D0138932

Springer Texts in Statistics

Springer
New York
Berlin
Heidelberg
Barcelona
Hong Kong
London
Milan
Paris
Singapore
Tokyo

Springer Texts in Statistics

Alfred	Elements of Statistics for the Life and Social Sciences
Berger	An Introduction to Probability and Stochastic Processes
Bilodeau and Brenner	Theory of Multivariate Statistics
Blom	Probability and Statistics: Theory and Applications
Brockwell and Davis	An Introduction to Time Series and Forecasting
Chow and Teicher	Probability Theory: Independence, Interchangeability, Martingales, Third Edition
Christensen	Plane Answers to Complex Questions: The Theory of Linear Models, Second Edition
Christensen	Linear Models for Multivariate, Time Series, and Spatial Data
Christensen	Log-Linear Models and Logistic Regression, Second Edition
Creighton	A First Course in Probability Models and Statistical Inference
Dean and Voss	Design and Analysis of Experiments
du Toit, Steyn, and Stumpf	Graphical Exploratory Data Analysis
Edwards	Introduction to Graphical Modelling
Finkelstein and Levin	Statistics for Lawyers
Flury	A First Course in Multivariate Statistics
Jobson	Applied Multivariate Data Analysis, Volume I: Regression and Experimental Design

Continued at end of book

Peter J. Brockwell
Richard A. Davis

Introduction to Time Series and Forecasting

With 122 Illustrations

Springer

Peter J. Brockwell
Colorado State University
Department of Statistics
Fort Collins, CO 80523
USA

Richard A. Davis
Colorado State University
Department of Statistics
Fort Collins, CO 80523
USA

Library of Congress Cataloging-in-Publication Data

Brockwell, Peter J.
Introduction to time series and forecasting / Peter J. Brockwell and Richard A. Davis.
p. cm. — (Springer texts in statistics)
Includes bibliographical references and index.
ISBN 0-387-94719-1 (hard : alk. paper)
1. Time-series analysis. I. Davis, Richard A. II. Title.
III. Series.
QA280.B757 1996
519.5′5—dc20 96-11743

Printed on acid-free paper.

Production managed by Bill Imbornoni; manufacturing supervised by Joe Quatela.
Typeset by The Bartlett Press, Inc.
Printed and bound by R.R. Donnelley and Sons, Harrisonburg, VA.
Printed in the United States of America.

9 8 7 6 5 4

ISBN 0-387-94719-1 Springer-Verlag New York Berlin Heidelberg SPIN 10755869

To Pam and Patti

Preface

This book is aimed at the reader who wishes to gain a working knowledge of time series and forecasting methods as applied in economics, engineering and the natural and social sciences. Unlike our earlier book, *Time Series: Theory and Methods*, referred to in the text as TSTM, this one has as prerequisites only a knowledge of basic calculus, matrix algebra and elementary statistics at the level (for example) of Mendenhall, Wackerly and Scheaffer (1990). It is intended for upper-level undergraduate students and beginning graduate students.

The emphasis is on methods and the analysis of data sets. A time series package enabling the reader to reproduce most of the calculations in the text (and to analyze further data sets of the reader's own choosing) is included on the CD-ROM which accompanies the book. The data sets used in the book are also included. The programs require an IBM-PC-compatible computer. Two versions of the package are included on the CD-ROM, but the new version, ITSM2000, should be installed on computers operating under Windows 95, 98 or NT. Installation instructions for both versions are provided on the jacket containing the CD-ROM.

A detailed introduction to the use of ITSM96 is contained in Appendix D of the book. Very little prior familiarity with computing is required. Detailed instructions on the use of ITSM2000 are found in the on-line help files which are accessed, when the program ITSM is running, by selecting the menu options `Help>Contents` and selecting the topic of interest. The book could also be used in conjunction with other computer packages for handling time series. Chapter 14 of the book by Venables and Ripley (1994) describes how to perform many of the calculations using S-plus.

There are numerous problems at the end of each chapter, many of which involve use of the programs to study the data sets provided.

To make the underlying theory accessible to a wider audience, we have stated some of the key mathematical results without proof, but have attempted to ensure that the logical structure of the development is otherwise complete. (References to proofs are provided for the interested reader.)

We have added a number of topics not considered in TSTM. These include harmonic regression (Chapter 1), the Burg and Hannan-Rissanen algorithms (Chapter 5), unit roots in time series models and regression with ARMA errors (Chapter 6), structural models, the EM algorithm and generalized state-space models, with applications to time series of count data (Chapter 8), and the Holt-Winters and ARAR forecasting algorithms (Chapter 9). The last chapter introduces the reader to transfer function models, intervention analysis, and nonlinear (including ARCH), continuous-time, and long-memory models.

There is sufficient material here for a full-year introduction to univariate and multivariate time series and forecasting. Chapters 1 through 6 have been used for several years as an introductory one-semester course in univariate time series at Royal Melbourne Institute of Technology. The chapter on spectral analysis can be excluded without loss of continuity by readers who are so inclined.

The book has benefited greatly from comments and corrections made by readers of TSTM, users of the computer package ITSM, and readers of preliminary versions of this book. To all these we express our thanks, particularly to two reviewers for their many valuable suggestions. We are especially indebted also to Anthony Brockwell and Joe Mandarino for their contributions to the development of the computer package, to Rob Hyndman for his collaboration in the writing of the manual, ITSM for Windows, to Carey and Michele Davis for their assistance in assembling the index, to the National Science Foundation for support of the research on which many of the algorithms are based, and to the Departments of Statistics at Colorado State University and Statistics and Operations Research at Royal Melbourne Institute of Technology for providing the excellent working environments in which the work was carried out.

Melbourne, Victoria
Fort Collins, Colorado
June, 1996

P.J. Brockwell
R.A. Davis

Contents

1 Introduction

In this chapter we introduce some basic ideas of time series analysis and stochastic processes. Of particular importance are the concepts of stationarity and the autocovariance and sample autocovariance functions. Some standard techniques are described for the estimation and removal of trend and seasonality (of known period) from an observed time series. These are illustrated with reference to the data sets in Section 1.1. The calculations in all the examples can be carried out using the programs supplied on the enclosed diskette. The data sets are contained in files with names ending in .DAT. For example, the Australian red wine sales are filed as WINE.DAT. Most of the topics covered in this chapter will be developed more fully in later sections of the book. The reader who is not already familiar with random variables and random vectors should first read Appendix A where a concise account of the required background is given.

1.1 Examples of Time Series

A **time series** is a set of observations x_t, each one being recorded at a specific time t. A *discrete-time time series* (the type to which this book is primarily devoted) is one in which the set T_0 of times at which observations are made is a discrete set, as is the case for example when observations are made at fixed time intervals. *Continuous-*

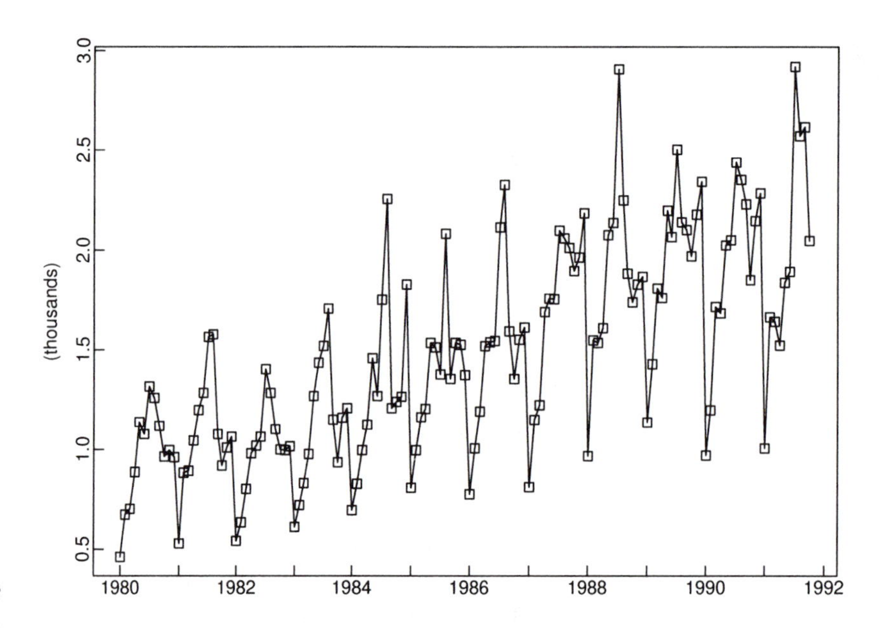

Figure 1-1
The Australian red wine sales, Jan. '80 – Oct. '91.

time time series are obtained when observations are recorded continuously over some time interval, e.g., when $T_0 = [0, 1]$.

Example 1.1.1 Australian red wine sales; WINE.DAT

Figure 1.1 shows the monthly sales (in kilolitres) of red wine by Australian winemakers from January 1980 through October 1991. In this case the set T_0 consists of the 142 times {(Jan. 1980), (Feb. 1980), ...,(Oct. 1991)}. Given a set of n observations made at uniformly spaced time intervals, it is often convenient to rescale the time axis in such a way that T_0 becomes the set of integers $\{1, 2, \ldots, n\}$. In the present example this amounts to measuring time in months with (Jan. 1980) as month 1. Then T_0 is the set $\{1, 2, \ldots, 142\}$. It appears from the graph that the sales have an upward trend and a seasonal pattern with a peak in July and a trough in January. □

Example 1.1.2 All-star baseball games, 1933–1995

Figure 1.2 shows the results of the all-star games by plotting x_t, where

$$x_t = \begin{cases} 1 & \text{if the National League won in year } t, \\ -1 & \text{if the American League won in year } t. \end{cases}$$

This is a series with only two possible values, ± 1. It also has some missing values since no game was played in 1945 and two games were scheduled for each of the years 1959–1962. □

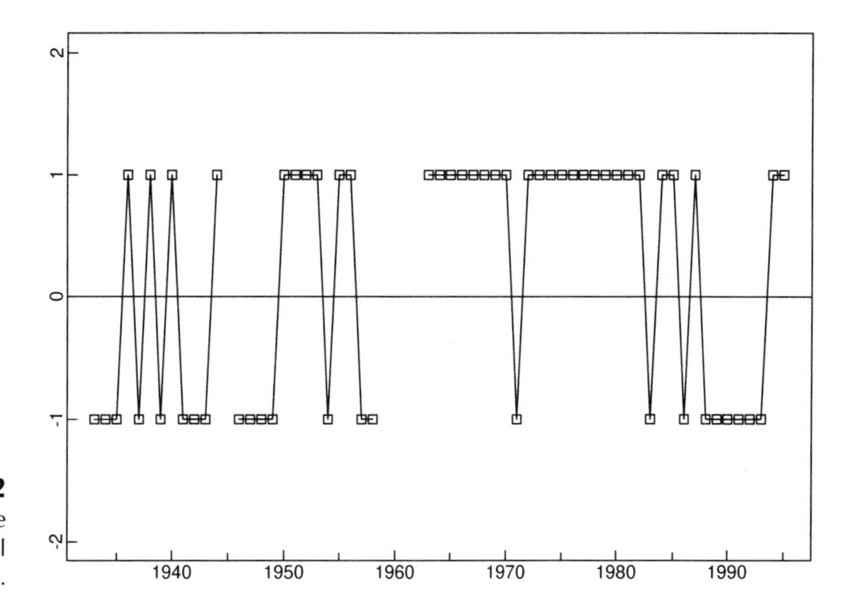

Figure 1-2 Results of the All-star baseball games, 1933–1995.

Example 1.1.3 Accidental deaths, U.S.A., 1973–1978; DEATHS.DAT

Like the red wine sales, the monthly accidental death figures show a strong seasonal pattern, with the maximum for each year occurring in July and the minimum for each year occurring in February. The presence of a trend in Figure 1.3 is much less apparent than in the wine sales. In Section 1.5 we shall consider the problem of representing the data as the sum of a trend, a seasonal component and a residual term. □

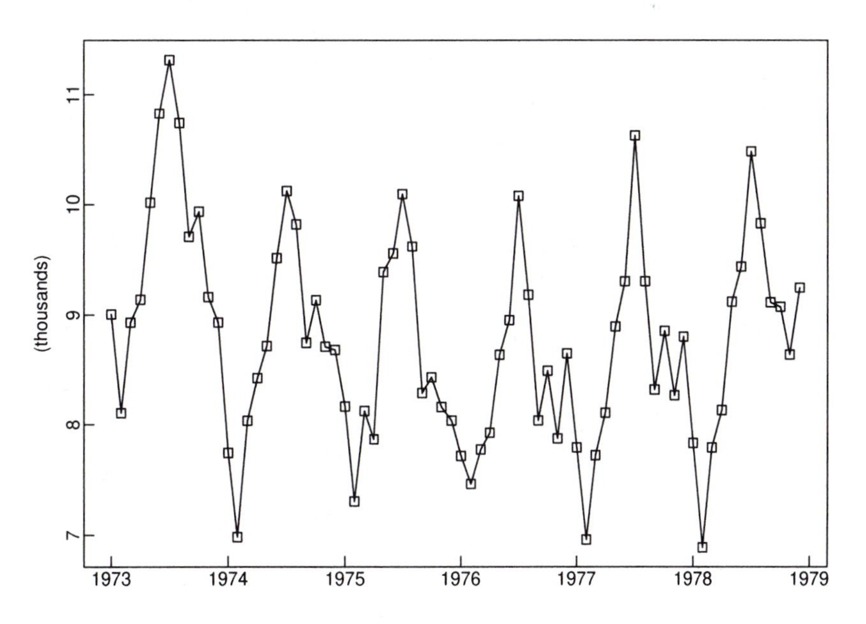

Figure 1-3 The monthly accidental deaths data, 1973–1978.

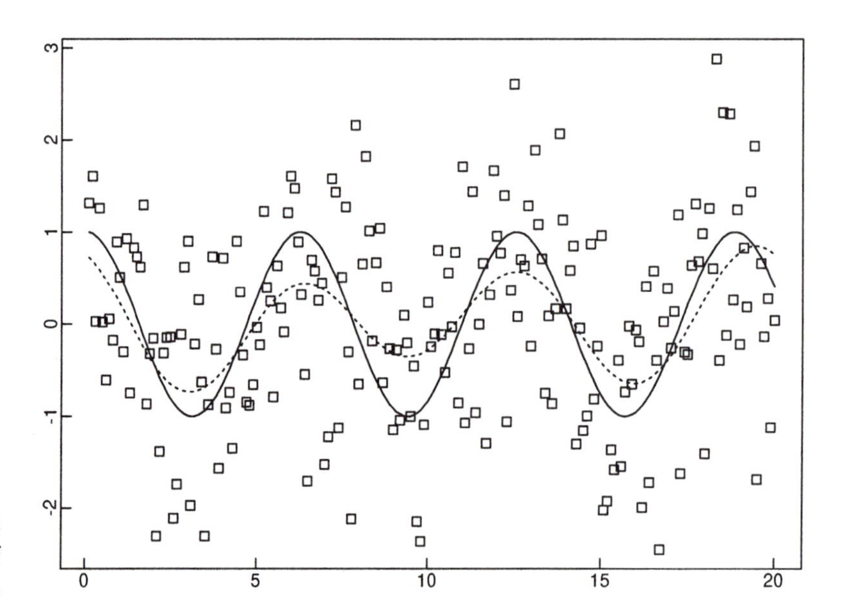

Figure 1-4 The series $\{X_t\}$ of Example 1.1.4.

Example 1.1.4 A signal detection problem; SIGNAL.DAT

Figure 1.4 shows simulated values of the series

$$X_t = \cos(t) + N_t, t = 0.1, 0.2, \ldots, 20.0,$$

where $\{N_t\}$ is a sequence of independent normal random variables, with mean 0 and variance 0.25. Such a series is often referred to as *signal plus noise*, the signal being the smooth function, $S_t = \cos(t)$ in this case. Given only the data X_t, how can we determine the unknown signal component? There are many approaches to this general problem under varying assumptions about the signal and the noise. One simple approach is to *smooth* the data by expressing X_t as a sum of sine waves of various frequencies (see Section 4.2) and eliminating the high frequency components. If we do this to the values of $\{X_t\}$ shown in Figure 1.4 and retain only the lowest 3.5% of the frequency components, we obtain the estimate of the signal also shown in Figure 1.4. The waveform of the signal is quite close to that of the true signal in this case, although its amplitude is somewhat smaller. □

Example 1.1.5 Population of the U.S.A., 1790–1990; USPOP.DAT

The population of the U.S.A., measured at ten-year intervals, is shown in Figure 1.5. The graph suggests the possibility of fitting a quadratic or exponential trend to the data. We shall explore this further in Section 1.3. □

Example 1.1.6 Strikes in the U.S.A., 1951–1980; STRIKE.DAT

The annual numbers of strikes in the U.S.A. for the years 1951–1980 are shown in Figure 1.6. They appear to fluctuate erratically about a slowly changing level. □

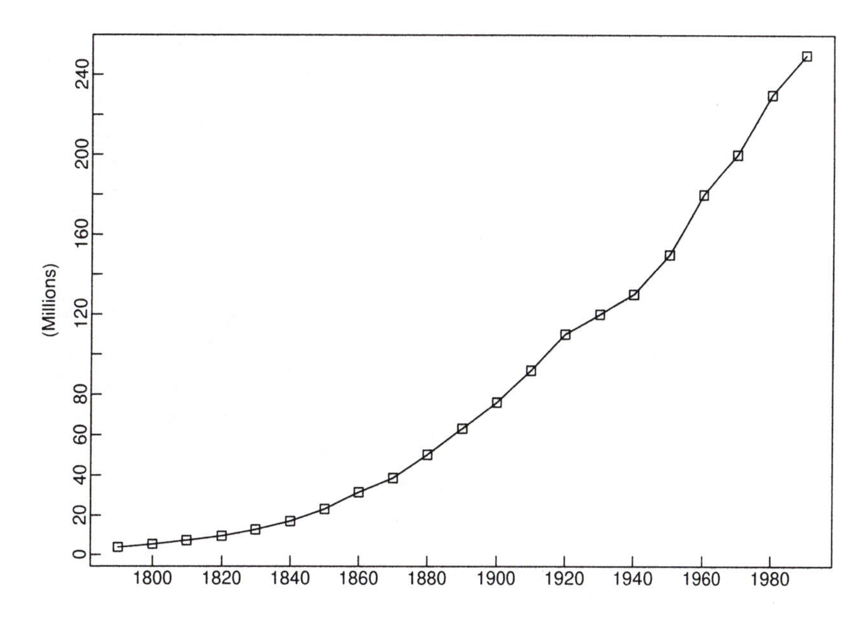

Figure 1-5
Population of the U.S.A. at ten-year intervals, 1790–1990.

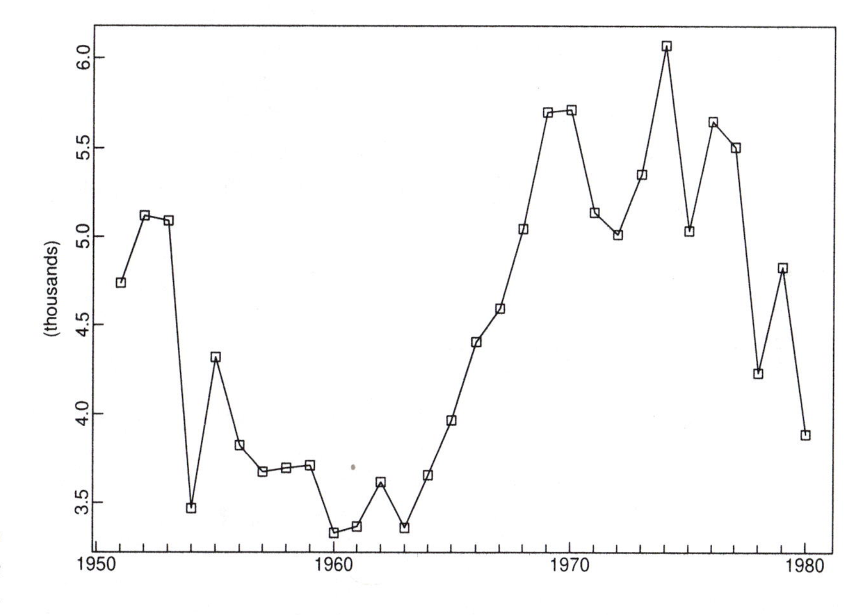

Figure 1-6
Strikes in the U.S.A., 1951–1980.

1.2 Objectives of Time Series Analysis

The examples considered in Section 1.1 are an extremely small sample from the multitude of time series encountered in the fields of engineering, science, sociology, and economics. Our purpose in this book is to study techniques for drawing inferences from such series. Before we can do this, however, it is necessary to set up a hypothetical probability model to represent the data. Having chosen a model (or family of models) it then becomes possible to estimate parameters, check for goodness of fit to the data, and possibly to use the fitted model to enhance our understanding of the mechanism generating the series. Once a satisfactory model has been developed, it may be used in a variety of ways depending on the particular field of application.

The model may be used simply to provide a compact description of the data. We may for example be able to represent the accidental deaths data of Example 1.1.3 as the sum of a specified trend, and seasonal and random terms. For the interpretation of economic statistics such as unemployment figures, it is important to recognize the presence of seasonal components and to remove them so as not to confuse them with long-term trends. This process is known as **seasonal adjustment**. Other applications of time series models include separation (or filtering) of noise from signals as in Example 1.1.4, prediction of future values of a series such as the red wine sales in Example 1.1.1 or the population data in Example 1.1.5, testing hypotheses such as global warming using recorded temperature data, predicting one series from observations of another, e.g., predicting future sales using advertising expenditure data, and controlling future values of a series by adjusting parameters. Time series models are also useful in simulation studies. For example, the performance of a reservoir depends heavily on the random daily inputs of water to the system. If these are modelled as a time series, then we can use the fitted model to simulate a large number of independent sequences of daily inputs. Knowing the size and mode of operation of the reservoir, we can determine the fraction of the simulated input sequences that cause the reservoir to run out of water in a given time period. This fraction will then be an estimate of the probability of emptiness of the reservoir at some time in the given period.

1.3 Some Simple Time Series Models

An important part of the analysis of a time series is the selection of a suitable probability model (or class of models) for the data. To allow for the possibly unpredictable nature of future observations it is natural to suppose that each observation x_t is a realized value of a certain random variable X_t.

Definition 1.3.1

A **time series model** for the observed data $\{x_t\}$ is a specification of the joint distributions (or possibly only the means and covariances) of a sequence of random variables $\{X_t\}$ of which $\{x_t\}$ is postulated to be a realization.

Remark. We shall frequently use the term *time series* to mean both the data and the process of which it is a realization. □

A complete probabilistic time series model for the sequence of random variables $\{X_1, X_2, \ldots,\}$ would specify all of the **joint distributions** of the random vectors $(X_1, \ldots, X_n)'$, $n = 1, 2, \ldots$, or equivalently all of the probabilities

$$P[X_1 \le x_1, \ldots, X_n \le x_n], \quad -\infty < x_1, \ldots, x_n < \infty, \quad n = 1, 2, \ldots.$$

Such a specification is rarely used in time series analysis (unless the data are generated by some well-understood simple mechanism), since in general it will contain far too many parameters to be estimated from the available data. Instead we specify only the **first-** and **second-order moments** of the joint distributions, i.e., the expected values EX_t and the expected products $E(X_{t+h}X_t)$, $t = 1, 2, \ldots$, $h = 0, 1, 2, \ldots$, focusing on properties of the sequence $\{X_t\}$ which depend only on these. Such properties of $\{X_t\}$ are referred to as **second-order properties**. In the particular case when all the joint distributions are multivariate normal, the second-order properties of $\{X_t\}$ completely determine the joint distributions and hence give a complete probabilistic characterization of the sequence. In general we shall lose a certain amount of information by looking at time series "through second-order spectacles"; however, as we shall see in Chapter 2, the theory of minimum mean squared error linear prediction depends only on the second-order properties, thus providing further justification for the use of the second-order characterization of time series models.

Figure 1.7 shows one of many possible realizations of $\{S_t, t = 1, \ldots, 200\}$, where $\{S_t\}$ is a sequence of random variables specified in Example 1.3.3 below. In most practical problems involving time series we see only *one* realization. For example, there is only one available realization of Fort Collins annual rainfall for the years 1900–1996, but we imagine it to be one of the many sequences that *might* have occurred. In the following examples we introduce some simple time series models. One of our goals will be to expand this repertoire so as to have at our disposal a broad range of models with which to try and match the observed behavior of given data sets.

1.3.1 Some Zero-Mean Models

Example 1.3.1 IID noise

Perhaps the simplest model for a time series is one in which there is no trend or seasonal component and in which the observations are simply independent and identically distributed (iid) random variables with zero mean. We refer to such a sequence of random variables $X_1, X_2, \ldots,$ as IID noise. By definition we can write, for any positive integer n and real numbers $x_1, \ldots, x_n$,

$$\begin{aligned} P[X_1 \le x_1, \ldots, X_n \le x_n] &= P[X_1 \le x_1] \cdots P[X_n \le x_n] \\ &= F(x_1) \cdots F(x_n) \end{aligned}$$

where $F(\cdot)$ is the cumulative distribution function (see Section A.1) of each of the identically distributed random variables $X_1, X_2, \ldots$. In this model there is no dependence between observations. In particular, for all $h \geq 1$ and all $x, x_1, \ldots, x_n$,

$$P[X_{n+h} \leq x | X_1 = x_1, \ldots, X_n = x_n] = P[X_{n+h} \leq x],$$

showing that knowledge of $X_1, \ldots, X_n$ is of no value for predicting the behavior of X_{n+h}. Given the values of $X_1, \ldots, X_n$, the function f that minimizes the mean squared error $E[(X_{n+h} - f(X_1, \ldots, X_n))^2]$ is in fact identically zero (see Problem 1.2). Although this means that IID noise is a rather uninteresting process for forecasters, it plays an important role as a building block for more complicated time series models. □

Example 1.3.2 A binary process

As an example of IID noise, consider the sequence of iid random variables $\{X_t, t = 1, 2, \ldots,\}$ with

$$P[X_t = 1] = p, \quad P[X_t = -1] = 1 - p,$$

where $p = 1/2$. The time series obtained by tossing a penny repeatedly and scoring $+1$ for each head and -1 for each tail is usually modelled as a realization of this process. A priori we might well consider the same process as a model for the all-star baseball games in Example 1.1.2. However, even a cursory inspection of the results from 1963–1982, which shows the National League winning 19 of 20 games, casts serious doubt on the hypothesis $P[X_t = 1] = 1/2$. □

Example 1.3.3 Random walk

The random walk $\{S_t, t = 0, 1, 2, \ldots,\}$ (starting at zero) is obtained by cumulatively summing (or "integrating") iid random variables. Thus a random walk with zero mean is obtained by defining $S_0 = 0$ and

$$S_t = X_1 + X_2 + \cdots + X_t, \quad \text{for } t = 1, 2, \ldots,$$

where $\{X_t\}$ is IID noise. If $\{X_t\}$ is the binary process of Example 1.3.2, then $\{S_t, t = 0, 1, 2, \ldots,\}$ is called a **simple symmetric random walk**. This walk can be viewed as the location of a pedestrian who starts at zero at time zero and at each integer time tosses a fair coin, stepping one unit to the right each time a head appears and one unit to the left for each tail. A realization of length 200 of a simple symmetric random walk is shown in Figure 1.7. Notice that the outcomes of the coin tosses can be recovered from $\{S_t, t = 0, 1, \ldots\}$ by differencing. Thus the result of the tth toss can be found from $S_t - S_{t-1} = X_t$. □

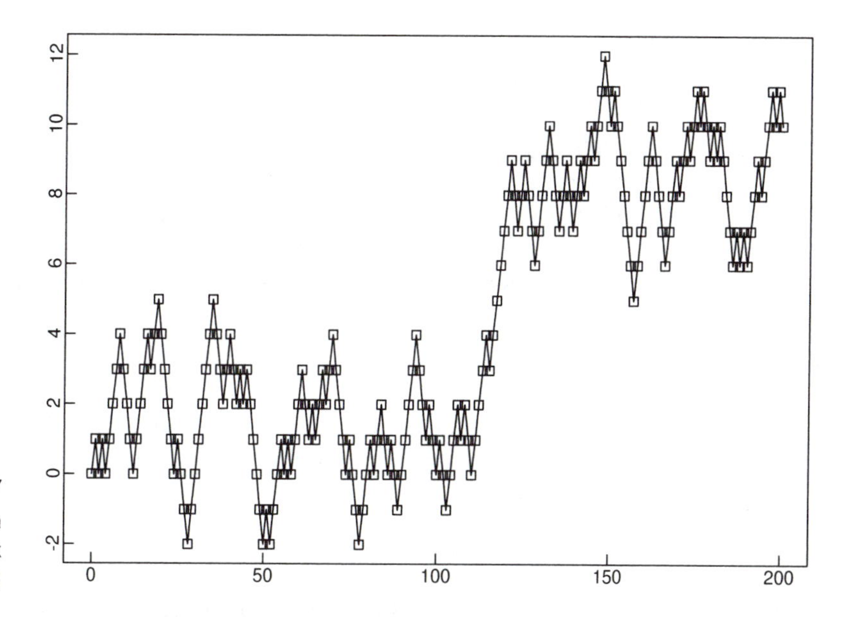

Figure 1-7 One realization of a simple random walk $\{S_t, t = 0, 1, 2, \ldots, 200\}$

1.3.2 Models with Trend and Seasonality

In several of the time series examples of Section 1.1, there is a clear trend in the data. An increasing trend is apparent in both the Australian red wine sales (Figure 1.1) and the population of the U.S.A. (Figure 1.5). In both cases a zero-mean model for the data is clearly inappropriate. The graph of the population data, which contains no apparent periodic component, suggests trying a model of the form

$$X_t = m_t + Y_t,$$

where m_t is a slowly changing function known as the **trend component** and Y_t has zero mean. A useful technique for estimating m_t is the method of least squares (some other methods are considered in Section 1.5).

In the least squares procedure we attempt to fit a parametric family of functions, e.g.,

$$m_t = a_0 + a_1 t + a_2 t^2, \tag{1.3.1}$$

to the data $\{x_1, \ldots, x_n\}$ by choosing the parameters, in this illustration a_0, a_1 and a_2, to minimize $\sum_{t=1}^{n}(x_t - m_t)^2$.

Example 1.3.4 Population of the U.S.A., 1790–1990

Fitting a function of the form (1.3.1) to the population data shown in Figure 1.5 for $1790 \le t \le 1990$, gives the estimated parameter values,

$$\hat{a}_0 = 2.1006 \times 10^{10},$$

$$\hat{a}_1 = -2.3379 \times 10^{7},$$

and

$$\hat{a}_2 = 6.5063 \times 10^3.$$

A graph of the fitted function is shown with the original data in Figure 1.8. The estimated values of the noise process Y_t, $1790 \le t \le 1990$, are the residuals obtained by subtraction of $\hat{m}_t = \hat{a}_0 + \hat{a}_1 t + \hat{a}_2 t^2$ from x_t.

The estimated trend component $\hat{m}_t$ furnishes us with a natural predictor of future values of X_t. For example, if we estimate Y_{2000} by its mean value (i.e., zero) we obtain the estimate

$$\hat{m}_{2000} = 2.732 \times 10^8$$

for the population of the U.S.A. in 2000. However, if the residuals $\{Y_t\}$ are highly correlated we may be able to use their values to give a better estimate of Y_{2000} and hence of X_{2000}. □

Example 1.3.5 Level of Lake Huron 1875–1972; LAKE.DAT

A graph of the level of Lake Huron (reduced by 570 feet) in the years 1875–1972 is displayed in Figure 1.9. Since the lake level appears to decline at a roughly linear rate, a model of the form

$$X_t = a_0 + a_1 t + Y_t, \tag{1.3.2}$$

was considered for this time series. The least squares estimates of the parameter values are

$$\hat{a}_0 = 10.202 \quad \text{and} \quad \hat{a}_1 = -.0242.$$

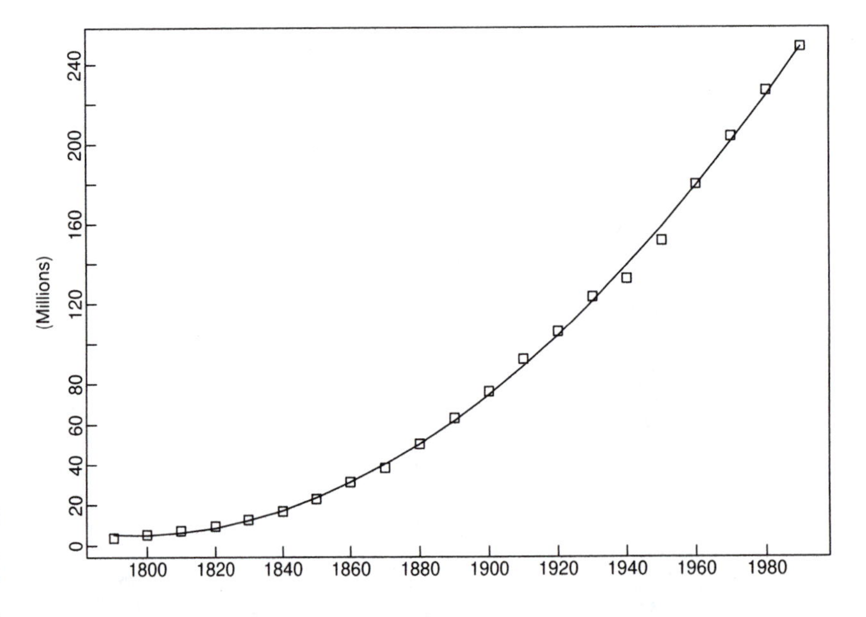

Figure 1-8
Population of the U.S.A. showing the quadratic trend fitted by least squares.

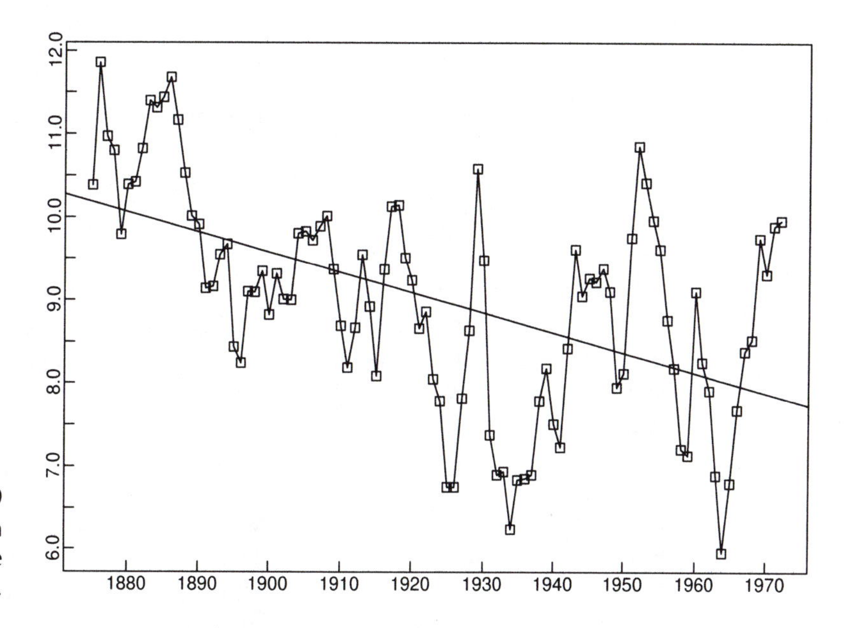

Figure 1-9
Level of Lake Huron 1875–1972 showing the line fitted by least squares.

(The resulting least squares line, $\hat{a}_0 + \hat{a}_1 t$, is also displayed in Figure 1.9.) The estimates of the noise, Y_t, in the model (1.3.2) are the residuals obtained by subtracting the least squares line from x_t and are plotted in Figure 1.10. There are two interesting features of the graph of the residuals. The first is the absence of any discernible trend. The second is the smoothness of the graph. (In particular there are long stretches of residuals that have the same sign. This would be very unlikely to occur if the residuals were observations of IID noise with zero mean.) Smoothness of the graph of a time

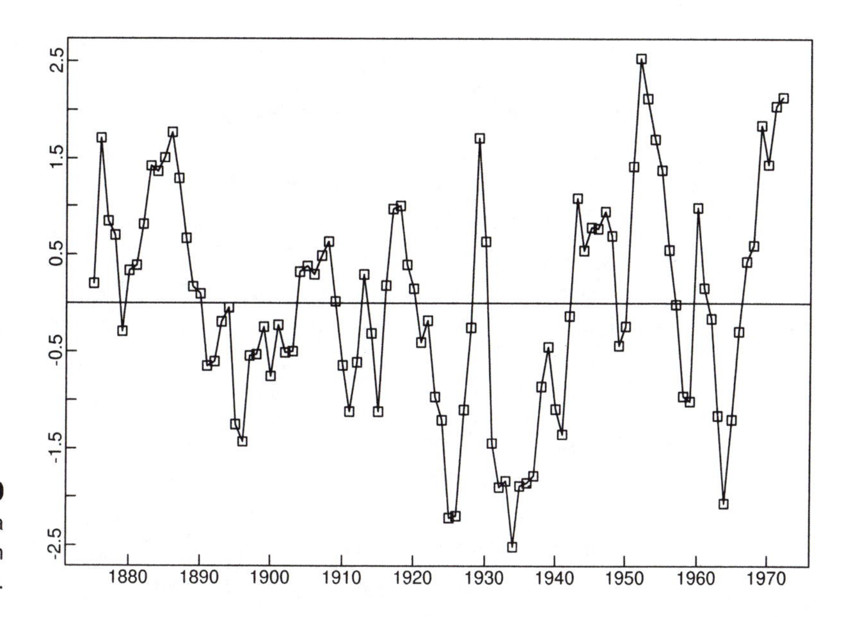

Figure 1-10
Residuals from fitting a line to the Lake Huron data in Figure 1.9.

series is generally indicative of the existence of some form of dependence between the observations.

Such dependence can be used to advantage in forecasting future values of the series. If we were to assume the validity of the fitted model with IID residuals $\{Y_t\}$, then the minimum mean squared error predictor of the next residual (Y_{1973}) would be zero (by Problem 1.2). However, Figure 1.10 strongly suggests that Y_{1973} will be positive.

How then do we quantify dependence and how do we construct models for forecasting that incorporate dependence of a particular type? To deal with these questions, Section 1.4 introduces the autocorrelation function as a measure of dependence, and stationary processes as a family of useful models exhibiting a wide variety of dependence structures. □

Harmonic Regression

Many time series are influenced by seasonally varying factors such as the weather, the effect of which can be modelled by a periodic component with fixed known period. For example the accidental deaths series (Figure 1.3) shows a repeating annual pattern with peaks in July and troughs in February, strongly suggesting a seasonal factor with period 12. In order to represent such a seasonal effect, allowing for noise but assuming no trend, we can use the simple model,

$$X_t = s_t + Y_t,$$

where s_t is a periodic function of t with period d ($s_{t-d} = s_t$). A convenient choice for s_t is a sum of harmonics given by

$$s_t = a_0 + \sum_{j=1}^{k}(a_j \cos(\lambda_j t) + b_j \sin(\lambda_j t))$$

where $a_0, a_1, \ldots, a_k$ and $b_1, \ldots, b_k$ are unknown parameters and $\lambda_1, \ldots, \lambda_k$ are fixed frequencies, each being some integer multiple of $2\pi/d$. If the number of observations n is an integer multiple of d then the frequencies λ_j, $j = 1, \ldots, k$ are integer multiples, $2\pi m_1/n, \ldots, 2\pi m_k/n$ of $2\pi/n$. The option *Remove harmonic trend* of the program PEST allows the estimation of s_t with $k \le 4$. It requires only that you specify the required multiples $m_1, \ldots, m_k$ of the frequency $2\pi/n$. These multiples are called *Fourier indices* (see Chapter 4). The period of a harmonic with Fourier index m is n/m. Thus to fit a single harmonic with period 365 to 365 daily observations we would choose a single Fourier index (i.e., $k = 1$) with $m_1 = 1$.

Example 1.3.6 Accidental deaths

To fit a sum of two harmonics with periods twelve months and six months to the monthly accidental deaths data $x_1, \ldots, x_{72}$ we choose $k = 2$, $m_1 = 6$, and $m_2 = 12$. (The indices $m_1 = 6$ and $m_2 = 12$ correspond to harmonics with periods $n/m_1 = 12$ and $n/m_2 = 6$ (months), respectively.) As can be seen in Figure 1.11, the periodic character of the series is captured reasonably well by this fitted function. In practice,

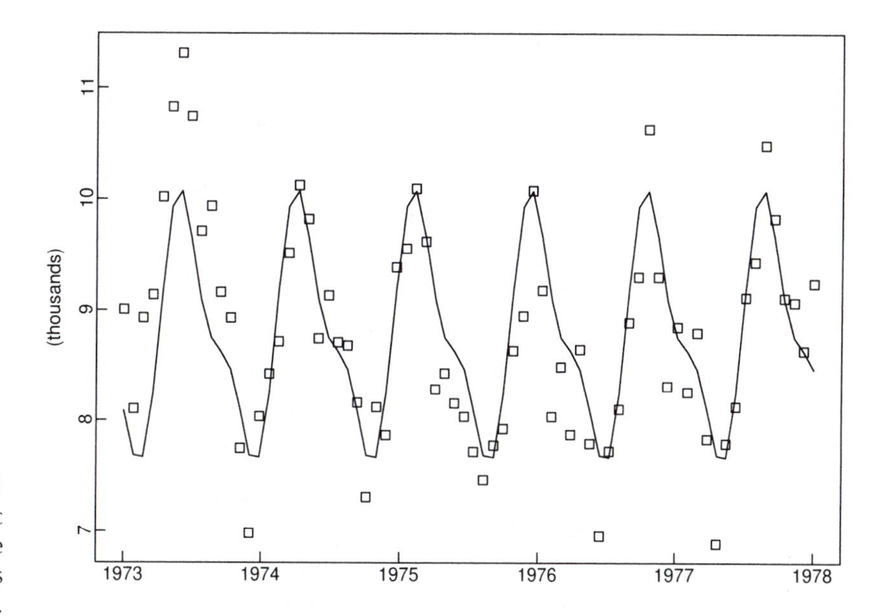

Figure 1-11 The estimated harmonic component of the accidental deaths data from PEST.

it is worth experimenting with several different combinations of harmonics, all with Fourier indices which are integer multiples of n/d, to find a satisfactory estimate of the seasonal component. A more systematic approach based on the periodogram is discussed in Chapter 4. Methods for estimating and eliminating seasonal components in the presence of trend are described in Section 1.5. □

1.3.3 A General Approach to Time Series Modelling

The examples of the previous section illustrate a general approach to time series analysis that will form the basis for much of what is done in this book. Before introducing the ideas of dependence and stationarity, we outline this approach to provide the reader with an overview of the way in which the various ideas of this chapter fit together.

- Plot the series and examine the main features of the graph, checking in particular if there is
  (a) a trend,
  (b) a seasonal component,
  (c) any apparent sharp changes in behavior,
  (d) any outlying observations.
- Remove the trend and seasonal components to get *stationary* residuals (as defined in Section 1.4). To achieve this goal it may sometimes be necessary to apply a preliminary transformation to the data. For example, if the magnitude of the fluctuations appears to grow roughly linearly with the level of the series, then the transformed series $\{\ln X_1, \ldots, \ln X_n\}$ will have fluctuations of more constant

magnitude. See, for example, Figures 1.1 and 1.17. (If some of the data are negative, add a positive constant to each of the data values to ensure that all values are positive before taking logarithms.) There are several ways in which trend and seasonality can be removed (see Section 1.5), some involving estimating the components and subtracting them from the data, and others depending on *differencing* the data, i.e., replacing the original series $\{X_t\}$ by $\{Y_t := X_t - X_{t-d}\}$ for some positive integer d. Whichever method is used, the aim is to produce a stationary series, whose values we shall refer to as residuals.

- Choose a model to fit the residuals, making use of various sample statistics including the sample autocorrelation function to be defined in Section 1.4.
- Forecasting will be achieved by forecasting the residuals and then inverting the transformations described above to arrive at forecasts of the original series $\{X_t\}$.
- An extremely useful alternative approach touched on only briefly in this book is to express the series in terms of its Fourier components, which are sinusoidal waves of different frequencies (cf. Example 1.1.4). This approach is especially important in engineering applications such as signal processing and structural design. It is important, for example, to ensure that the resonant frequency of a structure does not coincide with a frequency at which the loading forces on the structure have a particularly large component.

1.4 Stationary Models and the Autocorrelation Function

Loosely speaking, a time series $\{X_t, t = 0, \pm 1, \ldots\}$ is said to be stationary if it has statistical properties similar to those of the "time-shifted" series $\{X_{t+h}, t = 0, \pm 1, \ldots\}$, for each integer h. Restricting attention to those properties which depend only on the first- and second-order moments of $\{X_t\}$, we can make this idea precise with the following definitions.

Definition 1.4.1

Let $\{X_t\}$ be a time series with $EX_t^2 < \infty$. The **mean function** of $\{X_t\}$ is

$$\mu_X(t) = \mathrm{E}(X_t).$$

The **covariance function** of $\{X_t\}$ is

$$\gamma_X(r, s) = \mathrm{Cov}(X_r, X_s) = E[(X_r - \mu_X(r))(X_s - \mu_X(s))]$$

for all integers r and s.

Definition 1.4.2

$\{X_t\}$ is **(weakly) stationary** if

(i) $\mu_X(t)$ is independent of t,

and

(ii) $\gamma_X(t+h,t)$ is independent of t for each h.

Remark 1. Strict stationarity of a time series $\{X_t, t = 0, \pm 1, \ldots,\}$ is defined by the condition that $(X_1, \ldots, X_n)$ and $(X_{1+h}, \ldots, X_{n+h})$ have the same joint distributions for all integers h and $n > 0$. It is easy to check that if $\{X_t\}$ is strictly stationary and $EX_t^2 < \infty$ for all t, then $\{X_t\}$ is also weakly stationary (Problem 1.3). Whenever we use the term *stationary* we shall mean weakly stationary as in Definition 1.4.2, unless we specifically indicate otherwise. □

Remark 2. In view of condition (ii), whenever we use the term covariance function with reference to a *stationary* time series $\{X_t\}$ we shall mean the function γ_X of *one* variable, defined by

$$\gamma_X(h) := \gamma_X(h,0) = \gamma_X(t+h,t).$$

The function $\gamma_X(\cdot)$ will be referred to as the autocovariance function and $\gamma_X(h)$ as its value at *lag* h. □

Definition 1.4.3

Let $\{X_t\}$ be a stationary time series. The **autocovariance function** (ACVF) of $\{X_t\}$ is

$$\gamma_X(h) = \text{Cov}(X_{t+h}, X_t).$$

The **autocorrelation function** (ACF) of $\{X_t\}$ is

$$\rho_X(h) \equiv \frac{\gamma_X(h)}{\gamma_X(0)} = \text{Cor}(X_{t+h}, X_t).$$

In the following examples we shall frequently use the easily verified **linearity property of covariances**, that if $EX^2 < \infty$, $EY^2 < \infty$, $EZ^2 < \infty$ and a, b and c are any real constants, then

$$\text{Cov}(aX + bY + c, Z) = a\,\text{Cov}(X,Z) + b\,\text{Cov}(Y,Z).$$

Example 1.4.1 IID noise

If $\{X_t\}$ is IID noise and $E(X_t^2) = \sigma^2 < \infty$, then the first requirement of Definition 1.4.2 is obviously satisfied since $E(X_t) = 0$ for all t. By the assumed independence,

$$\gamma_X(t+h,t) = \begin{cases} \sigma^2, & \text{if } h = 0, \\ 0, & \text{if } h \neq 0, \end{cases}$$

which does not depend on t. Hence IID noise with finite second moment is stationary. We shall use the notation

$$\{X_t\} \sim \text{IID}(0, \sigma^2),$$

to indicate that the random variables X_t are independent and identically distributed random variables, each with mean 0 and variance σ^2. □

Example 1.4.2 White noise

If $\{X_t\}$ is a sequence of uncorrelated random variables, each with zero mean and variance σ^2, then clearly $\{X_t\}$ is stationary with the same covariance function as the IID noise in Example 1.4.1. Such a sequence is referred to as **white noise** (with mean 0 and variance σ^2). This is indicated by the notation

$$\{X_t\} \sim \text{WN}(0, \sigma^2).$$

Clearly every IID$(0, \sigma^2)$ sequence is WN$(0, \sigma^2)$ but not conversely (see Problem 1.8 and the ARCH(1) process of Section 10.3). □

Example 1.4.3 The random walk

If $\{S_t\}$ is the random walk defined in Example 1.3.3 with $\{X_t\}$ as in Example 1.4.1, then $\text{E}S_t = 0$, $\text{E}(S_t^2) = t\sigma^2 < \infty$ for all t and, for $h \geq 0$,

$$\begin{aligned} \gamma_S(t+h, t) &= \text{Cov}(S_{t+h}, S_t) \\ &= \text{Cov}(S_t + X_{t+1} + \cdots + X_{t+h}, S_t) \\ &= \text{Cov}(S_t, S_t) \\ &= t\sigma^2. \end{aligned}$$

Since $\gamma_S(t+h, t)$ depends on t, the series $\{S_t\}$ is *not* stationary. □

Example 1.4.4 First-order moving average or MA(1) process

Consider the series defined by the equation

$$X_t = Z_t + \theta Z_{t-1}, t = 0, \pm 1, \ldots, \tag{1.4.1}$$

where $\{Z_t\} \sim \text{WN}(0, \sigma^2)$ and θ is a real-valued constant. From (1.4.1) we see that $\text{E}X_t = 0$, $\text{E}X_t^2 = \sigma^2(1 + \theta^2) < \infty$, and

$$\gamma_X(t+h, t) = \begin{cases} \sigma^2(1+\theta^2), & \text{if } h = 0, \\ \sigma^2\theta, & \text{if } h = \pm 1, \\ 0, & \text{if } |h| > 1. \end{cases}$$

Thus the requirements of Definition 1.4.2 are satisfied and $\{X_t\}$ is stationary. The autocorrelation function of $\{X_t\}$ is

$$\rho_X(h) = \begin{cases} 1, & \text{if } h = 0, \\ \theta/(1+\theta^2), & \text{if } h = \pm 1, \\ 0, & \text{if } |h| > 1. \end{cases}$$ □

Example 1.4.5 First order autoregression or AR(1) process

Let us *assume* now that $\{X_t\}$ is a stationary series satisfying the equations

$$X_t = \phi X_{t-1} + Z_t, t = 0, \pm 1, \ldots, \tag{1.4.2}$$

where $\{Z_t\} \sim \mathrm{WN}(0, \sigma^2)$, $|\phi| < 1$ and Z_t is uncorrelated with X_s for each $s < t$. (We shall show in Section 2.2 that there is in fact exactly one such solution of (1.4.2).) By taking expectations on each side of (1.4.2) and using the fact that $\mathrm{E}Z_t = 0$, we see at once that

$$\mathrm{E}X_t = 0.$$

To find the autocorrelation function of $\{X_t\}$ we multiply each side of (1.4.2) by X_{t-h} ($h > 0$) and then take expectations to get

$$\begin{aligned} \gamma_X(h) &= \mathrm{Cov}(X_t, X_{t-h}) \\ &= \mathrm{Cov}(\phi X_{t-1}, X_{t-h}) + \mathrm{Cov}(Z_t, X_{t-h}) \\ &= \phi\gamma_X(h-1) + 0 = \cdots = \phi^h \gamma_X(0). \end{aligned}$$

Observing that $\gamma(h) = \gamma(-h)$ and using Definition 1.4.3, we find that

$$\rho_X(h) = \frac{\gamma_X(h)}{\gamma_X(0)} = \phi^{|h|}, h = 0, \pm 1, \ldots.$$

Since Z_t is uncorrelated with X_{t-1}, it follows from the linearity property of covariances that $\mathrm{Cov}(X_t, Z_t) = \mathrm{Cov}(\phi X_{t-1} + Z_t, Z_t) = \mathrm{Var}(Z_t) = \sigma^2$. Hence,

$$\gamma_X(0) = \mathrm{Cov}(X_t, X_t) = \mathrm{Cov}(X_t, \phi X_{t-1} + Z_t) = \phi\gamma_X(1) + \sigma^2 = \phi^2\gamma_X(0) + \sigma^2$$

and consequently $\gamma_X(0) = \sigma^2/(1-\phi^2)$. □

1.4.1 The Sample Autocorrelation Function

Although we have just seen how to compute the autocorrelation function for a few simple time series models, in practical problems we do not start with a model, but with *observed data* $\{x_1, x_2, \ldots, x_n\}$. To assess the degree of dependence in the data and to select a model for the data which reflects this, one of the important tools we use is the **sample autocorrelation function** (sample ACF) of the data. If we believe that the data are realized values of a stationary time series $\{X_t\}$, then the sample ACF will provide us with an estimate of the ACF of $\{X_t\}$. This estimate may suggest which of the many possible stationary time series models is a suitable candidate for representing the dependence in the data. For example, a sample ACF that is close to zero for all nonzero lags suggests that an appropriate model for the data might be IID noise. The following definitions are natural *sample* analogues of those for the autocovariance and autocorrelation functions given earlier for stationary time series *models*.

Definition 1.4.4

Let $x_1, \ldots, x_n$ be observations of a time series. The **sample mean** of $x_1, \ldots, x_n$ is

$$\bar{x} = \frac{1}{n}\sum_{t=1}^{n} x_t.$$

The **sample autocovariance function** is

$$\hat{\gamma}(h) := n^{-1}\sum_{t=1}^{n-|h|}(x_{t+|h|} - \bar{x})(x_t - \bar{x}), \qquad -n < h < n.$$

The **sample autocorrelation function** is

$$\hat{\rho}(h) = \frac{\hat{\gamma}(h)}{\hat{\gamma}(0)}, \qquad -n < h < n.$$

Remark 3. For $h \geq 0$, $\hat{\gamma}(h)$ is approximately equal to the sample covariance of the $n - h$ pairs of observations $(x_1, x_{1+h}), (x_2, x_{2+h}), \ldots, (x_{n-h}, x_n)$. The difference arises from use of the divisor n instead of $n - h$ and the subtraction of the *overall* mean, $\bar{x}$, from each factor of the summands. Use of the divisor n ensures that the sample covariance matrix $\hat{\Gamma}_n := [\hat{\gamma}(i-j)]_{i,j=1}^{n}$ is nonnegative definite (see Section 2.4.2). □

Remark 4. Like the sample covariance matrix defined in Remark 3, the sample correlation matrix $\hat{R}_n := [\hat{\rho}(i-j)]_{i,j=1}^{n}$ is nonnegative definite. Each of its diagonal elements is equal to 1 since $\hat{\rho}(0) = 1$. □

Example 1.4.6

Figure 1.12 shows 200 simulated values of normally distributed IID(0, 1), denoted IID N(0, 1), noise. Figure 1.13 shows the corresponding sample autocorrelation function at lags $0, 1, \ldots, 40$. Since $\rho(h) = 0$ for $h > 0$, one would also expect the corresponding sample autocorrelations to be near 0. It can be shown in fact that for IID noise with finite variance, the sample autocorrelations $\hat{\rho}(h)$, $h > 0$, are approximately IID N$(0, 1/n)$ for n large (see TSTM p. 222). Hence, approximately 95% of the sample autocorrelations should fall between the bounds $\pm 1.96/\sqrt{n}$ (since 1.96 is the .975

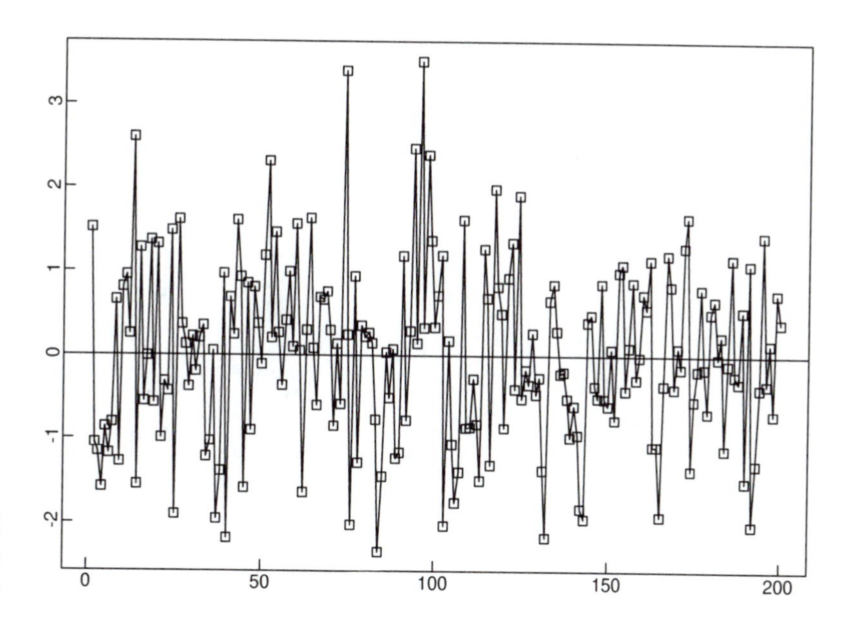

Figure 1-12 200 simulated values of IID N(0,1) noise.

quantile of the standard normal distribution). Therefore, in Figure 1.13 we would expect roughly $40(.05) = 2$ values to fall outside the bounds. □

Remark 5. The sample autocovariance and autocorrelation functions can be computed for *any* data set $\{x_1, \ldots, x_n\}$ and are not restricted to observations from a stationary time series. For data containing a trend, $|\hat{\rho}(h)|$ will exhibit slow decay as

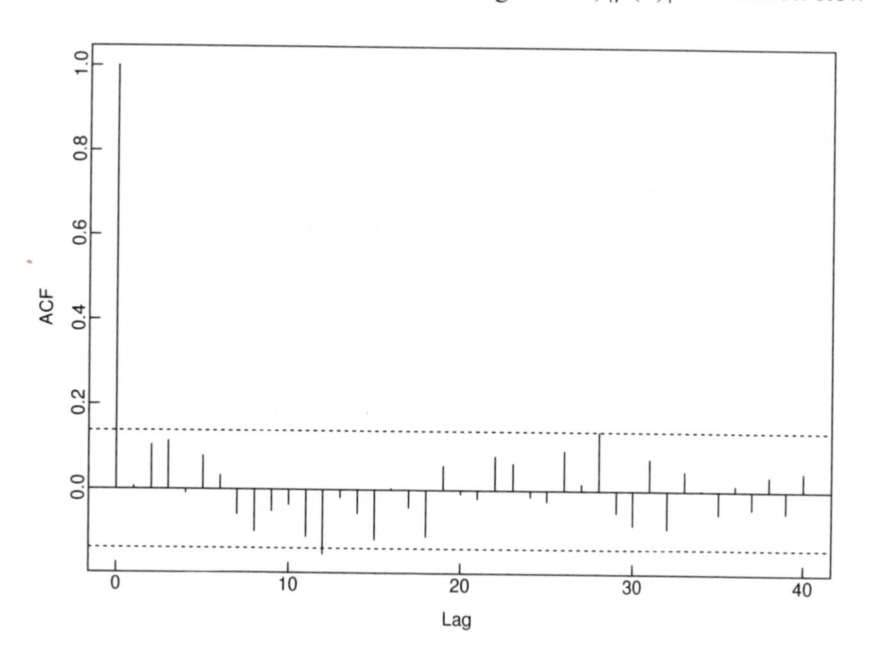

Figure 1-13 The sample autocorrelation function for the data of Figure 1.12 showing the bounds $\pm 1.96/\sqrt{n}$.

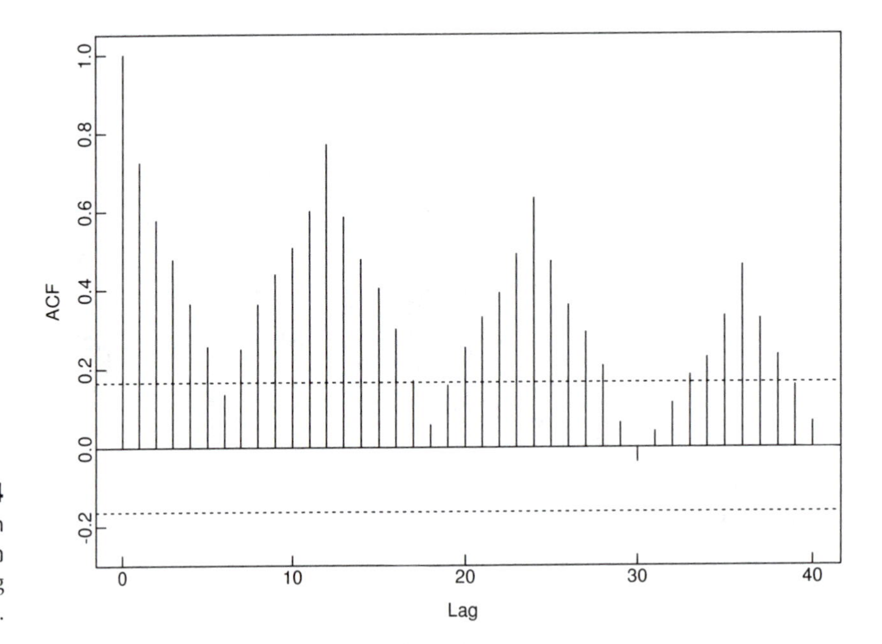

Figure 1-14
The sample autocorrelation function for the Australian red wine sales showing the bounds $\pm 1.96/\sqrt{n}$.

h increases, and for data with a substantial deterministic periodic component, $|\hat{\rho}(h)|$ will exhibit similar behavior with the same periodicity. (See the sample ACF of the Australian red wine sales in Figure 1.14 and Problem 1.9.) Thus $\hat{\rho}(\cdot)$ can be useful as an indicator of nonstationarity (see also Section 6.1). □

1.4.2 A Model for the Lake Huron Data

As noted earlier, an IID noise model for the residuals $\{y_1, \ldots, y_{98}\}$ obtained by fitting a straight line to the Lake Huron data in Example 1.3.5 appears to be inappropriate. This conclusion is confirmed by the sample ACF of the residuals (Figure 1.15), which has three of the first forty values well outside the bounds $\pm 1.96/\sqrt{98}$.

The roughly geometric decay of the first few sample autocorrelations (with $\hat{\rho}(h+1)/\hat{\rho}(h) \approx 0.7$) suggests that an AR(1) series (with $\phi \approx 0.7$) might provide a reasonable model for these residuals. (The form of the ACF for an AR(1) process was computed in Example 1.4.5.)

To explore the appropriateness of such a model, consider the points (y_1, y_2), $(y_2, y_3), \ldots, (y_{97}, y_{98})$ plotted in Figure 1.16. The graph does indeed suggest a linear relationship between y_t and y_{t-1}. Using simple least squares estimation to fit a straight line of the form $y_t = ay_{t-1}$, we obtain the model

$$Y_t = .791 Y_{t-1} + Z_t, \tag{1.4.3}$$

where $\{Z_t\}$ is IID noise with variance $\sum_{t=2}^{98}(y_t - .791 y_{t-1})^2/97 = .5024$. The sample ACF of the estimated noise sequence $z_t = y_t - .791 y_{t-1}, t = 2, \ldots, 98$ is slightly outside the bounds $\pm 1.96/\sqrt{97}$ at lag 1 ($\hat{\rho}(1) = .216$), but it is inside the bounds for

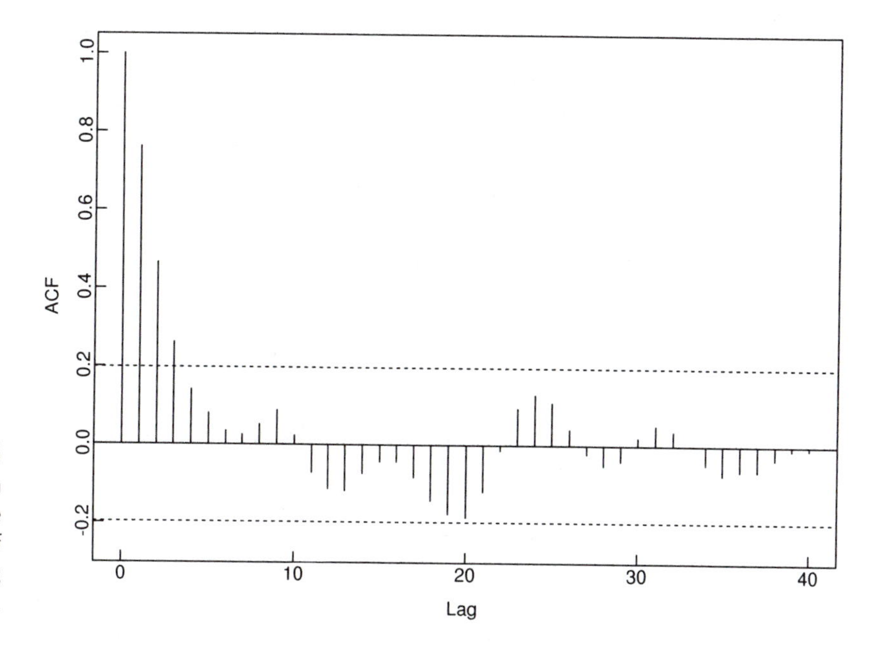

Figure 1-15 The sample autocorrelation function for the Lake Huron residuals of Figure 1.10 showing the bounds $\pm 1.96/\sqrt{n}$.

all other lags up to 40. This check that the estimated noise sequence is consistent with the IID assumption of (1.4.3) reinforces our belief in the fitted model. More *goodness of fit* tests for IID noise sequences are described in Section 1.6. The estimated noise sequence $\{z_t\}$ in this example passes them all, providing further support for the model (1.4.3).

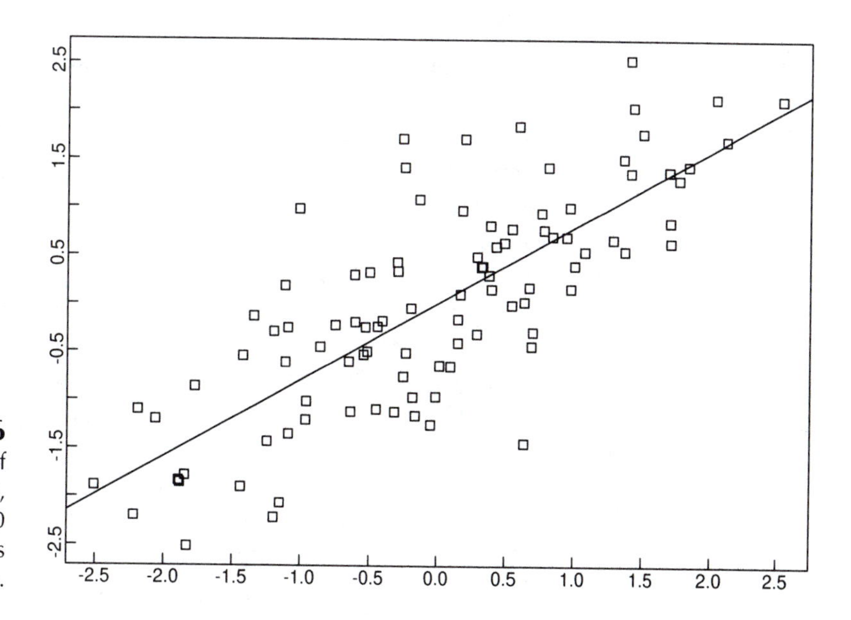

Figure 1-16 Scatter plot of $(y_{t-1}, y_t), t = 2, \ldots, 98$, for the data in Figure 1.10 showing the least squares regression line $y = .791x$.

A better fit to the residuals in equation (1.3.2) is provided by the second-order autoregression

$$Y_t = \phi_1 Y_{t-1} + \phi_2 Y_{t-2} + Z_t, \tag{1.4.4}$$

where $\{Z_t\}$ is IID noise with variance σ^2. This is analogous to a linear model in which Y_t is regressed on the previous *two* values Y_{t-1} and Y_{t-2} of the time series. The least squares estimates of the parameters ϕ_1 and ϕ_2, found by minimizing $\sum_{t=3}^{98}(y_t - \phi_1 y_{t-1} - \phi_2 y_{t-2})^2$, are $\hat{\phi}_1 = 1.002$ and $\hat{\phi}_2 = -.2834$. The estimate of σ^2 is $\hat{\sigma}^2 = \sum_{t=3}^{98}(y_t - \hat{\phi}_1 y_{t-1} - \hat{\phi}_2 y_{t-2})^2/96 = .4460$ which is approximately 11% smaller than the estimate of the noise variance for the AR(1) model (1.4.3). The improved fit is indicated by the sample ACF of the estimated residuals, $y_t - \hat{\phi}_1 y_{t-1} - \hat{\phi}_2 y_{t-2}$, which falls well within the bounds $\pm 1.96/\sqrt{96}$ for *all* lags up to 40.

1.5 Estimation and Elimination of Trend and Seasonal Components

The first step in the analysis of any time series is to plot the data. If there are any apparent discontinuities in the series, such as a sudden change of level, it may be advisable to analyze the series by first breaking it into homogeneous segments. If there are outlying observations, they should be studied carefully to check whether there is any justification for discarding them (as for example if an observation has been incorrectly recorded). Inspection of a graph may also suggest the possibility of representing the data as a realization of the process (the **classical decomposition** model),

$$X_t = m_t + s_t + Y_t, \tag{1.5.1}$$

where m_t is a slowly changing function known as a **trend component**, s_t is a function with known period d referred to as a **seasonal component**, and Y_t is a **random noise component** which is stationary in the sense of Definition 1.4.2. If the seasonal and noise fluctuations appear to increase with the level of the process, then a preliminary transformation of the data is often used to make the transformed data more compatible with the model (1.5.1). Compare for example the red wine sales in Figure 1.1 with the transformed data, Figure 1.17, obtained by applying a logarithmic transformation. The transformed data do not exhibit the increasing fluctuation with increasing level that was apparent in the original data. This suggests that the model (1.5.1) is more appropriate for the transformed than for the original series. In this section we shall assume that the model (1.5.1) is appropriate (possibly after a preliminary transformation of the data), and examine some techniques for estimating the components m_t, s_t, and Y_t in the model.

Our aim is to estimate and extract the deterministic components m_t and s_t in the hope that the residual or noise component Y_t will turn out to be a stationary time series. We can then use the theory of such processes to find a satisfactory probabilistic

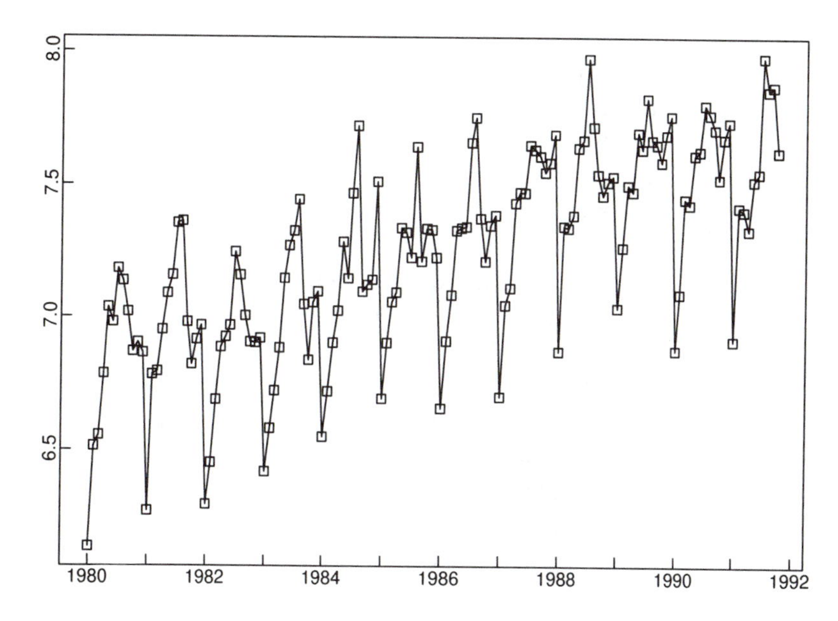

Figure 1-17 The natural logarithms of the red wine data.

model for the process Y_t, to analyze its properties, and to use it in conjunction with m_t and s_t for purposes of prediction and simulation of $\{X_t\}$.

Another approach, developed extensively by Box and Jenkins (1976), is to apply differencing operators repeatedly to the series $\{X_t\}$ until the differenced observations resemble a realization of some stationary time series $\{W_t\}$. We can then use the theory of stationary processes for the modelling, analysis, and prediction of $\{W_t\}$ and hence of the original process. The various stages of this procedure will be discussed in detail in Chapters 5 and 6.

The two approaches to trend and seasonality removal, (1) by estimation of m_t and s_t in (1.5.1) and (2) by differencing the series $\{X_t\}$, will now be illustrated with reference to the data introduced in Section 1.1.

1.5.1 Estimation and Elimination of Trend in the Absence of Seasonality

In the absence of a seasonal component the model (1.5.1) becomes the following.

Nonseasonal Model with Trend:

$$X_t = m_t + Y_t, t = 1, \ldots, n, \tag{1.5.2}$$

where $EY_t = 0$.

(If $EY_t \neq 0$, then we can replace m_t and Y_t in (1.5.2) with $m_t + EY_t$ and $Y_t - EY_t$, respectively.)

Method 1: Trend Estimation
Moving average and spectral smoothing are essentially nonparametric methods for trend (or signal) estimation and not for model building. Special smoothing filters can also be designed to remove periodic components as described under Method S1 below. The choice of smoothing filter requires a certain amount of subjective judgment and it is recommended that a variety of filters be tried in order to get a good idea of the underlying trend. Exponential smoothing, since it is based on a moving average of *past* values only, is often used for forecasting, the smoothed value at the present time being used as the forecast of the next value.

To construct a *model* for the data (with no seasonality) there are two recommended approaches, both available in PEST. One is to fit a polynomial trend (by least squares) as described in Method 1(d) below, then to subtract the fitted trend from the data and to find an appropriate stationary time series model for the residuals. The other is to eliminate the trend by differencing as described in Method 2 and then to find an appropriate stationary model for the differenced series. The latter method has the advantage that it usually requires the estimation of fewer parameters and does not rest on the assumption of a trend that remains fixed throughout the observation period. The study of the residuals (or of the differenced series) is taken up in Section 1.6.

(a) *Smoothing with a finite moving average filter* Let q be a nonnegative integer and consider the two-sided moving average,

$$W_t = (2q+1)^{-1} \sum_{j=-q}^{q} X_{t-j}, \tag{1.5.3}$$

of the process $\{X_t\}$ defined by (1.5.2). Then for $q+1 \le t \le n-q$,

$$W_t = (2q+1)^{-1} \sum_{j=-q}^{q} m_{t-j} + (2q+1)^{-1} \sum_{j=-q}^{q} Y_{t-j} \simeq m_t, \tag{1.5.4}$$

assuming that m_t is approximately linear over the interval $[t-q, t+q]$ and that the average of the error terms over this interval is close to zero (see Problem 1.11).

The moving average thus provides us with the estimates

$$\hat{m}_t = (2q+1)^{-1} \sum_{j=-q}^{q} X_{t-j}, q+1 \le t \le n-q. \tag{1.5.5}$$

Since X_t is not observed for $t \le 0$ or $t > n$ we cannot use (1.5.5) for $t \le q$ or $t > n-q$. The program SMOOTH deals with this problem by defining $X_t := X_1$ for $t < 1$ and $X_t := X_n$ for $t > n$.

Example 1.5.1 The result of applying the moving average filter (1.5.5) with $q = 2$ to the strike data of Figure 1.6 is shown in Figure 1.18. The estimated noise terms, $\hat{Y}_t = X_t - \hat{m}_t$, are shown in Figure 1.19. As expected, they show no apparent trend. □

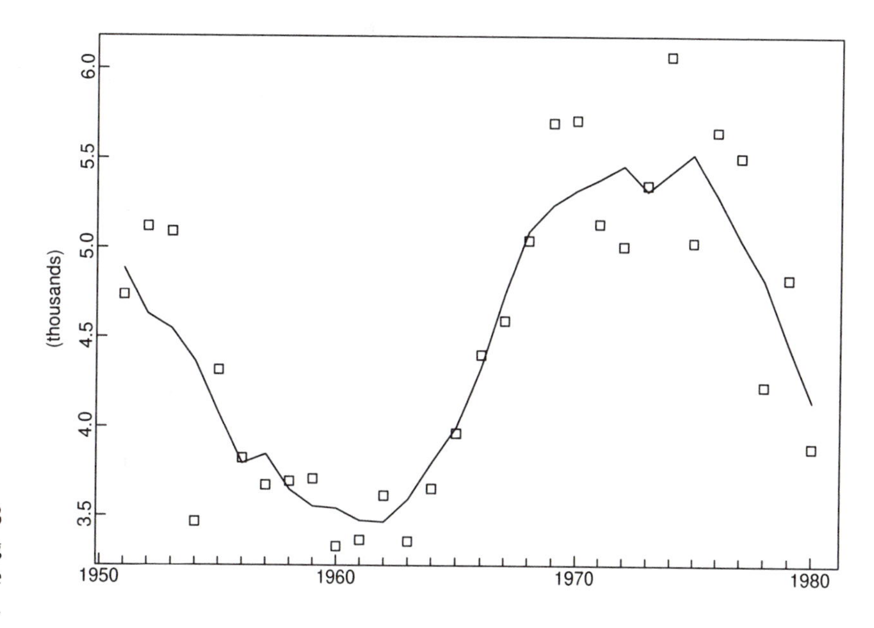

Figure 1-18 Simple 5-term moving average $\hat{m}_t$ of the strike data from Figure 1.6.

It is useful to think of $\{\hat{m}_t\}$ in (1.5.5) as a process obtained from $\{X_t\}$ by application of a linear operator or linear filter, $\hat{m}_t = \sum_{j=-\infty}^{\infty} a_j X_{t-j}$ with weights $a_j = (2q + 1)^{-1}$, $-q \le j \le q$. This particular filter is a **low-pass** filter in the sense that it takes the data $\{X_t\}$ and removes from it the rapidly fluctuating (or high frequency) component $\{\hat{Y}_t\}$ to leave the slowly varying estimated trend term $\{\hat{m}_t\}$ (see Figure 1.20).

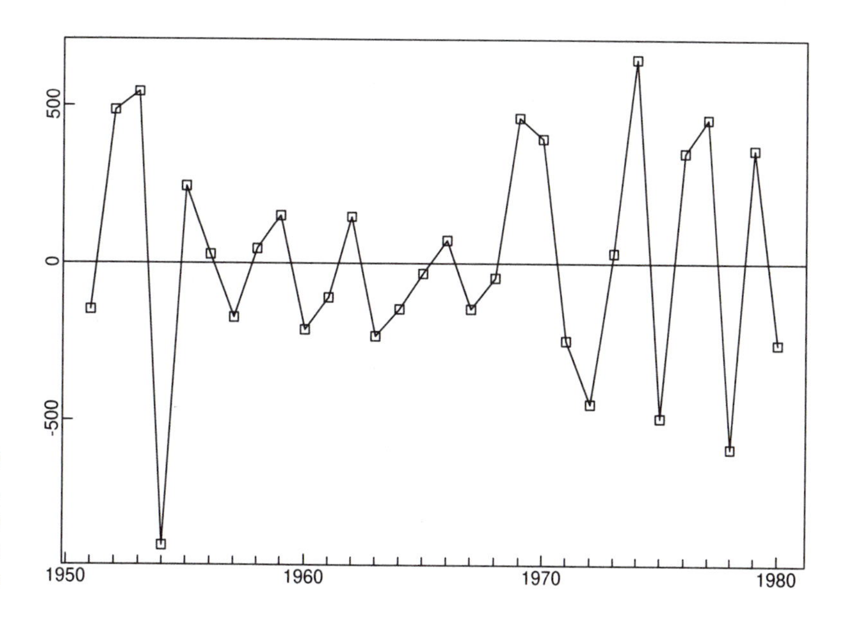

Figure 1-19 Residuals $\hat{Y}_t = X_t - \hat{m}_t$ after subtracting the 5-term moving average from the strike data

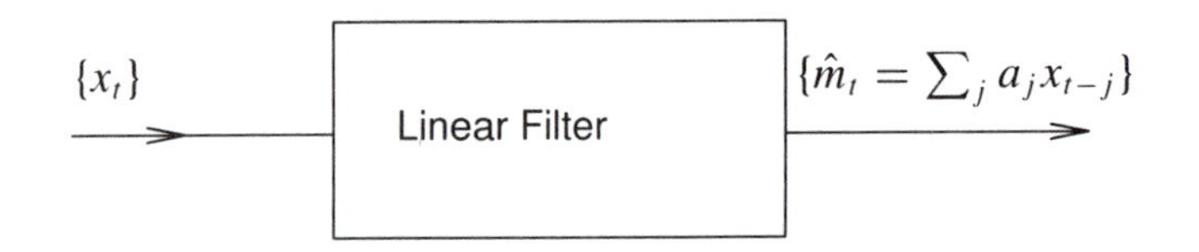

Figure 1-20 Smoothing with a low-pass linear filter.

The particular filter (1.5.5) is only one of many that could be used for smoothing. For large q, provided $(2q+1)^{-1}\sum_{j=-q}^{q} Y_{t-j} \simeq 0$, it will not only attenuate noise but at the same time will allow linear trend functions $m_t = c_0 + c_1 t$, to pass without distortion (see Problem 1.11). However we must beware of choosing q to be too large since if m_t is not linear, the filtered process, although smooth, will not be a good estimate of m_t. By clever choice of the weights $\{a_j\}$ it is possible (see Problems 1.12–1.14 and Section 4.3) to design a filter that will not only be effective in attenuating noise in the data, but which will also allow a larger class of trend functions (for example all polynomials of degree less than or equal to 3) to pass through without distortion. The Spencer 15-point moving average is a filter that passes polynomials of degree 3 without distortion. Its weights are

$$a_j = 0, |j| > 7,$$

with

$$a_j = a_{-j}, |j| \le 7,$$

and

$$[a_0, a_1, \ldots, a_7] = \frac{1}{320}[74, 67, 46, 21, 3, -5, -6, -3]. \tag{1.5.6}$$

Applied to the process (1.5.2) with $m_t = c_0 + c_1 t + c_2 t^2 + c_3 t^3$, it gives

$$\begin{aligned}\sum_{j=-7}^{7} a_j X_{t-j} &= \sum_{j=-7}^{7} a_j m_{t-j} + \sum_{j=-7}^{7} a_j Y_{t-j}\\ &\simeq \sum_{j=-7}^{7} a_j m_{t-j}\\ &= m_t,\end{aligned}$$

where the last step depends on the assumed form of m_t (Problem 1.12). Further details regarding this and other smoothing filters can be found in Kendall and Stuart (1976), Chapter 46.

(b) *Exponential smoothing* For any fixed $a \in [0, 1]$, the one-sided moving averages $\hat{m}_t$, $t = 1, \ldots, n$ defined by the recursions

$$\hat{m}_t = aX_t + (1-a)\hat{m}_{t-1}, t = 2, \ldots, n, \tag{1.5.7}$$

and

$$\hat{m}_1 = X_1, \tag{1.5.8}$$

can also be computed using the program SMOOTH. Application of (1.5.7) and (1.5.8) is often referred to as exponential smoothing, since the recursions imply that for $t \ge 2$, $\hat{m}_t = \sum_{j=0}^{t-2} a(1-a)^j X_{t-j} + (1-a)^{t-1} X_1$, a weighted moving average of $X_t, X_{t-1}, \ldots$, with weights decreasing exponentially (except for the last one).

(c) *Smoothing by elimination of high frequency components* The program SMOOTH allows us to smooth an arbitrary series either by application of a symmetric filter with coefficients $a_{-q}, a_{-q+1}, \ldots, a_q$ or by application of the exponential smoother defined by (1.5.7) and (1.5.8) with specified $a \in [0, 1]$. It also allows smoothing of a series $\{X_t\}$ by elimination of the uppermost frequency components of its Fourier series expansion (see Section 4.2). This option was used in Example 1.5.2, where we chose to retain the fraction $f = .04$ of the frequency components of the series in order to estimate the underlying signal. (The choice $f = 1$ would have left the series unchanged.)

Example 1.5.2 In Figures 1.21 and 1.22 we show the results of smoothing the strikes data by exponential smoothing with parameter $a = 0.4$ (see (1.5.7)) and by high frequency elimination with $f = 0.4$, i.e., by eliminating a fraction 0.6 of the Fourier components at the top of the frequency range. These should be compared with the simple 5-term moving average smoothing shown in Figure 1.18. Experimentation with different smoothing parameters can easily be carried out using the program SMOOTH. The exponentially smoothed value of the last observation is frequently used to forecast the next data value. The program can automatically select an optimal value of a for this purpose. □

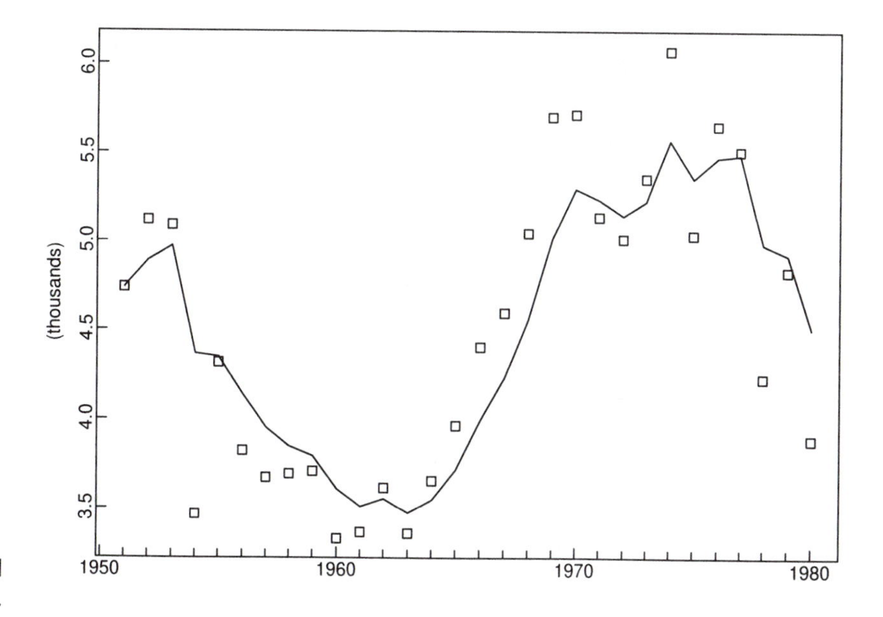

Figure 1-21 Exponentially smoothed strike data with $a = 0.4$.

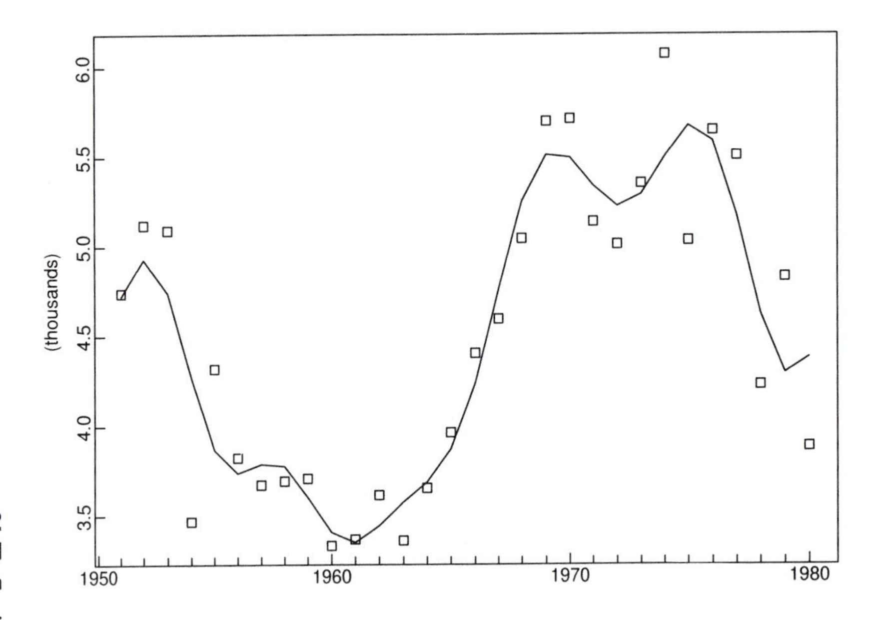

Figure 1-22
Strike data smoothed by elimination of high frequencies with $f = 0.4$.

(d) *Polynomial fitting* In Section 1.3.2 we showed how a trend of the form $m_t = a_0 + a_1 t + a_2 t^2$ can be fitted to the data $\{x_1, \ldots, x_n\}$, by choosing the parameters a_0, a_1, and a_2, to minimize the sum of squares, $\sum_{t=1}^{n}(x_t - m_t)^2$ (see Example 1.3.4). The method of least squares estimation can also be used to estimate higher order polynomial trends in the same way. The program PEST allows least squares fitting of either linear or quadratic trends.

Method 2: Trend Elimination by Differencing

Instead of attempting to remove the noise by smoothing as in Method 1, we now attempt to eliminate the trend term by differencing. We define the lag-1 difference operator ∇ by

$$\nabla X_t = X_t - X_{t-1} = (1 - B)X_t, \tag{1.5.9}$$

where B is the backward shift operator,

$$BX_t = X_{t-1}. \tag{1.5.10}$$

Powers of the operators B and ∇ are defined in the obvious way, i.e., $B^j(X_t) = X_{t-j}$ and $\nabla^j(X_t) = \nabla(\nabla^{j-1}(X_t))$, $j \geq 1$ with $\nabla^0(X_t) = X_t$. Polynomials in B and ∇ are manipulated in precisely the same way as polynomial functions of real variables. For example,

$$\begin{aligned}\nabla^2 X_t = \nabla(\nabla(X_t)) &= (1 - B)(1 - B)X_t = (1 - 2B + B^2)X_t \\ &= X_t - 2X_{t-1} + X_{t-2}.\end{aligned}$$

If the operator ∇ is applied to a linear trend function $m_t = c_0 + c_1 t$, then we obtain the constant function $\nabla m_t = m_t - m_{t-1} = c_0 + c_1 t - (c_0 + c_1(t-1)) = c_1$. In the same way any polynomial trend of degree k can be reduced to a constant by application of the operator ∇^k (Problem 1.10). For example, if $X_t = m_t + Y_t$, where $m_t = \sum_{j=0}^{k} c_j t^j$ and Y_t is stationary with mean zero, application of ∇^k gives

$$\nabla^k X_t = k! c_k + \nabla^k Y_t,$$

a stationary process with mean $k! c_k$. These considerations suggest the possibility, given any sequence $\{x_t\}$ of data, of applying the operator ∇ repeatedly until we find a sequence $\{\nabla^k x_t\}$ that can plausibly be modelled as a realization of a stationary process. It is often found in practice that the order k of differencing required is quite small, frequently one or two. (This relies on the fact that many functions can be well approximated, on an interval of finite length, by a polynomial of reasonably low degree.)

Example 1.5.3 Applying the operator ∇ to the population values $\{x_t, t = 1, \ldots, 20\}$ of Figure 1.5, we find that two differencing operations are sufficient to produce a series with no apparent trend. (The differencing can easily be carried out using the program PEST.) The differenced data, $\nabla^2 x_t = x_t - 2x_{t-1} + x_{t-2}$, are plotted in Figure 1.23. Notice that the magnitude of the fluctuations in $\nabla^2 x_t$ increases with the value of x_t. This effect can be suppressed by first taking natural logarithms, $y_t = \ln x_t$, and then applying the operator ∇^2 to the series $\{y_t\}$. (See also Figures 1.1 and 1.17.) □

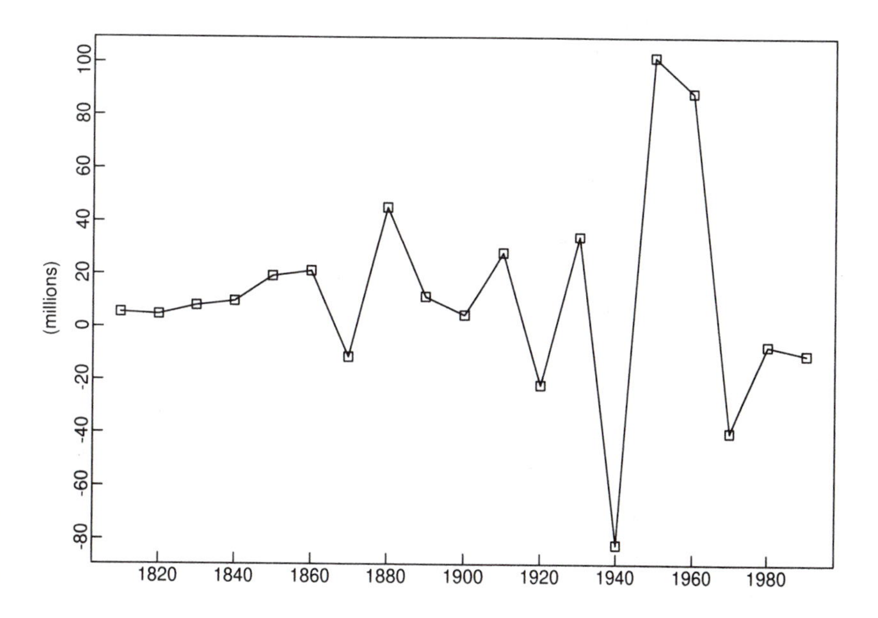

Figure 1-23 The twice-differenced series derived from the population data of Figure 1.5.

1.5.2 Estimation and Elimination of Both Trend and Seasonality

The methods described for the estimation and elimination of trend can be adapted in a natural way to eliminate both trend and seasonality in the general model, specified as follows.

Classical Decomposition Model

$$X_t = m_t + s_t + Y_t, t = 1, \ldots, n, \tag{1.5.11}$$

where $EY_t = 0$, $s_{t+d} = s_t$ and $\sum_{j=1}^{d} s_j = 0$.

We shall illustrate these methods with reference to the accidental deaths data of Example 1.1.3, for which the period d of the seasonal component is clearly 12.

Method S1: Estimation of Trend and Seasonal Components

The method we are about to describe is used in the "classical decomposition" option in the Data Menu of the program PEST.

Suppose we have observations $\{x_1, \ldots, x_n\}$. The trend is first estimated by applying a moving average filter specially chosen to eliminate the seasonal component and to dampen the noise. If the period d is even, say $d = 2q$, then we use

$$\hat{m}_t = (0.5x_{t-q} + x_{t-q+1} + \cdots + x_{t+q-1} + 0.5x_{t+q})/d, q < t \le n - q. \tag{1.5.12}$$

If the period is odd, say $d = 2q + 1$, then we use the simple moving average (1.5.5).

The second step is to estimate the seasonal component. For each $k = 1, \ldots, d$, we compute the average w_k of the deviations $\{(x_{k+jd} - \hat{m}_{k+jd}), q < k + jd \le n - q\}$. Since these average deviations do not necessarily sum to zero, we estimate the seasonal component s_k as

$$\hat{s}_k = w_k - d^{-1} \sum_{i=1}^{d} w_i, k = 1, \ldots, d, \tag{1.5.13}$$

and $\hat{s}_k = \hat{s}_{k-d}, k > d$.

The *deseasonalized* data is then defined to be the original series with the estimated seasonal component removed, i.e.,

$$d_t = x_t - \hat{s}_t, t = 1, \ldots, n. \tag{1.5.14}$$

Finally we reestimate the trend from the deseasonalized data $\{d_t\}$ using one of the methods already described. The program PEST allows you to fit either a linear or quadratic trend $\hat{m}_t$ by least squares to the deseasonalized series. In terms of this reestimated trend and the estimated seasonal component, the estimated noise series is then given by

$$\hat{Y}_t = x_t - \hat{m}_t - \hat{s}_t, t = 1, \ldots, n.$$

The reestimation of the trend is done in order to have a parametric form for the trend which can be extrapolated for the purposes of prediction and simulation.

Example 1.5.4 Figure 1.24 shows the deseasonalized accidental deaths data obtained from the program PEST by reading in the original series, and deseasonalizing by specifying the period as 12. The estimated seasonal component $\hat{s}_t$ is plotted in Figure 1.25. (After correcting for the mean, this estimate is similar to the one produced using harmonic regression with frequencies $2\pi/12$ and $2\pi/6$ displayed in Figure 1.11.) If a quadratic trend $\hat{m}_t$ is then fitted to the deseasonalized data, we find from the program that

$$\hat{m}_t = 9952 - 71.82t + 0.8260t^2, \; 1 \le t \le 72.$$

At this point the data stored in PEST is the estimated noise series

$$\hat{Y}_t = x_t - \hat{m}_t - \hat{s}_t, \; t = 1, \ldots, 72,$$

obtained by subtracting the estimated seasonal and trend components from the original data. □

Method S2: Elimination of Trend and Seasonal Components by Differencing
The technique of differencing that we applied earlier to nonseasonal data can be adapted to deal with seasonality of period d by introducing the lag-d differencing operator ∇_d defined by

$$\nabla_d X_t = X_t - X_{t-d} = (1 - B^d)X_t. \tag{1.5.15}$$

(This operator should not be confused with the operator $\nabla^d = (1 - B)^d$ defined earlier.)

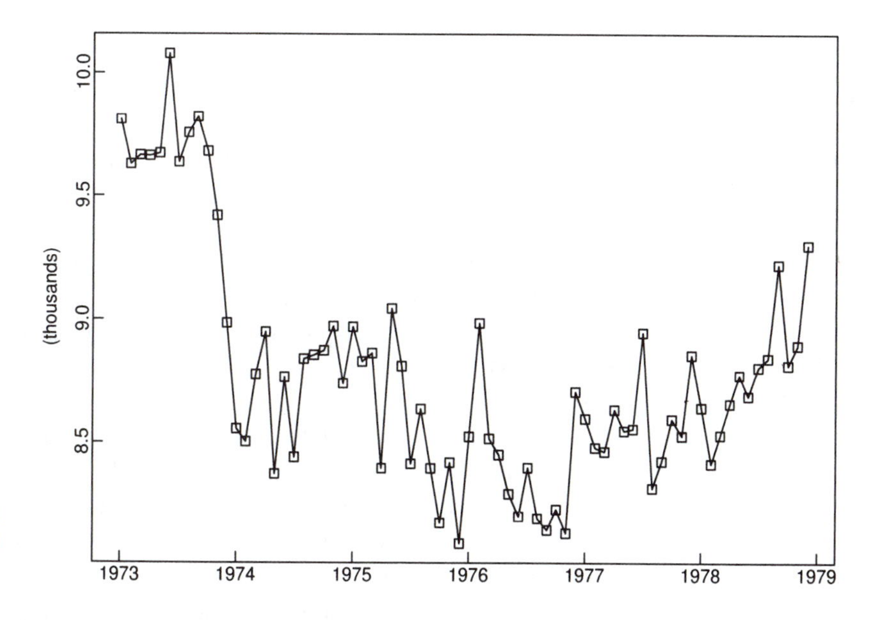

Figure 1-24 The deseasonalized accidental deaths data from PEST.

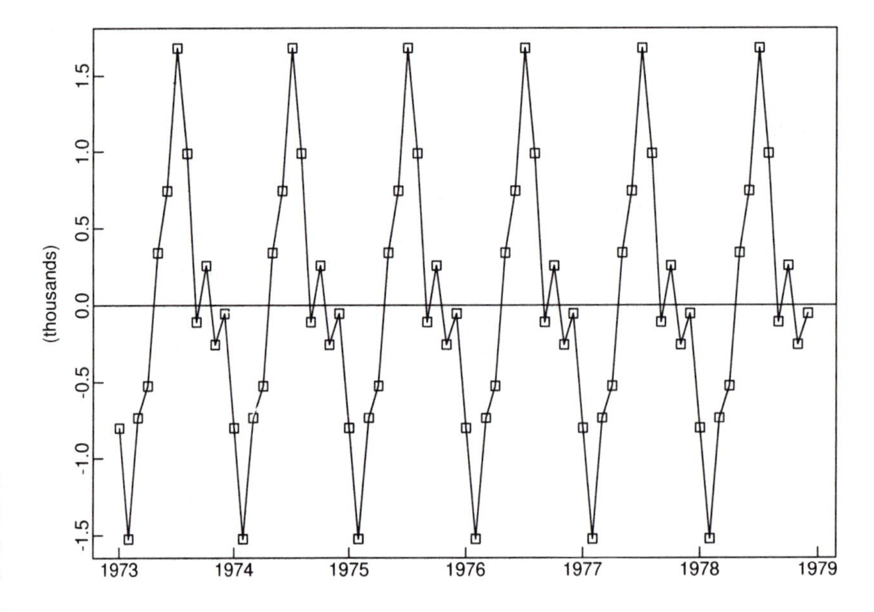

Figure 1-25 The estimated seasonal component of the accidental deaths data from PEST.

Applying the operator ∇_d to the model,

$$X_t = m_t + s_t + Y_t,$$

where $\{s_t\}$ has period d, we obtain

$$\nabla_d X_t = m_t - m_{t-d} + Y_t - Y_{t-d},$$

which gives a decomposition of the difference $\nabla_d X_t$ into a trend component $(m_t - m_{t-d})$ and a noise term $(Y_t - Y_{t-d})$. The trend, $m_t - m_{t-d}$, can then be eliminated using the methods already described, in particular by applying a power of the operator ∇.

Example 1.5.5 Figure 1.26 shows the result of applying the operator ∇_{12} to the accidental deaths data. The seasonal component evident in Figure 1.3 is absent from the graph of $\nabla_{12}x_t$, $13 \le t \le 72$. There still appears to be a nondecreasing trend, however. If we now apply the operator ∇ to $\{\nabla_{12}x_t\}$ and plot the resulting differences $\nabla\nabla_{12}x_t$, $14 \le t \le 72$, we obtain the graph shown in Figure 1.27, which has no apparent trend or seasonal component. In Chapter 5 we shall show that this doubly differenced series can in fact be well represented by a stationary time series model. □

In this section we have discussed a variety of methods for estimating and/or removing trend and seasonality. The particular method chosen for any given data set will depend on a number of factors including whether or not estimates of the components of the series are required and whether or not it appears that the data contain a seasonal component that does not vary with time. The program PEST allows two options, "classical decomposition," in which trend and/or seasonal components are

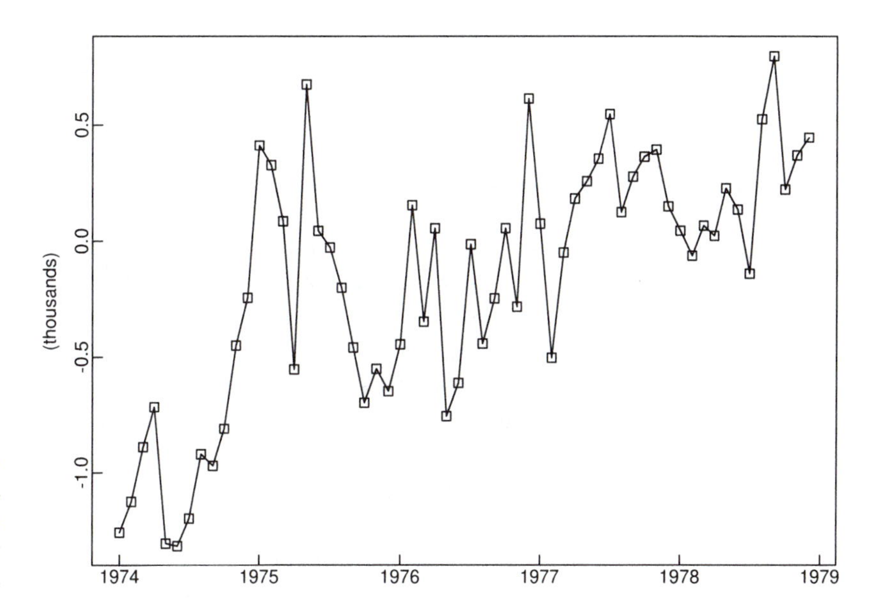

Figure 1-26 The differenced series $\{\nabla_{12}x_t, t = 13, \ldots, 72\}$ derived from the monthly accidental deaths $\{x_t, t = 1, \ldots, 72\}$.

estimated and subtracted from the data to generate a noise sequence, and "differencing," in which trend and/or seasonal components are removed from the data by repeated differencing at one or more lags in order to generate a noise sequence. In the next section we shall examine some techniques for deciding whether or not the noise sequence so generated differs significantly from IID noise. If the noise sequence *does* have sample autocorrelations significantly different from zero, then we can take

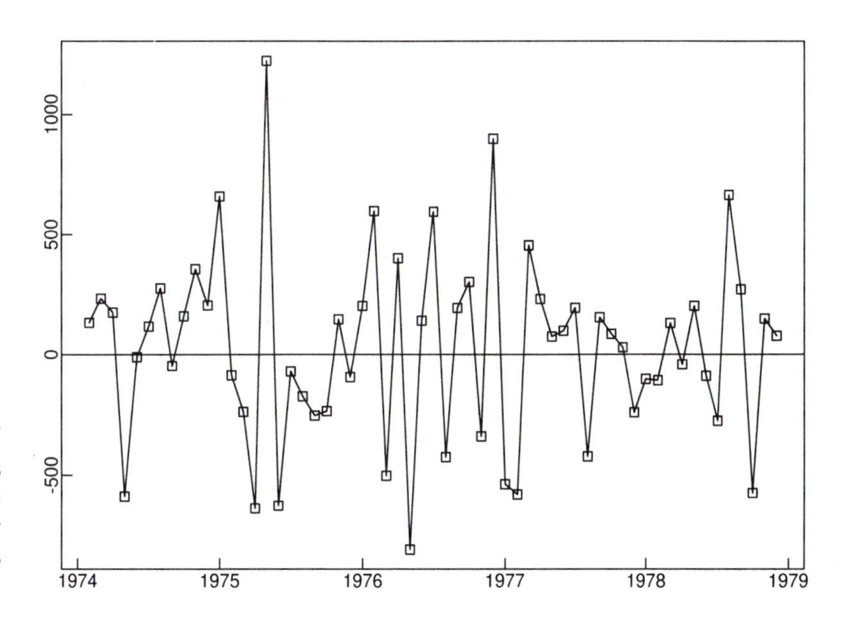

Figure 1-27 The differenced series $\{\nabla\nabla_{12}x_t, t = 14, \ldots, 72\}$ derived from the monthly accidental deaths $\{x_t, t = 1, \ldots, 72\}$.

advantage of this serial dependence to forecast future noise values in terms of past values by modelling the noise as a stationary time series.

1.6 Testing the Estimated Noise Sequence

The objective of the data transformations described in Section 1.5 is to produce a series with no apparent deviations from stationarity, and in particular with no apparent trend or seasonality. Assuming that this has been done, the next step is to model the estimated noise sequence (i.e., the **residuals** obtained either by differencing the data or by estimating and subtracting the trend and seasonal components). If there is no dependence between these residuals, then we can regard them as observations of independent random variables and there is no further modelling to be done except to estimate their mean and variance. However, if there is significant dependence between the residuals, then we need to look for a more complex stationary time series model for the noise that accounts for the dependence. This will be to our advantage since dependence means in particular that past observations of the noise sequence can assist in predicting future values.

In this section we examine some simple tests for checking the hypothesis that the residuals from Section 1.5 are observed values of independent and identically distributed random variables. If they are, then our work is done. If not, then we must use the theory of stationary processes to be developed in later chapters to find a more appropriate model.

(a) *The Sample Autocorrelation Function* For large n, the sample autocorrelations of an iid sequence $Y_1, \ldots, Y_n$ with finite variance are approximately iid with distribution $\mathrm{N}(0, 1/n)$ (see TSTM p. 222). Hence, if $y_1, \ldots, y_n$ is a realization of such an iid sequence, about 95% of the sample autocorrelations should fall between the bounds $\pm 1.96/\sqrt{n}$. If we compute the sample autocorrelations up to lag 40 and find that more than two or three values fall outside the bounds, or that one value falls far outside the bounds, we therefore reject the iid hypothesis. The bounds $\pm 1.96/\sqrt{n}$ are automatically plotted when the sample autocorrelation function is computed by the program PEST.

(b) *The Portmanteau Test* Instead of checking to see if each sample autocorrelation $\hat{\rho}(j)$ falls inside the bounds defined in (a) above, it is also possible to consider the single statistic

$$Q = n \sum_{j=1}^{h} \hat{\rho}^2(j).$$

If $Y_1, \ldots, Y_n$ is a finite variance iid sequence, then by the same result used in (a), Q is approximately distributed as the sum of squares of the independent $\mathrm{N}(0, 1)$ random variables, $\sqrt{n}\hat{\rho}(j)$, $j = 1, \ldots, h$, i.e., as chi-squared with h degrees of freedom. A large value of Q suggests that the sample autocorrelations of the data are too large for the data to be a sample from an iid sequence. We therefore reject the iid hypothesis at level α if $Q > \chi^2_{1-\alpha}(h)$, where $\chi^2_{1-\alpha}(h)$ is the $1-\alpha$ quantile of the chi-squared distribution with h degrees of freedom. The program PEST conducts a refinement of this test, formulated by Ljung and Box (1978), in which Q is replaced by

$$Q_{LB} = n(n+2)\sum_{j=1}^{h} \hat{\rho}^2(j)/(n-j),$$

whose distribution is better approximated by the chi-squared distribution with h degrees of freedom.

Another portmanteau test, formulated by McLeod and Li (1983), can be used for testing whether the data are observations from an iid sequence of *normally distributed* random variables. It is based on the same statistic used for the Ljung-Box test, except that the sample autocorrelations of the data are replaced by the sample autocorrelations of the *squared* data, $\hat{\rho}_{WW}(h)$, giving

$$\tilde{Q} = n(n+2)\sum_{k=1}^{h} \hat{\rho}^2_{WW}(k)/(n-k).$$

The hypothesis of iid normal data is then rejected at level α if the observed value of $\tilde{Q}$ is larger than the (1-α) quantile of the $\chi^2(h)$ distribution.

(c) *The Turning Point Test* If $y_1, \ldots, y_n$ is a sequence of observations, we say that there is a turning point at time i, $1 < i < n$, if $y_{i-1} < y_i$ and $y_i > y_{i+1}$ or if $y_{i-1} > y_i$ and $y_i < y_{i+1}$. If T is the number of turning points of an iid sequence of length n, then, since the probability of a turning point at time i is $\frac{2}{3}$, the expected value of T is

$$\mu_T = E(T) = 2(n-2)/3.$$

It can also be shown for an iid sequence that the variance of T is

$$\sigma_T^2 = \text{Var}(T) = (16n - 29)/90.$$

A large value of $T - \mu_T$ indicates that the series is fluctuating more rapidly than expected for an iid sequence. On the other hand a value of $T - \mu_T$ much smaller than zero indicates a positive correlation between neighboring observations. For an iid sequence with n large, it can be shown that

$$T \text{ is approximately } \text{N}(\mu_T, \sigma_T^2).$$

This means we can carry out a test of the iid hypothesis, rejecting it at level α if $|T - \mu_T|/\sigma_T > \Phi_{1-\alpha/2}$, where $\Phi_{1-\alpha/2}$ is the $1-\alpha/2$ quantile of the standard normal distribution. (A commonly used value of α is .05, for which the corresponding value of $\Phi_{1-\alpha/2}$ is 1.96.)

(d) *The Difference-Sign Test* For this test we count the number, S, of values of i such that $y_i > y_{i-1}$, $i = 2, \ldots, n$ or equivalently the number of times the differenced series $y_i - y_{i-1}$ is positive. For an iid sequence it is clear that

$$\mu_S = ES = \frac{1}{2}(n-1).$$

It can also be shown, under the same assumption, that

$$\sigma_S^2 = \text{Var}(S) = (n+1)/12,$$

and that for large n,

$$S \text{ is approximately } \mathrm{N}(\mu_S, \sigma_S^2).$$

A large positive (or negative) value of $S - \mu_S$ indicates the presence of an increasing (or decreasing) trend in the data. We therefore reject the assumption of no trend in the data if $|S - \mu_S|/\sigma_S > \Phi_{1-\alpha/2}$.

The difference-sign test must be used with caution. A set of observations exhibiting a strong cyclic component will pass the difference-sign test for randomness since roughly half of the observations will be points of increase.

(e) *The Rank Test* The rank test is particularly useful for detecting a linear trend in the data. Define P to be the number of pairs (i, j) such that $y_j > y_i$, and $j > i, i = 1, \ldots, n-1$. There is a total of $\binom{n}{2} = \frac{1}{2}n(n-1)$ pairs (i, j) such that $j > i$. For an iid sequence $\{Y_1, \ldots, Y_n\}$, each event $\{Y_j > Y_i\}$ has probability $\frac{1}{2}$ and the mean of P is therefore

$$\mu_P = \frac{1}{4}n(n-1).$$

It can also be shown for an iid sequence that the variance of P is

$$\sigma_P^2 = n(n-1)(2n+5)/72$$

and that for large n,

$$P \text{ is approximately } \mathrm{N}(\mu_P, \sigma_P^2)$$

(see Kendall and Stuart, 1976). A large positive (negative) value of $P - \mu_P$ indicates the presence of an increasing (decreasing) trend in the data. The assumption that $\{y_j\}$ is a sample from an iid sequence is therefore rejected at level $\alpha = 0.05$ if $|P - \mu_P|/\sigma_P > \Phi_{1-\alpha/2} = 1.96$.

(f) *Checking for Normality* If the noise process is Gaussian, i.e., if all of its joint distributions are normal, then stronger conclusions can be drawn when a model is fitted to the data. The following test enables us to check whether it is reasonable to assume that observations from an iid sequence are also Gaussian.

Let $Y_{(1)} < Y_{(2)} < \cdots < Y_{(n)}$ be the order statistics of a random sample $Y_1, \ldots, Y_n$ from the distribution $\mathrm{N}(\mu, \sigma^2)$. If $X_{(1)} < X_{(2)} < \cdots < X_{(n)}$ are the order statistics from a $\mathrm{N}(0, 1)$ sample of size n, then

$$EY_{(j)} = \mu + \sigma m_j,$$

where $m_j = EX_{(j)}$, $j = 1, \ldots, n$. Thus a graph of the points $(m_1, Y_{(1)}), \ldots, (m_n, Y_{(n)})$ (known as a **q-q plot**) should be approximately linear. However, if the sample values Y_i are not normally distributed, then the graph should be nonlinear. Consequently the squared correlation of the points $(m_i, Y_{(i)})$, $i = 1, \ldots, n$ should be near 1 if the normal assumption is correct. The assumption of normality is therefore rejected if the squared correlation R^2 is sufficiently small. If we approximate m_i by $\Phi^{-1}((i - .5)/n)$ (see Mage, 1982 for some alternative approximations), then R^2 reduces to

$$R^2 = \frac{\left(\sum_{i=1}^n (Y_{(i)} - \overline{Y})\Phi^{-1}\left(\frac{i-.5}{n}\right)\right)^2}{\sum_{i=1}^n (Y_{(i)} - \overline{Y})^2 \sum_{i=1}^n \left(\Phi^{-1}\left(\frac{i-.5}{n}\right)\right)^2},$$

where $\overline{Y} = n^{-1}(Y_1 + \cdots + Y_n)$. Percentage points for the distribution of R^2, assuming normality of the sample values, are given by Shapiro and Francia (1972) for sample sizes $n < 100$. For $n = 200$, $P(R^2 < .987) = .05$ and $P(R^2 < .989) = .10$.

Example 1.6.1 If we did not know in advance how the signal plus noise data of Example 1.1.4 were generated, we might suspect that it came from an iid sequence. To check this hypothesis we can apply the tests (a)–(e) introduced above using the program PEST.

(a) Figure 1.28 shows the sample autocorrelation function of the data. Since over 25% of the autocorrelations are outside the bounds $\pm 1.96/\sqrt{200}$, we reject the hypothesis that the series is iid.

(b) The sample value of Q with $h = 20$ is 50.40. Comparing this with the .95 quantile of $\chi^2(20)$, i.e., 31.4, we reject the iid hypothesis at level .05.

(c) The sample value of T is 138 and the asymptotic distribution under the iid hypothesis (with $n = 200$) is N(132, 35.3). Thus $|T - \mu_T|/\sigma_T = 1.01$, which is

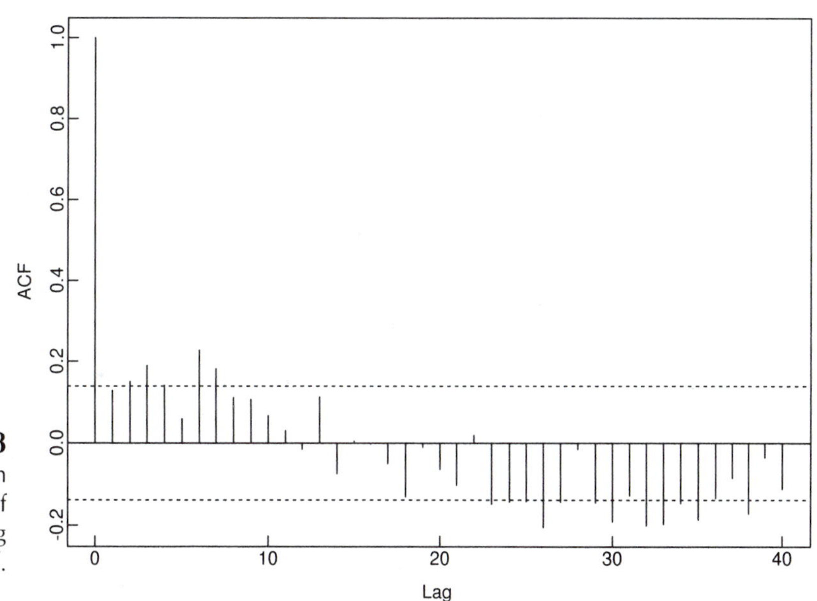

Figure 1-28 The sample autocorrelation function for the data of Example 1.1.4 showing the bounds $\pm 1.96/\sqrt{n}$.

substantially less than the cutoff value of 1.96 for a test of level .05. On the basis of the value of T there is not sufficient evidence to reject the iid hypothesis.

(d) The sample value of S is 101 and the asymptotic distribution under the iid hypothesis (with $n = 200$) is N(99.5, 16.7). Thus $|S - \mu_S|/\sigma_S = 0.38$, which is substantially less than the cutoff value of 1.96 for a test of level .05. On the basis of the value of S there is not sufficient evidence to reject the iid hypothesis.

(e) The sample value of P is 10310 and the asymptotic distribution under the iid hypothesis (with $n = 200$) is $\text{N}(9950, 2.015 \times 10^6)$. Thus $|P - \mu_P|/\sigma_P = 0.23$, which is substantially less than the cutoff value of 1.96 for a test of level .05. On the basis of the value of P there is not sufficient evidence to reject the iid hypothesis.

Thus, although tests (c), (d), and (e) are not able to detect any deviation from iid behavior, the tests (a) and (b) do so quite strongly, causing us to (correctly) reject the iid hypothesis in this example. We shall return to the tests for normality in Chapter 5. □

The general strategy in applying the tests described in this section is to check them all and to proceed with caution if any of them suggests a serious deviation from the iid hypothesis. (Remember that as you increase the number of tests, the probability that *at least one* rejects the null hypothesis when it is true increases. You should therefore not necessarily reject the null hypothesis on the basis of one test result only.)

Problems

1.1. Let X and Y be two random variables with $E(Y) = \mu$ and $EY^2 < \infty$.

a. Show that the constant c that minimizes $E(Y - c)^2$ is $c = \mu$.

b. Deduce that the random variable $f(X)$ that minimizes $E[(Y - f(X))^2|X]$ is

$$f(X) = E[Y|X].$$

c. Deduce that the random variable $f(X)$ that minimizes $E(Y - f(X))^2$ is also

$$f(X) = E[Y|X].$$

1.2. (Generalization of Problem 1.1.) Suppose that $X_1, X_2, \ldots,$ is a sequence of random variables with $E(X_t^2) < \infty$ and $E(X_t) = \mu$.

a. Show that the random variable $f(X_1, \ldots, X_n)$ that minimizes $E[(X_{n+1} - f(X_1, \ldots, X_n))^2|X_1, \ldots, X_n]$ is

$$f(X_1, \ldots, X_n) = E[X_{n+1}|X_1, \ldots, X_n].$$

b. Deduce that the random variable $f(X_1, \ldots, X_n)$ that minimizes $E[(X_{n+1} - f(X_1, \ldots, X_n))^2]$ is also

$$f(X_1, \ldots, X_n) = E[X_{n+1}|X_1, \ldots, X_n].$$

c. If $X_1, X_2, \ldots$ is iid with $E(X_i^2) < \infty$ and $EX_i = \mu$, where μ is known, what is the minimum mean squared error predictor of X_{n+1} in terms of $X_1, \ldots, X_n$?

d. Under the conditions of part (c) show that the best linear unbiased estimator of μ in terms of $X_1, \ldots, X_n$ is $\bar{X} = \frac{1}{n}(X_1 + \cdots + X_n)$. ($\hat{\mu}$ said to be an unbiased estimator of μ if $E\hat{\mu} = \mu$ for all μ.)

e. Under the conditions of part (c) show that $\bar{X}$ is the best linear predictor of X_{n+1} which is unbiased for μ.

f. If $X_1, X_2, \ldots$ is iid with $E(X_i^2) < \infty$ and $EX_i = \mu$, and if $S_0 = 0$, $S_n = X_1 + \cdots + X_n$, $n = 1, 2, \ldots$, what is the minimum mean squared error predictor of S_{n+1} in terms of $S_1, \ldots, S_n$?

1.3. Show that a strictly stationary process with $E(X_i^2) < \infty$ is weakly stationary.

1.4. Let $\{Z_t\}$ be a sequence of independent normal random variables, each with mean 0 and variance σ^2 and let a, b, and c be constants. Which, if any, of the following processes are stationary? For each stationary process specify the mean and autocovariance function.

a. $X_t = a + bZ_t + cZ_{t-2}$

b. $X_t = Z_1 \cos(ct) + Z_2 \sin(ct)$

c. $X_t = Z_t \cos(ct) + Z_{t-1} \sin(ct)$

d. $X_t = a + bZ_0$

e. $X_t = Z_0 \cos(ct)$

f. $X_t = Z_t Z_{t-1}$

1.5. Let $\{X_t\}$ be the moving average process of order 2 given by

$$X_t = Z_t + \theta Z_{t-2}$$

where $\{Z_t\}$ is WN(0, 1).

a. Find the autocovariance and autocorrelation function for this process when $\theta = .8$.

b. Compute the variance of the sample mean $(X_1 + X_2 + X_3 + X_4)/4$ when $\theta = .8$.

c. Repeat (b) when $\theta = -.8$ and compare your answer with the result obtained in (b).

1.6. Let $\{X_t\}$ be the AR(1) process defined in Example 1.4.5.

a. Compute the variance of the sample mean $(X_1 + X_2 + X_3 + X_4)/4$ when $\phi = .9$ and $\sigma^2 = 1$.

b. Repeat (a) when $\phi = -.9$ and compare your answer with the result obtained in (a).

1.7. If $\{X_t\}$ and $\{Y_t\}$ are uncorrelated stationary sequences, i.e., if X_r and Y_s are uncorrelated for every r and s, show that $\{X_t + Y_t\}$ is stationary with autocovariance function equal to the sum of the autocovariance functions of $\{X_t\}$ and $\{Y_t\}$.

1.8. Let $\{Z_t\}$ be IID N(0,1) noise and define

$$X_t = \begin{cases} Z_t, & \text{if } t \text{ is even,} \\ (Z_{t-1}^2 - 1)/\sqrt{2}, & \text{if } t \text{ is odd.} \end{cases}$$

a. Show that $\{X_t\}$ is WN(0,1) but not IID(0,1) noise.

b. Find $E(X_{n+1}|X_1, \ldots, X_n)$ for n odd and n even and compare the results.

1.9. Let $\{x_1, \ldots, x_n\}$ be observed values of a time series at times $1, \ldots, n$, and let $\hat{\rho}(h)$ be the sample ACF at lag h as in Definition 1.4.4.

a. If $x_t = a + bt$, where a and b are constants and $b \neq 0$, show that for each fixed $h \geq 1$,

$$\hat{\rho}(h) \to 1, \text{ as } n \to \infty.$$

b. If $x_t = c\cos(\omega t)$, where c and ω are constants ($c \neq 0$ and $\omega \in (-\pi, \pi]$), show that for each fixed h,

$$\hat{\rho}(h) \to \cos(\omega h) \text{ as } n \to \infty.$$

1.10. If $m_t = \sum_{k=0}^{p} c_k t^k$, $t = 0, \pm 1, \ldots,$ show that ∇m_t is a polynomial of degree $p - 1$ in t and hence that $\nabla^{p+1} m_t = 0$.

1.11. Consider the simple moving average filter with weights $a_j = (2q+1)^{-1}$, $-q \leq j \leq q$.

a. If $m_t = c_0 + c_1 t$, show that $\sum_{j=-q}^{q} a_j m_{t-j} = m_t$.

b. If Z_t, $t = 0, \pm 1, \pm 2, \ldots,$ are independent random variables with mean 0 and variance σ^2, show that the moving average $A_t = \sum_{j=-q}^{q} a_j Z_{t-j}$ is "small" for large q in the sense that $EA_t = 0$ and $\text{Var}(A_t) = \sigma^2/(2q+1)$.

1.12. a. Show that a linear filter $\{a_j\}$ passes an arbitrary polynomial of degree k without distortion, i.e., that

$$m_t = \sum_j a_j m_{t-j}$$

for all k^{th} degree polynomials $m_t = c_0 + c_1 t + \cdots + c_k t^k$, if and only if

$$\begin{cases} \sum_j a_j = 1 & \text{and} \\ \sum_j j^r a_j = 0, & \text{for } r = 1, \ldots, k. \end{cases}$$

b. Deduce that the Spencer 15-point moving average filter $\{a_j\}$ defined by (1.5.6) passes arbitrary third-degree polynomial trends without distortion.

1.13. Find a filter of the form $1 + \alpha B + \beta B^2 + \gamma B^3$ (i.e., find α, β and γ) that passes linear trends without distortion and that eliminates arbitrary seasonal components of period 2.

1.14. Show that the filter with coefficients $[a_{-2}, a_{-1}, a_0, a_1, a_2] = \frac{1}{9}[-1, 4, 3, 4, -1]$ passes third degree polynomials and eliminates seasonal components with period 3.

1.15. Let $\{Y_t\}$ be a stationary process with mean zero and let a and b be constants.

a. If $X_t = a + bt + s_t + Y_t$, where s_t is a seasonal component with period 12, show that $\nabla\nabla_{12}X_t = (1 - B)(1 - B^{12})X_t$ is stationary and express its autocovariance function in terms of that of $\{Y_t\}$.

b. If $X_t = (a + bt)s_t + Y_t$, where s_t is a seasonal component with period 12, show that $\nabla_{12}^2 X_t = (1 - B^{12})^2 X_t$ is stationary and express its autocovariance function in terms of that of $\{Y_t\}$.

1.16. USING SMOOTH TO SMOOTH THE STRIKES DATA. Double click on the SMOOTH icon. When you see the title, PROGRAM SMOOTH, press `<Enter>` and follow the program prompts, selecting options and data from the menus as they appear. These should be self-explanatory. After choosing *Enter Data* you will see a list of files in the ITSM directory. To access the file STRIKES.DAT, move the highlight bar to the file name STRIKES.DAT and either press `<Enter>` or click with the mouse. Not all the data files are visible in the data display box simultaneously. To access the others, use the arrow keys to scroll up or down (after making sure that the mouse pointer is outside the data display box). Once the data have been read in you will be given a choice of three different smoothing options, simple moving average smoothing, exponential smoothing and removal of high frequency components. Try using each of them to reproduce the results shown in Figures 1.18, 1.21, and 1.22. Do not use the filename STRIKES.DAT to store the smoothed data produced by the program or it will attempt to overwrite the original datafile with that name.

1.17. USING PEST TO PLOT THE DEATHS DATA. Select the module PEST in the same way we selected SMOOTH in the previous problem. When you see the boxed title, PROGRAM PEST press `<Enter>` and you will see the PEST menu. Choose the first option and you will then see a smaller menu. Choose the first option again. Choose YES and you will be asked to select a dataset. Use the arrow keys to highlight the file name DEATHS.DAT in the data display box and then type `<Enter>`. The program will then read the contents of the file DEATHS.DAT and give you another menu. This time choose the second option to plot the data. A histogram of the datafile DEATHS will appear on the screen. To continue with the program, type `<Enter>` and a graph of DEATHS vs time will appear on the screen. Typing <Enter> and selecting the option Return will return you to the data menu. Then choose the last option to return to the main menu and the last option of the main menu to exit from PEST.

1.18. USING PEST TO ANALYZE THE DEATHS DATA. Proceed as in the previous problem to run PEST and to read in the datafile DEATHS.DAT. Plot the data as before, type <Enter> and choose the Return option to return to the data menu. Choose option 6 for classical decomposition and type 12<Enter> to specify period 12. The seasonal components will then appear on the screen. Type <Enter> to return to the data menu. At this stage the data have been deseasonalized by subtraction of the seasonal components. Use option 7 to fit a quadratic trend $m(i) = a(0) + ia(1) + i^2a(2)$, $i = 1, \ldots, 72$. Type <Enter> to return to the data menu. The data have now been deseasonalized and detrended. Use option 2 of the data menu to plot the residuals and option 3 to plot their sample autocorrelation function. The sample autocorrelation function of the residuals suggests there is dependence between them. (The tests for dependence described in Section 1.6 can also be performed by returning to the main menu of PEST, selecting the option *ARMA parameter estimation*, then *Likelihood of model*, then *File and analyze residuals*, and finally *Tests of randomness*.) To forecast the data without allowing for dependence between the residuals, return to the main menu of PEST and select the option *Forecasting*. Specify 24 for the number of values to be forecast and the program will compute forecasts based on the assumption that the estimated seasonal and trend components are true values and that the residuals constitute a white noise sequence with zero mean. You will first see the forecasts of the residuals (which are all zero). You will then be given the option of forecasting the original series and of plotting the data together with the forecasts. Later we shall see how to improve on these forecasts by taking into account the evidence of dependence between successive residuals.

1.19. Use a text editor or WORD6 to construct a data file with filename TEST.DAT, which consists of a single column of 30 numbers, $\{x_1, \ldots, x_{30}\}$, defined by

$$x_1, \ldots, x_{10} : 486\ 474\ 434\ 441\ 435\ 401\ 414\ 414\ 386\ 405$$

$$x_{11}, \ldots, x_{20} : 411\ 389\ 414\ 426\ 410\ 441\ 459\ 449\ 486\ 510$$

$$x_{21}, \ldots, x_{30} : 506\ 549\ 579\ 581\ 630\ 666\ 674\ 729\ 771\ 785$$

This series is in fact the sum of a quadratic trend and a period-three seasonal component. Use the program SMOOTH to apply the filter in Problem 1.13 to this time series and discuss the results.

2 Stationary Processes

A key role in time series analysis is played by processes whose properties, or some of them, do not vary with time. If we wish to make predictions, then clearly we must assume that *something* does not vary with time. In extrapolating deterministic functions it is common practice to assume that either the function itself or one of its derivatives is constant. The assumption of a constant first derivative leads to linear extrapolation as a means of prediction. In time series analysis our goal is to predict a series that typically is not deterministic but contains a random component. If this random component is stationary, in the sense of Definition 1.4.2, then we can develop powerful techniques to forecast its future values. These techniques will be developed and discussed in this and subsequent chapters.

2.1 Basic Properties

In Section 1.4 we introduced the concept of stationarity and defined the autocovariance function (ACVF) of a stationary time series $\{X_t\}$ as

$$\gamma(h) = \text{Cov}(X_{t+h}, X_t), h = 0, \pm1, \pm2, \ldots .$$

The autocorrelation function (ACF) of $\{X_t\}$ was defined similarly as the function $\rho(\cdot)$, whose value at lag h is

$$\rho(h) = \frac{\gamma(h)}{\gamma(0)}.$$

The ACVF and ACF provide a useful measure of the degree of dependence between the values of a time series at different times and for this reason play an important role when we consider the prediction of future values of the series in terms of the past and present values. They can be estimated from observations of $X_1, \ldots, X_n$ by computing the *sample* ACVF and ACF as described in Section 1.4.1.

The role of the autocorrelation function in prediction is illustrated by the following simple example. Suppose that $\{X_t\}$ is a stationary Gaussian time series (see Definition A.3.2) and that we have observed X_n. We would like to find the function of X_n which gives us the best predictor of X_{n+h}, the value of the series after another h time units have elapsed. To define the problem we must first say what we mean by "best." A natural and computationally convenient definition is to specify our required predictor to be the function of X_n with minimum mean squared error. In this illustration, and indeed throughout the remainder of this book, we shall use this as our criterion for "best." Now by Proposition A.3.1 the conditional distribution of X_{n+h} given that $X_n = x_n$ is

$$N(\mu + \rho(h)(x_n - \mu), \sigma^2(1 - \rho(h)^2)),$$

where μ and σ^2 are the mean and variance of $\{X_t\}$. It was shown in Problem 1.1 that the value of the constant c that minimizes $E(X_{n+h} - c)^2$ is $c = E(X_{n+h})$ and that the function m of X_n that minimizes $E(X_{n+h} - m(X_n))^2$ is the conditional mean,

$$m(X_n) = E(X_{n+h}|X_n) = \mu + \rho(h)(X_n - \mu). \tag{2.1.1}$$

The corresponding mean squared error is

$$E(X_{n+h} - m(X_n))^2 = \sigma^2(1 - \rho(h)^2). \tag{2.1.2}$$

This calculation shows that, at least for stationary Gaussian time series, prediction of X_{n+h} in terms of X_n is more accurate as $|\rho(h)|$ becomes closer to 1, and in the limit as $\rho \to \pm 1$ the best predictor approaches $\mu \pm (X_n - \mu)$ and the corresponding mean squared error approaches 0.

In the preceding calculation the assumption of joint normality of X_{n+h} and X_n played a crucial role. For time series with nonnormal joint distributions the corresponding calculations are in general much more complicated. However, if instead of looking for the best function of X_n for predicting X_{n+h}, we look for the best **linear predictor**, i.e., the best predictor of the form $\ell(X_n) = aX_n + b$, then our problem becomes that of finding a and b to minimize $E(X_{n+h} - aX_n - b)^2$. An elementary calculation (Problem 2.1), shows that the best predictor of this form is

$$\ell(X_n) = \mu + \rho(h)(X_n - \mu) \tag{2.1.3}$$

with corresponding mean squared error,

$$E(X_{n+h} - \ell(X_n))^2 = \sigma^2(1 - \rho(h)^2). \tag{2.1.4}$$

Comparison with (2.1.1) and (2.1.3) shows that for Gaussian processes, $\ell(X_n)$ and $m(X_n)$ are the same. In general, of course, $m(X_n)$ will give smaller mean squared error than $\ell(X_n)$ since it is the best of a larger class of predictors (see Problem 1.8). However, the fact that the best linear predictor depends only on the mean and ACF of the series $\{X_t\}$ means that it can be calculated without more detailed knowledge of the joint distributions. This is extremely important in practice because of the difficulty of estimating all of the joint distributions and because of the difficulty of computing the required conditional expectations even if the distributions were known.

As we shall see later in this chapter, similar conclusions apply when we consider the more general problem of predicting X_{n+h} as a function not only of X_n, but also of $X_{n-1}, X_{n-2}, \ldots$. Before pursuing this question we need to examine in more detail the properties of the autocovariance and autocorrelation functions of a stationary time series.

Basic Properties of $\gamma(\cdot)$:

$$\gamma(0) \geq 0,$$

$$|\gamma(h)| \leq \gamma(0) \text{ for all } h,$$

and $\gamma(\cdot)$ is even, i.e.,

$$\gamma(h) = \gamma(-h) \text{ for all } h.$$

Proof The first property is simply the statement that $\text{Var}(X_t) \geq 0$, the second is an immediate consequence of the fact that correlations are less than or equal to 1 in absolute value (or the Cauchy-Schwarz inequality), and the third is established by observing that

$$\gamma(h) = \text{Cov}(X_{t+h}, X_t) = \text{Cov}(X_t, X_{t+h}) = \gamma(-h). \quad \blacksquare$$

Autocovariance functions have another fundamental property, namely that of nonnegative definiteness.

Definition 2.1.1 A real-valued function κ defined on the integers is **nonnegative definite** if

$$\sum_{i,j=1}^{n} a_i \kappa(i-j) a_j \geq 0 \tag{2.1.5}$$

for all positive integers n and vectors $\mathbf{a} = (a_1, \ldots, a_n)'$ with real-valued components, a_i.

Theorem 2.1.1 *A real-valued function defined on the integers is the autocovariance function of a stationary time series if and only if it is even and nonnegative definite.*

Proof To show that the autocovariance function $\gamma(\cdot)$ of any stationary time series $\{X_t\}$ is nonnegative definite, let $\mathbf{a}$ be any $n \times 1$ vector with real components $a_1, \ldots, a_n$ and let $\mathbf{X}_n = (X_n, \ldots, X_1)'$. Then by equation (A.2.5) and the nonnegativity of variances,

$$\begin{aligned}\operatorname{Var}(\mathbf{a}'\mathbf{X}_n) &= \mathbf{a}'\Gamma_n\mathbf{a} \\ &= \sum_{i,j=1}^{n} a_i\gamma(i-j)a_j \\ &\geq 0,\end{aligned}$$

where Γ_n is the covariance matrix of the random vector $\mathbf{X}_n$. The last inequality, however, is precisely the statement that $\gamma(\cdot)$ is nonnegative definite. The converse result, that there exists a stationary time series with autocovariance function κ if κ is even, real-valued, and nonnegative definite, is more difficult to establish (see TSTM, Theorem 1.5.1 for a proof). A slightly stronger statement can be made, namely that under the specified conditions there exists a stationary *Gaussian* time series $\{X_t\}$ with mean 0 and autocovariance function $\kappa(\cdot)$. ■

Remark 1. An autocorrelation function $\rho(\cdot)$ has all the properties of an autocovariance function and satisfies the additional condition $\rho(0) = 1$. In particular we can say that $\rho(\cdot)$ is the autocorrelation function of a stationary process if and only if $\rho(\cdot)$ is an ACVF with $\rho(0) = 1$. □

Remark 2. To verify that a given function is nonnegative definite it is often simpler to find a stationary process which has the given function as its ACVF than to verify the conditions (2.1.5) directly. For example, the function $\kappa(h) = \cos(\omega h)$ is nonnegative definite since (see Problem 2.2) it is the ACVF of the stationary process,

$$X_t = A\cos(\omega t) + B\sin(\omega t),$$

where A and B are uncorrelated random variables, both with mean 0 and variance 1. Another illustration is provided by the following example. □

Example 2.1.1 We shall show now that the function defined on the integers by

$$\kappa(h) = \begin{cases} 1, & \text{if } h = 0, \\ \rho, & \text{if } h = \pm 1, \\ 0, & \text{otherwise,} \end{cases}$$

is the ACVF of a stationary time series if and only if $|\rho| \leq \frac{1}{2}$. Inspection of the ACVF of the MA(1) process of Example 1.4.4 shows that κ is the ACVF of such a process provided we can find real θ and nonnegative σ^2 such that

$$\sigma^2(1+\theta^2) = 1$$

and

$$\sigma^2\theta = \rho.$$

If $|\rho| \le \frac{1}{2}$ these equations give solutions $\theta = (2\rho)^{-1}(1 \pm \sqrt{1-4\rho^2})$ and $\sigma^2 = (1+\theta^2)^{-1}$. However, if $|\rho| > \frac{1}{2}$ there is no real solution for θ and hence no MA(1) process with ACVF κ. To show that there is no *stationary* process with ACVF κ, we need to show that κ is not nonnegative definite. We shall do this directly from the definition (2.1.5). First, if $\rho > \frac{1}{2}$, $K = [\kappa(i-j)]_{i,j=1}^n$ and $\mathbf{a}$ is the n-component vector $\mathbf{a} = (1, -1, 1, -1, \ldots)'$, then

$$\mathbf{a}'K\mathbf{a} = n - 2(n-1)\rho < 0 \text{ for } n > 2\rho/(2\rho - 1),$$

showing that $\kappa(\cdot)$ is not nonnegative definite and therefore, by Theorem 2.1.1, is not an autocovariance function. If $\rho < -\frac{1}{2}$, the same argument with $\mathbf{a} = (1, 1, 1, 1, \ldots)'$ again shows that $\kappa(\cdot)$ is not nonnegative definite.

If $\{X_t\}$ is a (weakly) stationary time series, then the vector $(X_1, \ldots, X_n)'$ and the time-shifted vector $(X_{1+h}, \ldots, X_{n+h})'$ have the same mean vectors and covariance matrices for every integer h and positive integer n. A strictly stationary sequence is one in which the joint distributions of these two vectors (and not just the means and covariances) are the same. The precise definition is given below. □

Definition 2.1.2

$\{X_t\}$ is a **strictly stationary time series** if

$$(X_1, \ldots, X_n)' \stackrel{d}{=} (X_{1+h}, \ldots, X_{n+h})'$$

for all all integers h and $n \ge 1$. (Here $\stackrel{d}{=}$ is used to indicate that the two random vectors have the same joint distribution function.)

For reference, we record some of the elementary properties of strictly stationary time series.

Properties of a Strictly Stationary Time Series $\{X_t\}$:

a. The random variables X_t are identically distributed.

b. $(X_t, X_{t+h})' \stackrel{d}{=} (X_1, X_{1+h})'$ for all integers t and h.

c. $\{X_t\}$ is weakly stationary if $E(X_t^2) < \infty$ for all t.

d. Weak stationarity does not imply strict stationarity.

e. An iid sequence is strictly stationary.

Proof Properties (a) and (b) follow at once from Definition 2.1.2. If $EX_t^2 < \infty$, then by (a) and (b) EX_t is independent of t and $\text{Cov}(X_t, X_{t+h}) = \text{Cov}(X_1, X_{1+h})$, which is also independent of t, proving (c). For (d) see Problem 1.8. If $\{X_t\}$ is an iid sequence of random variables with common distribution function F, then the joint distribution function of $(X_{1+h}, \ldots, X_{n+h})'$ evaluated at $(x_1, \ldots, x_n)'$ is $F(x_1)\cdots F(x_n)$, which is independent of h. ■

One of the simplest ways to construct a time series $\{X_t\}$ that is strictly stationary (and hence stationary if $EX_t^2 < \infty$) is to "filter" an iid sequence of random variables. Let $\{Z_t\}$ be an iid sequence, which by (e) is strictly stationary, and define

$$X_t = g(Z_t, Z_{t-1}, \cdots, Z_{t-q}) \tag{2.1.6}$$

for some real-valued function $g(\cdot, \cdots, \cdot)$. Then $\{X_t\}$ is strictly stationary, since $(Z_{t+h}, \ldots, Z_{t+h-q})' \stackrel{d}{=} (Z_t, \ldots, Z_{t-q})'$ for all integers h. It follows also from the defining equation (2.1.6) that $\{X_t\}$ is ***q*-dependent**, i.e., that X_s and X_t are independent whenever $|t - s| > q$. (An iid sequence is 0-dependent.) In the same way, adopting a second-order viewpoint, we say that a stationary time series is ***q*-correlated** if $\gamma(h) = 0$ whenever $|h| > q$. A white noise sequence is then 0-correlated while the MA(1) process of Example 1.4.4 is 1-correlated. The moving average process of order q defined below is q-correlated and, perhaps surprisingly, the converse is also true (Proposition 2.1.1).

The MA(q) Process:

$\{X_t\}$ is a **moving average process of order *q*** if

$$X_t = Z_t + \theta_1 Z_{t-1} + \cdots + \theta_q Z_{t-q} \tag{2.1.7}$$

where $\{Z_t\} \sim \text{WN}(0, \sigma^2)$ and $\theta_1, \ldots, \theta_q$ are constants.

It is a simple matter to check that (2.1.7) defines a stationary time series that is strictly stationary if $\{Z_t\}$ is IID noise. In the latter case, (2.1.7) is a special case of (2.1.6) with g a linear function.

The importance of MA(q) processes derives from the fact that *every* q-correlated process is an MA(q) process. This is the content of the following proposition whose proof can be found in TSTM, Section 3.2. The extension of this result to the case $q = \infty$ is essentially Wold's decomposition (see Section 2.6).

Proposition 2.1.1 *If $\{X_t\}$ is a stationary q-correlated time series with mean 0, then it can be represented as the MA(q) process in (2.1.7).*

2.2 Linear Processes

The class of linear time series models, which includes the class of autoregressive-moving average (ARMA) models, provides a general framework for studying stationary processes. In fact, every second-order stationary process is either a linear process or can be transformed to a linear process by subtracting a *deterministic* component. This result is known as Wold's decomposition and is discussed in Section 2.6.

Definition 2.2.1 The time series $\{X_t\}$ is a **linear process** if it has the representation

$$X_t = \sum_{j=-\infty}^{\infty} \psi_j Z_{t-j}, \tag{2.2.1}$$

for all t, where $\{Z_t\} \sim \mathrm{WN}(0, \sigma^2)$ and $\{\psi_j\}$ is a sequence of constants with $\sum_{j=-\infty}^{\infty} |\psi_j| < \infty$.

In terms of the backward shift operator B, (2.2.1) can be written more compactly as

$$X_t = \psi(B) Z_t, \tag{2.2.2}$$

where $\psi(B) = \sum_{j=-\infty}^{\infty} \psi_j B^j$. A linear process is called a **moving average** or **MA(∞)** if $\psi_j = 0$ for all $j < 0$, i.e., if

$$X_t = \sum_{j=0}^{\infty} \psi_j Z_{t-j}.$$

Remark 1. The condition $\sum_{j=-\infty}^{\infty} |\psi_j| < \infty$ ensures that the infinite sum in (2.2.1) converges (with probability one) since $E|Z_t| \le \sigma$ and

$$E|X_t| \le \sum_{j=-\infty}^{\infty} \left(|\psi_j| E|Z_{t-j}| \right) \le \left(\sum_{j=-\infty}^{\infty} |\psi_j| \right) \sigma < \infty.$$

It also ensures that $\sum_{j=-\infty}^{\infty} \psi_j^2 < \infty$ and hence (see Appendix C, Example C.1.1) that the series in (2.2.1) converges in mean square, i.e., that X_t is the mean square limit of the partial sums $\sum_{j=-n}^{n} \psi_j Z_{t-j}$. The condition $\sum_{j=-n}^{n} |\psi_j| < \infty$ also ensures convergence in both senses of the more general series (2.2.3) considered in Proposition 2.2.1 below. In Section 10.5 we consider a more general class of linear processes, the fractionally integrated ARMA processes, for which the coefficents are not absolutely summable but only square summable. □

The operator $\psi(B)$ can be thought of as a linear filter, which when applied to the white noise "input" series $\{Z_t\}$ produces the "output" $\{X_t\}$ (see Section 4.3). As established in the following proposition, a linear filter, when applied to any stationary input series, produces a stationary output series.

Proposition 2.2.1 *Let $\{Y_t\}$ be a stationary time series with mean 0 and covariance function γ_Y. If $\sum_{j=-\infty}^{\infty} |\psi_j| < \infty$, then the time series*

$$X_t = \sum_{j=-\infty}^{\infty} \psi_j Y_{t-j} = \psi(B) Y_t \tag{2.2.3}$$

is stationary with mean 0 and autocovariance function

$$\gamma_X(h) = \sum_{j=-\infty}^{\infty} \sum_{k=-\infty}^{\infty} \psi_j \psi_k \gamma_Y(h + k - j). \tag{2.2.4}$$

In the special case when $\{X_t\}$ is a linear process,

$$\gamma_X(h) = \sum_{j=-\infty}^{\infty} \psi_j \psi_{j+h} \sigma^2. \tag{2.2.5}$$

Proof The argument used in Remark 1, with σ replaced by $\sqrt{\gamma_Y(0)}$, shows that the series in (2.2.3) is convergent. Since $EY_t = 0$, we have

$$E(X_t) = E\left(\sum_{j=-\infty}^{\infty} \psi_j Y_{t-j}\right) = \sum_{j=-\infty}^{\infty} \psi_j E(Y_{t-j}) = 0$$

and

$$\begin{aligned} E(X_{t+h} X_t) &= E\left[\left(\sum_{j=-\infty}^{\infty} \psi_j Y_{t+h-j}\right)\left(\sum_{k=-\infty}^{\infty} \psi_k Y_{t-k}\right)\right] \\ &= \sum_{j=-\infty}^{\infty} \sum_{k=-\infty}^{\infty} \psi_j \psi_k E(Y_{t+h-j} Y_{t-k}) \\ &= \sum_{j=-\infty}^{\infty} \sum_{k=-\infty}^{\infty} \psi_j \psi_k \gamma_Y(h - j + k), \end{aligned}$$

which shows that $\{X_t\}$ is stationary with covariance function (2.2.4). (The interchange of summation and expectation operations in the above calculations can be justified by the absolute summability of ψ_j.) Finally if $\{Y_t\}$ is the white noise sequence $\{Z_t\}$ in (2.2.1), then $\gamma_Y(h - j + k) = \sigma^2$ if $k = j - h$ and 0 otherwise, from which (2.2.5) follows. ■

Remark 2. The absolute convergence of (2.2.3) implies (Problem 2.6) that filters of the form $\alpha(B) = \sum_{j=-\infty}^{\infty} \alpha_j B^j$ and $\beta(B) = \sum_{j=-\infty}^{\infty} \beta_j B^j$ with absolutely summable coefficients can be applied successively to a stationary series $\{Y_t\}$ to generate a new stationary series,

$$W_t = \sum_{j=-\infty}^{\infty} \psi_j Y_{t-j},$$

where

$$\psi_j = \sum_{k=-\infty}^{\infty} \alpha_k \beta_{j-k} = \sum_{k=-\infty}^{\infty} \beta_k \alpha_{j-k}. \tag{2.2.6}$$

These relations can be expressed in the equivalent form,

$$W_t = \psi(B) Y_t,$$

where

$$\psi(B) = \alpha(B)\beta(B) = \beta(B)\alpha(B), \tag{2.2.7}$$

and the products are defined by (2.2.6) or equivalently by multiplying the series $\sum_{j=-\infty}^{\infty} \alpha_j B^j$ and $\sum_{j=-\infty}^{\infty} \beta_j B^j$ term by term and collecting powers of B. It is clear from (2.2.6) and (2.2.7) that the order of application of the filters $\alpha(B)$ and $\beta(B)$ is immaterial. □

Example 2.2.1 An AR(1) process

In Example 1.4.5, an AR(1) process was defined as a stationary solution $\{X_t\}$ of the equations

$$X_t - \phi X_{t-1} = Z_t, \tag{2.2.8}$$

where $\{Z_t\} \sim \mathrm{WN}(0, \sigma^2)$, $|\phi| < 1$ and Z_t is uncorrelated with X_s for each $s < t$. To show that such a solution exists and is the unique stationary solution of (2.2.8), we consider the linear process defined by

$$X_t = \sum_{j=0}^{\infty} \phi^j Z_{t-j}. \tag{2.2.9}$$

(The coefficients ϕ^j for $j \geq 0$ are absolutely summable since $|\phi| < 1$.) It is easy to verify directly that the process (2.2.9) is a solution of (2.2.8) and by Proposition 2.2.1 it is also stationary with mean 0 and ACVF

$$\gamma_X(h) = \sum_{j=0}^{\infty} \phi^j \phi^{j+h} \sigma^2$$
$$= \frac{\sigma^2 \phi^h}{1 - \phi^2},$$

for $h \geq 0$.

To show that (2.2.9) is the *only* stationary solution of (2.2.8) let $\{Y_t\}$ be *any* stationary solution. Then, iterating (2.2.8), we obtain

$$\begin{aligned} Y_t &= \phi Y_{t-1} + Z_t \\ &= Z_t + \phi Z_{t-1} + \phi^2 Y_{t-2} \\ &= \cdots \\ &= Z_t + \phi Z_{t-1} + \cdots + \phi^k Z_{t-k} + \phi^{k+1} Y_{t-k-1}. \end{aligned}$$

If $\{Y_t\}$ is stationary, then $EY_t^2 \le \gamma_Y(0)$ for all t so that

$$\begin{aligned} E(Y_t - \sum_{j=0}^{k} \phi^j Z_{t-j})^2 &= \phi^{2k+2} E(Y_{t-k})^2 \\ &\le \phi^{2k+2}\gamma_Y(0) \\ &\to 0 \text{ as } k \to \infty. \end{aligned}$$

This implies that Y_t is equal to the mean square limit $\sum_{j=0}^{\infty} \phi^j Z_{t-j}$ and hence that the process defined by (2.2.9) is the unique stationary solution of the equations (2.2.8).

It the case $|\phi| > 1$, the series in (2.2.9) does not converge. However, we can rewrite (2.2.8) in the form

$$X_t = -\phi^{-1} Z_{t+1} + \phi^{-1} X_{t+1}. \tag{2.2.10}$$

Iterating (2.2.10) gives

$$\begin{aligned} X_t &= -\phi^{-1} Z_{t+1} - \phi^{-2} Z_{t+2} + \phi^{-2} X_{t+2} \\ &= \cdots \\ &= -\phi^{-1} Z_{t+1} - \cdots - \phi^{-k-1} Z_{t+k+1} + \phi^{-k-1} X_{t+k+1}, \end{aligned}$$

which shows, by the same arguments used above, that

$$X_t = -\sum_{j=1}^{\infty} \phi^{-j} Z_{t+j} \tag{2.2.11}$$

is the unique stationary solution of (2.2.8). This solution should not be confused with the nonstationary solution $\{X_t\}$ of (2.2.8) obtained when X_0 is any specified random variable that is uncorrelated with $\{Z_t\}$.

The solution (2.2.11) is frequently regarded as unnatural since X_t as defined by (2.2.11) is correlated with *future* values of Z_s, contrasting with the solution (2.2.9), which has the property that X_t is uncorrelated with Z_s for all $s > t$. It is customary therefore when modelling stationary time series to restrict attention to AR(1) processes with $|\phi| < 1$. Then X_t has the representation (2.2.8) in terms of $\{Z_s, s \le t\}$ and we say that $\{X_t\}$ is a **causal** or **future-independent** function of $\{Z_t\}$, or more concisely that $\{X_t\}$ is a causal autoregressive process. It should be noted that every AR(1) process with $|\phi| > 1$ can be reexpressed as an AR(1) process with $|\phi| < 1$ and a new white noise sequence (Problem 3.8). From a second-order point of view therefore, nothing is lost by eliminating AR(1) processes with $|\phi| > 1$ from consideration.

If $\phi = \pm 1$, there is no stationary solution of (2.2.8) (see Problem 2.8). □

Remark 3. It is worth remarking that when $|\phi| < 1$ the unique stationary solution (2.2.9) can be found immediately with the aid of (2.2.7). To do this let $\phi(B) = 1 - \phi B$ and $\pi(B) = \sum_{j=0}^{\infty} \phi^j B^j$. Then

$$\psi(B) := \phi(B)\pi(B) = 1.$$

Applying the operator $\pi(B)$ to both sides of (2.2.8), we obtain

$$X_t = \pi(B)Z_t = \sum_{j=0}^{\infty} \phi^j Z_{t-j}$$

as claimed. □

2.3 Introduction to ARMA Processes

In this section we introduce, through an example, some of the key properties of an important class of linear processes known as ARMA (autoregressive moving average) processes. These are defined by linear difference equations with constant coefficients. As our example, we shall consider the ARMA(1,1) process. Higher-order ARMA processes will be discussed in Chapter 3.

Definition 2.3.1

The time series $\{X_t\}$ is an **ARMA(1,1) process** if it is stationary and satisfies (for every t)

$$X_t - \phi X_{t-1} = Z_t + \theta Z_{t-1}, \tag{2.3.1}$$

where $\{Z_t\} \sim \mathrm{WN}(0, \sigma^2)$ and $\phi + \theta \neq \theta$.

Using the backward shift operator B, (2.3.1) can be written more concisely as

$$\phi(B)X_t = \theta(B)Z_t \tag{2.3.2}$$

where $\phi(B)$ and $\theta(B)$ are the linear filters

$$\phi(B) = 1 - \phi B \text{ and } \theta(B) = 1 + \theta B,$$

respectively.

We first investigate the range of values of ϕ and θ for which a stationary solution of (2.3.1) exists. If $|\phi| < 1$, let $\chi(z)$ denote the power series expansion of $\frac{1}{\phi(z)}$, i.e., $\sum_{j=0}^{\infty} \phi^j z^j$, which has absolutely summable coefficients. Then from (2.2.7) we conclude that $\chi(B)\phi(B) = 1$. Applying $\chi(B)$ to each side of (2.3.2) therefore gives

$$X_t = \chi(B)\theta(B)Z_t = \psi(B)Z_t,$$

where

$$\psi(B) = \sum_{j=0}^{\infty} \psi_j B^j = \left(1 + \phi B + \phi^2 B^2 + \cdots\right)(1 + \theta B).$$

By multiplying out the right-hand side or using (2.2.6), we find that

$$\psi_0 = 1 \text{ and } \psi_j = (\phi + \theta)\phi^{j-1} \text{ for } j \geq 1.$$

As in Example 2.2.1, we conclude that the MA(∞) process

$$X_t = Z_t + (\phi + \theta) \sum_{j=1}^{\infty} \phi^{j-1} Z_{t-j} \tag{2.3.3}$$

is the unique stationary solution of (2.3.1).

Now suppose that $|\phi| > 1$. We first represent $\frac{1}{\phi(z)}$ as a series of powers of z with absolutely summable coefficients by expanding in powers of z^{-1}, giving (Problem 2.7)

$$\frac{1}{\phi(z)} = -\sum_{j=1}^{\infty} \phi^{-j} z^{-j}.$$

Then we can apply the same argument as in the case when $|\phi| < 1$ to obtain the unique stationary solution of (2.3.1). We let $\chi(B) = -\sum_{j=1}^{\infty} \phi^{-j} B^{-j}$ and apply $\chi(B)$ to each side of (2.3.2) to obtain

$$X_t = \chi(B)\theta(B)Z_t = -\theta\phi^{-1}Z_t - (\theta + \phi) \sum_{j=1}^{\infty} \phi^{-j-1} Z_{t+j}. \tag{2.3.4}$$

If $\phi = \pm 1$, there is no stationary solution of (2.3.1). Consequently, there is no such thing as an ARMA(1,1) process with $\phi = \pm 1$ according to our definition.

We can now summarize our findings about the existence and nature of the stationary solutions of the ARMA(1,1) recursions (2.3.2) as follows:

- A stationary solution of the ARMA(1,1) equations exists if and only if $\phi \neq \pm 1$.
- If $|\phi| < 1$, then the unique stationary solution is given by (2.3.3). In this case we say that $\{X_t\}$ is **causal** or a causal function of $\{Z_t\}$ since X_t can be expressed in terms of the current and past values, Z_s, $s \leq t$.
- If $|\phi| > 1$, then the unique stationary solution is given by (2.3.4). The solution is **noncausal** since X_t is then a function of Z_s, $s \geq t$.

Just as causality means that X_t is expressible in terms of $Z_s, s \le t$, the dual concept of invertibility means that Z_t is expressible in terms of $X_s, s \le t$. We show now that the ARMA(1,1) process defined by (2.3.1) is invertible if $|\theta| < 1$. To demonstrate this, let $\xi(z)$ denote the power series expansion of $\frac{1}{\theta(z)}$, i.e., $\sum_{j=0}^{\infty}(-\theta)^j z^j$, which has absolutely summable coefficients. From (2.2.7) it therefore follows that $\xi(B)\theta(B) = 1$ and applying $\xi(B)$ to each side of (2.3.2) gives

$$Z_t = \xi(B)\phi(B)X_t = \pi(B)X_t$$

where

$$\pi(B) = \sum_{j=0}^{\infty} \pi_j B^j = \left(1 - \theta B + (-\theta)^2 B^2 + \cdots\right)(1 - \phi B).$$

By multiplying out the right-hand side or using (2.2.6), we find that

$$Z_t = X_t - (\phi + \theta)\sum_{j=1}^{\infty}(-\theta)^{j-1}X_{t-j}. \tag{2.3.5}$$

Thus the ARMA(1,1) process is **invertible** since Z_t can be expressed in terms of the present and past values of the process, $X_s,\ s \le t$. An argument like the one used to show noncausality when $|\phi| > 1$ shows that the ARMA(1,1) process is **noninvertible** when $|\theta| > 1$ since then

$$Z_t = -\phi\theta^{-1}X_t + (\theta + \phi)\sum_{j=1}^{\infty}(-\theta)^{-j-1}X_{t+j}. \tag{2.3.6}$$

We summarize these results as follows:

- If $|\theta| < 1$, then the ARMA(1,1) process is **invertible** and Z_t is expressed in terms of $X_s, s \le t$, by (2.3.5).
- If $|\theta| > 1$, then the ARMA(1,1) process is **noninvertible** and Z_t is expressed in terms of $X_s, s \ge t$, by (2.3.6).

Remark 1. In the cases $\theta = \pm 1$, the ARMA(1,1) process is invertible in the more general sense that Z_t is a mean square limit of finite linear combinations of $X_s,\ s \le t$, although it cannot be expressed explicitly as an infinite linear combination of $X_s,\ s \le t$ (see Section 4.4 of TSTM). In this book the term *invertible* will always be used in the more restricted sense that $Z_t = \sum_{j=0}^{\infty}\pi_j X_{t-j}$, where $\sum_{j=0}^{\infty}|\pi_j| < \infty$. □

Remark 2. If the ARMA(1,1) process $\{X_t\}$ is noncausal or noninvertible with $|\theta| > 1$, then it is possible to find a new white noise sequence $\{W_t\}$ such that $\{X_t\}$ is a causal and noninvertible ARMA(1,1) process relative to $\{W_t\}$ (Problem 4.10). Therefore, from a second-order point of view, nothing is lost by restricting attention to causal and invertible ARMA(1,1) models. The last sentence is also valid for higher-order ARMA models. □

2.4 Properties of the Sample Mean and Autocorrelation Function

A stationary process $\{X_t\}$ is characterized, at least from a second-order point of view, by its mean μ and its autocovariance function $\gamma(\cdot)$. The estimation of μ, $\gamma(\cdot)$ and the autocorrelation function $\rho(\cdot) = \gamma(\cdot)/\gamma(0)$ from observations $X_1, \ldots, X_n$ therefore plays a crucial role in problems of inference and in particular in the problem of constructing an appropriate model for the data. In this section, we examine some of the properties of the sample estimates $\bar{x}$ and $\hat{\rho}(\cdot)$ of μ and $\rho(\cdot)$, respectively.

2.4.1 Estimation of μ

The moment estimator of the mean μ of a stationary process is the sample mean

$$\bar{X}_n = n^{-1}(X_1 + X_2 + \cdots + X_n). \tag{2.4.1}$$

It is an unbiased estimator of μ since

$$E(\bar{X}_n) = n^{-1}(EX_1 + \cdots + EX_n) = \mu.$$

The mean squared error of $\bar{X}_n$ is

$$\begin{aligned} E(\bar{X}_n - \mu)^2 &= \mathrm{Var}(\bar{X}_n) \\ &= n^{-2} \sum_{i=1}^{n} \sum_{j=1}^{n} \mathrm{Cov}(X_i, X_j) \\ &= n^{-2} \sum_{i-j=-n}^{n} (n - |i-j|)\gamma(i-j) \\ &= n^{-1} \sum_{h=-n}^{n} \left(1 - \frac{|h|}{n}\right) \gamma(h). \end{aligned} \tag{2.4.2}$$

Now if $\gamma(h) \to 0$ as $h \to \infty$, the right-hand side of (2.4.2) converges to zero so that $\bar{X}_n$ converges in mean square to μ. If $\sum_{h=-\infty}^{\infty} |\gamma(h)| < \infty$, then (2.4.2) gives $\lim_{n\to\infty} n\mathrm{Var}(\bar{X}_n) = \sum_{|h|<\infty} \gamma(h)$. We record these results in the following proposition.

Proposition 2.4.1 *If $\{X_t\}$ is a stationary time series with mean μ and autocovariance function $\gamma(\cdot)$, then as $n \to \infty$*

$$\operatorname{Var}(\bar{X}_n) = E(\bar{X}_n - \mu)^2 \to 0 \qquad \text{if } \gamma(n) \to 0,$$

$$nE(\bar{X}_n - \mu)^2 \to \sum_{|h|<\infty} \gamma(h) \qquad \text{if } \sum_{h=-\infty}^{\infty} |\gamma(h)| < \infty.$$

To make inferences about μ using the sample mean $\bar{X}_n$, it is necessary to know the distribution or an approximation to the distribution of $\bar{X}_n$. If the time series is Gaussian (see Definition A.3.2) then by Remark 2 of Section A.3 and (2.4.2),

$$n^{1/2}(\bar{X}_n - \mu) \sim \mathrm{N}\left(0, \sum_{|h|<n} \left(1 - \frac{|h|}{n}\right) \gamma(h)\right).$$

It is easy to construct exact confidence bounds for μ using this result if $\gamma(\cdot)$ is known, and approximate confidence bounds if it is necessary to estimate $\gamma(\cdot)$ from the observations.

For many time series, in particular for linear and ARMA models, $\bar{X}_n$ is approximately normal with mean μ and variance $n^{-1} \sum_{|h|<\infty} \gamma(h)$ for large n (see TSTM, p. 219). An approximate 95% confidence interval for μ is then

$$(\bar{X}_n - 1.96v^{1/2}/\sqrt{n},\ \bar{X}_n + 1.96v^{1/2}/\sqrt{n}), \tag{2.4.3}$$

where $v = \sum_{|h|<\infty} \gamma(h)$. Of course v is not generally known so it must be estimated from the data. The estimator computed in the program PEST is $\hat{v} = \sum_{|h|<\sqrt{n}} (1 - \frac{|h|}{n})\hat{\gamma}(h)$. For ARMA processes this is a good approximation to v for large n.

Example 2.4.1 An AR(1) model

Let $\{X_t\}$ be an AR(1) process with mean μ, defined by the equations

$$X_t - \mu = \phi(X_{t-1} - \mu) + Z_t,$$

where $|\phi| < 1$ and $\{Z_t\} \sim \mathrm{WN}(0, \sigma^2)$. From Example 2.2.1 we have $\gamma(h) = \phi^{|h|}\sigma^2/(1-\phi^2)$ and hence $v = (1 + 2\sum_{h=1}^{\infty} \phi^h)\sigma^2/(1-\phi^2) = \sigma^2/(1-\phi)^2$. Approximate 95% confidence bounds for μ are therefore given by $\bar{x}_n \pm 1.96\sigma n^{-1/2}/(1-\phi)$. Since ϕ and σ are unknown in practice, they must be replaced in these bounds by estimated values. □

2.4.2 Estimation of $\gamma(\cdot)$ and $\rho(\cdot)$

Recall from Section 1.4.1 that the sample autocovariance and autocorrelation functions are defined by

$$\hat{\gamma}(h) = n^{-1} \sum_{t=1}^{n-|h|} (X_{t+|h|} - \bar{X}_n)(X_t - \bar{X}_n) \tag{2.4.4}$$

and

$$\hat{\rho}(h) = \frac{\hat{\gamma}(h)}{\hat{\gamma}(0)}. \tag{2.4.5}$$

Both the estimators $\hat{\gamma}(h)$ and $\hat{\rho}(h)$ are biased even if the factor n^{-1} in (2.4.4) is replaced by $(n-h)^{-1}$. Nevertheless, under general assumptions, they are nearly unbiased for large sample sizes. The sample ACVF has the desirable property that for each $k \geq 1$ the k-dimensional sample covariance matrix

$$\hat{\Gamma}_k = \begin{bmatrix} \hat{\gamma}(0) & \hat{\gamma}(1) & \cdots & \hat{\gamma}(k-1) \\ \hat{\gamma}(1) & \hat{\gamma}(0) & \cdots & \hat{\gamma}(k-2) \\ \vdots & \vdots & \cdots & \vdots \\ \hat{\gamma}(k-1) & \hat{\gamma}(k-2) & \cdots & \hat{\gamma}(0) \end{bmatrix} \tag{2.4.6}$$

is nonnegative definite. To see this, first note that if $\hat{\Gamma}_m$ is nonnegative definite, then $\hat{\Gamma}_k$ is nonnegative definite for all $k < m$. So assume $k \geq n$ and write

$$\hat{\Gamma}_k = n^{-1} T T'$$

where T is the $k \times 2k$ matrix

$$T = \begin{bmatrix} 0 & \cdots & 0 & 0 & Y_1 & Y_2 & \cdots & Y_k \\ 0 & \cdots & 0 & Y_1 & Y_2 & \cdots & Y_k & 0 \\ \vdots & & & & & & & \vdots \\ 0 & Y_1 & Y_2 & \cdots & Y_k & 0 \cdots & 0 & \end{bmatrix},$$

$Y_i = X_i - \bar{X}_n$, $i = 1, \ldots, n$, and $Y_i = 0$ for $i = n+1, \ldots, k$. Then for any real $k \times 1$ vector $\mathbf{a}$ we have

$$\mathbf{a}'\hat{\Gamma}_k \mathbf{a} = n^{-1}(\mathbf{a}'T)(T'\mathbf{a}) \geq 0, \tag{2.4.7}$$

and consequently the sample autocovariance matrix $\hat{\Gamma}_k$ and sample autocorrelation matrix

$$\hat{R}_k = \hat{\Gamma}_k / \gamma(0) \tag{2.4.8}$$

are nonnegative definite. Sometimes the factor n^{-1} is replaced by $(n-h)^{-1}$ in the definition of $\hat{\gamma}(h)$, but the resulting covariance and correlation matrices $\hat{\Gamma}_n$ and $\hat{R}_n$ may not then be nonnegative definite. We shall therefore use the definitions (2.4.4) and (2.4.5) of $\hat{\gamma}(h)$ and $\hat{\rho}(h)$.

Remark 1. The matrices $\hat{\Gamma}_k$ and $\hat{R}_k$ are in fact nonsingular if there is at least one nonzero Y_i, or equivalently if $\hat{\gamma}(0) > 0$. To establish this result, suppose that $\hat{\gamma}(0) > 0$ and $\hat{\Gamma}_k$ is singular. Then there is equality in (2.4.7) for some nonzero vector $\mathbf{a}$, implying that $\mathbf{a}'T = 0$ and hence that the rank of T is less than k. Let Y_i be the first nonzero value of $Y_1, Y_2, \ldots, Y_k$, and consider the $k \times k$ submatrix of T consisting of columns $(i+1)$ through $(i+k)$. Since this matrix is lower right triangular with each diagonal element equal to Y_i, its determinant is $Y_i^k \neq 0$. Consequently the submatrix is nonsingular and T must have rank k, a contradiction. □

Without further information beyond the observed data, $X_1, \ldots, X_n$, it is impossible to give reasonable estimates of $\gamma(h)$ and $\rho(h)$ for $h \geq n$. Even for h slightly smaller than n, the estimates $\hat{\gamma}(h)$ and $\hat{\rho}(h)$ are unreliable since there are so few pairs (X_{t+h}, X_t) available (only one if $h = n - 1$). A useful guide is provided by Box and Jenkins (1976), p. 33, who suggest that n should be at least about 50 and $h \leq n/4$.

The sample ACF plays an important role in the selection of suitable models for the data. We have already seen in Example 1.4.6 and Section 1.6 how the sample ACF can be used to test for IID noise. For systematic inference concerning $\rho(h)$, we need the sampling distribution of the estimator $\hat{\rho}(h)$. Although the distribution of $\hat{\rho}(h)$ is intractable for samples from even the simplest time series models, it can usually be well approximated by a normal distribution for large sample sizes. For linear models and in particular for ARMA models (see Theorem 7.2.2 of TSTM for exact conditions) $\hat{\boldsymbol{\rho}}_k = (\hat{\rho}(1), \ldots, \hat{\rho}(k))'$ is approximately distributed for large n as $N(\boldsymbol{\rho}_k, n^{-1}W)$, i.e.,

$$\hat{\boldsymbol{\rho}} \approx N(\boldsymbol{\rho}, n^{-1}W), \tag{2.4.9}$$

where $\boldsymbol{\rho} = (\rho(1), \ldots, \rho(k))'$, and W is the covariance matrix whose (i, j)-element is given by **Bartlett's formula**,

$$\begin{aligned} w_{ij} = & \sum_{k=-\infty}^{\infty} \{\rho(k+i)\rho(k+j) + \rho(k-i)\rho(k+j) + 2\rho(i)\rho(j)\rho^2(k) \\ & - 2\rho(i)\rho(k)\rho(k+j) - 2\rho(j)\rho(k)\rho(k+i)\}. \end{aligned}$$

Simple algebra shows that

$$\begin{aligned} w_{ij} = & \sum_{k=1}^{\infty} \{\rho(k+i) + \rho(k-i) - 2\rho(i)\rho(k)\} \\ & \times \{\rho(k+j) + \rho(k-j) - 2\rho(j)\rho(k)\}, \end{aligned} \tag{2.4.10}$$

which is a more convenient form of w_{ij} for computational purposes.

Example 2.4.2 IID Noise

If $\{X_t\} \sim \text{IID}(0, \sigma^2)$, then $\rho(h) = 0$ for $|h| > 0$, so from (2.4.10), we obtain

$$w_{ij} = \begin{cases} 1 & \text{if } i = j, \\ 0 & \text{otherwise.} \end{cases}$$

For large n, therefore, $\hat{\rho}(1), \ldots, \hat{\rho}(h)$ are approximately independent and identically distributed normal random variables with mean 0 and variance n^{-1}. This result is the basis for the test that data are generated from IID noise using the sample ACF described in Section 1.6. (See also Example 1.4.6.) □

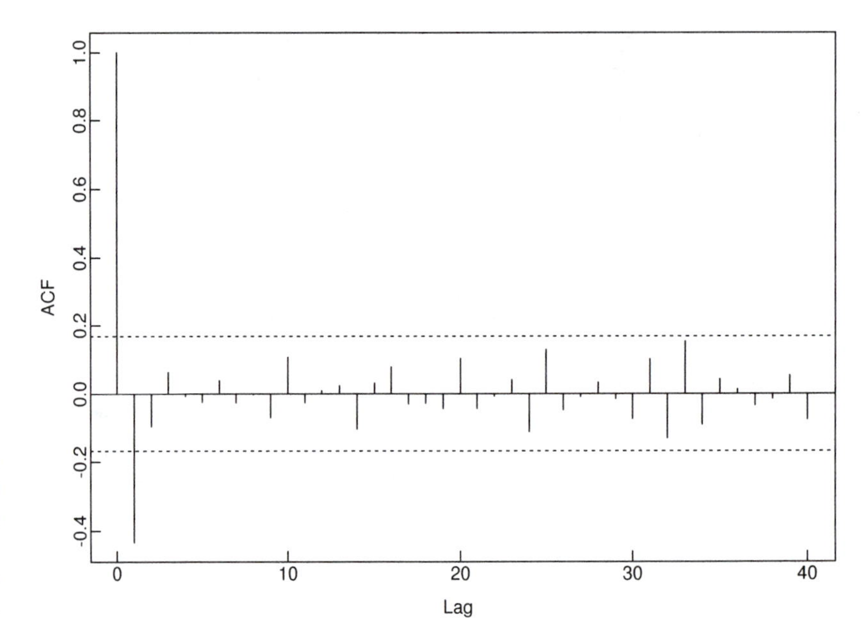

Figure 2-1 The sample autocorrelation function of $n = 200$ observations of the MA(1) process in Example 2.4.3, showing the bounds $\pm 1.96 n^{-1/2}(1 + 2\hat{\rho}^2(1))^{1/2}$.

Example 2.4.3 An MA(1) process

If $\{X_t\}$ is the MA(1) process of Example 1.4.4, i.e., if

$$X_t = Z_t + \theta Z_{t-1}, t = 0, \pm 1, \ldots,$$

where $\{Z_t\} \sim \text{WN}(0, \sigma^2)$, then from (2.4.10)

$$w_{ii} = \begin{cases} 1 - 3\rho^2(1) + 4\rho^4(1), & \text{if } i = 1, \\ 1 + 2\rho^2(1), & \text{if } i > 1, \end{cases}$$

is the approximate variance of $n^{-1/2}(\hat{\rho}(i) - \rho(i))$ for large n. In Figure 2.1, we have plotted the sample autocorrelation function $\hat{\rho}(k)$, $k = 0, \ldots, 40$ for 200 observations from the MA(1) model,

$$X_t = Z_t - .8Z_{t-1}, \tag{2.4.11}$$

where $\{Z_t\}$ is a sequence of IID N(0, 1) random variables. Here $\rho(1) = -.8/1.64 = -.4878$ and $\rho(h) = 0$ for $h > 1$. The lag-one sample ACF is found to be $\hat{\rho}(1) = -.4333 = -6.128n^{-1/2}$, which would cause us (in the absence of our prior knowledge of $\{X_t\}$) to reject the hypothesis that the data are a sample from an IID noise sequence. The fact that $|\hat{\rho}(h)| \leq 1.96n^{-1/2}$ for $h = 2, \ldots, 40$ strongly suggests that the data are from a model in which observations are uncorrelated past lag 1. In Figure 2.1, we have plotted the bounds $\pm 1.96n^{-1/2}(1 + 2\rho^2(1))^{1/2}$, indicating the compatibility of the data with the model (2.4.11). Since, however, $\rho(1)$ is not normally known in advance, the autocorrelations $\hat{\rho}(2), \ldots, \hat{\rho}(40)$ would in practice have been compared with the more stringent bounds $\pm 1.96n^{-1/2}$ or with the bounds $\pm 1.96n^{-1/2}(1 + 2\hat{\rho}^2(1))^{1/2}$ in order to check the hypothesis that the data are generated by a moving average process of order 1. Finally, it is worth noting that the lag-one correlation, $-.4878$, is well inside the 95% confidence bounds for $\rho(1)$ given by $\hat{\rho}(1) \pm 1.96 n^{-1/2}(1 - 3\hat{\rho}^2(1) + 4\hat{\rho}^4(1))^{1/2} = -.4333 \pm .1510$. This further supports the compatibility of the data with the model $X_t = Z_t - 0.8 Z_{t-1}$. □

Example 2.4.4 An AR(1) process

For the AR(1) process of Example 2.2.1,

$$X_t = \phi X_{t-1} + Z_t,$$

where $\{Z_t\}$ is IID noise and $|\phi| < 1$, we have from (2.4.10) with $\rho(h) = \phi^{|h|}$

$$\begin{aligned} w_{ii} &= \sum_{k=1}^{i} \phi^{2i}(\phi^{-k} - \phi^{k})^2 + \sum_{k=i+1}^{\infty} \phi^{2k}(\phi^{-i} - \phi^{i})^2 \\ &= (1 - \phi^{2i})(1 + \phi^2)(1 - \phi^2)^{-1} - 2i\phi^{2i}, \end{aligned} \tag{2.4.12}$$

$i = 1, 2, \ldots$. In Figure 2.2, we have plotted the sample ACF of the Lake Huron residuals, $y_1, \ldots, y_{98}$, from Figure 1.10 together with 95% confidence bounds for $\rho(i)$, $i = 1, \ldots, 40$ assuming the data are generated from the AR(1) model

$$Y_t = .791 Y_{t-1} + Z_t \tag{2.4.13}$$

(see equation (1.4.3)). The confidence bounds are computed from $\hat{\rho}(i) \pm 1.96 n^{-1/2} w_{ii}^{1/2}$, where w_{ii} is given in (2.4.12) with $\phi = .791$. The model ACF, $\rho(i) = (.791)^i$,

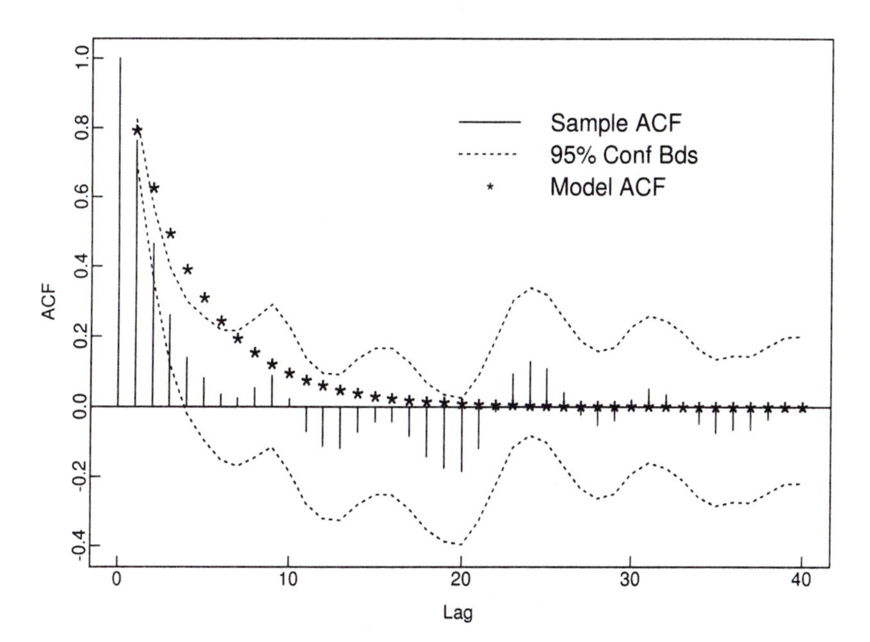

Figure 2-2 The sample autocorrelation function of the Lake Huron residuals of Figure 1.10 showing the bounds $\hat{\rho}(i) \pm 1.96 n^{-1/2} w_{ii}^{1/2}$ and the model ACF $\rho(i) = (.791)^i$.

is also plotted in Figure 2.2. Notice that the model ACF lies just outside the confidence bounds at lags 2–6. This suggests some incompatibility of the data with the model (2.4.13). A much better fit to the residuals is provided by the second-order autoregression defined by (1.4.4). □

2.5 Forecasting Stationary Time Series

We now consider the problem of predicting the values X_{n+h}, $h > 0$, of a stationary time series with known mean μ and autocovariance function γ in terms of the values $\{X_n, \ldots, X_1\}$, up to time n. Our goal is to find the *linear combination* of $1, X_n, X_{n-1}, \ldots, X_1$, which forecasts X_{n+h} with minimum mean squared error. The best linear predictor in terms of $1, X_n, \ldots, X_1$ will be denoted by $P_n X_{n+h}$ and clearly has the form

$$P_n X_{n+h} = a_0 + a_1 X_n + \cdots + a_n X_1. \tag{2.5.1}$$

It remains only to determine the coefficients $a_0, a_1, \ldots, a_n$, by finding the values that minimize

$$S(a_0, \ldots, a_n) = E(X_{n+h} - a_0 - a_1 X_n - \cdots - a_n X_1)^2. \tag{2.5.2}$$

(We already know from Problem 1.1 that $P_0 Y = E(Y)$.) Since S is a quadratic function of $a_0, \ldots, a_n$ and is bounded below by zero, it is clear that there is at least one value of $(a_0, \ldots, a_n)$ which minimizes S and that at the minimum $(a_0, \ldots, a_n)$ satisfies the equations,

$$\frac{\partial S(a_0, \ldots, a_n)}{\partial a_j} = 0, \ j = 0, \ldots, n. \tag{2.5.3}$$

Evaluation of the derivatives in equations (2.5.3) gives the equivalent equations,

$$E[X_{n+h} - a_0 - \sum_{i=1}^{n} a_i X_{n+1-i}] = 0, \tag{2.5.4}$$

$$E[(X_{n+h} - a_0 - \sum_{i=1}^{n} a_i X_{n+1-i}) X_{n+1-j}] = 0, \ j = 1, \ldots, n. \tag{2.5.5}$$

These equations can be written more neatly in vector notation as

$$a_0 = \mu(1 - \sum_{i=1}^{n} a_i), \tag{2.5.6}$$

and

$$\Gamma_n \mathbf{a}_n = \gamma_n(h), \tag{2.5.7}$$

where

$$\mathbf{a}_n = (a_1, \ldots, a_n)',$$
$$\Gamma_n = [\gamma(i-j)]_{i,j=1}^n,$$

and

$$\gamma_n(h) = (\gamma(h), \gamma(h+1), \ldots, \gamma(h+n-1))'.$$

Hence,

$$P_n X_{n+h} = \mu + \sum_{i=1}^n a_i (X_{n+1-i} - \mu), \tag{2.5.8}$$

where $\mathbf{a}_n$ satisfies (2.5.7). From (2.5.8) the expected value of the prediction error $X_{n+h} - P_n X_{n+h}$ is zero and the mean square prediction error is therefore

$$\begin{aligned} E(X_{n+h} - P_n X_{n+h})^2 &= \gamma(0) - 2\sum_{i=1}^n a_i \gamma(h+i-1) + \sum_{i=1}^n \sum_{j=1}^n a_i \gamma(i-j) a_j \\ &= \gamma(0) - \mathbf{a}_n' \gamma_n(h), \end{aligned} \tag{2.5.9}$$

where the last line follows from (2.5.7).

Remark 1. To show that equations (2.5.4) and (2.5.5) determine $P_n X_{n+h}$ uniquely, let $\{a_j^{(1)}, j = 0, \ldots, n\}$ and $\{a_j^{(2)}, j = 0, \ldots, n\}$ be two solutions and let Z be the difference between the corresponding predictors, i.e.,

$$Z = a_0^{(1)} - a_0^{(2)} + \sum_{j=1}^n (a_j^{(1)} - a_j^{(2)}) X_{n+1-j}.$$

Then

$$Z^2 = Z\left(a_0^{(1)} - a_0^{(2)} + \sum_{j=1}^n (a_j^{(1)} - a_j^{(2)}) X_{n+1-j}\right).$$

But from (2.5.4) and (2.5.5) we have $EZ = 0$ and $E(ZX_{n+1-j}) = 0$ for $j = 1, \ldots, n$. Consequently $E(Z^2) = 0$ and hence $Z = 0$. □

Properties of $P_n X_{n+h}$:

1. $P_n X_{n+h} = \mu + \sum_{i=1}^n a_i (X_{n+1-i} - \mu)$, where $\mathbf{a}_n = (a_1, \ldots, a_n)'$ satisfies (2.5.7)
2. $E(X_{n+h} - P_n X_{n+h})^2 = \gamma(0) - \mathbf{a}_n' \gamma_n(h)$, where $\gamma_n(h) = (\gamma(h), \ldots, \gamma(h+n))'$.
3. $E(X_{n+h} - P_n X_{n+h}) = 0$.
4. $E[(X_{n+h} - P_n X_{n+h}) X_j] = 0$, $j = 1, \ldots, n$.

Remark 2. Notice that properties 3 and 4 are exactly equivalent to (2.5.4) and (2.5.5). They can be written more succinctly in the form

$$E[(Error) \times (Predictor Variable)] = 0. \tag{2.5.10}$$

The equations (2.5.10) therefore uniquely determine $P_n X_{n+h}$. □

Example 2.5.1 One-step prediction of an AR(1) series

Consider now the stationary time series defined in Example 2.2.1 by the equations

$$X_t = \phi X_{t-1} + Z_t, t = 0, \pm 1, \ldots,$$

where $|\phi| < 1$ and $\{Z_t\} \sim \mathrm{WN}(0, \sigma^2)$. From (2.5.7) and (2.5.8), the best linear predictor of X_{n+1} in terms of $\{1, X_n, \ldots, X_1\}$ is (for $n \geq 1$)

$$P_n X_{n+1} = \mathbf{a}_n' \mathbf{X}_n,$$

where $\mathbf{X}_n = (X_n, \ldots, X_1)'$ and

$$\begin{bmatrix} 1 & \phi & \phi^2 & \cdots & \phi^{n-1} \\ \phi & 1 & \phi & \cdots & \phi^{n-2} \\ \vdots & \vdots & \vdots & \vdots & \vdots \\ \phi^{n-1} & \phi^{n-2} & \phi^{n-3} & \cdots & 1 \end{bmatrix} \begin{bmatrix} a_1 \\ a_2 \\ \vdots \\ a_n \end{bmatrix} = \begin{bmatrix} \phi \\ \phi^2 \\ \vdots \\ \phi^n \end{bmatrix}. \tag{2.5.11}$$

A solution of (2.5.11) is clearly

$$\mathbf{a}_n = (\phi, 0, \ldots, 0)',$$

and hence the best linear predictor of X_{n+1} in terms of $\{X_1, \ldots, X_n\}$ is

$$P_n X_{n+1} = \mathbf{a}_n' \mathbf{X}_n = \phi X_n,$$

with mean squared error

$$E(X_{n+1} - P_n X_{n+1})^2 = \gamma(0) - \mathbf{a}_n' \gamma_n(1) = \frac{\sigma^2}{1 - \phi^2} - \phi \gamma(1) = \sigma^2.$$

A simpler approach to this problem is to guess, by inspection of the equation defining X_{n+1}, that the best predictor is ϕX_n. Then to verify this conjecture, it suffices to check (2.5.10) for each of the predictor variables $1, X_n, \ldots, X_1$. The prediction error of the predictor ϕX_n is clearly $X_{n+1} - \phi X_n = Z_{n+1}$. But $E(Z_{n+1} Y) = 0$ for $Y = 1$ and for $Y = X_j, j = 1, \ldots, n$. Hence, by (2.5.10), ϕX_n is the required best linear predictor in terms of $1, X_1, \ldots, X_n$. □

Prediction of Second-Order Random Variables

Suppose now that Y and $W_n, \ldots, W_1$ are *any* random variables with finite second moments and that the means $\mu = EY$, $\mu_i = EW_i$ and covariances $\mathrm{Cov}(Y, Y)$, $\mathrm{Cov}(Y, W_i)$, and $\mathrm{Cov}(W_i, W_j)$ are all known. It is convenient to introduce the random vector $\mathbf{W} = (W_n, \ldots, W_1)'$, the corresponding vector of means, $\boldsymbol{\mu}_W = (\mu_n, \ldots, \mu_1)'$, the vector of covariances,

$$\boldsymbol{\gamma} = \text{Cov}(Y, \mathbf{W}) = (\text{Cov}(Y, W_n), \text{Cov}(Y, W_{n-1}), \ldots, \text{Cov}(Y, W_1))',$$

and the covariance matrix,

$$\Gamma = \text{Cov}(\mathbf{W}, \mathbf{W}) = \left[\text{Cov}(W_{n+1-i}, W_{n+1-j})\right]_{i,j=1}^{n}.$$

Then by the same arguments used in the calculation of $P_n X_{n+h}$, the best linear predictor of Y in terms of $\{1, W_n, \ldots, W_1\}$ is found to be

$$P(Y|\mathbf{W}) = \mu_Y + \mathbf{a}'(\mathbf{W} - \boldsymbol{\mu}_W), \tag{2.5.12}$$

where $\mathbf{a} = (a_1, \ldots, a_n)'$ is any solution of

$$\Gamma\mathbf{a} = \boldsymbol{\gamma}. \tag{2.5.13}$$

The mean squared error of the predictor is

$$E[(Y - P(Y|\mathbf{W}))^2] = \text{Var}(Y) - \mathbf{a}'\boldsymbol{\gamma}. \tag{2.5.14}$$

Example 2.5.2 Estimation of a missing value

Consider again the stationary series defined in Example 2.2.1 by the equations,

$$X_t = \phi X_{t-1} + Z_t, t = 0, \pm 1, \ldots,$$

where $|\phi| < 1$ and $\{Z_t\} \sim \text{WN}(0, \sigma^2)$. Suppose that we observe the series at times 1 and 3 and wish to use these observations to find the linear combination of 1, X_1, and X_3 that estimates X_2 with minimum mean squared error. The solution to this problem can be obtained directly from (2.5.12) and (2.5.13) by setting $Y = X_2$ and $\mathbf{W} = (X_1, X_3)'$. This gives the equations

$$\begin{bmatrix} 1 & \phi^2 \\ \phi^2 & 1 \end{bmatrix} \mathbf{a} = \begin{bmatrix} \phi \\ \phi \end{bmatrix},$$

with solution

$$\mathbf{a} = \frac{1}{1 + \phi^2} \begin{bmatrix} \phi \\ \phi \end{bmatrix}.$$

The best estimator of X_2 is thus

$$P(X_2|\mathbf{W}) = \frac{\phi}{1 + \phi^2}(X_1 + X_3),$$

with mean squared error

$$E[(X_2 - P(X_2|\mathbf{W}))^2] = \frac{\sigma^2}{1-\phi^2} - \mathbf{a}' \begin{bmatrix} \dfrac{\sigma^2}{1-\phi^2} \\ \dfrac{\sigma^2}{1-\phi^2} \end{bmatrix}.$$

$$= \frac{\sigma^2}{1+\phi^2}.$$

□

The Prediction Operator, $P(\cdot|\mathbf{W})$

For any given $\mathbf{W} = (W_n, \dots, W_1)'$ and Y with finite second moments, we have seen how to compute the best linear predictor $P(Y|\mathbf{W})$ of Y in terms of $1, W_n, \dots, W_1'$ from (2.5.12) and (2.5.13). The function $P(\cdot|\mathbf{W})$, which converts Y into $P(Y|\mathbf{W})$, is called a **prediction operator**. (The operator P_n defined by equations (2.5.7) and (2.5.8) is an example with $\mathbf{W} = (X_n, X_{n-1}, \dots, X_1)'$.) Prediction operators have a number of useful properties that can sometimes be used to simplify the calculation of best linear predictors. We list some of these below.

Properties of the Prediction Operator $P(\cdot|\mathbf{W})$:

Suppose that $EU^2 < \infty$, $EV^2 < \infty$, $\Gamma = \text{cov}(\mathbf{W}, \mathbf{W})$ and $\beta, \alpha_1, \dots, \alpha_n$ are constants.

1. $P(U|\mathbf{W}) = EU + \mathbf{a}'(\mathbf{W} - E\mathbf{W})$ where $\Gamma\mathbf{a} = \text{cov}(U, \mathbf{W})$.
2. $E[(U - P(U|\mathbf{W}))\mathbf{W}] = \mathbf{0}$ and $E[U - P(U|\mathbf{W})] = 0$
3. $E[(U - P(U|\mathbf{W}))^2] = \text{var}(U) - \mathbf{a}'\text{cov}(U, \mathbf{W})$.
4. $P(\alpha_1 U + \alpha_2 V + \beta|\mathbf{W}) = \alpha_1 P(U|\mathbf{W}) + \alpha_2 P(V|\mathbf{W}) + \beta$.
5. $P(\sum_{i=1}^n \alpha_i W_i + \beta|\mathbf{W}) = \sum_{i=1}^n \alpha_i W_i + \beta$.
6. $P(U|\mathbf{W}) = EU$ if $\text{cov}(U, \mathbf{W}) = \mathbf{0}$.
7. $P(U|\mathbf{W}) = P(P(U|\mathbf{W}, \mathbf{V})|\mathbf{W})$ if the components of $E(\mathbf{V}\mathbf{V}')$ are all finite.

Example 2.5.3 One-step prediction of an AR(p) series

Suppose now that $\{X_t\}$ is a stationary time series satisfying the equations

$$X_t = \phi_1 X_{t-1} + \cdots + \phi_p X_{t-p} + Z_t, t = 0, \pm 1, \dots,$$

where $\{Z_t\} \sim \text{WN}(0, \sigma^2)$ and Z_t is uncorrelated with X_s for each $s < t$. Then if $n > p$ we can apply the prediction operator P_n to each side of the defining equations, using properties (4), (5), and (6) to get

$$P_n X_{n+1} = \phi_1 X_n + \cdots + \phi_p X_{n+1-p}.$$

□

Example 2.5.4 An AR(1) series with nonzero mean

The time series $\{Y_t\}$ is said to be an AR(1) process with mean μ if $\{X_t = Y_t - \mu\}$ is a zero-mean AR(1) process. Defining $\{X_t\}$ as in Example 2.5.1 and letting $Y_t = X_t + \mu$, we see that Y_t satisfies the equation

$$Y_t - \mu = \phi(Y_{t-1} - \mu) + Z_t. \tag{2.5.15}$$

If $P_n Y_{n+h}$ is the best linear predictor of Y_{n+h} in terms of $\{1, Y_n, \ldots, Y_1\}$, then application of P_n to (2.5.15) with $t = n+1, n+2, \ldots$ gives the recursions

$$P_n Y_{n+h} - \mu = \phi(P_n Y_{n+h-1} - \mu), h = 1, 2, \ldots.$$

Noting that $P_n Y_n = Y_n$, we can solve these equations recursively for $P_n Y_{n+h}$, $h = 1, 2, \ldots$, to obtain

$$P_n Y_{n+h} = \mu + \phi^h(Y_n - \mu). \tag{2.5.16}$$

The corresponding mean squared error is (from (2.5.14))

$$E(Y_{n+h} - P_n Y_{n+h})^2 = \gamma(0)[1 - \mathbf{a}_n' \rho_n(h)]. \tag{2.5.17}$$

From Example 2.2.1 we know that $\gamma(0) = \sigma^2/(1-\phi^2)$ and $\rho(h) = \phi^h$, $h \geq 0$. Hence, substituting $\mathbf{a}_n = (\phi^h, 0, \ldots, 0)'$ (from (2.5.16)) into (2.5.17) gives

$$E(Y_{n+h} - P_n Y_{n+h})^2 = \sigma^2(1 - \phi^{2h})/(1 - \phi^2). \tag{2.5.18}$$

□

Remark 3. In general, if $\{Y_t\}$ is a stationary time series with mean μ and if $\{X_t\}$ is the zero-mean series defined by $X_t = Y_t - \mu$, then since the collection of all linear combinations of $1, Y_n, \ldots, Y_1$ is the same as the collection of all linear combinations of $1, X_n, \ldots, X_1$, the linear predictor of any random variable W in terms of $1, Y_n, \ldots, Y_1$ is the same as the linear predictor in terms of $1, X_n, \ldots, X_1$. Denoting this predictor by $P_n W$ and applying P_n to the equation $Y_{n+h} = X_{n+h} + \mu$ gives

$$P_n Y_{n+h} = \mu + P_n X_{n+h}. \tag{2.5.19}$$

Thus the best linear predictor of Y_{n+h} can be determined by finding the best linear predictor of X_{n+h} and then adding μ. Note from (2.5.8) that since $E(X_t) = 0$, $P_n X_{n+h}$ is the same as the best linear predictor of X_{n+h} in terms of $X_n, \ldots, X_1$ only. □

2.5.1 The Durbin-Levinson Algorithm

In view of Remark 3 above, we can restrict attention from now on to zero-mean stationary time series, making the necessary adjustments for the mean if we wish to predict a stationary series with nonzero mean. If $\{X_t\}$ is a zero-mean stationary series with autocovariance function $\gamma(\cdot)$, then in principle the equations (2.5.12) and (2.5.13) completely solve the problem of determining the best linear predictor

$P_n X_{n+h}$ of X_{n+h} in terms of $\{X_n, \ldots, X_1\}$. However, the direct approach requires the determination of a solution of a system of n linear equations, which for large n may be difficult and time-consuming. In cases when the process is defined by a system of linear equations (as in Examples 2.5.2 and 2.5.3) we have seen how the linearity of P_n can be used to great advantage. For more general stationary processes it would be helpful if the one-step predictor $P_n X_{n+1}$ based on n previous observations could be used to simplify the calculation of $P_{n+1} X_{n+2}$, the one-step predictor based on $n+1$ previous observations. Prediction algorithms that utilize this idea are said to be **recursive**. Two important examples are the Durbin-Levinson algorithm, discussed in this section, and the innovations algorithm, discussed in Section 2.5.2 below.

We know from (2.5.12) and (2.5.13) that if the matrix Γ_n is nonsingular, then

$$P_n X_{n+1} = \phi_n' \mathbf{X}_n = \phi_{n1} X_n + \cdots + \phi_{nn} X_1,$$

where

$$\phi_n = \Gamma_n^{-1} \gamma_n,$$

$\gamma_n = (\gamma(1), \ldots, \gamma(n))'$, and the corresponding mean squared error is

$$v_n := E(X_{n+1} - P_n X_{n+1})^2 = \gamma(0) - \phi_n' \gamma_n.$$

A useful sufficient condition for nonsingularity of *all* the autocovariance matrices, $\Gamma_1, \Gamma_2, \ldots$ is $\gamma(0) > 0$ and $\gamma(h) \to 0$ as $h \to \infty$. (For a proof of this result see TSTM, Proposition 5.1.1.)

The Durbin-Levinson Algorithm:

The coefficients $\phi_{n1}, \ldots, \phi_{nn}$ can be computed recursively from the equations

$$\phi_{nn} = \left[\gamma(n) - \sum_{j=1}^{n-1} \phi_{n-1,j}\gamma(n-j)\right] v_{n-1}^{-1}, \tag{2.5.20}$$

$$\begin{bmatrix} \phi_{n1} \\ \vdots \\ \phi_{n,n-1} \end{bmatrix} = \begin{bmatrix} \phi_{n-1,1} \\ \vdots \\ \phi_{n-1,n-1} \end{bmatrix} - \phi_{nn} \begin{bmatrix} \phi_{n-1,n-1} \\ \vdots \\ \phi_{n-1,1} \end{bmatrix} \tag{2.5.21}$$

and

$$v_n = v_{n-1}[1 - \phi_{nn}^2], \tag{2.5.22}$$

where $\phi_{11} = \gamma(1)/\gamma(0)$ and $v_0 = \gamma(0)$.

Proof The definition of ϕ_{11} ensures that the equation

$$R_n \phi_n = \rho_n \tag{2.5.23}$$

(where $\boldsymbol{\rho}_n = (\rho(1), \ldots, \rho(n))'$) is satisfied for $n = 1$. The first step in the proof is to show that $\boldsymbol{\phi}_n$, defined recursively by (2.5.20) and (2.5.21), satisfies (2.5.23) for all n. Suppose this is true for $n = k$. Then, partitioning R_{k+1} and defining

$$\boldsymbol{\rho}_k^{(r)} := (\rho(k), \rho(k-1), \ldots, \rho(1))'$$

and

$$\boldsymbol{\phi}_k^{(r)} := (\phi_{kk}, \phi_{k,k-1}, \ldots, \phi_{k1})',$$

we see that the recursions imply

$$\begin{aligned} R_{k+1}\boldsymbol{\phi}_{k+1} &= \begin{bmatrix} R_k & \boldsymbol{\rho}_k^{(r)} \\ \boldsymbol{\rho}_k^{(r)\prime} & 1 \end{bmatrix} \begin{bmatrix} \boldsymbol{\phi}_k - \phi_{k+1,k+1}\boldsymbol{\phi}_k^{(r)} \\ \phi_{k+1,k+1} \end{bmatrix} \\ &= \begin{bmatrix} \boldsymbol{\rho}_k - \phi_{k+1,k+1}\boldsymbol{\rho}_k^{(r)} + \phi_{k+1,k+1}\boldsymbol{\rho}_k^{(r)} \\ \boldsymbol{\rho}_k^{(r)\prime}\boldsymbol{\phi}_k - \phi_{k+1,k+1}\boldsymbol{\rho}_k^{(r)\prime}\boldsymbol{\phi}_k^{(r)} + \phi_{k+1,k+1} \end{bmatrix} \\ &= \boldsymbol{\rho}_{k+1}, \end{aligned}$$

as required. Here we have used the fact that if $R_k\boldsymbol{\phi}_k = \boldsymbol{\rho}_k$, then $R_k\boldsymbol{\phi}_k^{(r)} = \boldsymbol{\rho}_k^{(r)}$. This is easily checked by writing out the component equations in reverse order. Since (2.5.23) is satisfied for $n = 1$, it follows by induction that the coefficient vectors $\boldsymbol{\phi}_n$ defined recursively by (2.5.20) and (2.5.21) satisfy (2.5.23) for all n.

It remains only to establish that the mean squared errors,

$$v_n := E(X_{n+1} - \boldsymbol{\phi}_n'\mathbf{X}_n)^2$$

satisfy $v_0 = \gamma(0)$ and (2.5.22). The fact that $v_0 = \gamma(0)$ is an immediate consequence of the definition $P_0X_1 := E(X_1) = 0$. Since we have shown that $\boldsymbol{\phi}_n'\mathbf{X}_n$ is the best linear predictor of X_{n+1}, we can write, from (2.5.9) and (2.5.21),

$$v_n = \gamma(0) - \boldsymbol{\phi}_n'\boldsymbol{\gamma}_n = \gamma(0) - \boldsymbol{\phi}_{n-1}'\boldsymbol{\gamma}_{n-1} + \phi_{nn}\boldsymbol{\phi}_{n-1}^{(r)\prime}\boldsymbol{\gamma}_{n-1} - \phi_{nn}\gamma(n).$$

Applying (2.5.9) again gives

$$v_n = v_{n-1} + \phi_{nn}(\boldsymbol{\phi}_{n-1}^{(r)\prime}\boldsymbol{\gamma}_{n-1} - \gamma(n)),$$

and hence, by (2.5.20),

$$v_n = v_{n-1} - \phi_{nn}^2(\gamma(0) - \boldsymbol{\phi}_{n-1}'\boldsymbol{\gamma}_{n-1}) = v_{n-1}(1 - \phi_{nn}^2).$$ ■

Remark 4. Under the conditions of the proposition, the function defined by $\alpha(0) = 1$ and $\alpha(n) = \phi_{nn}$, $n = 1, 2, \ldots$, is known as the **partial autocorrelation function** (PACF) of $\{X_t\}$. It will be discussed further in Section 3.2. Of particular interest is equation (2.5.22), which shows the relation between $\alpha(n)$ and the reduction in the one-step mean squared error as the number of predictors is increased from $n - 1$ to n. □

2.5.2 The Innovations Algorithm

The recursive algorithm to be discussed in this section is applicable to *all* series with finite second moments, regardless of whether they are stationary or not. Its application, however, can be simplified in certain special cases.

Suppose then that $\{X_t\}$ is a zero-mean series with $E|X_t|^2 < \infty$ for each t and

$$E(X_i X_j) = \kappa(i, j). \tag{2.5.24}$$

It will be convenient to introduce the following notation for the best one-step predictors and their mean squared errors:

$$\hat{X}_n = \begin{cases} 0, & \text{if } n = 1 \\ P_{n-1}X_n, & \text{if } n = 2, 3, \ldots, \end{cases}$$

and

$$v_n = E(X_{n+1} - P_n X_{n+1})^2.$$

We shall also introduce the **innovations**, or one-step prediction errors,

$$U_n = X_n - \hat{X}_n.$$

In terms of the vectors $\mathbf{U}_n = (U_1, \ldots, U_n)'$ and $\mathbf{X}_n = (X_1, \ldots, X_n)'$ the last equations can be written as

$$\mathbf{U}_n = A_n \mathbf{X}_n, \tag{2.5.25}$$

where A_n has the form

$$A_n = \begin{bmatrix} 1 & 0 & 0 & \cdots & 0 \\ a_{11} & 1 & 0 & \cdots & 0 \\ a_{22} & a_{21} & 1 & \cdots & 0 \\ \vdots & \vdots & \vdots & \ddots & 0 \\ a_{n-1,n-1} & a_{n-1,n-2} & a_{n-1,n-3} & \cdots & 1 \end{bmatrix}.$$

(If $\{X_t\}$ is stationary, then $a_{ij} = -a_j$ with a_j as in (2.5.7) with $h = 1$.) This implies that A_n is nonsingular, with inverse C_n of the form,

$$C_n = \begin{bmatrix} 1 & 0 & 0 & \cdots & 0 \\ \theta_{11} & 1 & 0 & \cdots & 0 \\ \theta_{22} & \theta_{21} & 1 & \cdots & 0 \\ \vdots & \vdots & \vdots & \ddots & 0 \\ \theta_{n-1,n-1} & \theta_{n-1,n-2} & \theta_{n-1,n-3} & \cdots & 1 \end{bmatrix}.$$

The vector of one-step predictors, $\hat{\mathbf{X}}_n := (X_1, P_1X_2, \ldots, P_{n-1}X_n)'$ can therefore be expressed as

$$\hat{\mathbf{X}}_n = \mathbf{X}_n - \mathbf{U}_n = C_n\mathbf{U}_n - \mathbf{U}_n = \Theta_n(\mathbf{X}_n - \hat{\mathbf{X}}_n), \tag{2.5.26}$$

where

$$\mathbf{\Theta}_n = \begin{bmatrix} 0 & 0 & 0 & \cdots & 0 \\ \theta_{11} & 0 & 0 & \cdots & 0 \\ \theta_{22} & \theta_{21} & 0 & \cdots & 0 \\ \vdots & \vdots & \vdots & \ddots & 0 \\ \theta_{n-1,n-1} & \theta_{n-1,n-2} & \theta_{n-1,n-3} & \cdots & 0 \end{bmatrix}$$

and $\mathbf{X}_n$ itself satisfies

$$\mathbf{X}_n = C_n(\mathbf{X}_n - \hat{\mathbf{X}}_n). \tag{2.5.27}$$

The equations (2.5.26) can be rewritten as

$$\hat{X}_{n+1} = \begin{cases} 0, & \text{if } n = 0, \\ \displaystyle\sum_{j=1}^{n} \theta_{nj}(X_{n+1-j} - \hat{X}_{n+1-j}), & \text{if } n = 1, 2, \ldots, \end{cases} \tag{2.5.28}$$

from which the one-step predictors $\hat{X}_1$, $\hat{X}_2$, ..., can be computed recursively once the coefficients θ_{ij} have been determined. The following algorithm generates these coefficients and the mean squared errors $v_i = E(X_{i+1} - \hat{X}_{i+1})^2$, starting from the covariances $\kappa(i, j)$.

The Innovations Algorithm:

The coefficients $\theta_{n1}, \ldots, \theta_{nn}$ can be computed recursively from the equations

$$v_0 = \kappa(1, 1),$$

$$\theta_{n,n-k} = v_k^{-1}\left(\kappa(n+1, k+1) - \sum_{j=0}^{k-1} \theta_{k,k-j}\theta_{n,n-j}v_j\right), \quad 0 \le k < n,$$

and

$$v_n = \kappa(n+1, n+1) - \sum_{j=0}^{n-1} \theta_{n,n-j}^2 v_j.$$

(It is a trivial matter to solve first for v_0, then successively for $\theta_{11}, v_1; \theta_{22}, \theta_{21}, v_2; \theta_{33}, \theta_{32}, \theta_{31}, v_3; \ldots$.)

Proof See TSTM, Proposition 5.2.2. ∎

Remark 5. While the Durbin-Levinson recursion gives the coefficients of $X_n, \ldots, X_1$ in the representation $\hat{X}_{n+1} = \sum_{j=1}^{n} \phi_{nj} X_{n+1-j}$, the innovations algorithm gives the coefficients of $(X_n - \hat{X}_n), \ldots, (X_1 - \hat{X}_1)$, in the expansion $\hat{X}_{n+1} =$

$\sum_{j=1}^{n} \theta_{nj}(X_{n+1-j} - \hat{X}_{n+1-j})$. The latter expansion has a number of advantages deriving from the fact that the innovations are uncorrelated (see Problem 2.20). It can also be greatly simplified in the case of ARMA(p, q) series, as we shall see in Section 3.3. An immediate consequence of (2.5.28) is the innovations representation of X_{n+1} itself. Thus (defining $\theta_{n0} := 1$),

$$X_{n+1} = X_{n+1} - \hat{X}_{n+1} + \hat{X}_{n+1} = \sum_{j=0}^{n} \theta_{nj}(X_{n+1-j} - \hat{X}_{n+1-j}), n = 0, 1, 2, \ldots \square$$

Example 2.5.5 Recursive prediction of an MA(1)

If $\{X_t\}$ is the time series defined by

$$X_t = Z_t + \theta Z_{t-1}, \{Z_t\} \sim \text{WN}(0, \sigma^2),$$

then $\kappa(i, j) = 0$ for $|i-j| > 1, \kappa(i, i) = \sigma^2(1+\theta^2)$ and $\kappa(i, i+1) = \theta\sigma^2$. Application of the innovations algorithm leads at once to the recursions,

$$\theta_{nj} = 0, 2 \leq j \leq n,$$
$$\theta_{n1} = v_{n-1}^{-1}\theta\sigma^2,$$
$$v_0 = (1 + \theta^2)\sigma^2,$$

and

$$v_n = [1 + \theta^2 - v_{n-1}^{-1}\theta^2\sigma^2]\sigma^2.$$

For the particular case

$$X_t = Z_t - 0.9Z_{t-1}, \{Z_t\} \sim \text{WN}(0, 1),$$

the mean squared errors v_n, of $\hat{X}_{n+1}$ and coefficients θ_{nj}, $1 \leq j \leq n$, in the innovations representation,

$$\hat{X}_{n+1} = \sum_{j=1}^{n} \theta_{nj}(X_{n+1-j} - \hat{X}_{n+1-j}) = \theta_{n1}(X_n - \hat{X}_n),$$

are found from the recursions to be as follows:

$v_0 = 1.8100$

$\theta_{11} = -.4972$ $\quad v_1 = 1.3625$

$\theta_{21} = -.6606$ $\quad \theta_{22} = 0$ $\quad v_2 = 1.2155$

$\theta_{31} = -.7404$ $\quad \theta_{32} = 0$ $\quad \theta_{33} = 0$ $\quad v_3 = 1.1436$

$\theta_{41} = -.7870$ $\quad \theta_{42} = 0$ $\quad \theta_{43} = 0$ $\quad \theta_{44} = 0$ $\quad v_4 = 1.1017.$

If we apply the Durbin-Levinson algorithm to the same problem, we find that the mean squared errors v_n, of $\hat{X}_{n+1}$ and coefficients ϕ_{nj}, $1 \leq j \leq n$, in the representation

$$\hat{X}_{n+1} = \sum_{j=1}^{n} \phi_{nj} X_{n+1-j}$$

are as follows:

$$
\begin{array}{lllll}
v_0 = 1.8100 & & & & \\
\phi_{11} = -.4972 & v_1 = 1.3625 & & & \\
\phi_{21} = -.6606 & \phi_{22} = -.3285 & v_2 = 1.2155 & & \\
\phi_{31} = -.7404 & \phi_{32} = -.4892 & \phi_{33} = -.2433 & v_3 = 1.1436 & \\
\phi_{41} = -.7870 & \phi_{42} = -.5828 & \phi_{43} = -.3850 & \phi_{44} = -.1914 & v_4 = 1.1017.
\end{array}
$$

Notice that as n increases, v_n approaches the white noise variance and θ_{n1} approaches θ. These results hold for any MA(1) process with $|\theta| < 1$. The innovations algorithm is particularly well suited to forecasting MA(q) processes since for them $\theta_{nj} = 0$ for $n - j > q$. For AR(p) processes the Durbin-Levinson algorithm is usually more convenient since $\phi_{nj} = 0$ for $n - j > p$. □

Recursive Calculation of the h-step Predictors

For h-step prediction we use the result

$$P_n(X_{n+k} - P_{n+k-1}X_{n+k}) = 0, k \geq 1. \tag{2.5.29}$$

This follows from (2.5.10) and the fact that

$$E[(X_{n+k} - P_{n+k-1}X_{n+k} - 0)X_{n+j-1}] = 0, j = 1, \ldots, n.$$

Hence,

$$
\begin{aligned}
P_n X_{n+h} &= P_n P_{n+h-1} X_{n+h} \\
&= P_n \hat{X}_{n+h} \\
&= P_n \left(\sum_{j=1}^{n+h-1} \theta_{n+h-1,j}(X_{n+h-j} - \hat{X}_{n+h-j}) \right).
\end{aligned}
$$

Applying (2.5.29) again and using the linearity of P_n we find that

$$P_n X_{n+h} = \sum_{j=h}^{n+h-1} \theta_{n+h-1,j}(X_{n+h-j} - \hat{X}_{n+h-j}), \tag{2.5.30}$$

where the coefficients θ_{nj} are determined as before by the innovations algorithm. Moreover, the mean squared error can be expressed as

$$
\begin{aligned}
E(X_{n+h} - P_n X_{n+h})^2 &= EX_{n+h}^2 - E(P_n X_{n+h})^2 \\
&= \kappa(n+h, n+h) - \sum_{j=h}^{n+h-1} \theta_{n+h-1,j}^2 v_{n+h-j-1}.
\end{aligned} \tag{2.5.31}
$$

2.5.3 Prediction of a Stationary Process in Terms of Infinitely Many Past Values

It is often useful, when many past observations $X_m, \ldots, X_0, X_1, \ldots, X_n$ $(m < 0)$ are available, to evaluate the best linear predictor of X_{n+h} in terms of $1, X_m, \ldots, X_0,$

$\ldots, X_n$. This predictor, which we shall denote by $P_{m,n}X_{n+h}$, can easily be evaluated by the methods described above. If $|m|$ is large, this predictor can be approximated by the sometimes more easily calculated mean square limit,

$$\tilde{P}_n X_{n+h} = \lim_{m \to -\infty} P_{m,n} X_{n+h}.$$

We shall refer to $\tilde{P}_n$ as the **prediction operator based on the infinite past, $\{X_t, -\infty < t \le n\}$**. Analogously we shall refer to P_n as the **prediction operator based on the finite past**, $\{X_1, \ldots, X_n\}$. (Mean square convergence of random variables is discussed in Appendix C.)

Determination of $\tilde{P}_n X_{n+h}$

Like $P_n X_{n+h}$, the best linear predictor $\tilde{P}_n X_{n+h}$ when $\{X_n\}$ is a zero-mean stationary process with autocovariance function $\gamma(\cdot)$, is characterized by the equations

$$E[(X_{n+h} - \tilde{P}_n X_{n+h})X_{n+1-i}] = 0, i = 1, 2, \ldots.$$

Provided we can find a solution to these equations, it will necessarily be the uniquely defined predictor $\tilde{P}_n X_{n+h}$. An approach to this problem that is often effective is to *assume* that $\tilde{P}_n X_{n+h}$ can be expressed in the form

$$\tilde{P}_n X_{n+h} = \sum_{j=1}^{\infty} \alpha_j X_{n+1-j},$$

in which case the preceding equations reduce to

$$E[(X_{n+h} - \sum_{j=1}^{\infty} \alpha_j X_{n+1-j})X_{n+1-i}] = 0, i = 1, 2, \ldots,$$

or equivalently,

$$\sum_{j=0}^{\infty} \gamma(i-j)\alpha_j = \gamma(h+i-1), i = 1, 2, \ldots.$$

This is an infinite set of linear equations for the unknown coefficients α_i which determine $\tilde{P}_n X_{n+h}$ *provided the resulting series converges.*

Properties of $\tilde{P}_n$:

Suppose that $EU^2 < \infty, EV^2 < \infty, a, b,$ and c are constants, and $\Gamma = \text{Cov}(\mathbf{W}, \mathbf{W})$.

1. $E[(U - \tilde{P}_n(U))X_j] = 0, j \le n.$
2. $\tilde{P}_n(aU + bV + c) = a\tilde{P}_n(U) + b\tilde{P}_n(V) + c.$
3. $\tilde{P}_n(U) = U$ if U is a limit of linear combinations of $X_j, j \le n.$
4. $\tilde{P}_n(U) = EU$ if $\text{Cov}(U, X_j) = 0$ for all $j \le n.$

These properties can sometimes be used to simplify the calculation of $\tilde{P}_n X_{n+h}$, notably when the process $\{X_t\}$ is an ARMA process.

Example 2.5.7 Consider the ARMA(1,1) process

$$X_t - \phi X_{t-1} = Z_t + \theta Z_{t-1}, \{Z_t\} \sim \mathrm{WN}(0, \sigma^2).$$

We know from (2.3.3) and (2.3.5) that we have the representations

$$X_{n+1} = Z_{n+1} + (\phi + \theta) \sum_{j=1}^{\infty} \phi^{j-1} Z_{n+1-j}$$

and

$$Z_{n+1} = X_{n+1} - (\phi + \theta) \sum_{j=1}^{\infty} (-\theta)^{j-1} X_{n+1-j}.$$

Applying the operator $\tilde{P}_n$ to the second equation and using the properties of $\tilde{P}_n$ gives

$$\tilde{P}_n X_{n+1} = (\phi + \theta) \sum_{j=1}^{\infty} (-\theta)^{j-1} X_{n+1-j}.$$

Applying the operator $\tilde{P}_n$ to the first equation and using the properties of $\tilde{P}_n$ gives

$$\tilde{P}_n X_{n+1} = (\phi + \theta) \sum_{j=1}^{\infty} \phi^{j-1} Z_{n+1-j}.$$

Hence,

$$X_{n+1} - \tilde{P}_n X_{n+1} = Z_{n+1}$$

and so the mean squared error of the predictor $\tilde{P}_n X_{n+1}$ is $E Z_{n+1}^2 = \sigma^2$. □

2.6 The Wold Decomposition

Consider the stationary process

$$X_t = A \cos(\omega t) + B \sin(\omega t)$$

where $\omega \in (0, \pi)$ is constant and A, B are uncorrelated random variables with mean 0 and variance σ^2. Notice that

$$X_n = (2 \cos \omega) X_{n-1} - X_{n-2} = \tilde{P}_{n-1} X_n, n = 0, \pm 1, \ldots$$

so that $X_n - \tilde{P}_{n-1} X_n = 0$ for all n. Processes with the latter property are said to be **deterministic**.

The Wold Decomposition:

If $\{X_t\}$ is a nondeterministic stationary time series, then

$$X_t = \sum_{j=0}^{\infty} \psi_j Z_{t-j} + V_t, \tag{2.6.1}$$

where

1. $\psi_0 = 1$ and $\sum_{j=0}^{\infty} \psi_j^2 < \infty$,
2. $\{Z_t\} \sim \mathrm{WN}(0, \sigma^2)$,
3. $\mathrm{Cov}(Z_s, V_t) = 0$ for all s and t,
4. Z_t is the limit of linear combinations of X_s, $s \le t$, and
5. $\{V_t\}$ is deterministic.

If $\tilde{P}_t Y$ denotes the best predictor of Y in terms of linear combinations or limits of linear combinations of X_s, $-\infty < s \le t$, then the one-step mean squared error is given by

$$\sigma^2 = E(X_t - \tilde{P}_{t-1} X_t)^2.$$

The process is then deterministic if and only if $\sigma^2 = 0$. For the case $\sigma^2 > 0$, the decomposition (2.6.1) holds with one-step prediction error given by $Z_t = X_t - \tilde{P}_{t-1} X_t$. Moreover, the process $\{V_t\}$ is perfectly predictable from the process $\{X_t\}$, i.e., $\tilde{P}_t V_s = V_s$ for all s and t. For most of the time series dealt with in this book (in particular for all ARMA processes) the deterministic component V_t is 0 for all t. In such cases, the process is said to be **purely nondeterministic**.

Example 2.6.1 The series $X_t = Z_t + Y$, where $\{Z_t\} \sim \mathrm{WN}(0, \sigma^2)$, $EZ_tY = 0$ for all t, and Y has mean 0 and variance σ^2, has Wold decomposition $X_t = Z_t + V_t$ with $V_t = Y$ for all t. In this example it is easy to see that $\tilde{P}_t Y = Y$ since

$$\frac{1}{n} \sum_{j=0}^{n-1} X_{t-j} = \frac{1}{n} \sum_{j=0}^{n-1} Z_{t-j} + Y,$$

which converges in mean square to Y. □

Problems

2.1. Suppose that $X_1, X_2, \ldots$, is a stationary time series with mean μ and ACF $\rho(\cdot)$. Show that the best predictor of X_{n+h} of the form $aX_n + b$ is obtained by choosing $a = \rho(h)$ and $b = \mu(1 - \rho(h))$.

2.2. Show that the process

$$X_t = A\cos(\omega t) + B\sin(\omega t),\ t = 0, \pm 1, \ldots,$$

(where A and B are uncorrelated random variables with mean 0 and variance 1 and ω is a fixed frequency in the interval $[0, \pi]$), is stationary and find its mean and autocovariance function. Deduce that the function $\kappa(h) = \cos(\omega h)$, $h = 0, \pm 1, \ldots$, is nonnegative definite.

2.3. a. Find the ACVF of the time series $X_t = Z_t + .3Z_{t-1} - .4Z_{t-2}$ where $\{Z_t\} \sim \text{WN}(0, 1)$.

b. Find the ACVF of the time series $Y_t = \tilde{Z}_t - 1.2\tilde{Z}_{t-1} - 1.6\tilde{Z}_{t-2}$ where $\{\tilde{Z}_t\} \sim \text{WN}(0, .25)$. Compare with the answer found in (a).

2.4. It is clear that the function $\kappa(h) = 1$, $h = 0, \pm 1, \ldots$ is an autocovariance function since it is the autocovariance function of the process $X_t = Z$, $t = 0, \pm 1, \ldots$, where Z is a random variable with mean 0 and variance 1. By identifying appropriate sequences of random variables, show that the following functions are also autocovariance functions.

(a) $\kappa(h) = (-1)^{|h|}$

(b) $\kappa(h) = 1 + \cos(\frac{\pi h}{2}) + \cos(\frac{\pi h}{4})$

(c) $\kappa(h) = \begin{cases} 1, & \text{if } h = 0, \\ 0.4, & \text{if } h = \pm 1, \\ 0, & \text{otherwise.} \end{cases}$

2.5. Suppose that $\{X_t, t = 0, \pm 1, \ldots\}$ is stationary and that $|\theta| < 1$. Show that for each fixed n the sequence

$$S_m = \sum_{j=1}^{m} \theta^j X_{n-j}$$

is convergent absolutely and in mean square (see Appendix C) as $m \to \infty$.

2.6. Verify the equations (2.2.6).

2.7. Show, using the geometric series $\frac{1}{1-x} = \sum_{j=0}^{\infty} x^j$ for $|x| < 1$, that $\frac{1}{1-\phi z} = -\sum_{j=1}^{\infty} \phi^{-j} z^{-j}$ for $|\phi| > 1$ and $|z| \geq 1$.

2.8. Show that the autoregressive equations

$$X_t = \phi_1 X_{t-1} + Z_t, t = 0, \pm 1, \ldots,$$

where $\{Z_t\} \sim \mathrm{WN}(0, \sigma^2)$ and $|\phi| = 1$, have *no stationary solution*. HINT: Suppose there does exist a stationary solution $\{X_t\}$ and use the autoregressive equation to derive an expression for the variance of $X_t - \phi_1^{n+1} X_{t-n-1}$ which contradicts the stationarity assumption.

2.9. Let $\{Y_t\}$ be the AR(1) plus noise time series defined by

$$Y_t = X_t + W_t,$$

where $\{W_t\} \sim \mathrm{WN}(0, \sigma_w^2)$, $\{X_t\}$ is the AR(1) process of Example 2.2.1, i.e.,

$$X_t - \phi X_{t-1} = Z_t, \{Z_t\} \sim \mathrm{WN}(0, \sigma_z^2),$$

and $E(W_s Z_t) = 0$ for all s and t.

a. Show that $\{Y_t\}$ is stationary and find its autocovariance function.

b. Show that the time series $U_t := Y_t - \phi Y_{t-1}$ is 1-correlated and hence, by Proposition 2.1.1, is an MA(1) process.

c. Conclude from (b) that $\{Y_t\}$ is an ARMA(1,1) process and express the three parameters of this model in terms of ϕ, σ_w^2, and σ_z^2.

2.10. Use the program PEST to compute the coefficients ψ_j and π_j, $j = 1, \ldots, 5$ in the expansions

$$X_t = \sum_{j=0}^{\infty} \psi_j Z_{t-j}$$

and

$$Z_t = \sum_{j=0}^{\infty} \pi_j X_{t-j}$$

for the ARMA(1,1) process defined by the equations

$$X_t - 0.5X_{t-1} = Z_t + 0.5Z_{t-1}, \{Z_t\} \sim \mathrm{WN}(0, \sigma^2).$$

(To do this you will need to enter the model in PEST and then choose the option *Model ACF/PACF, AR/MA(∞) representations*.) Check the results with those obtained in Section 2.3.

2.11. Suppose that in a sample of size 100 from an AR(1) process with mean μ, $\phi = .6$, and $\sigma^2 = 2$, we obtain $\bar{x}_{100} = .271$. Construct an approximate 95% confidence interval for μ. Do the data suggest that $\mu = 0$?

2.12. Suppose that in a sample of size 100 from an MA(1) process with mean μ, $\theta = -.6$, and $\sigma^2 = 1$, we obtain $\bar{x}_{100} = .157$. Construct an approximate 95% confidence interval for μ. Do the data suggest that $\mu = 0$?

2.13. Suppose that in a sample of size 100, we obtain $\hat{\rho}(1) = .438$ and $\hat{\rho}(2) = .145$.

a. Assuming that the data were generated from an AR(1) model, construct approximate 95% confidence intervals for both $\rho(1)$ and $\rho(2)$. Based on these two confidence intervals, are the data consistent with an AR(1) model with $\phi = .8$?

b. Assuming that the data were generated from an MA(1) model, construct approximate 95% confidence intervals for both $\rho(1)$ and $\rho(2)$. Based on these two confidence intervals, are the data consistent with an MA(1) model with $\theta = .6$?

2.14. Let $\{X_t\}$ be the process defined in Problem 2.2.

a. Find $P_1 X_2$ and its mean squared error.

b. Find $P_2 X_3$ and its mean squared error.

c. Find $\tilde{P}_n X_{n+1}$ and its mean squared error.

2.15. Suppose that $\{X_t, t = 0, \pm 1, \ldots\}$ is a stationary process satisfying the equations

$$X_t = \phi_1 X_{t-1} + \cdots + \phi_p X_{t-p} + Z_t,$$

where $\{Z_t\} \sim \text{WN}(0, \sigma^2)$ and Z_t is uncorrelated with X_s for each $s < t$. Show that the best linear predictor $P_n X_{n+1}$ of X_{n+1} in terms of $1, X_1, \ldots, X_n$, assuming $n > p$, is

$$P_n X_{n+1} = \phi_1 X_n + \cdots + \phi_p X_{n+1-p}.$$

What is the mean squared error of $P_n X_{n+1}$?

2.16. Use the program PEST to plot and print out the sample ACF and PACF up to lag 40 of the sunspot series, $D_t, t = 1, 100$, (contained in the ITSM data file SUNSPOTS.DAT). By making use of the option *Preliminary Estimation* and suboption *Yule-Walker*, fit an AR(2) model to the mean-corrected data. This will give you a model of the form

$$X_t = \phi_1 X_{t-1} + \phi_2 X_{t-2} + Z_t, \text{ where } \{Z_t\} \sim \text{WN}(0, \sigma^2),$$

for the mean-corrected series $X_t = D_t - 46.93$. Record the values of the estimated parameters ϕ_1, ϕ_2, and σ^2. Plot and print out the *model* ACF and PACF and compare them with the corresponding sample quantities. This is most conveniently done by superposing the graphs using option *Model ACF/PACF* and suboption *Sample ACF/PACF with model ACF/PACF*.

2.17. Without exiting from PEST use the model found in the preceding problem to compute forecasts of the next nine values of the sunspot series as follows. Select the option *Forecasting* from the main menu of PEST and specify that you wish to forecast nine future values. You will see the nine forecasts $P_{100} X_{100+h}$, $h = 1, \ldots, 9$ tabulated on the screen in the column labelled XHAT. The corresponding forecasts of the original data, i.e., of $D_{101}, \ldots, D_{109}$, are listed in the column labelled `XHAT + MEAN`. You will also see estimates of the mean squared errors beside each forecast. The details of the calculations will be taken up in Chapter 3 when we discuss ARMA models in detail.

2.18. Let $\{X_t\}$ be the stationary process defined by the equations

$$X_t = Z_t - \theta Z_{t-1}, t = 0, \pm 1, \ldots,$$

where $|\theta| < 1$ and $\{Z_t\} \sim \mathrm{WN}(0, \sigma^2)$. Show that the best linear predictor $\tilde{P}_n X_{n+1}$ of X_{n+1} based on $\{X_j, -\infty < j \leq n\}$ is

$$\tilde{P}_n X_{n+1} = -\sum_{j=1}^{\infty} \theta^j X_{n+1-j}.$$

What is the mean squared error of the predictor $\tilde{P}_n X_{n+1}$?

2.19. If $\{X_t\}$ is defined as in Problem 2.18 and $\theta = 1$, find the best linear predictor $P_n X_{n+1}$ of X_{n+1} in terms of $X_1, \ldots, X_n$. What is the corresponding mean squared error?

2.20. In the innovations algorithm, show that for each $n \geq 2$, the innovation $X_n - \hat{X}_n$ is uncorrelated with $X_1, \ldots, X_{n-1}$. Conclude that $X_n - \hat{X}_n$ is uncorrelated with the innovations $X_1 - \hat{X}_1, \ldots, X_{n-1} - \hat{X}_{n-1}$.

2.21. Let X_1, X_2, X_4, X_5 be observations from the MA(1) model

$$X_t = Z_t + \theta Z_{t-1}, \{Z_t\} \sim \mathrm{WN}(0, \sigma^2).$$

a. Find the best linear estimate of the missing value X_3 in terms of X_1 and X_2.

b. Find the best linear estimate of the missing value X_3 in terms of X_4 and X_5.

c. Find the best linear estimate of the missing value X_3 in terms of X_1, X_2, X_4, and X_5.

d. Compute the mean squared errors for each of the estimates in (a), (b), and (c).

2.22. Repeat parts (a)–(d) of Problem 2.21 assuming now that the observations X_1, X_2, X_4, X_5 are from the causal AR(1) model,

$$X_t = \phi X_{t-1} + Z_t, \{Z_t\} \sim \mathrm{WN}(0, \sigma^2).$$

3 ARMA Models

In this chapter, we introduce an important parametric family of stationary time series, the autoregressive moving average or ARMA processes. For a large class of autocovariance functions $\gamma(\cdot)$, it is possible to find an ARMA process $\{X_t\}$ with ACVF $\gamma_X(\cdot)$ such that $\gamma(\cdot)$ is well approximated by $\gamma_X(\cdot)$. In particular, for any positive integer K, there exists an ARMA process $\{X_t\}$ such that $\gamma_X(h) = \gamma(h)$ for $h = 0, 1, \ldots, K$. For this (and other) reasons, the family of ARMA processes plays a key role in the modelling of time series data. The linear structure of ARMA processes also leads to a substantial simplification of the general methods for linear prediction discussed earlier in Section 2.5.

3.1 ARMA(p, q) Processes

In Section 2.3, we introduced an ARMA(1,1) process and discussed some of its key properties. These included existence and uniqueness of stationary solutions of the defining equations and the concepts of causality and invertibility. In this section we extend these notions to the general ARMA(p, q) process.

Definition 3.1.1 $\{X_t\}$ is an **ARMA(p, q) process** if $\{X_t\}$ is stationary and if for every t

$$X_t - \phi_1 X_{t-1} - \cdots - \phi_p X_{t-p} = Z_t + \theta_1 Z_{t-1} + \cdots + \theta_q Z_{t-q}, \qquad (3.1.1)$$

where $\{Z_t\} \sim \mathrm{WN}(0, \sigma^2)$ and the polynomials $(1 - \phi_1 z - \cdots - \phi_p z^p)$ and $(1 + \theta_1 z + \cdots + \theta_q z^q)$ have no common factors.

The process $\{X_t\}$ is said to be an **ARMA(p,q) process with mean μ** if $\{X_t-\mu\}$ is an ARMA(p,q) process.

It is convenient to use the more concise form of (3.1.1),

$$\phi(B)X_t=\theta(B)Z_t, \tag{3.1.2}$$

where $\phi(\cdot)$ and $\theta(\cdot)$ are the pth and q^{th} degree polynomials,

$$\phi(z)=1-\phi_1 z-\cdots-\phi_p z^p$$

and

$$\theta(z)=1+\theta_1 z+\cdots+\theta_q z^q,$$

and B is the backward shift operator ($B^jX_t=X_{t-j},\ B^jZ_t=Z_{t-j},\ j=0,\pm1,\ldots$). The time series $\{X_t\}$ is said to be an **autoregressive process of order p** (or AR(p)) if $\theta(z)\equiv1$ and a **moving average process of order q** (or MA(q)) if $\phi(z)\equiv1$.

An important part of Definition 3.1.1 is the requirement that $\{X_t\}$ be stationary. In Section 2.3, we showed, for the ARMA(1,1) equations (2.3.1), that a stationary solution exists (and is unique) if and only if $\phi_1\neq\pm1$. The latter is equivalent to the condition that the autoregressive polynomial $\phi(z)=1-\phi_1z\neq0$ for $z=\pm1$. The analogous condition for the general ARMA(p,q) process is $\phi(z)=1-\phi_1z-\cdots-\phi_pz^p\neq0$ for all complex z with $|z|=1$. (Complex z is used here since the zeroes of a polynomial of degree $p>1$ may be either real or complex. The region defined by the set of complex z such that $|z|=1$ is referred to as the unit circle.) If $\phi(z)\neq0$ for all z on the unit circle, then there exists $\delta>0$ such that

$$\frac{1}{\phi(z)}=\sum_{j=-\infty}^{\infty}\chi_jz^j \quad\text{for } 1-\delta<|z|<1+\delta,$$

and $\sum_{j=-\infty}^{\infty}|\chi_j|<\infty$. We can then define $\frac{1}{\phi(B)}$ as the linear filter with absolutely summable coefficients

$$\frac{1}{\phi(B)}=\sum_{j=-\infty}^{\infty}\chi_jB^j.$$

Applying the operator $\chi(B):=1/\phi(B)$ to both sides of (3.1.2), we obtain

$$X_t=\chi(B)\phi(B)X_t=\chi(B)\theta(B)Z_t=\psi(B)Z_t=\sum_{j=-\infty}^{\infty}\psi_jZ_{t-j}, \tag{3.1.3}$$

where $\psi(z)=\chi(z)\theta(z)=\sum_{j=-\infty}^{\infty}\psi_jz^j$. Using the argument given in Section 2.3 for the ARMA(1,1) process, it follows that $\psi(B)Z_t$ is the unique stationary solution of (3.1.1).

Existence and Uniqueness:

A stationary solution $\{X_t\}$ of the equations (3.1.1) exists (and is also the unique stationary solution) if and only if

$$\phi(z) = 1 - \phi_1 z - \cdots - \phi_p z^p \neq 0 \quad \text{for all } |z| = 1. \tag{3.1.4}$$

In Section 2.3, we saw that the ARMA(1,1) process is causal, i.e., that X_t can be expressed in terms of Z_s, $s \leq t$, if and only if $|\phi_1| < 1$. For a general ARMA(p, q) process, the analogous condition is that $\phi(z) \neq 0$ for $|z| \leq 1$, i.e., the zeroes of the autoregressive polynomial must all be greater than 1 in absolute value.

Causality:

An ARMA(p, q) process $\{X_t\}$ is **causal**, or a **causal function of** $\{Z_t\}$, if there exist constants $\{\psi_j\}$ such that $\sum_{j=0}^{\infty} |\psi_j| < \infty$ and

$$X_t = \sum_{j=0}^{\infty} \psi_j Z_{t-j} \text{ for all } t. \tag{3.1.5}$$

Causality is equivalent to the condition

$$\phi(z) = 1 - \phi_1 z - \cdots - \phi_p z^p \neq 0 \text{ for all } |z| \leq 1. \tag{3.1.6}$$

The proof of the equivalence between causality and (3.1.6) follows from elementary properties of power series. From (3.1.3), we see that $\{X_t\}$ is causal if and only if $\chi(z) := 1/\phi(z) = \sum_{j=0}^{\infty} \chi_j z^j$ (assuming $\phi(z)$ and $\theta(z)$ have no common factors). But this in turn is equivalent to (3.1.6).

The sequence $\{\psi_j\}$ in (3.1.5) is determined by the relation $\psi(z) = \sum_{j=0}^{\infty} \psi_j z^j = \theta(z)/\phi(z)$, or equivalently by the identity

$$(1 - \phi_1 z - \cdots - \phi_p z^p)(\psi_0 + \psi_1 z + \cdots) = 1 + \theta_1 z + \cdots + \theta_q z^q.$$

Equating coefficients of z^j, $j = 0, 1, \ldots$, we find that

$$\begin{aligned}
1 &= \psi_0 \\
\theta_1 &= \psi_1 - \psi_0 \phi_1 \\
\theta_2 &= \psi_2 - \psi_1 \phi_1 - \psi_0 \phi_2 \\
&\vdots \quad \vdots \quad\quad \vdots,
\end{aligned}$$

or equivalently,

$$\psi_j - \sum_{k=1}^{p} \phi_k \psi_{j-k} = \theta_j, \quad j = 0, 1, \ldots, \tag{3.1.7}$$

where $\theta_0 := 1, \theta_j := 0$ for $j > q$ and $\psi_j := 0$ for $j < 0$.

Invertibility, which allows Z_t to be expressed in terms of X_s, $s \le t$, has a similar characterization in terms of the moving average polynomial.

Invertibility:

An ARMA(p, q) process $\{X_t\}$ is **invertible** if there exist constants $\{\pi_j\}$ such that $\sum_{j=0}^{\infty} |\pi_j| < \infty$ and

$$Z_t = \sum_{j=0}^{\infty} \pi_j X_{t-j} \text{ for all } t.$$

Invertibility is equivalent to the condition

$$\theta(z) = 1 + \theta_1 z + \cdots + \theta_q z^q \neq 0 \text{ for all } |z| \le 1.$$

Interchanging the roles of the AR and MA polynomials, we find from (3.1.7) that the sequence $\{\pi_j\}$ is determined by the equations

$$\pi_j + \sum_{k=1}^{q} \theta_k \pi_{j-k} = -\phi_j, \quad j = 0, 1, \ldots, \tag{3.1.8}$$

where $\phi_0 := -1, \phi_j := 0$ for $j > p$ and $\pi_j := 0$ for $j < 0$.

Example 3.1.1 An ARMA(1,1) process

Consider the ARMA(1,1) process $\{X_t\}$ satisfying the equations

$$X_t - .5X_{t-1} = Z_t + .4Z_{t-1}, \quad \{Z_t\} \sim \text{WN}(0, \sigma^2). \tag{3.1.9}$$

Since the autoregressive polynomial $\phi(z) = 1 - .5z$ has a zero at $z = 2$, which is located outside the unit circle, we conclude from (3.1.4) and (3.1.6) that there exists a unique ARMA process satisfying (3.1.9) that is also causal. The coefficients $\{\psi_j\}$ in the MA(∞) representation of $\{X_t\}$ are found directly from (3.1.7),

$$\begin{aligned}
\psi_0 &= 1 \\
\psi_1 &= .4 + .5 \\
\psi_2 &= .5(.4 + .5) \\
\psi_j &= .5^{j-1}(.4 + .5), \quad j = 1, 2, \ldots.
\end{aligned}$$

The MA polynomial, $\theta(z) = 1 + .4z$, has a zero at $z = -1/.4 = 2.5$, which is also located outside the unit circle. This implies that $\{X_t\}$ is invertible with coefficients

$\{\pi_j\}$ given by (see (3.1.8))

$$\pi_0 = 1$$
$$\pi_1 = -(.4 + .5)$$
$$\pi_2 = -(.4 + .5)(-.4)$$
$$\pi_j = -(.4 + .5)(-.4)^{j-1}, \quad j = 1, 2, \ldots.$$

(A direct derivation of these formulas for $\{\psi_j\}$ and $\{\pi_j\}$ was given in Section 2.3 without appealing to the recursions (3.1.7) and (3.1.8).) □

Example 3.1.2 An AR(2) process

Let $\{X_t\}$ be the AR(2) process,

$$X_t = .7X_{t-1} - .1X_{t-2} + Z_t, \quad \{Z_t\} \sim \mathrm{WN}(0, \sigma^2).$$

The autoregressive polynomial for this process has the factorization $\phi(z) = 1 - .7z + .1z^2 = (1 - .5z)(1 - .2z)$, and is therefore zero at $z = 2$ and $z = 5$. Since these zeroes lie outside the unit circle, we conclude that $\{X_t\}$ is a causal AR(2) process with coefficients $\{\psi_j\}$ given by

$$\psi_0 = 1$$
$$\psi_1 = .7$$
$$\psi_2 = .7^2 - .1$$
$$\psi_j = .7\psi_{j-1} - .1\psi_{j-2}, \quad j = 2, 3, \ldots.$$

While it is a simple matter to calculate ψ_j numerically for any j, it is possible also to give an explicit solution of these difference equations using the theory of linear difference equations (see TSTM, Section 3.6). □

The option *Entry of an ARMA(p,q) model* of the program PEST will accept any causal ARMA model. This option contains causality and invertibility checks and will immediately let you know if the entered model is either noncausal or noninvertible. If the entered model is noncausal, then all the AR coefficients can be set to .001 to ensure a causal model. Once a causal model has been entered, the coefficients ψ_j in the MA(∞) representation of the process can then be computed using the the option *Model ACF/PACF, AR/MA Infinity Representations*. This option can also be used to compute the AR(∞) coefficients π_j provided the model is invertible.

Example 3.1.3 An ARMA(2,1) process

Consider the ARMA(2,1) process defined by the equations

$$X_t - .75X_{t-1} + .5625X_{t-2} = Z_t + 1.25Z_{t-1}, \quad \{Z_t\} \sim \mathrm{WN}(0, \sigma^2).$$

The AR polynomial $\phi(z) = 1 - .75z + .5625z^2$ has zeroes at $z = 2(1 \pm i\sqrt{3})/3$, which lie outside the unit circle. The process is therefore causal. On the other hand, the MA polynomial $\theta(z) = 1 + 1.25z$ has a zero at $z = -.8$ and hence $\{X_t\}$ is not invertible. □

Remark 1. It should be noted that causality and invertibility are properties not of $\{X_t\}$ alone, but rather of the relationship between the two processes $\{X_t\}$ and $\{Z_t\}$ appearing in the defining ARMA equations (3.1.1). □

Remark 2. If $\{X_t\}$ is an ARMA process defined by $\phi(B)X_t = \theta(B)Z_t$, where $\theta(z) \neq 0$ if $|z| = 1$, then it is always possible (see TSTM, p. 127) to find polynomials $\tilde{\phi}(z)$ and $\tilde{\theta}(z)$ and a white noise sequence $\{W_t\}$ such that $\tilde{\phi}(B)X_t = \tilde{\theta}(B)W_t$ and $\tilde{\theta}(z)$ and $\tilde{\phi}(z)$ are non-zero for $|z| \leq 1$. However, if the original white noise sequence $\{Z_t\}$ is iid, then the new white noise sequence will not be iid unless $\{Z_t\}$ is Gaussian. □

In view of the preceding remark, we will focus our attention principally on causal and invertible ARMA processes.

3.2 The ACF and PACF of an ARMA(p, q) Process

In this section we discuss three methods for computing the autocovariance function $\gamma(\cdot)$ of a causal ARMA process $\{X_t\}$. The autocorrelation function is readily found from the ACVF on dividing by $\gamma(0)$. The partial autocorrelation function (PACF) is also found from the function $\gamma(\cdot)$.

3.2.1 Calculation of the ACVF

First we determine the ACVF $\gamma(\cdot)$ of the causal ARMA(p, q) process defined by

$$\phi(B)X_t = \theta(B)Z_t, \quad \{Z_t\} \sim \mathrm{WN}(0, \sigma^2), \tag{3.2.1}$$

where $\phi(z) = 1 - \phi_1 z - \cdots - \phi_p z^p$ and $\theta(z) = 1 + \theta_1 z + \cdots + \theta_q z^q$. The causality assumption implies that

$$X_t = \sum_{j=0}^{\infty} \psi_j Z_{t-j}, \tag{3.2.2}$$

where $\sum_{j=0}^{\infty} \psi_j z^j = \theta(z)/\phi(z)$, $|z| \leq 1$. The calculation of the sequence $\{\psi_j\}$ was discussed in Section 3.1.

First Method. From Proposition 2.2.1 and the representation (3.2.2), we obtain

$$\gamma(h) = E(X_{t+h}X_t) = \sigma^2 \sum_{j=0}^{\infty} \psi_j \psi_{j+|h|}. \tag{3.2.3}$$

Example 3.2.1 The ARMA(1,1) process

Substituting from (2.3.3) into (3.2.3), we find that the ACVF of the process defined by

$$X_t - \phi X_{t-1} = Z_t + \theta Z_{t-1}, \quad \{Z_t\} \sim \mathrm{WN}(0, \sigma^2) \tag{3.2.4}$$

with $|\phi| < 1$ is given by

$$\begin{aligned}
\gamma(0) &= \sigma^2 \sum_{j=0}^{\infty} \psi_j^2 \\
&= \sigma^2 \left[1 + (\theta + \phi)^2 \sum_{j=0}^{\infty} \phi^{2j} \right] \\
&= \sigma^2 \left[1 + \frac{(\theta + \phi)^2}{1 - \phi^2} \right], \\
\gamma(1) &= \sigma^2 \sum_{j=0}^{\infty} \psi_{j+1} \psi_j \\
&= \sigma^2 \left[\theta + \phi + (\theta + \phi)^2 \phi \sum_{j=0}^{\infty} \phi^{2j} \right] \\
&= \sigma^2 \left[\theta + \phi + \frac{(\theta + \phi)^2 \phi}{1 - \phi^2} \right],
\end{aligned}$$

and

$$\gamma(h) = \phi^{h-1} \gamma(1), \quad h \geq 2.$$

□

Example 3.2.2 The MA(q) process

For the process

$$X_t = Z_t + \theta_1 Z_{t-1} + \cdots + \theta_q Z_{t-q}, \quad \{Z_t\} \sim \mathrm{WN}(0, \sigma^2),$$

equation (3.2.3) immediately gives the result

$$\gamma(h) = \begin{cases} \sigma^2 \sum_{j=0}^{q-|h|} \theta_j \theta_{j+|h|}, & \text{if } |h| \leq q, \\ 0, & \text{if } |h| > q, \end{cases}$$

where θ_0 is defined to be 1. The ACVF of the MA(q) process thus has the distinctive feature of vanishing at lags greater than q. Data for which the sample ACVF is small for lags greater than q therefore suggest that an appropriate model might be a moving average of order q (or less). Recall from Proposition 2.1.1 that *every* zero-mean stationary process with correlations vanishing at lags greater than q can be represented as a moving average process of order q or less. □

Second Method. If we multiply each side of the equations

$$X_t - \phi_1 X_{t-1} - \cdots - \phi_p X_{t-p} = Z_t + \theta_1 Z_{t-1} + \cdots + \theta_q Z_{t-q},$$

by X_{t-k}, $k = 0, 1, 2, \ldots$ and take expectations on each side, we find that

$$\gamma(k) - \phi_1\gamma(k-1) - \cdots - \phi_p\gamma(k-p) = \sigma^2 \sum_{j=0}^{\infty} \theta_{k+j}\psi_j, \quad 0 \le k < m, \tag{3.2.5}$$

and

$$\gamma(k) - \phi_1\gamma(k-1) - \cdots - \phi_p\gamma(k-p) = 0, \quad k \ge m, \tag{3.2.6}$$

where $m = \max(p, q+1)$, $\psi_j := 0$ for $j < 0$, $\theta_0 := 1$, and $\theta_j := 0$ for $j \notin \{0, \ldots, q\}$. In calculating the right-hand side of (3.2.5) we have made use of the expansion (3.2.2). The equations (3.2.6) are a set of homogeneous linear difference equations with constant coefficients, for which the solution is well known (see e.g., TSTM, Section 3.6) to be of the form

$$\gamma(h) = \alpha_1\xi_1^{-h} + \alpha_2\xi_2^{-h} + \cdots + \alpha_p\xi_p^{-h}, \quad h \ge m - p, \tag{3.2.7}$$

where $\xi_1, \ldots, \xi_p$ are the roots (assumed to be distinct) of the equation $\phi(z) = 0$ and $\alpha_1, \ldots, \alpha_p$ are arbitrary constants. (For further details, and for the treatment of the case when the roots are not distinct, see TSTM, Section 3.6.) Of course we are looking for the solution of (3.2.6) that also satisfies (3.2.5). We therefore substitute the solution (3.2.7) into (3.2.5) to obtain a set of m linear equations that then uniquely determine the constants $\alpha_1, \ldots, \alpha_p$ and the $m - p$ autocovariances $\gamma(h)$, $0 \le h < m - p$.

Example 3.2.3 The ARMA(1,1) process

For the causal ARMA(1,1) process defined in Example 3.2.1, the equations (3.2.5) are

$$\gamma(0) - \phi\gamma(1) = \sigma^2(1 + \theta(\theta + \phi)) \tag{3.2.8}$$

and

$$\gamma(1) - \phi\gamma(0) = \sigma^2\theta. \tag{3.2.9}$$

Equation (3.2.6) takes the form

$$\gamma(k) - \phi\gamma(k-1) = 0, \quad k \ge 2. \tag{3.2.10}$$

The solution of (3.2.10) is

$$\gamma(h) = \alpha\phi^h, \quad h \ge 1.$$

Substituting this expression for $\gamma(h)$ into the two preceding equations (3.2.8) and (3.2.9) gives two linear equations for α and the unknown autocovariance $\gamma(0)$. These equations are easily solved, giving the autocovariances already found for this process in Example 3.2.1. □

Example 3.2.4 The general AR(2) process

For the causal AR(2) process defined by

$$(1 - \xi_1^{-1}B)(1 - \xi_2^{-1}B)X_t = Z_t, \quad |\xi_1|, |\xi_2| > 1, \xi_1 \neq \xi_2,$$

we easily find from (3.2.7) and (3.2.5) using the relations

$$\phi_1 = \xi_1^{-1} + \xi_2^{-1}$$

and

$$\phi_2 = -\xi_1^{-1}\xi_2^{-1}$$

that

$$\gamma(h) = \frac{\sigma^2\xi_1^2\xi_2^2}{(\xi_1\xi_2 - 1)(\xi_2 - \xi_1)}[(\xi_1^2 - 1)^{-1}\xi_1^{1-h} - (\xi_2^2 - 1)^{-1}\xi_2^{1-h}]. \tag{3.2.11}$$

Figures 3.1–3.4 illustrate some of the possible forms of $\gamma(\cdot)$ for different values of ξ_1 and ξ_2. Notice that in the case of complex conjugate roots, $\xi_1 = re^{i\theta}$ and $\xi_2 = re^{-i\theta}$, $0 < \theta < \pi$, we can write (3.2.11) in the more illuminating form

$$\gamma(h) = \frac{\sigma^2 r^4 \cdot r^{-h}\sin(h\theta + \psi)}{(r^2 - 1)(r^4 - 2r^2\cos 2\theta + 1)^{1/2}\sin\theta}, \tag{3.2.12}$$

where

$$\tan\psi = \frac{r^2 + 1}{r^2 - 1}\tan\theta \tag{3.2.13}$$

and $\cos\psi$ has the same sign as $\cos\theta$. Thus in this case $\gamma(\cdot)$ has the form of a damped sinusoidal function with damping factor r^{-1} and period $2\pi/\theta$. If the roots are close to the unit circle, then r is close to 1, the damping is slow, and we obtain a nearly sinusoidal autocovariance function. □

Third Method. The autocovariances can also be found by solving the first $p+1$ equations of (3.2.5) and (3.2.6) for $\gamma(0), \ldots, \gamma(p)$ and then using the subsequent equations to solve successively for $\gamma(p + 1), \gamma(p + 2), \ldots$. This is an especially convenient method for numerical determination of the autocovariances $\gamma(h)$ and is used in the option *Model ACF/PACF* of the program PEST.

Example 3.2.5 Consider again the causal ARMA(1,1) process of Example 3.2.1. To apply the third method we simply solve (3.2.8) and (3.2.9) for $\gamma(0)$ and $\gamma(1)$. Then $\gamma(2), \gamma(3), \ldots$,

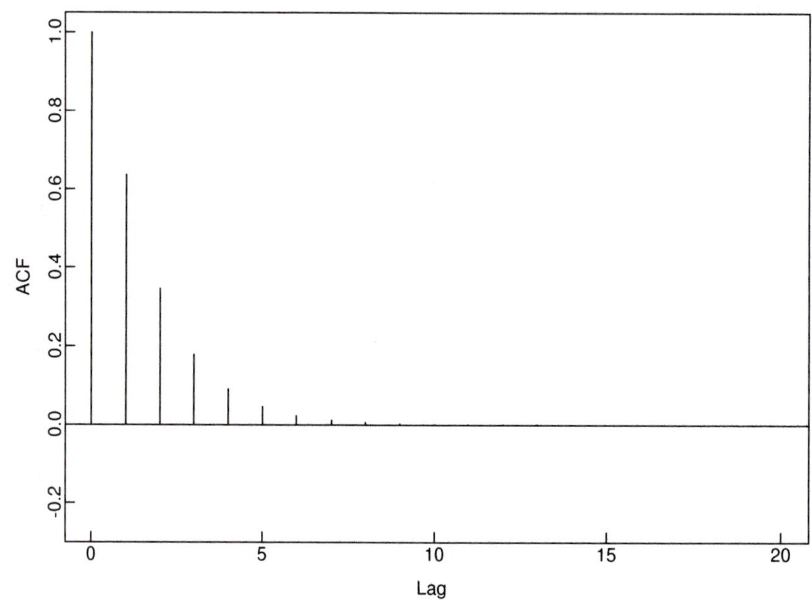

Figure 3-1
The model ACF of the AR(2) series of Example 3.2.4 with $\xi_1 = 2$ and $\xi_2 = 5$.

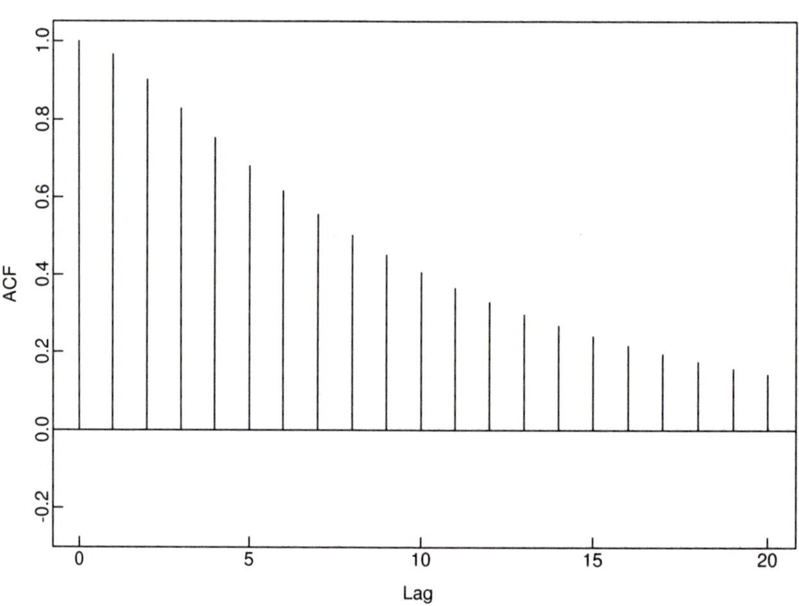

Figure 3-2
The model ACF of the AR(2) series of Example 3.2.4 with $\xi_1 = 10/9$ and $\xi_2 = 2$.

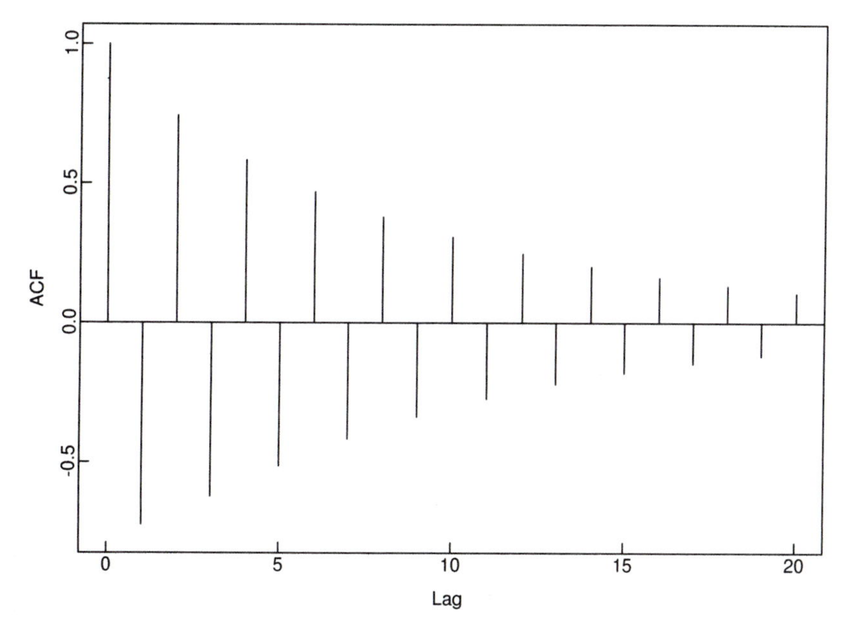

Figure 3-3
The model ACF of the AR(2) series of Example 3.2.4 with $\xi_1 = -10/9$ and $\xi_2 = 2$.

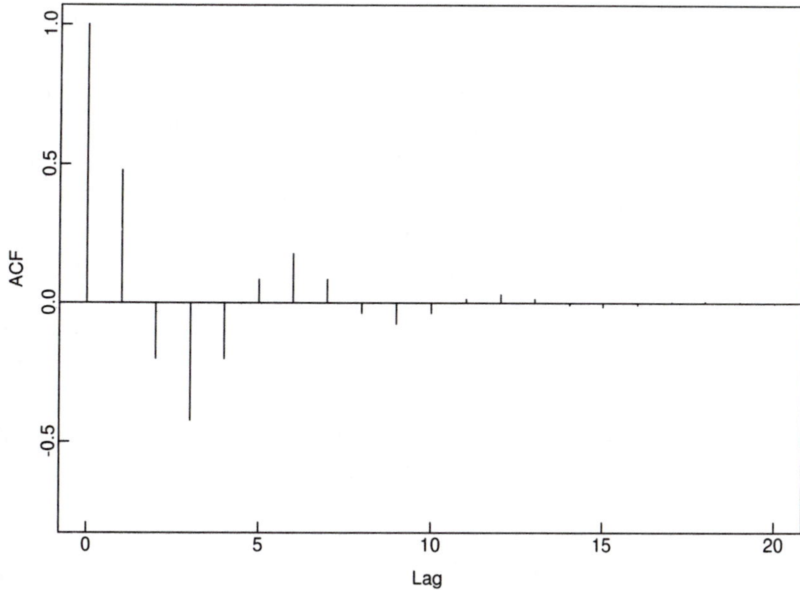

Figure 3-4
The model ACF of the AR(2) series of Example 3.2.4 with $\xi_1 = 2(1 + i\sqrt{3})/3$ and $\xi_2 = 2(1 - i\sqrt{3})/3$.

can be found successively from (3.2.10). It is easy to check that this procedure gives the same results as those obtained in Examples 3.2.1 and 3.2.3. □

3.2.2 The Autocorrelation Function

Recall that the ACF of an ARMA process $\{X_t\}$ is the function $\rho(\cdot)$ found immediately from the ACVF $\gamma(\cdot)$ as

$$\rho(h) = \frac{\gamma(h)}{\gamma(0)}.$$

Likewise for any set of observations $\{x_1, \ldots, x_n\}$, the sample ACF $\hat{\rho}(\cdot)$ is computed as

$$\hat{\rho}(h) = \frac{\hat{\gamma}(h)}{\hat{\gamma}(0)}.$$

The Sample ACF of an MA(q) Series. Given observations $\{x_1, \ldots, x_n\}$ of a time series, one approach to the fitting of a model to the data is to match the sample ACF of the data with the ACF of the model. In particular, if the sample ACF $\hat{\rho}(h)$ is significantly different from zero for $0 \le h \le q$ and negligible for $h > q$, Example 3.2.2 suggests that an MA(q) model might provide a good representation of the data. In order to apply this criterion we need to take into account the random variation expected in the sample autocorrelation function before we can classify ACF values as "negligible." To resolve this problem we can use Bartlett's formula (Section 2.4), which implies that for a large sample of size n from an MA(q) process, the sample ACF values at lags greater than q are approximately normally distributed with means 0 and variances $w_{hh}/n = (1+2\rho^2(1)+\cdots+2\rho^2(q))/n$. This means that if the sample is from an MA(q) process and if $h > q$, then $\hat{\rho}(h)$ should fall between the bounds $\pm 1.96\sqrt{w_{hh}/n}$ with probability approximately 0.95. In practice we frequently use the more stringent values $\pm 1.96/\sqrt{n}$ as the bounds between which sample autocovariances are considered "negligible." A more effective and systematic approach to the problem of model selection, which also applies to ARMA(p, q) models with $p > 0$ and $q > 0$, will be discussed in Section 5.5.

3.2.3 The Partial Autocorrelation Function

The **partial autocorrelation function (PACF)** of an ARMA process $\{X_t\}$ is the function $\alpha(\cdot)$ defined by the equations

$$\alpha(0) = 1$$

and

$$\alpha(h) = \phi_{hh}, \quad h \ge 1,$$

where ϕ_{hh} is the last component of

$$\phi_h = \Gamma_h^{-1}\gamma_h, \tag{3.2.14}$$

$\Gamma_h = [\gamma(i-j)]_{i,j=1}^{h}$ and $\gamma_h = [\gamma(1), \gamma(2), \ldots, \gamma(h)]'$.

For any set of observations $\{x_1, \ldots, x_n\}$ with $x_i \neq x_j$ for some i and j, the **sample PACF**, $\hat{\alpha}(h)$, is given by

$$\hat{\alpha}(0) = 1$$

and

$$\hat{\alpha}(h) = \hat{\phi}_{hh}, \quad h \geq 1,$$

where $\hat{\phi}_{hh}$ is the last component of

$$\hat{\phi}_h = \hat{\Gamma}_h^{-1}\hat{\gamma}_h. \tag{3.2.15}$$

We show in the next example that the PACF of a causal AR(p) process is zero for lags greater than p. Both sample and model partial autocorrelation functions can be computed numerically using the program PEST. Algebraic calculation of the PACF is quite complicated except when q is zero or p and q are both small.

It can be shown (TSTM, p. 171) that ϕ_{hh} is the correlation between the prediction errors $X_h - P(X_h|X_1, \ldots, X_{h-1})$ and $X_0 - P(X_0|X_1, \ldots, X_{h-1})$.

Example 3.2.6 The PACF of an AR(p) process

For the causal AR(p) process defined by

$$X_t - \phi_1 X_{t-1} - \cdots - \phi_p X_{t-p} = Z_t, \quad \{Z_t\} \sim \mathrm{WN}(0, \sigma^2),$$

we know (Problem 2.15) that for $h \geq p$ the best linear predictor of X_{h+1} in terms of $1, X_1, \ldots, X_h$ is

$$\hat{X}_{h+1} = \phi_1 X_h + \phi_2 X_{h-1} + \cdots + \phi_p X_{h+1-p}.$$

Since the coefficient ϕ_{hh} of X_1 is ϕ_p if $h = p$ and 0 if $h > p$, we conclude that the PACF $\alpha(\cdot)$ of the process $\{X_t\}$ has the properties

$$\alpha(p) = \phi_p$$

and

$$\alpha(h) = 0 \text{ for } h > p.$$

For $h < p$ the values of $\alpha(h)$ can easily be computed from (3.2.14). For any specified ARMA model the PACF can be evaluated numerically using the option *Model ACF/PACF* of the program PEST. □

Example 3.2.7 The PACF of an MA(1) process

For the MA(1) process, it can be shown from (3.2.14) (see Problem 3.12) that the PACF at lag h is

$$\alpha(h) = \phi_{hh} = -(-\theta)^h/(1 + \theta^2 + \cdots + \theta^{2h}). \qquad \square$$

The Sample PACF of an AR(p) Series. If $\{X_t\}$ is an AR(p) series, then the sample PACF based on observations $\{x_1, \ldots, x_n\}$ should reflect (with sampling variation) the properties of the PACF itself. In particular, if the sample PACF $\hat{\alpha}(h)$ is significantly different from zero for $0 \le h \le p$ and negligible for $h > p$, Example 3.2.6 suggests that an AR(p) model might provide a good representation of the data. To decide what is meant by "negligible" we can use the result that for an AR(p) process the sample PACF values at lags greater than p are approximately independent $N(0, 1/n)$ random variables. This means that roughly 95% of the sample PACF values beyond lag p should fall within the bounds $\pm 1.96/\sqrt{n}$. If we observe a sample PACF satisfying $|\hat{\alpha}(h)| > 1.96/\sqrt{n}$ for $0 \le h \le p$ and $|\hat{\alpha}(h)| < 1.96/\sqrt{n}$ for $h > p$, this suggests an AR(p) model for the data. For a more systematic approach to model selection, see Section 5.5.

3.2.4 Examples

Example 3.2.8 The time series plotted in Figure 3.5 consists of 57 consecutive daily *overshorts* from an underground gasoline tank at a filling station in Colorado. If y_t is the measured amount of fuel in the tank at the end of the tth day and a_t is the measured amount sold minus the amount delivered during the course of the tth day, then the overshort at the end of day t is defined as $x_t = y_t - y_{t-1} + a_t$. Due to the error in measuring the current amount of fuel in the tank, the amount sold, and the amount delivered to the station, we view y_t, a_t, and x_t as observed values from some set of random variables Y_t, A_t, and X_t for $t = 1, \ldots, 57$. (In the absence of any measurement error and any leak in the tank, each x_t would be zero.) The data and their ACF are plotted in Figures 3.5 and 3.6. To check the plausibility of an MA(1) model, the bounds $\pm 1.96(1 + 2\hat{\rho}^2(1))^{1/2}/n^{1/2}$ are also plotted in Figure 3.6. Since $\hat{\rho}(h)$ is well within these bounds for $h > 1$, the data appear to be compatible with the model

$$X_t = \mu + Z_t + \theta Z_{t-1}, \quad \{Z_t\} \sim \text{WN}(0, \sigma^2). \tag{3.2.16}$$

The mean μ may be estimated by the sample mean $\bar{x}_{57} = -4.035$ and the parameters θ, σ^2 may be estimated by equating the sample ACVF with the model ACVF at lags 0 and 1, and solving the resulting equations for θ and σ^2. This estimation procedure is known as the method of moments, and in this case gives the equations,

$$(1 + \theta^2)\sigma^2 = \hat{\gamma}(0) = 3415.72,$$

$$\theta\sigma^2 = \hat{\gamma}(1) = -1719.95.$$

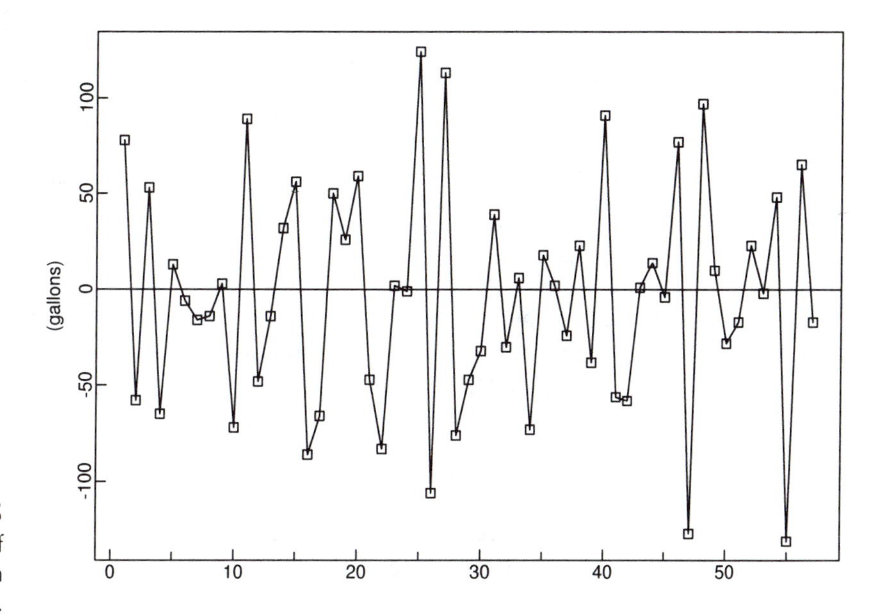

Figure 3-5 Time series of the overshorts in Example 3.2.8.

Using the approximate solution $\theta = -1$ and $\sigma^2 = 1708$, we obtain the noninvertible MA(1) model

$$X_t = -4.035 + Z_t - Z_{t-1}, \ \{Z_t\} \sim \text{WN}(0, 1708).$$

Typically in time series modelling, we have little or no knowledge of the underlying physical mechanism generating the data and the choice of a suitable class of

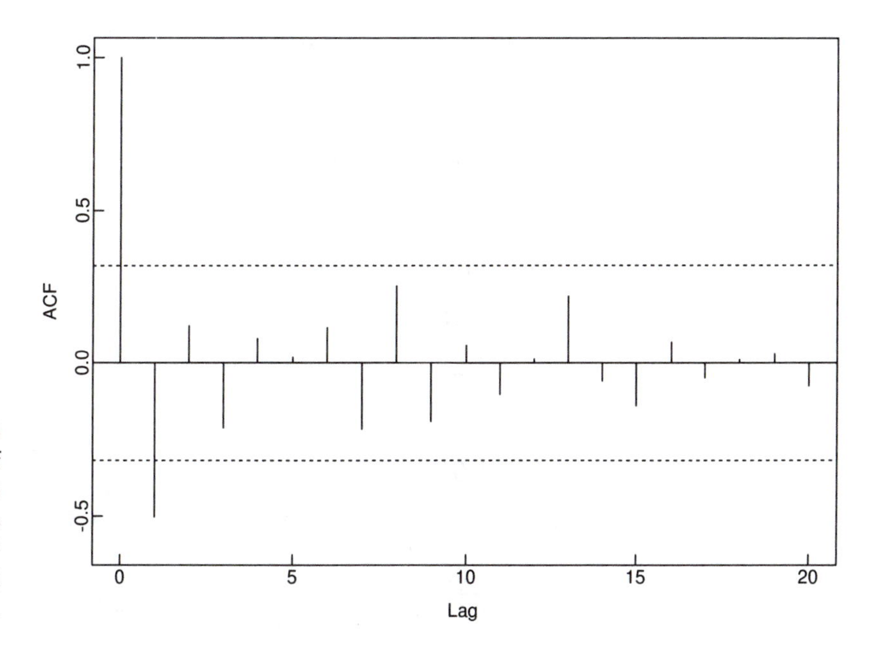

Figure 3-6 The sample ACF of the data in Figure 3.5 showing the bounds $\pm 1.96 n^{-1/2}(1 + 2\hat{\rho}^2(1))^{1/2}$ assuming an MA(1) model for the data.

models is entirely data driven. For the time series of overshorts, the data, through the graph of the ACF, lead us to the MA(1) model. Alternatively, we can attempt to model the mechanism generating the time series of overshorts using a *structural model*. As we will see, the structural model formulation leads us again to the MA(1) model. In the structural model setup, write Y_t, the observed amount of fuel in the tank at time t, as

$$Y_t = y_t^* + U_t, \tag{3.2.17}$$

where y_t^* is the true (or actual) amount of fuel in the tank at time t (not to be confused with y_t above) and U_t is the resulting *measurement error*. The variable y_t^* is an idealized quantity that in principle cannot be observed even with the most sophisticated measurement devices. Similarly, we assume that

$$A_t = a_t^* + V_t, \tag{3.2.18}$$

where a_t^* is the actual amount of fuel sold minus the actual amount delivered during day t and V_t is the associated measurement error. We further assume that $\{U_t\} \sim \mathrm{WN}(0, \sigma_U^2)$, $\{V_t\} \sim \mathrm{WN}(0, \sigma_V^2)$, and that two sequences $\{U_t\}$ and $\{V_t\}$ are uncorrelated with one another ($E(U_t V_s) = 0$ for all s and t). If the change of level per day due to leakage is μ gallons ($\mu < 0$ indicates leakage), then

$$y_t^* = \mu + y_{t-1}^* - a_t^*. \tag{3.2.19}$$

This equation relates the actual amount of fuel in the tank at the end of days t and $t-1$, adjusted for the actual amounts that have been sold and delivered during the day. Using (3.2.17)–(3.2.19), the model for the time series of overshorts is given by

$$X_t = Y_t - Y_{t-1} + A_t = \mu + U_t - U_{t-1} + V_t.$$

This model is stationary and 1-correlated since

$$EX_t = E(\mu + U_t - U_{t-1} + V_t) = \mu$$

and

$$\begin{aligned}
\gamma(h) &= E[(X_{t+h} - \mu)(X_t - \mu)] \\
&= E[(U_{t+h} - U_{t+h-1} + V_{t+h})(U_t - U_{t-1} + V_t)] \\
&= \begin{cases} 2\sigma_U^2 + \sigma_V^2, & \text{if } h = 0, \\ -\sigma_U^2, & \text{if } |h| = 1, \\ 0, & \text{otherwise.} \end{cases}
\end{aligned}$$

It follows from Proposition 2.1.1 that $\{X_t\}$ is the MA(1) model (3.2.16) with

$$\rho(1) = \frac{\theta_1}{1+\theta_1^2} = \frac{-\sigma_U^2}{2\sigma_U^2 + \sigma_V^2}.$$

From this equation, we see that the measurement error associated with the adjustment $\{A_t\}$ is zero (i.e., $\sigma_V^2 = 0$) if and only if $\rho(1) = -.5$ or, equivalently, if and only if $\theta_1 = -1$. From the analysis above, the moment estimator of θ_1 for the overshort data is in fact -1, so that we conclude there is relatively little measurement error associated with the amount of fuel sold and delivered.

We shall return to a more general discussion of structural models in Chapter 8. □

Example 3.2.9 The sunspot numbers

Figure 3.7 shows the sample PACF of the sunspot numbers $S_1, \ldots, S_{100}$ (for the years 1770 – 1869) as computed using the first option in the main menu of the program PEST. The graph also show the bounds $\pm 1.96/\sqrt{100}$. The fact that all of the PACF values beyond lag 2 fall within the bounds suggests the possible suitability of an AR(2) model for the mean-corrected data set $X_t = S_t - 46.93$. One simple way to estimate the parameters ϕ_1, ϕ_2, and σ^2 of such a model is to require that the ACVF of the model at lags 0, 1, and 2 should match the sample ACVF at those lags. Substituting the sample ACVF values

$$\hat{\gamma}(0) = 1382.2, \quad \hat{\gamma}(1) = 1114.4, \quad \hat{\gamma}(2) = 591.73,$$

for $\gamma(0)$, $\gamma(1)$, and $\gamma(2)$ in the first three equations of (3.2.5) and (3.2.6) and solving for ϕ_1, ϕ_2, and σ^2 give the fitted model,

$$X_t - 1.318X_{t-1} + 0.634X_{t-2} = Z_t, \quad \{Z_t\} \sim \text{WN}(0, 289.2). \tag{3.2.20}$$

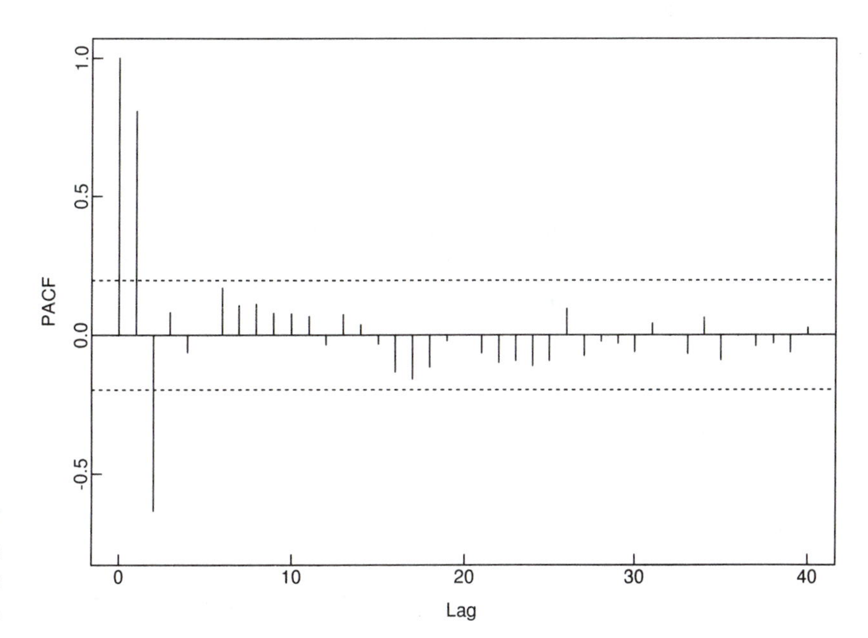

Figure 3-7 The sample PACF of the sunspot numbers with the bounds $\pm 1.96/\sqrt{100}$.

(This method of model-fitting is called Yule-Walker estimation and will be discussed more fully in Section 5.1.1.) □

3.3 Forecasting ARMA Processes

The innovations algorithm (see Section 2.5.2) provided us with a recursive method for forecasting second-order zero-mean processes that are not necessarily stationary. For the causal ARMA process

$$\phi(B)X_t = \theta(B)Z_t, \quad \{Z_t\} \sim \mathrm{WN}(0, \sigma^2),$$

it is possible to simplify the application of the algorithm drastically. The idea is to apply it not to the process $\{X_t\}$ itself, but to the transformed process (cf. Ansley, 1979),

$$\begin{cases} W_t = \sigma^{-1}X_t, & t = 1, \ldots, m, \\ W_t = \sigma^{-1}\phi(B)X_t, & t > m, \end{cases} \tag{3.3.1}$$

where

$$m = \max(p, q). \tag{3.3.2}$$

For notational convenience we define $\theta_0 := 1$ and $\theta_j := 0$ for $j > q$. We shall also assume that $p \geq 1$ and $q \geq 1$. (There is no loss of generality in these assumptions since in the analysis that follows we may take any of the coefficients ϕ_i and θ_i to be zero.)

The autocovariance function $\gamma_X(\cdot)$ of $\{X_t\}$ can easily be computed using any of the methods described in Section 3.2.1. The autocovariances $\kappa(i, j) = E(W_i W_j)$, $i, j \geq 1$, are then found from

$$\kappa(i, j) = \begin{cases} \sigma^{-2}\gamma_X(i-j), & 1 \leq i, j \leq m \\ \sigma^{-2}\left[\gamma_X(i-j) - \sum_{r=1}^{p}\phi_r\gamma_X(r - |i-j|)\right], & \min(i, j) \leq m < \max(i, j) \leq 2m, \\ \sum_{r=0}^{q}\theta_r\theta_{r+|i-j|}, & \min(i, j) > m, \\ 0, & \text{otherwise.} \end{cases} \tag{3.3.3}$$

Applying the innovations algorithm to the process $\{W_t\}$ we obtain

$$\begin{cases} \hat{W}_{n+1} = \sum_{j=1}^{n} \theta_{nj}(W_{n+1-j} - \hat{W}_{n+1-j}), & 1 \le n < m, \\ \hat{W}_{n+1} = \sum_{j=1}^{q} \theta_{nj}(W_{n+1-j} - \hat{W}_{n+1-j}), & n \ge m, \end{cases} \tag{3.3.4}$$

where the coefficients θ_{nj} and the mean squared errors $r_n = E(W_{n+1} - \hat{W}_{n+1})^2$ are found recursively from the innovations algorithm with κ defined as in (3.3.3). The notable feature of the predictors (3.3.4) is the vanishing of θ_{nj} when both $n \ge m$ and $j > q$. This is a consequence of the innovations algorithm and the fact that $\kappa(r, s) = 0$ if $r > m$ and $|r - s| > q$.

Observe now that the equations (3.3.1) allow each X_n, $n \ge 1$, to be written as a linear combination of W_j, $1 \le j \le n$, and conversely each W_n, $n \ge 1$, to be written as a linear combination of X_j, $1 \le j \le n$. This means that the best linear predictor of any random variable Y in terms of $\{1, X_1, \dots, X_n\}$ is the same as the best linear predictor of Y in terms of $\{1, W_1, \dots, W_n\}$. We shall denote this predictor by $P_n Y$. In particular the one-step predictors of W_{n+1} and X_{n+1} are given by

$$\hat{W}_{n+1} = P_n W_{n+1}$$

and

$$\hat{X}_{n+1} = P_n X_{n+1}.$$

Using the linearity of P_n and equations (3.3.1) we see that

$$\begin{cases} \hat{W}_t = \sigma^{-1}\hat{X}_t, & t = 1, \dots, m, \\ \hat{W}_t = \sigma^{-1}[\hat{X}_t - \phi_1 X_{t-1} - \cdots - \phi_p X_{t-p}], & t > m, \end{cases} \tag{3.3.5}$$

which, together with (3.3.1), shows that

$$X_t - \hat{X}_t = \sigma[W_t - \hat{W}_t] \quad \text{for all } t \ge 1. \tag{3.3.6}$$

Replacing $(W_j - \hat{W}_j)$ by $\sigma^{-1}(X_j - \hat{X}_j)$ in (3.3.3) and then substituting into (3.3.4), we finally obtain

$$\hat{X}_{n+1} = \begin{cases} \sum_{j=1}^{n} \theta_{nj}(X_{n+1-j} - \hat{X}_{n+1-j}), & 1 \le n < m, \\ \phi_1 X_n + \cdots + \phi_p X_{n+1-p} + \sum_{j=1}^{q} \theta_{nj}(X_{n+1-j} - \hat{X}_{n+1-j}), & n \ge m, \end{cases} \tag{3.3.7}$$

and

$$E(X_{n+1} - \hat{X}_{n+1})^2 = \sigma^2 E(W_{n+1} - \hat{W}_{n+1})^2 = \sigma^2 r_n, \tag{3.3.8}$$

where θ_{nj} and r_n are found from the innovations algorithm with κ as in (3.3.3). Equations (3.3.7) determine the one-step predictors $\hat{X}_2, \hat{X}_3, \ldots$, recursively.

Remark 1. It can be shown (see TSTM, Problem 5.6) that if $\{X_t\}$ is invertible, then as $n \to \infty$,

$$E(X_n - \hat{X}_n - Z_n)^2 \to 0,$$

$$\theta_{nj} \to \theta_j, \ j = 1, \ldots, q,$$

and

$$r_n \to 1.$$

Algebraic calculation of the coefficients θ_{nj} and r_n is not feasible except for very simple models, such as those considered in the following examples. However, numerical implementation of the recursions is quite straightforward and is used to compute predictors in the program PEST. □

Example 3.3.1 Prediction of an AR(p) process

Applying (3.3.7) to the ARMA(p, 1) process with $\theta_1 = 0$, we easily find that

$$\hat{X}_{n+1} = \phi_1 X_n + \ldots + \phi_p X_{n+1-p}, \quad n \geq p.$$ □

Example 3.3.2 Prediction of an MA(q) process

Applying (3.3.7) to the ARMA(1, q) process with $\phi_1 = 0$, gives

$$\hat{X}_{n+1} = \sum_{j=1}^{\min(n,q)} \theta_{nj}(X_{n+1-j} - \hat{X}_{n+1-j}), \quad n \geq 1,$$

where the coefficients θ_{nj} are found by applying the innovations algorithm to the covariances $\kappa(i, j)$ defined in (3.3.3). Since in this case the processes $\{X_t\}$ and $\{\sigma^{-1} W_t\}$ are identical, these covariances are simply

$$\kappa(i, j) = \sigma^{-2}\gamma_X(i - j) = \sum_{r=0}^{q-|i-j|} \theta_r \theta_{r+|i-j|}.$$ □

Example 3.3.3 Prediction of an ARMA(1,1) process

If

$$X_t - \phi X_{t-1} = Z_t + \theta Z_{t-1}, \quad \{Z_t\} \sim \text{WN}(0, \sigma^2),$$

and $|\phi| < 1$, then equations (3.3.7) reduce to the single equation,

$$\hat{X}_{n+1} = \phi X_n + \theta_{n1}(X_n - \hat{X}_n), \quad n \geq 1.$$

To compute θ_{n1} we first use Example 3.2.1 to find that $\gamma_X(0) = \sigma^2(1+2\theta\phi+\theta^2)/(1-\phi^2)$. Substituting in (3.3.3) then gives, for $i, j \geq 1$,

$$\kappa(i,j) = \begin{cases} (1+2\theta\phi+\theta^2)/(1-\phi^2), & i=j=1, \\ 1+\theta^2, & i=j\geq 2, \\ \theta, & |i-j|=1, i\geq 1, \\ 0, & \text{otherwise.} \end{cases}$$

With these values of $\kappa(i, j)$, the recursions of the innovations algorithm reduce to

$$\begin{aligned} r_0 &= (1+2\theta\phi+\theta^2)/(1-\phi^2), \\ \theta_{n1} &= \theta/r_{n-1}, \\ r_n &= 1+\theta^2-\theta^2/r_{n-1}, \end{aligned} \tag{3.3.9}$$

which can be solved quite explicitly (see Problem 3.13). □

Example 3.3.4 Numerical prediction of an ARMA(2,3) process

In this example we illustrate the steps involved in numerical prediction of an ARMA(2,3) process. Of course these steps are shown for illustration only. The calculations are all carried out automatically by PEST in the course of computing predictors for any specified data set and ARMA model. The process we shall consider is the ARMA process defined by the equations

$$X_t - X_{t-1} + 0.24X_{t-2} = Z_t + 0.4Z_{t-1} + 0.2Z_{t-2} + 0.1Z_{t-3}, \tag{3.3.10}$$

where $\{Z_t\} \sim \mathrm{WN}(0, 1)$. Ten values of $X_1, \ldots, X_{10}$ simulated by the program PEST are shown in Table 3.1. (These were produced using the option *Entry of an ARMA(p,q) Model* to specify the model and the option *Generation of Simulated Data* to generate the data.)

The first step is to compute the covariances $\gamma_X(h)$, $h = 0, 1, 2$, which are easily found from equations (3.2.5) with $k = 0, 1, 2$ to be

$$\gamma_X(0) = 7.17133,\ \gamma_X(1) = 6.44139,\ \text{and}\ \gamma_X(2) = 5.0603.$$

From (3.3.3), we find that the symmetric matrix $K = [\kappa(i, j)]_{i,j=1,2,\ldots}$ is given by

$$K = \begin{bmatrix} 7.1713 \\ 6.4414 & 7.1713 \\ 5.0603 & 6.4414 & 7.1713 \\ 0.10 & 0.34 & 0.816 & 1.21 \\ 0 & 0.10 & 0.34 & 0.50 & 1.21 \\ 0 & 0 & 0.10 & 0.24 & 0.50 & 1.21 \\ \cdot & 0 & 0 & 0.10 & 0.24 & 0.50 & 1.21 \\ \cdot & \cdot & 0 & 0 & 0.10 & 0.24 & 0.50 & 1.21 \\ \cdot & \cdot & \cdot & \cdot & \cdot & \cdot & \cdot & \cdot & \cdot \end{bmatrix}$$

The next step is to solve the recursions of the innovations algorithm for θ_{nj} and r_n using these values for $\kappa(i, j)$. Then

$$\hat{X}_{n+1} = \begin{cases} \sum_{j=1}^{n} \theta_{nj}(X_{n+1-j} - \hat{X}_{n+1-j}), & n = 1, 2, \\ X_n - 0.24X_{n-1} + \sum_{j=1}^{3} \theta_{nj}(X_{n+1-j} - \hat{X}_{n+1-j}), & n = 3, 4, \ldots, \end{cases}$$

and

$$E(X_{n+1} - \hat{X}_{n+1})^2 = \sigma^2 r_n = r_n.$$

The results are shown in Table 3.1. □

h-step Prediction of an ARMA(p, q) Process

As in Section 2.5, we use $P_n Y$ to denote the best linear predictor of Y in terms of $X_1, \ldots, X_n$ (which, as pointed out after (3.3.4), is the same as the best linear predictor of Y in terms of $W_1, \ldots, W_n$). Then from (2.5.30) we have

$$P_n W_{n+h} = \sum_{j=h}^{n+h-1} \theta_{n+h-1,j}(W_{n+h-j} - \hat{W}_{n+h-j})$$

$$= \sigma^2 \sum_{j=h}^{n+h-1} \theta_{n+h-1,j}(X_{n+h-j} - \hat{X}_{n+h-j}).$$

Table 3.1 $\hat{X}_{n+1}$ for the ARMA(2,3) Process of Example 3.3.4.

n	X_{n+1}	r_n	θ_{n1}	θ_{n2}	θ_{n3}	$\hat{X}_{n+1}$
0	1.704	7.1713				0
1	0.527	1.3856	0.8982			1.5305
2	1.041	1.0057	1.3685	0.7056		−0.1710
3	0.942	1.0019	0.4008	0.1806	0.0139	1.2428
4	0.555	1.0019	0.3998	0.2020	0.0732	0.7443
5	−1.002	1.0005	0.3992	0.1995	0.0994	0.3138
6	−0.585	1.0000	0.4000	0.1997	0.0998	−1.7293
7	0.010	1.0000	0.4000	0.2000	0.0998	−0.1688
8	−0.638	1.0000	0.4000	0.2000	0.0999	0.3193
9	0.525	1.0000	0.4000	0.2000	0.1000	−0.8731
10		1.0000	0.4000	0.2000	0.1000	1.0638
11		1.0000	0.4000	0.2000	0.1000	
12		1.0000	0.4000	0.2000	0.1000	

Using this result and applying the operator P_n to each side of the equations (3.3.1), we conclude that the h-step predictors $P_n X_{n+h}$ satisfy

$$P_n X_{n+h} = \begin{cases} \sum_{j=h}^{n+h-1} \theta_{n+h-1,j}(X_{n+h-j} - \hat{X}_{n+h-j}), & 1 \le h \le m-n, \\ \sum_{i=1}^{p} \phi_i P_n X_{n+h-i} + \sum_{j=h}^{n+h-1} \theta_{n+h-1,j}(X_{n+h-j} - \hat{X}_{n+h-j}), & h > m-n. \end{cases} \tag{3.3.11}$$

If, as is almost always the case, $n > m = \max(p, q)$, then for all $h \ge 1$,

$$P_n X_{n+h} = \sum_{i=1}^{p} \phi_i P_n X_{n+h-i} + \sum_{j=h}^{q} \theta_{n+h-1,j}(X_{n+h-j} - \hat{X}_{n+h-j}). \tag{3.3.12}$$

Once the predictors $\hat{X}_1, \ldots \hat{X}_n$ have been computed from (3.3.7), it is a straightforward calculation, with n fixed, to determine the predictors $P_n X_{n+1}, P_n X_{n+2}, P_n X_{n+3}, \ldots$, recursively from (3.3.12) (or (3.3.11) if $n \le m$). The calculations are performed automatically in the *Forecasting* option of the program PEST.

Example 3.3.5 h-step prediction of an ARMA(2,3) process

To compute h-step predictors, $h = 1, \ldots, 10$, for the data of Example 3.3.4 using the model (3.3.10) we must read both the data set and the model into the program PEST. After reading in the data file and entering the model (3.3.10) into PEST (be careful *not* to subtract the sample mean from the data), select the option *Forecasting* from the main menu. You will notice that the white noise variance is automatically set by PEST to an estimate based on the sample. To retain the model value of 1 you must reset the white noise variance to this value. Then, if you specify that ten future predicted values are required, the program will print out the ten predictors and the square roots of their mean squared errors. (These are calculated as described below.) The results from PEST are shown in the following table. Notice how the predictors converge fairly rapidly to the mean of the process (i.e., zero) as the lead time h increases. Correspondingly the one-step mean squared error increases from the white noise variance (i.e., 1) at $h = 1$ to the variance of X_t (i.e., 7.1713), which is virtually reached at $h = 10$. □

The Mean Squared Error of $P_n X_{n+h}$

The mean squared error of $P_n X_{n+h}$ is easily computed by PEST from the formula

$$\sigma_n^2(h) := E(X_{n+h} - P_n X_{n+h})^2 = \sum_{j=0}^{h-1} \left(\sum_{r=0}^{j} \chi_r \theta_{n+h-r-1,j-r} \right)^2 v_{n+h-j-1}, \tag{3.3.13}$$

Table 3.2 h-step predictors for the ARMA(2,3) Series of Example 3.3.4.

h	$P_{10}X_{10+h}$	$\sqrt{MSE}$
1	1.0638	1.0000
2	1.1217	1.7205
3	1.0062	2.1931
4	0.7370	2.4643
5	0.4955	2.5902
6	0.3186	2.6434
7	0.1997	2.6648
8	0.1232	2.6730
9	0.0753	2.6761
10	0.0457	2.6773

where the coefficients χ_j are computed recursively from the equations $\chi_0 = 1$ and

$$\chi_j = \sum_{k=1}^{\min(p,j)} \phi_k \chi_{j-k}, \quad j = 1, 2, \ldots. \tag{3.3.14}$$

Example 3.3.6 h-step prediction of an ARMA(2,3) process

We now illustrate the use of (3.3.12) and (3.3.13) for the h-step predictors and their mean squared errors by manually reproducing the output of PEST shown in Table 3.2. From (3.3.12) and Table 3.1 we obtain

$$\begin{aligned} P_{10}X_{12} &= \sum_{i=1}^{2} \phi_i P_{10}X_{12-i} + \sum_{j=2}^{3} \theta_{11,j}(X_{12-j} - \hat{X}_{12-j}) \\ &= \phi_1 \hat{X}_{11} + \phi_2 X_{10} + 0.2(X_{10} - \hat{X}_{10}) + 0.1(X_9 - \hat{X}_9) \\ &= 1.1217 \end{aligned}$$

and

$$\begin{aligned} P_{10}X_{13} &= \sum_{i=1}^{2} \phi_i P_{10}X_{13-i} + \sum_{j=3}^{3} \theta_{12,j}(X_{13-j} - \hat{X}_{13-j}) \\ &= \phi_1 P_{10}X_{12} + \phi_2 \hat{X}_{11} + 0.1(X_{10} - \hat{X}_{10}) \\ &= 1.0062. \end{aligned}$$

For $k > 13$, $P_{10}X_k$ is easily found recursively from

$$P_{10}X_k = \phi_1 P_{10}X_{k-1} + \phi_2 P_{10}X_{k-2}.$$

To find the mean squared errors we use (3.3.13) with $\chi_0 = 1$, $\chi_1 = \phi_1 = 1$, and $\chi_2 = \phi_1\chi_1 + \phi_2 = 0.76$. Using the values of θ_{nj} and $v_j (= r_j)$ in Table 3.1, we obtain

$$\sigma_{10}^2(2) = E(X_{12} - P_{10}X_{12})^2 = 2.960$$

and

$$\sigma_{10}^2(3) = E(X_{13} - P_{10}X_{13})^2 = 4.810,$$

in accordance with the results shown in Table 3.2. □

Large-Sample Approximations

Assuming as usual that the ARMA(p, q) process defined by $\phi(B)X_t = \theta(B)Z_t$, $\{Z_t\} \sim \text{WN}(0, \sigma^2)$, is causal and invertible, we have the representations

$$X_{n+h} = \sum_{j=0}^{\infty} \psi_j Z_{n+h-j} \tag{3.3.15}$$

and

$$Z_{n+h} = X_{n+h} + \sum_{j=1}^{\infty} \pi_j X_{n+h-j}, \tag{3.3.16}$$

where $\{\psi_j\}$ and $\{\pi_j\}$ are uniquely determined by the equations (3.1.7) and (3.1.8), respectively. Let $\tilde{P}_nY$ denote the best (i.e., minimum mean squared error) approximation to Y which is a linear combination or limit of linear combinations of X_t, $-\infty < t \le n$ or equivalently (by (3.3.15) and (3.3.16)) of Z_t, $-\infty < t \le n$. The properties of the operator $\tilde{P}_n$ were discussed in Section 2.5.3. Applying $\tilde{P}_n$ to each side of the equations (3.3.15) and (3.3.16) gives

$$\tilde{P}_nX_{n+h} = \sum_{j=h}^{\infty} \psi_j Z_{n+h-j} \tag{3.3.17}$$

and

$$\tilde{P}_nX_{n+h} = -\sum_{j=1}^{\infty} \pi_j \tilde{P}_n X_{n+h-j}. \tag{3.3.18}$$

For $h = 1$ the jth term on the right of (3.3.18) is just X_{n+1-j}. Once $\tilde{P}_nX_{n+1}$ has been evaluated, $\tilde{P}_nX_{n+2}$ can then be computed from (3.3.18). The predictors $\tilde{P}_nX_{n+3}$, $\tilde{P}_nX_{n+4}, \ldots$, can then be computed successively in the same way. Subtracting (3.3.17) from (3.3.15) gives the h-step prediction error as

$$X_{n+h} - \tilde{P}_nX_{n+h} = \sum_{j=0}^{h-1} \psi_j Z_{n+h-j},$$

from which we see that the mean squared error is

$$\tilde{\sigma}^2(h) = \sigma^2 \sum_{j=0}^{h-1} \psi_j^2. \tag{3.3.19}$$

The predictors obtained in this way have the form

$$\tilde{P}_nX_{n+h} = \sum_{j=0}^{\infty} c_j X_{n-j}. \tag{3.3.20}$$

In practice of course we only have observations $X_1, \ldots, X_n$ available so we must truncate the series (3.3.20) after n terms. The resulting predictor is a useful approximation to $P_n X_{n+h}$ if n is large and the coefficients c_j converge to zero rapidly as j increases. It can be shown that the mean squared error (3.3.19) of $\tilde{P}_n X_{n+h}$ can also be obtained by letting $n \to \infty$ in the expression (3.3.13) for the mean squared error of $P_n X_{n+h}$ so that $\tilde{\sigma}^2(h)$ is an easily calculated approximation to $\sigma_n^2(h)$ for large n.

Prediction Bounds for Gaussian Processes

If the ARMA process $\{X_t\}$ is driven by Gaussian white noise (i.e., if $\{Z_t\} \sim$ IID $\mathrm{N}(0, \sigma^2)$), then for each $h \geq 1$ the prediction error $X_{n+h} - P_n X_{n+h}$ is normally distributed with mean 0 and variance $\sigma_n^2(h)$ given by (3.3.19).

Consequently if $\Phi_{1-\alpha/2}$ denotes the $(1-\alpha/2)$ quantile of the standard normal distribution function, it follows that X_{n+h} lies between the bounds $P_n X_{n+h} \pm \Phi_{1-\alpha/2}\sigma_n(h)$ with probability $(1-\alpha)$. These bounds are therefore called $(1-\alpha)$ prediction bounds for X_{n+h}.

Problems

3.1. Determine which of the following ARMA processes are causal and which of them are invertible. (In each case $\{Z_t\}$ denotes white noise.)

a. $X_t + 0.2X_{t-1} - 0.48X_{t-2} = Z_t$.

b. $X_t + 1.9X_{t-1} + 0.88X_{t-2} = Z_t + 0.2Z_{t-1} + 0.7Z_{t-2}$.

c. $X_t + 0.6X_{t-1} = Z_t + 1.2Z_{t-1}$.

d. $X_t + 1.8X_{t-1} + 0.81X_{t-2} = Z_t$.

e. $X_t + 1.6X_{t-1} = Z_t - 0.4Z_{t-1} + 0.04Z_{t-2}$.

3.2. For those processes in Problem 3.1 that are causal, compute and graph their ACF and PACF using the program PEST.

3.3. For those processes in Problem 3.1 that are causal, compute the first six coefficients $\psi_0, \psi_1, \ldots, \psi_5$ in the causal representation $X_t = \sum_{j=0}^{\infty} \psi_j Z_{t-j}$, of $\{X_t\}$.

3.4. Compute the ACF and PACF of the AR(2) process

$$X_t = .8X_{t-2} + Z_t, \quad \{Z_t\} \sim \mathrm{WN}(0, \sigma^2).$$

3.5. Let $\{Y_t\}$ be the ARMA plus noise time series defined by

$$Y_t = X_t + W_t,$$

where $\{W_t\} \sim \mathrm{WN}(0, \sigma_w^2)$, $\{X_t\}$ is the ARMA(p, q) process satisfying

$$\phi(B)X_t = \theta(B)Z_t, \quad \{Z_t\} \sim \mathrm{WN}(0, \sigma_z^2),$$

and $E(W_s Z_t) = 0$ for all s and t.

a. Show that $\{Y_t\}$ is stationary and find its autocovariance in terms of σ_W^2 and the ACVF of $\{X_t\}$.

b. Show that the process $U_t := \phi(B)Y_t$ is r-correlated where $r = \max(p, q)$ and hence, by Proposition 2.1.1, is an MA(r) process. Conclude that $\{Y_t\}$ is an ARMA(p, r) process.

3.6. Show that the two MA(1) processes

$$X_t = Z_t + \theta Z_{t-1}, \quad \{Z_t\} \sim \text{WN}(0, \sigma^2)$$

$$Y_t = \tilde{Z}_t + \frac{1}{\theta}\tilde{Z}_{t-1}, \quad \{\tilde{Z}_t\} \sim \text{WN}(0, \sigma^2\theta^2),$$

where $0 < |\theta| < 1$, have the same autocovariance functions.

3.7. Suppose that $\{X_t\}$ is the noninvertible MA(1) process

$$X_t = Z_t + \theta Z_{t-1}, \quad \{Z_t\} \sim \text{WN}(0, \sigma^2)$$

where $|\theta| > 1$. Define a new process $\{W_t\}$ as

$$W_t = \sum_{j=0}^{\infty} (-\theta)^{-j} X_{t-j}$$

and show that $\{W_t\} \sim \text{WN}(0, \sigma_W^2)$. Express σ_W^2 in terms of θ and σ^2 and show that $\{X_t\}$ has the *invertible* representation (in terms of $\{W_t\}$),

$$X_t = W_t + \frac{1}{\theta} W_{t-1}.$$

3.8. Let $\{X_t\}$ denote the unique stationary solution of the autoregressive equations

$$X_t = \phi X_{t-1} + Z_t, \ t = 0, \pm 1, \ldots,$$

where $\{Z_t\} \sim \text{WN}(0, \sigma^2)$ and $|\phi| > 1$. Then X_t is given by the expression (2.2.11). Define the new sequence

$$W_t = X_t - \frac{1}{\phi} X_{t-1},$$

show that $\{W_t\} \sim \text{WN}(0, \sigma_W^2)$, and express σ_W^2 in terms of σ^2 and ϕ. These calculations show that $\{X_t\}$ is the (unique stationary) solution of the *causal* AR equations

$$X_t = \frac{1}{\phi} X_{t-1} + W_t, \ t = 0, \pm 1, \ldots.$$

3.9. a. Calculate the autocovariance function $\gamma(\cdot)$ of the stationary time series

$$Y_t = \mu + Z_t + \theta_1 Z_{t-1} + \theta_{12} Z_{t-12}, \quad \{Z_t\} \sim \text{WN}(0, \sigma^2).$$

b. Use the program PEST to compute the sample mean and sample autocovariances $\hat{\gamma}(h)$, $0 \le h \le 20$, of $\{\nabla\nabla_{12}X_t\}$, where $\{X_t, t = 1, \ldots, 72\}$ is the accidental deaths series DEATHS.DAT of Example 1.1.3.

c. By equating $\hat{\gamma}(1)$, $\hat{\gamma}(11)$, and $\hat{\gamma}(12)$ from part (b) to $\gamma(1)$, $\gamma(11)$, and $\gamma(12)$, respectively, from part (a), find a model of the form defined in (a) to represent $\{\nabla\nabla_{12}X_t\}$.

3.10. By matching the autocovariances and sample autocovariances at lags 0 and 1, fit a model of the form

$$X_t - \mu = \phi(X_{t-1} - \mu) + Z_t, \quad \{Z_t\} \sim \mathrm{WN}(0, \sigma^2),$$

to the data STRIKES.DAT of Example 1.1.6. Use the fitted model to compute the best predictor of the number of strikes in 1981. Estimate the mean squared error of your predictor and construct 95% prediction bounds for the number of strikes in 1981 assuming that $\{Z_t\} \sim \mathrm{IID\ N}(0, \sigma^2)$.

3.11. Show that the value at lag 2 of the partial ACF of the MA(1) process

$$X_t = Z_t + \theta Z_{t-1}, \; t = 0, \pm 1, \ldots,$$

where $\{Z_t\} \sim \mathrm{WN}(0, \sigma^2)$, is

$$\alpha(2) = -\theta^2/(1 + \theta^2 + \theta^4).$$

3.12. For the MA(1) process of Problem 3.11, the best linear predictor of X_{n+1} based on $X_1, \ldots, X_n$ is

$$\hat{X}_{n+1} = \phi_{n,1}X_n + \cdots + \phi_{n,n}X_1,$$

where $\phi_n = (\phi_{n1}, \ldots, \phi_{nn})'$ satisfies $R_n\phi_n = \rho_n$ (equation (2.5.23)). By substituting the appropriate correlations into R_n and ρ_n and solving the resulting equations (starting with the last and working up), show that for $1 \le j < n$, $\phi_{n,n-j} = (-\theta)^{-j}(1+\theta^2+\cdots+\theta^{2j})\phi_{nn}$ and hence that the PACF $\alpha(n) := \phi_{nn} = -(-\theta)^n/(1 + \theta^2 + \cdots + \theta^{2n})$.

3.13. The coefficients θ_{nj} and one-step mean squared errors $v_n = r_n\sigma^2$ for the general causal ARMA(1,1) process in Example 3.3.3 can be found as follows:

a. Show that if $y_n := r_n/(r_n - 1)$, then the last of equations (3.3.9) can be rewritten in the form

$$y_n = \theta^{-2}y_{n-1} + 1, \quad n \ge 1.$$

b. Deduce that $y_n = \theta^{-2n}y_0 + \sum_{j=1}^{n} \theta^{-2(j-1)}$ and hence determine r_n and θ_{n1}, $n = 1, 2, \ldots$.

c. Evaluate the limits as $n \to \infty$ of r_n and θ_{n1} in the two cases $|\theta| < 1$ and $|\theta| \ge 1$.

4 Spectral Analysis

This chapter can be omitted without any loss of continuity. The reader with no background in Fourier or complex analysis should go straight to Chapter 5. The spectral representation of a stationary time series $\{X_t\}$ essentially decomposes $\{X_t\}$ into a sum of sinusoidal components with uncorrelated random coefficients. In conjunction with this decomposition there is a corresponding decomposition into sinusoids of the autocovariance function of $\{X_t\}$. The spectral decomposition is thus an analogue for stationary processes of the more familiar Fourier representation of deterministic functions. The analysis of stationary processes by means of their spectral representation is often referred to as the "frequency domain analysis" of time series or "spectral analysis." It is equivalent to "time domain" analysis based on the autocovariance function, but provides an alternative way of viewing the process, which for some applications may be more illuminating. For example, in the design of a structure subject to a randomly fluctuating load, it is important to be aware of the presence in the loading force of a large sinusoidal component with a particular frequency to ensure that this is not a resonant frequency of the structure. The spectral point of view is also particularly useful in the analysis of multivariate stationary processes and in the analysis of linear filters. In Section 4.1 we introduce the spectral density of a stationary process $\{X_t\}$, which specifies the frequency decomposition of the autocovariance function, and the closely related spectral representation (or frequency decomposition) of the process $\{X_t\}$ itself. Section 4.2 deals with the periodogram, a sample-based function from which we obtain estimators of the spectral density. In Section 4.3 we discuss time-invariant linear filters from a spectral point of view and in Section 4.4 we use the results to derive the spectral density of an arbitrary ARMA process.

4.1 Spectral Densities

Suppose that $\{X_t\}$ is a zero-mean stationary time series with autocovariance function $\gamma(\cdot)$ satisfying $\sum_{h=-\infty}^{\infty} |\gamma(h)| < \infty$. The **spectral density** of $\{X_t\}$ is the function $f(\cdot)$ defined by

$$f(\lambda) = \frac{1}{2\pi} \sum_{h=-\infty}^{\infty} e^{-ih\lambda}\gamma(h), \quad -\infty < \lambda < \infty, \tag{4.1.1}$$

where $e^{i\lambda} = \cos(\lambda) + i\sin(\lambda)$ and $i = \sqrt{-1}$. The summability of $|\gamma(\cdot)|$ implies that the series in (4.1.1) converges absolutely (since $|e^{ih\lambda}|^2 = \cos^2(h\lambda) + \sin^2(h\lambda) = 1$). Since cos and sin have period 2π, so also does f and it suffices to confine attention to the values of f on the interval $(-\pi, \pi]$.

Basic Properties of f:

$$\text{(a) } f \text{ is even, i.e., } f(\lambda) = f(-\lambda), \tag{4.1.2}$$

$$\text{(b) } f(\lambda) \geq 0 \text{ for all } \lambda \in (-\pi, \pi], \tag{4.1.3}$$

and

$$\text{(c) } \gamma(k) = \int_{-\pi}^{\pi} e^{ik\lambda} f(\lambda)\, d\lambda = \int_{-\pi}^{\pi} \cos(k\lambda) f(\lambda)\, d\lambda. \tag{4.1.4}$$

Proof Since $\sin(\cdot)$ is an odd function and $\cos(\cdot)$ and $\gamma(\cdot)$ are even functions, we have

$$\begin{aligned} f(\lambda) &= \frac{1}{2\pi} \sum_{h=-\infty}^{\infty} (\cos(h\lambda) - i\sin(h\lambda))\gamma(h) \\ &= \frac{1}{2\pi} \sum_{h=-\infty}^{\infty} \cos(-h\lambda)\gamma(h) + 0 \\ &= f(-\lambda). \end{aligned}$$

For each positive integer N define

$$f_N(\lambda) = \frac{1}{2\pi N} E\left(\left|\sum_{r=1}^{N} X_r e^{-ir\lambda}\right|^2\right)$$

$$= \frac{1}{2\pi N} E\left(\sum_{r=1}^{N} X_r e^{-ir\lambda} \sum_{s=1}^{N} X_s e^{is\lambda}\right)$$

$$= \frac{1}{2\pi N} \sum_{|h|<N} (N - |h|) e^{-ih\lambda} \gamma(h),$$

where $\Gamma_N = [\gamma(i-j)]_{i,j=1}^{N}$. Clearly the function f_N is nonnegative for each N, and since $f_N(\lambda) \to \frac{1}{2\pi} \sum_{h=-\infty}^{\infty} e^{-ih\lambda}\gamma(h) = f(\lambda)$ as $N \to \infty$, f must also be nonnegative. This proves (4.1.3). Turning to (4.1.4),

$$\int_{-\pi}^{\pi} e^{ik\lambda} f(\lambda)\, d\lambda = \int_{-\pi}^{\pi} \frac{1}{2\pi} \sum_{h=-\infty}^{\infty} e^{i(k-h)\lambda} \gamma(h)\, d\lambda$$

$$= \frac{1}{2\pi} \sum_{h=-\infty}^{\infty} \gamma(h) \int_{-\pi}^{\pi} e^{i(k-h)\lambda}\, d\lambda$$

$$= \gamma(k)$$

since the only nonzero summand in the second line is the one for which $h = k$ (see Problem 4.1). ■

Equation (4.1.4) expresses the autocovariances of a stationary time series with absolutely summable ACVF as the Fourier coefficients of the nonnegative even function on $(-\pi, \pi]$ defined by (4.1.1). However, even if $\sum_{h=-\infty}^{\infty} |\gamma(h)| = \infty$, there may exist a corresponding spectral density defined as follows.

Definition 4.1.1

A function f is the **spectral density** of a stationary time series $\{X_t\}$ with ACVF $\gamma(\cdot)$ if

(i) $f(\lambda) \geq 0$ for all $\lambda \in (0, \pi]$

(ii) $\gamma(h) = \int_{-\pi}^{\pi} e^{ih\lambda} f(\lambda)\, d\lambda$ for all integers h.

Remark 1. Spectral densities are essentially unique. That is, if f and g are two spectral densities corresponding to the autocovariance function $\gamma(\cdot)$, i.e., $\gamma(h) = \int_{-\pi}^{\pi} e^{ih\lambda} f(\lambda)\, d\lambda = \int_{-\pi}^{\pi} e^{ih\lambda} g(\lambda)\, d\lambda$ for all integers h, then f and g have the same Fourier coefficients and hence are equal (see for example TSTM, Section 2.8). □

The following proposition characterizes spectral densities.

Proposition 4.1.1 *A real-valued function f defined on $(-\pi, \pi]$ is the spectral density of a stationary process if and only if*

(i) $f(\lambda) = f(-\lambda)$,

(ii) $f(\lambda) \geq 0$, *and*

(iii) $\int_{-\pi}^{\pi} f(\lambda)\, d\lambda < \infty$.

Proof If $\gamma(\cdot)$ is absolutely summable, then (i)–(iii) follow from the basic properties of f, (4.1.2)–(4.1.4). For the argument in the general case, see TSTM, Section 4.3.

Conversely suppose f satisfies (i)–(iii). Then it is easy to check, using (i), that the function defined by

$$\gamma(h) = \int_{-\pi}^{\pi} e^{ih\lambda} f(\lambda)\, d\lambda$$

is even. Moreover, if $a_r \in \mathbb{R}$, $r = 1, \ldots, n$, then

$$\begin{aligned}\sum_{r,s=1}^{n} a_r \gamma(r-s) a_s &= \int_{-\pi}^{\pi} \sum_{r,s=1}^{n} a_r a_s e^{i\lambda(r-s)} f(\lambda)\, d\lambda \\ &= \int_{-\pi}^{\pi} \left| \sum_{r=1}^{n} a_r e^{i\lambda r} \right|^2 f(\lambda)\, d\lambda \\ &\geq 0,\end{aligned}$$

so that $\gamma(\cdot)$ is also nonnegative definite and therefore, by Theorem 2.1.1, is an autocovariance function. ■

Corollary 4.1.1 *An absolutely summable function $\gamma(\cdot)$ is the autocovariance function of a stationary time series if and only if it is even and*

$$f(\lambda) = \frac{1}{2\pi} \sum_{h=-\infty}^{\infty} e^{-ih\lambda} \gamma(h) \geq 0, \quad \textit{for all } \lambda \in (-\pi, \pi], \tag{4.1.5}$$

in which case $f(\cdot)$ is the spectral density of $\gamma(\cdot)$.

Proof We have already established the necessity of (4.1.5). Now suppose (4.1.5) holds. Applying Proposition 4.1.1 (the assumptions are easily checked) we conclude that f is the spectral density of some autocovariance function. But this ACVF must be $\gamma(\cdot)$ since $\gamma(k) = \int_{-\pi}^{\pi} e^{ik\lambda} f(\lambda)\, d\lambda$ for all integers k. ■

Example 4.1.1 Using Corollary 4.1.1, it is a simple matter to show that the function defined by

$$\kappa(h) = \begin{cases} 1, & \text{if } h = 0, \\ \rho, & \text{if } h = \pm 1, \\ 0, & \text{otherwise,} \end{cases}$$

is the ACVF of a stationary time series if and only if $|\rho| \leq \frac{1}{2}$ (see Example 2.1.1). Since $\kappa(\cdot)$ is even and nonzero only at lags $0, \pm 1$, it follows from the corollary that

κ is an ACVF if and only if the function

$$f(\lambda) = \frac{1}{2\pi} \sum_{h=-\infty}^{\infty} e^{-ih\lambda} \gamma(h)$$
$$= \frac{1}{2\pi} [1 + 2\rho \cos \lambda]$$

is nonnegative for *all* $\lambda \in [-\pi, \pi]$. But this occurs if and only if $|\rho| \le \frac{1}{2}$. □

As illustrated in the previous example, Corollary 4.1.1 provides us with a powerful tool for checking whether or not an absolutely summable function on the integers is an autocovariance function. It is much simpler and much more informative than direct verification of nonnegative definiteness as required in Theorem 2.1.1.

Not all autocovariance functions have a spectral density. For example, the stationary time series

$$X_t = A \cos(\omega t) + B \sin(\omega t), \tag{4.1.6}$$

where A and B are uncorrelated random variables with mean 0 and variance 1, has ACVF $\gamma(h) = \cos(\omega h)$ (Problem 2.2), which is not expressible as $\int_{-\pi}^{\pi} e^{ih\lambda} f(\lambda) d\lambda$, with f a function on $(-\pi, \pi]$. Nevertheless, $\gamma(\cdot)$ can be written as the Fourier transform of the discrete distribution function

$$F(\lambda) = \begin{cases} 0 & \text{if } \lambda < -\omega, \\ 0.5 & \text{if } -\omega \le \lambda < \omega, \\ 1.0 & \text{if } \lambda \ge \omega, \end{cases}$$

i.e.,

$$\cos(\omega h) = \int_{(-\pi,\pi]} e^{ih\lambda} dF(\lambda),$$

where the integral is as defined in Section A.1. As the following theorem states (see TSTM, p. 117), every ACVF is the Fourier transform of a (generalized) distribution function on $[-\pi, \pi]$. This representation is called the **spectral representation of the ACVF**.

Theorem 4.1.1 (Spectral Representation of the ACVF) *A function $\gamma(\cdot)$ defined on the integers is the ACVF of a stationary time series if and only if there exists a right-continuous, nondecreasing, bounded function F on $[-\pi,\pi]$ with $F(-\pi) = 0$ such that*

$$\gamma(h) = \int_{(-\pi,\pi]} e^{ih\lambda} dF(\lambda) \tag{4.1.7}$$

for all integers h. (For real-valued time series, F is symmetric in the sense that $\int_{(a,b]} dF(x) = \int_{[-b,-a)} dF(x)$ for all a and b such that $0 < a < b$.)

Remark 2. The function F is a **generalized distribution function** on $[-\pi, \pi]$ if $G(\lambda) = F(\lambda)/F(\pi)$ is a probability distribution function on $[-\pi, \pi]$. Note that since $F(\pi) = \gamma(0) = \text{Var}(X_1)$, the ACF of $\{X_t\}$ has spectral representation

$$\rho(h) = \int_{(-\pi,\pi]} e^{ih\lambda} dG(\lambda).$$

The function F in (4.1.7) is called the **spectral distribution function** of $\gamma(\cdot)$. If $F(\lambda)$ can be expressed as $F(\lambda) = \int_{-\pi}^{\lambda} f(y)\,dy$ for all $\lambda \in [-\pi, \pi]$, then f is the **spectral density function** and the time series is said to have a **continuous spectrum**. If F is a discrete distribution (i.e., if G is a discrete probability distribution), then the time series is said to have a **discrete spectrum**. The time series (4.1.6) has a discrete spectrum. □

Example 4.1.2 Linear combination of sinusoids

Consider now the process obtained by adding uncorrelated processes of the type defined in (4.1.6), i.e.,

$$X_t = \sum_{j=1}^{k} (A_j \cos(\omega_j t) + B_j \sin(\omega_j t)), \quad 0 < \omega_1 < \cdots < \omega_k < \pi, \tag{4.1.8}$$

where $A_1, B_1, \ldots, A_k, B_k$ are uncorrelated random variables with $E(A_j) = 0$ and $\text{Var}(A_j) = \text{Var}(B_j) = \sigma_j^2$, $j = 1, \ldots, k$. By Problem 4.5, the ACVF of this time series is $\gamma(h) = \sum_{j=1}^{k} \sigma_j^2 \cos(\omega_j h)$ and its spectral distribution function is $F(\lambda) = \sum_{j=1}^{k} \sigma_j^2 F_j(\lambda)$ where

$$F_j(\lambda) = \begin{cases} 0 & \text{if } \lambda < -\omega_j, \\ 0.5 & \text{if } -\omega_j \le \lambda < \omega_j, \\ 1.0 & \text{if } \lambda \ge \omega_j. \end{cases}$$

A sample path of this time series with $k = 2$, $\omega_1 = \pi/4$, $\omega_2 = \pi/6,$, $\sigma_1^2 = 9$ and $\sigma_2^2 = 1$ is plotted in Figure 4.1. Not surprisingly, the sample path closely approximates a sinusoid with frequency $\omega_1 = \pi/4$ (and period $2\pi/\omega_1 = 8$). The general features of this sample path could have been deduced from the spectral distribution function (see Figure 4.2), which places 90% of its total mass at the frequencies $\pm\pi/4$. This means that 90% of the variance of X_t is contributed by the term $A_1 \cos(\omega_1 t) + B_1 \cos(\omega_1 t)$, which is a sinusoid with period 8. □

The remarkable feature of Example 4.1.2 is that *every* zero-mean stationary process can be expressed as a superposition of uncorrelated sinusoids with frequencies $\omega \in [0, \pi]$. In general, however, a stationary process is a superposition of infinitely many sinusoids rather than a finite number as in (4.1.8). The required generalization

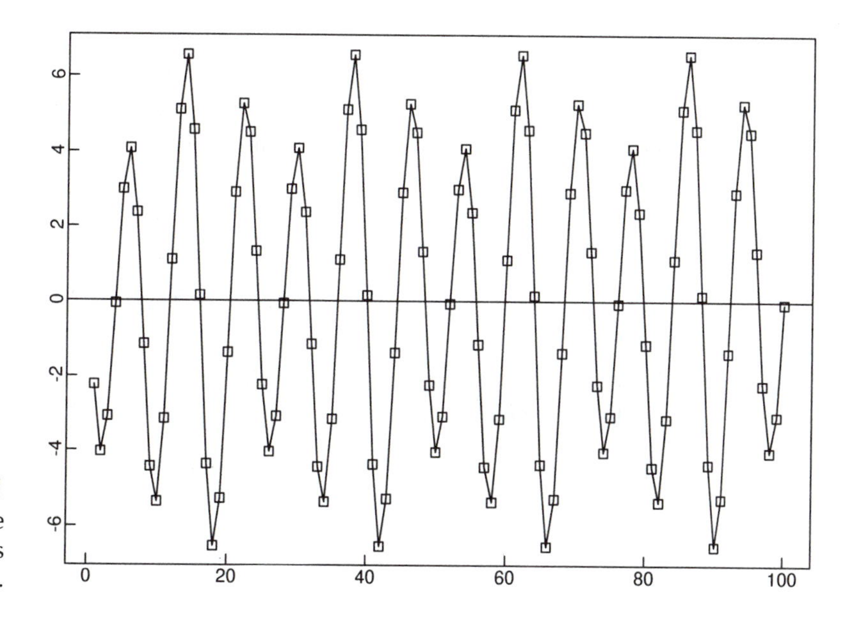

Figure 4-1
A sample path of size 100 from the time series in Example 4.1.2.

of (4.1.8) that allows for this is called a stochastic integral, written as

$$X_t = \int_{(-\pi,\pi]} e^{ih\lambda} dZ(\lambda), \tag{4.1.9}$$

where $\{Z(\lambda), -\pi < \lambda \le \pi\}$ is a complex-valued process with orthogonal (or uncorrelated) increments. The representation (4.1.9) of a zero-mean stationary process $\{X_t\}$ is called the **spectral representation of the process** and should be compared

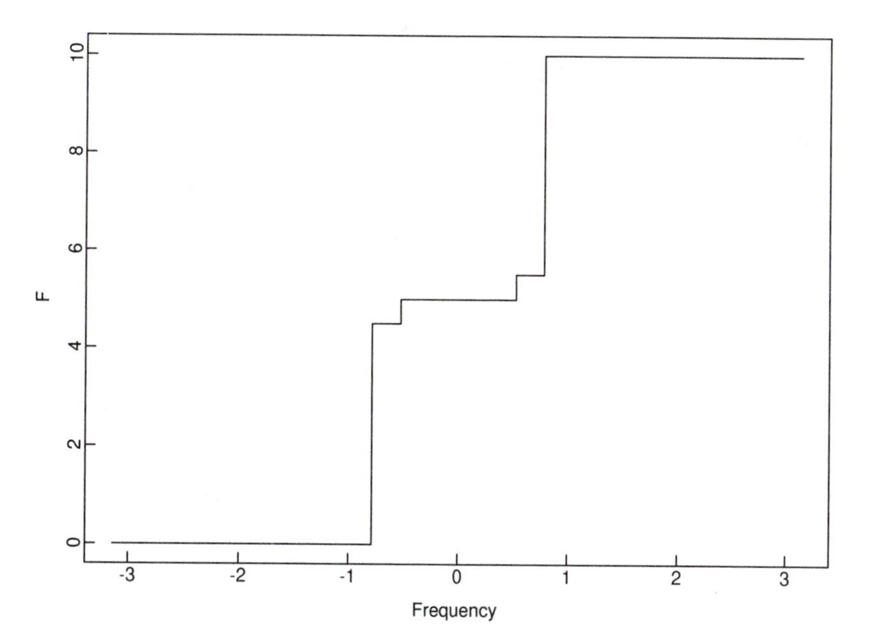

Figure 4-2
The spectral distribution function $F(\lambda)$, $-\pi \le \lambda \le \pi$, of the time series in Example 4.1.2.

with the corresponding spectral representation (4.1.7) of the autocovariance function $\gamma(\cdot)$. The underlying technical aspects of stochastic integration are beyond the scope of this book; however, in the simple case of the process (4.1.8) it is not difficult to see that it can be reexpressed in the form (4.1.9) by choosing

$$dZ(\lambda) = \begin{cases} \dfrac{A_j + iB_j}{2}, & \text{if } \lambda = -\omega_j \text{ and } j \in \{1, \ldots, k\} \\ \dfrac{A_j - iB_j}{2}, & \text{if } \lambda = \omega_j \text{ and } j \in \{1, \ldots, k\}, \\ 0, & \text{otherwise.} \end{cases}$$

For this example it is also clear that

$$E(dZ(\lambda)\overline{dZ(\lambda)}) = \begin{cases} \dfrac{\sigma_j^2}{2}, & \text{if } \lambda = \pm\omega_j, \\ 0, & \text{otherwise.} \end{cases}$$

In general, the connection between $dZ(\lambda)$ and the spectral distribution function of the process can be expressed symbolically as

$$E(dZ(\lambda)\overline{dZ(\lambda)}) = \begin{cases} F(\lambda) - F(\lambda-), & \text{for a discrete spectrum,} \\ f(\lambda)d\lambda, & \text{for a continuous spectrum.} \end{cases} \tag{4.1.10}$$

These relations show that a large jump in the spectral distribution function (or a large peak in the spectral density) at frequency $\pm\omega$ indicates the presence in the time series of strong sinusoidal components at (or near) ω. The period of a sinusoid with frequency ω radians per unit time is $2\pi/\omega$.

Example 4.1.3 White noise

If $\{X_t\} \sim \text{WN}(0, \sigma^2)$, then $\gamma(0) = \sigma^2$ and $\gamma(h) = 0$ for all $|h| > 0$. This process has a flat spectral density (see Problem 4.2),

$$f(\lambda) = \frac{\sigma^2}{2\pi}, \quad -\pi \le \lambda \le \pi.$$

A process with this spectral density is called **white noise** since each frequency in the spectrum contributes equally to the variance of the process. □

Example 4.1.4 The spectral density of an AR(1) process

If

$$X_t = \phi X_{t-1} + Z_t,$$

where $\{Z_t\} \sim \mathrm{WN}(0, \sigma^2)$, then from (4.1.1) $\{X_t\}$ has spectral density

$$\begin{aligned} f(\lambda) &= \frac{\sigma^2}{2\pi(1-\phi^2)}\left(1 + \sum_{h=1}^{\infty} \phi^h (e^{-ih\lambda} + e^{ih\lambda})\right) \\ &= \frac{\sigma^2}{2\pi(1-\phi^2)}\left(1 + \frac{\phi e^{i\lambda}}{1-\phi e^{i\lambda}} + \frac{\phi e^{-i\lambda}}{1-\phi e^{-i\lambda}}\right) \\ &= \frac{\sigma^2}{2\pi}(1 - 2\phi\cos\lambda + \phi^2)^{-1}. \end{aligned}$$

Graphs of $f(\lambda)$, $0 \le \lambda \le \pi$, are displayed in Figures 4.3 and 4.4 for $\phi = .7$ and $\phi = -.7$. Observe that for $\phi = .7$ the density is large for low frequencies and small for high frequencies. This is not unexpected since when $\phi = .7$ the process has a positive ACF with a large value at lag one (see Figure 4.5), making the series smooth with relatively few high frequency components. On the other hand, for $\phi = -.7$ the ACF has a large negative value at lag one (see Figure 4.6) producing a series that fluctuates rapidly about its mean value. In this case the series has a large contribution from high frequency components as reflected by the size of the spectral density near frequency π. □

Example 4.1.5 Spectral density of an MA(1) process

If

$$X_t = Z_t + \theta Z_{t-1},$$

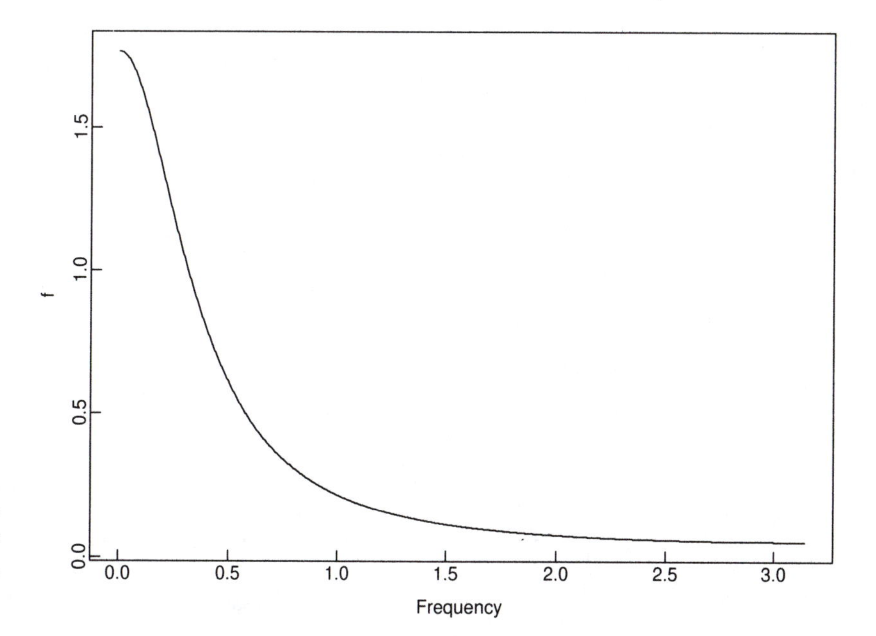

Figure 4-3 The spectral density $f(\lambda)$, $0 \le \lambda \le \pi$, of $X_t = .7X_{t-1} + Z_t$, where $\{Z_t\} \sim \mathrm{WN}(0, \sigma^2)$.

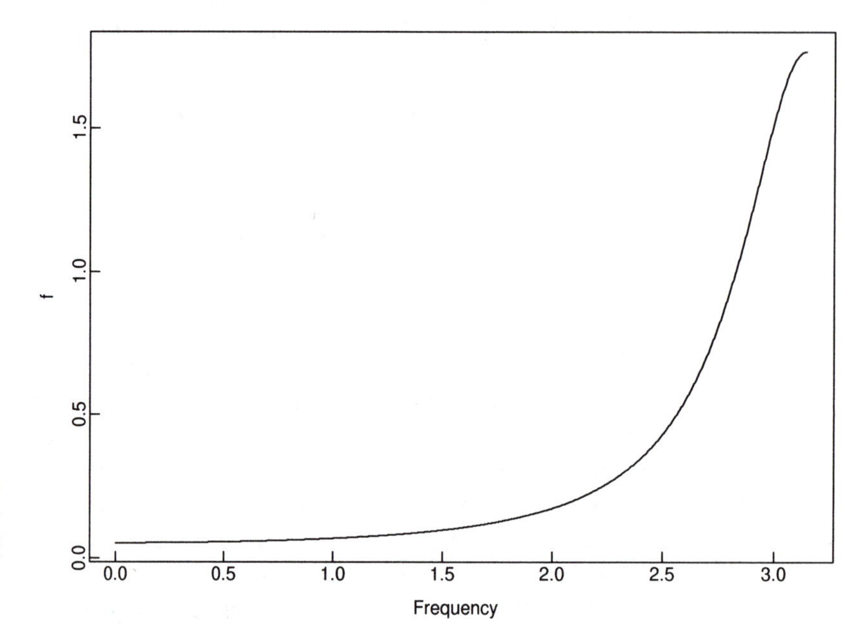

Figure 4-4 The spectral density $f(\lambda)$, $0 \le \lambda \le \pi$, of $X_t = -.7X_{t-1} + Z_t$, where $\{Z_t\} \sim \text{WN}(0, \sigma^2)$.

where $\{Z_t\} \sim \text{WN}(0, \sigma^2)$, then from (4.1.1),

$$
\begin{aligned}
f(\lambda) &= \frac{\sigma^2}{2\pi}\left(1 + \theta^2 + \theta(e^{-i\lambda} + e^{i\lambda})\right) \\
&= \frac{\sigma^2}{2\pi}(1 + 2\theta\cos\lambda + \theta^2).
\end{aligned}
$$

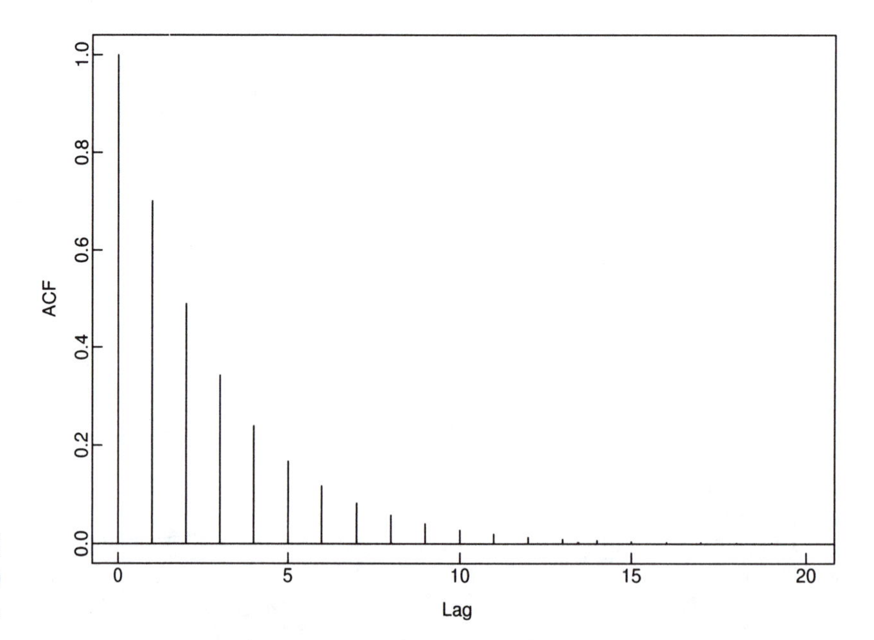

Figure 4-5 The ACF of the AR(1) process $X_t = .7X_{t-1} + Z_t$.

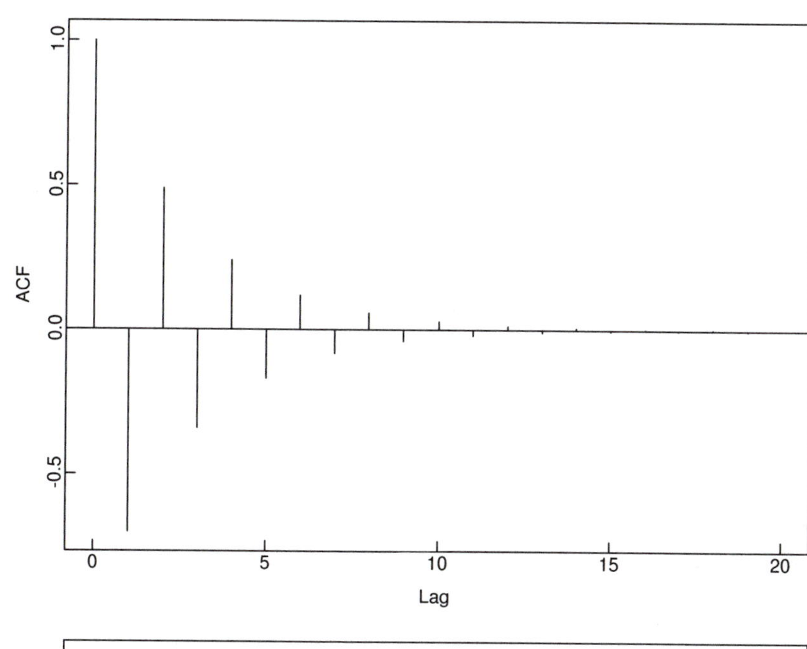

Figure 4-6
The ACF of the AR(1) process $X_t = -.7X_{t-1} + Z_t$.

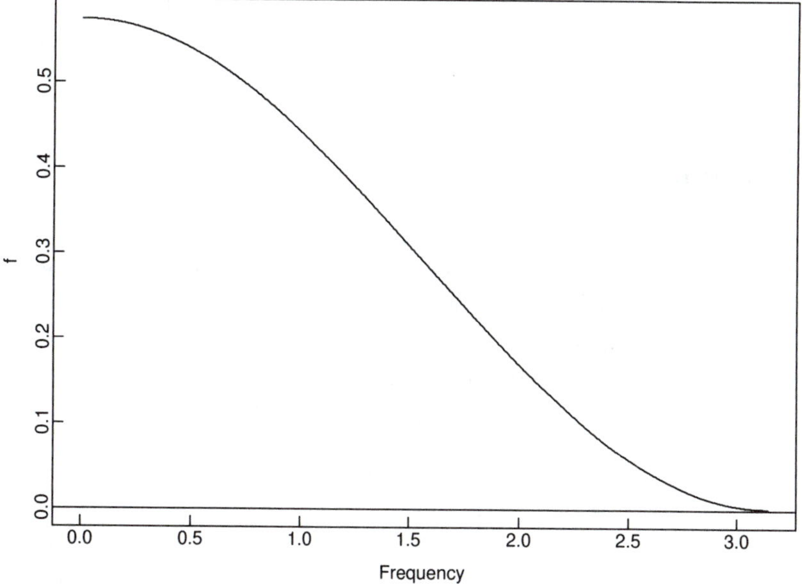

Figure 4-7
The spectral density $f(\lambda)$, $0 \le \lambda \le \pi$, of $X_t = Z_t + .9Z_{t-1}$ where $\{Z_t\} \sim \mathrm{WN}(0, \sigma^2)$.

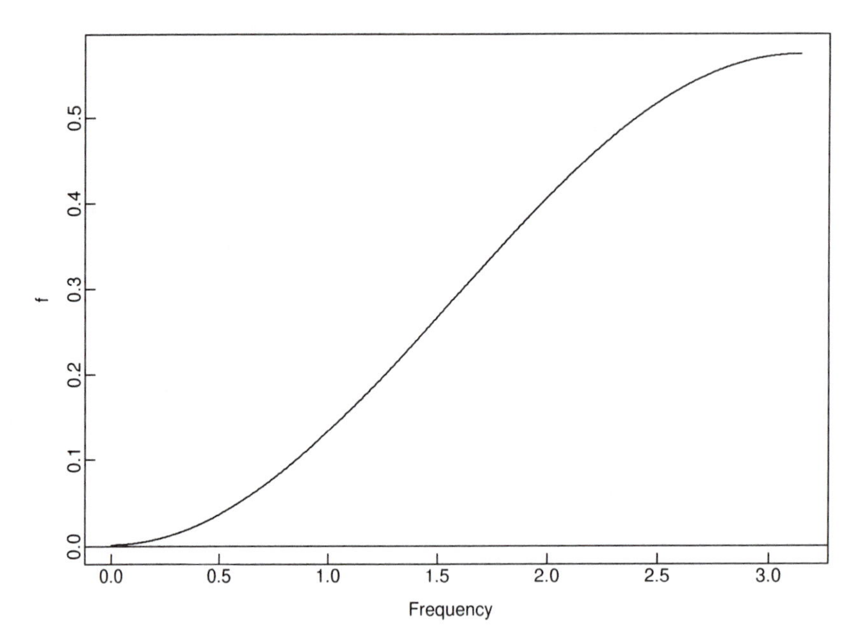

Figure 4-8 The spectral density $f(\lambda)$, $0 \le \lambda \le \pi$, of $X_t = Z_t - .9Z_{t-1}$ where $\{Z_t\} \sim \text{WN}(0, \sigma^2)$.

This function is shown in Figures 4.7 and 4.8 for the values $\theta = .9$ and $\theta = -.9$. Interpretations of the graphs analogous to those in Example 4.1.4 can again be made. □

4.2 The Periodogram

If $\{X_t\}$ is a stationary time series $\{X_t\}$ with ACVF $\gamma(\cdot)$ and spectral density $f(\cdot)$, then just as the sample ACVF $\hat{\gamma}(\cdot)$ of the observations $\{x_1, \ldots, x_n\}$ can be regarded as a sample analogue of $\gamma(\cdot)$, so also can the periodogram $I_n(\cdot)$ of the observations be regarded as a sample analogue of $2\pi f(\cdot)$.

To introduce the periodogram, we consider the vector of complex numbers,

$$\mathbf{x} = \begin{bmatrix} x_1 \\ x_2 \\ \vdots \\ x_n \end{bmatrix} \in \mathbb{C}^n,$$

where $\mathbb{C}^n$ denotes the set of all column vectors with complex-valued components. Now let $\omega_k = 2\pi k/n$, where k is any integer between $-(n-1)/2$ and $n/2$ (inclusive), i.e.,

$$\omega_k = \frac{2\pi k}{n}, \quad k = -\left[\frac{n-1}{2}\right], \ldots, \left[\frac{n}{2}\right], \tag{4.2.1}$$

where $[y]$ denotes the largest integer less than or equal to y. We shall refer to the set F_n of these values as the **Fourier frequencies** associated with sample size n, noting that

F_n is a subset of the interval $(-\pi, \pi]$. Correspondingly we introduce the n vectors,

$$\mathbf{e}_k = \frac{1}{\sqrt{n}} \begin{bmatrix} e^{i\omega_k} \\ e^{2i\omega_k} \\ \vdots \\ e^{ni\omega_k} \end{bmatrix}, \qquad k = -\left[\frac{n-1}{2}\right], \ldots, \left[\frac{n}{2}\right]. \tag{4.2.2}$$

Now $\mathbf{e}_1, \ldots, \mathbf{e}_n$ are **orthonormal** in the sense that

$$\mathbf{e}_j^* \mathbf{e}_k = \begin{cases} 1, & \text{if } j = k, \\ 0, & \text{if } j \neq k \end{cases}, \tag{4.2.3}$$

where $\mathbf{e}_j^*$ denotes the row vector whose kth component is the complex conjugate of the kth component of $\mathbf{e}_j$ (see Problem 4.3). This implies that $\{\mathbf{e}_1, \ldots, \mathbf{e}_n\}$ is a basis for $\mathbb{C}^n$ so that any $\mathbf{x} \in \mathbb{C}^n$ can be expressed as the sum of n components,

$$\mathbf{x} = \sum_{k=-[(n-1)/2]}^{[n/2]} a_k \mathbf{e}_k. \tag{4.2.4}$$

The coefficients a_k are easily found by multiplying (4.2.4) on the left by $\mathbf{e}_k^*$ and using (4.2.3). Thus,

$$a_k = \mathbf{e}_k^* \mathbf{x} = \frac{1}{\sqrt{n}} \sum_{t=1}^{n} x_t e^{-it\omega_k}. \tag{4.2.5}$$

The sequence $\{a_k\}$ is called the **discrete Fourier transform** of the sequence $\{x_1, \ldots, x_n\}$.

Remark 1. The tth component of (4.2.4) can be written as

$$x_t = \sum_{k=-[(n-1)/2]}^{[n/2]} a_k[\cos(\omega_k t) + i \sin(\omega_k t)], \quad t = 1, \ldots, n, \tag{4.2.6}$$

showing that (4.2.4) is just a way of representing x_t as a linear combination of sine waves with frequencies $\omega_k \in F_n$. □

Definition 4.2.1

The **periodogram** of $\{x_1, \ldots, x_n\}$ is the function

$$I_n(\lambda) = \frac{1}{n} \left| \sum_{t=1}^{n} x_t e^{-it\lambda} \right|^2. \tag{4.2.7}$$

Remark 2. If λ is one of the Fourier frequencies ω_k, then $I_n(\omega_k) = |a_k|^2$ and so from (4.2.4) and (4.2.3), we find at once that the squared length of $\mathbf{x}$ is

$$\sum_{t=1}^{n} |x_t|^2 = \mathbf{x}^* \mathbf{x} = \sum_{k=-[(n-1)/2]}^{[n/2]} |a_k|^2 = \sum_{k=-[(n-1)/2]}^{[n/2]} I_n(\omega_k).$$

The value of the periodogram at frequency ω_k is thus the contribution to this sum of squares from the "frequency ω_k" term $a_k\mathbf{e}_k$ in (4.2.4). □

The next proposition shows that $I_n(\lambda)$ can be regarded as a sample analogue of $2\pi f(\lambda)$. Recall that if $\sum_{h=-\infty}^{\infty}|\gamma(h)| < \infty$, then

$$2\pi f(\lambda) = \sum_{h=-\infty}^{\infty} \gamma(h)e^{-ih\lambda}, \quad \lambda \in (-\pi, \pi]. \tag{4.2.8}$$

Proposition 4.2.1 *If $x_1, \ldots, x_n$ are any real numbers and ω_k is any of the nonzero Fourier frequencies $2\pi k/n$ in $(-\pi, \pi]$, then*

$$I_n(\omega_k) = \sum_{|h|<n} \hat{\gamma}(h)e^{-ih\omega_k}, \tag{4.2.9}$$

where $\hat{\gamma}(h)$ is the sample ACVF of $x_1, \ldots, x_n$.

Proof Since $\sum_{t=1}^{n} e^{-it\omega_k} = 0$ if $\omega_k \neq 0$, we can subtract the sample mean $\bar{x}$ from x_t in the defining equation (4.2.7) of $I_n(\omega_k)$. Hence,

$$\begin{aligned} I_n(\omega_k) &= n^{-1}\sum_{s=1}^{n}\sum_{t=1}^{n}(x_s - \bar{x})(x_t - \bar{x})e^{-i(s-t)\omega_k} \\ &= \sum_{|h|<n} \hat{\gamma}(h)e^{-ih\omega_k}. \end{aligned}$$

■

In view of the similarity between (4.2.8) and (4.2.9), a natural estimate of the spectral density $f(\lambda)$ is $I_n(\lambda)/(2\pi)$. For a very large class of stationary time series $\{X_t\}$ with strictly positive spectral density, it can be shown that for any fixed frequencies $\lambda_1, \ldots, \lambda_m$ such that $0 < \lambda_1 < \cdots < \lambda_m < \pi$, the joint distribution function $F_n(x_1, \ldots, x_m)$ of the periodogram values $(I_n(\lambda_1), \ldots, I_n(\lambda_m))$ converges, as $n \to \infty$, to $F(x_1, \ldots, x_m)$, where

$$F(x_1, \ldots, x_m) = \begin{cases} \prod_{i=1}^{m}\left(1 - \exp\left\{\dfrac{-x_i}{2\pi f(\lambda_i)}\right\}\right), & \text{if } x_1, \ldots, x_m > 0, \\ 0, & \text{otherwise.} \end{cases} \tag{4.2.10}$$

Thus for large n the periodogram ordinates $(I_n(\lambda_1), \ldots, I_n(\lambda_m))$ are approximately distributed as independent exponential random variables with means $2\pi f(\lambda_1), \ldots, 2\pi f(\lambda_m)$, respectively. In particular, for each fixed $\lambda \in (0, \pi)$ and $\epsilon > 0$,

$$P[|I_n(\lambda) - 2\pi f(\lambda)| > \epsilon] \to p > 0, \text{ as } n \to \infty,$$

so the probability of an estimation error larger than ϵ cannot be made arbitrarily small by choosing a sufficiently large sample size n. Thus, $I_n(\lambda)$ is not a *consistent* estimator of $2\pi f(\lambda)$.

Since for large n the periodogram ordinates at fixed frequencies are approximately independent with variances changing only slightly over small frequency intervals, we might hope to construct a consistent estimator of $f(\lambda)$ by averaging the periodogram estimates in a small frequency interval containing λ, provided we can choose the interval in such a way that its width decreases to zero while at the same time the number of Fourier frequencies in the interval increases to ∞ as $n \to \infty$. This can indeed be done since the number of Fourier frequencies in any *fixed* frequency interval increases approximately linearly with n. Consider for example the estimator,

$$\tilde{f}(\lambda) = \frac{1}{2\pi} \sum_{|j| \le m} (2m+1)^{-1} I_n(g(n,\lambda) + 2\pi j/n), \tag{4.2.11}$$

where $m = \sqrt{n}$ and $g(n, \lambda)$ is the multiple of $2\pi/n$ closest to λ. The number of periodogram ordinates being averaged is approximately $2\sqrt{n}$ and the width of the frequency interval over which the average is taken is approximately $4\pi/\sqrt{n}$. It can be shown (see TSTM, Section 10.4) that this estimator is consistent for the spectral density f. The argument in fact establishes the consistency of a whole class of estimators defined as follows.

Definition 4.2.2

A **discrete spectral average estimator** of the spectral density $f(\lambda)$ has the form

$$\hat{f}(\lambda) = \frac{1}{2\pi} \sum_{|j| \le m_n} W_n(j) I_n(g(n,\lambda) + 2\pi j/n), \tag{4.2.12}$$

where the **bandwidths** m_n satisfy

$$m_n \to \infty \text{ and } m_n/n \to 0 \text{ as } n \to \infty, \tag{4.2.13}$$

and the **weight functions** $W_n(\cdot)$ satisfy

$$W_n(j) = W_n(-j),\ W_n(j) \ge 0 \text{ for all } j, \tag{4.2.14}$$

$$\sum_{|j| \le m_n} W_n(j) = 1, \tag{4.2.15}$$

and

$$\sum_{|j| \le m_n} W_n^2(j) \to 0 \text{ as } n \to \infty. \tag{4.2.16}$$

Remark 3. The conditions imposed on the sequences $\{m_n\}$ and $\{W_n(\cdot)\}$ ensure consistency of $\hat{f}(\lambda)$ for $f(\lambda)$ for a very large class of stationary processes (see TSTM, Theorem 10.4.1) including all the ARMA processes considered in this book. The conditions (4.2.13) simply mean that the number of terms in the weighted average (4.2.12) goes to ∞ as $n \to \infty$ while at the same time the width of the *frequency* interval over which the average is taken goes to zero. The conditions on $\{W_n(\cdot)\}$

ensure that the mean and variance of $\hat{f}(\lambda)$ converge as $n \to \infty$ to $f(\lambda)$ and 0, respectively. Under the conditions of TSTM, Theorem 10.4.1, it can be shown in fact that

$$\lim_{n\to\infty} E\hat{f}(\lambda) = f(\lambda)$$

and

$$\lim_{n\to\infty} \left(\sum_{|j|\le m_n} W_n^2(j) \right)^{-1} \operatorname{Cov}(\hat{f}(\lambda), \hat{f}(\nu)) = \begin{cases} 2f^2(\lambda) & \text{if } \lambda = \nu = 0 \text{ or } \pi, \\ f^2(\lambda) & \text{if } 0 < \lambda = \nu < \pi, \\ 0 & \text{if } \lambda \ne \nu. \end{cases}$$

□

Example 4.2.1 For the simple moving average estimator with $m_n = \sqrt{n}$ and $W_n(j) = (2m_n + 1)^{-1}$, $|j| \le m_n$, Remark 3 gives

$$(2\sqrt{n} + 1)\operatorname{Var}(\hat{f}(\lambda)) \to \begin{cases} 2f^2(\lambda) & \text{if } \lambda = 0 \text{ or } \pi, \\ f^2(\lambda) & \text{if } 0 < \lambda < \pi. \end{cases}$$

□

In practice, when the sample size n is a fixed finite number, the choice of m and $\{W(\cdot)\}$ involves a compromise between achieving small bias and small variance for the estimator $\hat{f}(\lambda)$. A weight function that assigns roughly equal weights to a broad band of frequencies will produce an estimate of $f(\lambda)$ that, although smooth, may have a large bias, since the estimate of $f(\lambda)$ depends on the values of I_n at frequencies distant from λ. On the other hand a weight function which assigns most of its weight to a narrow frequency band centered at zero will give an estimator with relatively

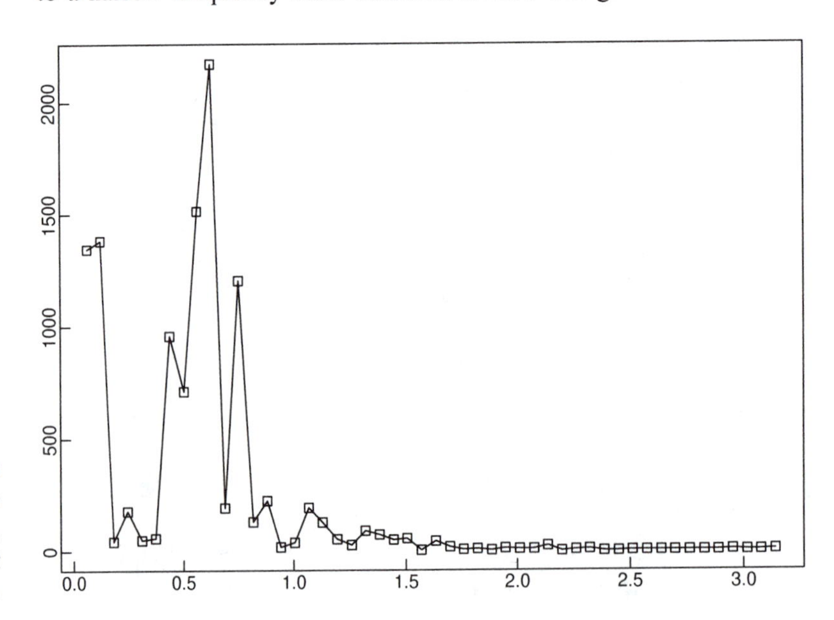

Figure 4-9 The spectral density estimate, $I_{100}(\lambda)/(2\pi)$, $0 < \lambda \le \pi$, of the sunspot numbers, 1770–1869.

small bias, but with a larger variance. In practice it is advisable to experiment with a range of weight functions and to select the one that appears to strike a satisfactory balance between bias and variance.

The option *Nonparametric Spectral Estimation* in the main menu of the program PEST allows the user to specify arbitrary values for m and for the weights $\{W(j)\}$, $j = 0, \ldots, m$ and computes the resulting discrete spectral average estimators $\hat{f}(\lambda)$, $0 \leq \lambda \leq \pi$.

Example 4.2.2 The sunspot numbers, 1770–1869

Figure 4.9 displays a plot of $(2\pi)^{-1}$ times the periodogram of the annual sunspot numbers (obtained by applying PEST to the data file SUNSPOTS.DAT). Figure 4.10 shows the result (also from PEST) of applying the discrete spectral weights, $\{\frac{1}{3}, \frac{1}{3}, \frac{1}{3}\}$ (corresponding to $m = 1$, $W(j) = 1/(2m+1)$, $|j| \leq m$). As expected, with such a small value of m, not much smoothing of the periodogram occurs. Next we use a more dispersed set of weights, $W(0) = W(1) = \frac{3}{15}$, $W(2) = \frac{2}{15}$, $W(3) = \frac{1}{15}$, producing the smoother spectral estimate shown in Figure 4.11. All three spectral density estimates show a well-defined peak at the frequency $\omega_{10} = 2\pi/10$ radians per year, in keeping with the suggestion from the graph of the data itself that the sunspot series contains an approximate cycle with period around 10 or 11 years. □

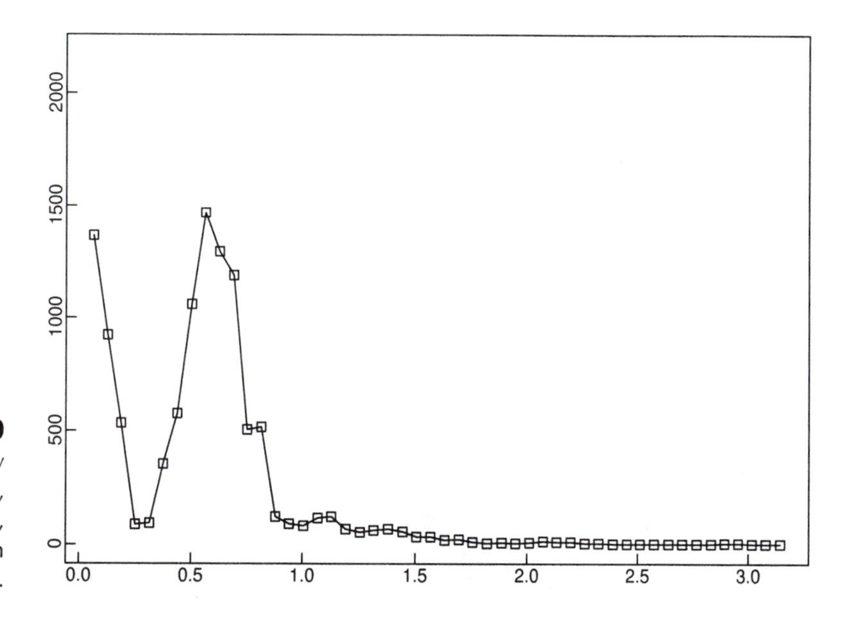

Figure 4-10 The spectral density estimate, $\hat{f}(\lambda)$, $0 < \lambda \leq \pi$, of the sunspot numbers, 1770–1869, with weights $\{\frac{1}{3}, \frac{1}{3}, \frac{1}{3}\}$.

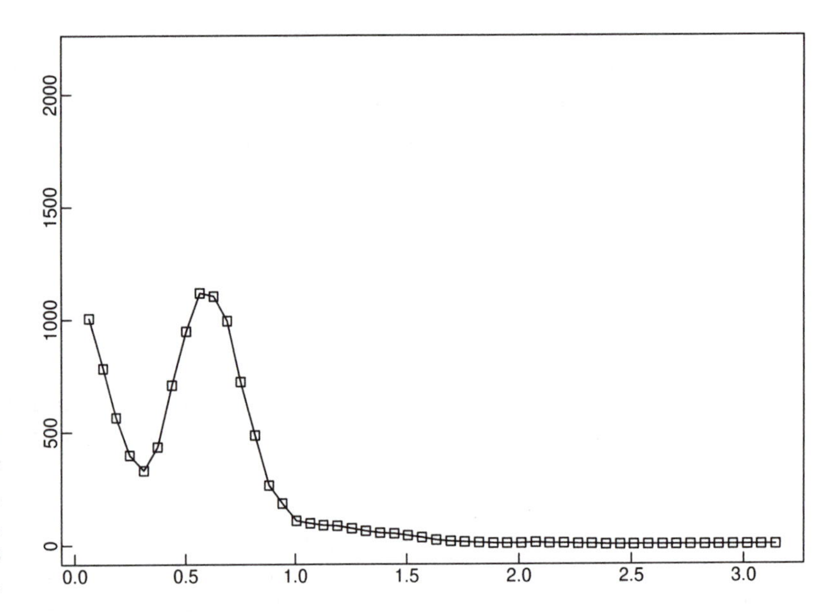

Figure 4-11 The spectral density estimate, $\hat{f}(\lambda)$, $0 < \lambda \leq \pi$, of the sunspot numbers, 1770–1869, with weights $\{\frac{1}{15}, \frac{2}{15}, \frac{3}{15}, \frac{3}{15}, \frac{3}{15}, \frac{2}{15}, \frac{1}{15}\}$.

4.3 Time-Invariant Linear Filters

In Section 1.5, we saw the utility of time-invariant linear filters for smoothing the data, estimating the trend, eliminating the seasonal and/or trend components of the data, etc. A linear process is the output of a time-invariant linear filter (TLF) applied to a white noise input series. More generally, we say that the process $\{Y_t\}$ is the output of a linear filter $C = \{c_{t,k}, t, k = 0 \pm 1, \ldots, \}$ applied to an input process $\{X_t\}$ if

$$Y_t = \sum_{k=-\infty}^{\infty} c_{t,k} X_k, \quad t = 0, \pm 1, \ldots. \tag{4.3.1}$$

The filter is said to be **time-invariant** if the weights $c_{t,t-k}$ are independent of t, i.e., if

$$c_{t,t-k} = \psi_k.$$

In this case,

$$Y_t = \sum_{k=-\infty}^{\infty} \psi_k X_{t-k}$$

and

$$Y_{t-s} = \sum_{k=-\infty}^{\infty} \psi_k X_{t-s-k}$$

so that the time-shifted process $\{Y_{t-s}, t = 0, \pm 1, \ldots, \}$ is obtained from $\{X_{t-s}, t = 0, \pm 1, \ldots, \}$ by application of the same linear filter $\Psi = \{\psi_j, j = 0, \pm 1, \ldots\}$. The TLF Ψ is said to be **causal** if

$$\psi_j = 0 \text{ for } j < 0,$$

since then Y_t is expressible in terms only of X_s, $s \le t$.

Example 4.3.1 The filter defined by

$$Y_t = aX_{-t}, \quad t = 0, \pm 1, \ldots,$$

is linear but not time-invariant since $c_{t,t-k} = 0$ except when $2t = k$. Thus, $c_{t,t-k}$ depends on the value of t. □

Example 4.3.2 The simple moving average

The filter

$$Y_t = (2q+1)^{-1} \sum_{|j| \le q} X_{t-j}$$

is a TLF with $\psi_j = (2q+1)^{-1}$, $j = -q, \ldots, q$ and $\psi_j = 0$ otherwise. □

Spectral methods are particularly valuable in describing the behavior of time-invariant linear filters as well as in designing filters for particular purposes such as the suppression of high frequency components. The following proposition shows how the spectral density of the output of a TLF is related to the spectral density of the input—a fundamental result in the study of time-invariant linear filters.

Proposition 4.3.1 *Let $\{X_t\}$ be a stationary time series with mean zero and spectral density $f_X(\lambda)$. Suppose that $\Psi = \{\psi_j, j = 0, \pm 1, \ldots\}$ is an absolutely summable TLF (i.e., $\sum_{j=-\infty}^{\infty} |\psi_j| < \infty$). Then the time series*

$$Y_t = \sum_{j=-\infty}^{\infty} \psi_j X_{t-j}$$

is stationary with mean zero and spectral density

$$f_Y(\lambda) = |\Psi(e^{-i\lambda})|^2 f_X(\lambda) = \Psi(e^{-i\lambda})\Psi(e^{i\lambda}) f_X(\lambda)$$

where $\Psi(e^{-i\lambda}) = \sum_{j=-\infty}^{\infty} \psi_j e^{-ij\lambda}$. (The function $\Psi(e^{-i\cdot})$ is called the **transfer function** *of the filter and the squared modulus $|\Psi(e^{-i\cdot})|^2$ is referred to as the* **power transfer function** *of the filter.)*

Proof Applying Proposition 2.2.1, we see that $\{Y_t\}$ is stationary with mean 0 and ACVF

$$\gamma_Y(h) = \sum_{j,k=-\infty}^{\infty} \psi_j \psi_k \gamma_X(h + k - j). \tag{4.3.2}$$

Since $\{X_t\}$ has spectral density $f_X(\lambda)$, we have

$$\gamma_X(h + k - j) = \int_{-\pi}^{\pi} e^{i(h-j+k)\lambda} f_X(\lambda)\, d\lambda, \tag{4.3.3}$$

which, by substituting (4.3.3) into (4.3.2), gives

$$\begin{aligned}\gamma_Y(h) &= \sum_{j,k=-\infty}^{\infty} \psi_j \psi_k \int_{-\pi}^{\pi} e^{i(h-j+k)\lambda} f_X(\lambda)\, d\lambda \\ &= \int_{-\pi}^{\pi} \left(\sum_{j=-\infty}^{\infty} \psi_j e^{-ij\lambda} \right) \left(\sum_{k=-\infty}^{\infty} \psi_j e^{ik\lambda} \right) e^{ih\lambda} f_X(\lambda)\, d\lambda \\ &= \int_{-\pi}^{\pi} e^{ih\lambda} \left| \sum_{j=-\infty}^{\infty} \psi_j e^{-ij\lambda} \right|^2 f_X(\lambda)\, d\lambda.\end{aligned}$$

The last expression immediately identifies the spectral density function of $\{Y_t\}$ as

$$f_Y(\lambda) = |\Psi(e^{-i\lambda})|^2 f_X(\lambda) = \Psi(e^{-i\lambda})\Psi(e^{i\lambda}) f_X(\lambda). \quad \blacksquare$$

Remark 1. Proposition 4.3.1 allows us to analyze the net effect of applying one or more filters in succession. For example, if the input process $\{X_t\}$ with spectral density f_X is operated on sequentially by two absolutely summable TLFs Ψ_1 and Ψ_2, then the net effect is the same as that of a TLF with transfer function $\Psi_1(e^{-i\lambda})\Psi_2(e^{-i\lambda})$ and the spectral density of the output process

$$W_t = \Psi_1(B)\Psi_2(B)X_t$$

is $|\Psi_1(e^{-i\lambda})\Psi_2(e^{-i\lambda})|^2 f_X(\lambda)$. (See also Remark 2 of Section 2.2.) □

As we saw in Section 1.5, differencing at lag s is one method for removing a seasonal component with period s from a time series. The transfer function for this filter is $1 - e^{-is\lambda}$, which is zero for all frequencies that are integer multiples of $2\pi/s$ radians per unit time. Consequently, this filter has the desired effect of removing all components with period s.

The simple moving average filter in Example 4.3.2 has transfer function

$$\Psi(e^{-i\lambda}) = D_q(\lambda)$$

where $D_q(\lambda)$ is the Dirichlet kernel

$$D_q(\lambda) = (2q+1)^{-1} \sum_{|j|\le q} e^{-ij\lambda} = \begin{cases} \dfrac{\sin[(q+.5)\lambda]}{(2q+1)\sin(\lambda/2)}, & \text{if } \lambda \ne 0, \\ 1, & \text{if } \lambda = 0. \end{cases}$$

A graph of D_q is given in Figure 4.12. Notice that $|D_q(\lambda)|$ is near 1 in a neighborhood of 0 and tapers off to 0 for large frequencies. This is an example of a low-pass filter. The ideal low-pass filter would have a transfer function of the form

$$\Psi(e^{-i\lambda}) = \begin{cases} 1, & \text{if } |\lambda| \le \omega_c, \\ 0, & \text{if } |\lambda| > \omega_c, \end{cases}$$

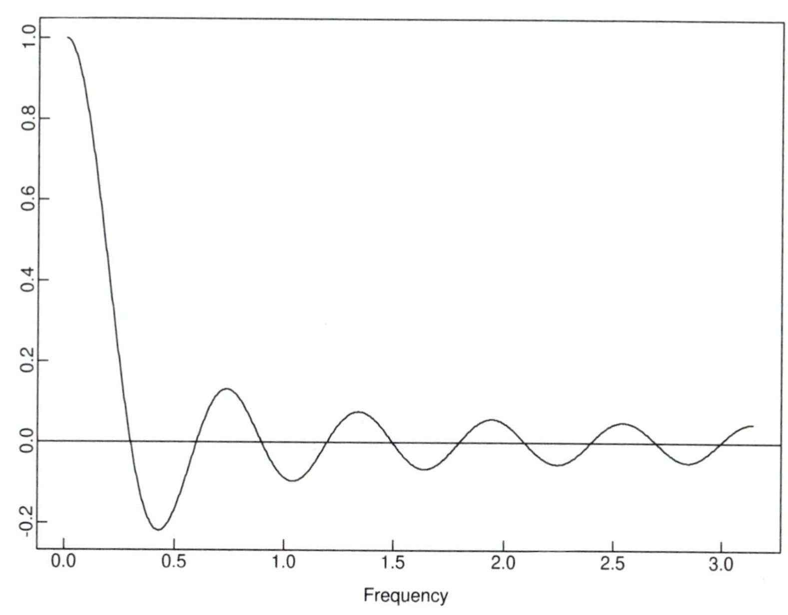

Figure 4-12 The transfer function $D_{10}(\lambda)$ for the simple moving average filter.

where ω_c is a predetermined cutoff value. To determine the corresponding linear filter, we expand $\Psi(e^{-i\lambda})$ as a Fourier series

$$\Psi(e^{-i\lambda}) = \sum_{j=-\infty}^{\infty} \psi_j e^{-ij\lambda} \tag{4.3.4}$$

with coefficients

$$\begin{aligned}\psi_j &= \frac{1}{2\pi}\int_{-\omega_c}^{\omega_c} e^{ij\lambda}\,d\lambda \\ &= \begin{cases} \dfrac{\omega_c}{\pi}, & \text{if } j = 0, \\ \dfrac{\sin(j\omega_c)}{j\pi}, & \text{if } |j| > 0. \end{cases}\end{aligned}$$

We can approximate the ideal low-pass filter by truncating the series in (4.3.4) at some large value q, which may depend on the length of the observed input series. In Figure 4.13, the transfer function of the ideal low-pass filter with $w_c = \pi/4$ is plotted with the approximations $\Psi^{(q)}(e^{-i\lambda}) = \sum_{j=-q}^{q} \psi_j e^{-ij\lambda}$ for $q = 2$ and $q = 10$. As can be seen in the figure, the approximations do not mirror Ψ very well near the cutoff value ω_c and behave like damped sinusoids for frequencies greater than ω_c. The poor approximation in the neighborhood of ω_c is typical of Fourier series approximations to functions with discontinuities—an effect known as the **Gibbs phenomenon**. Convergence factors may be employed to help mitigate the overshoot

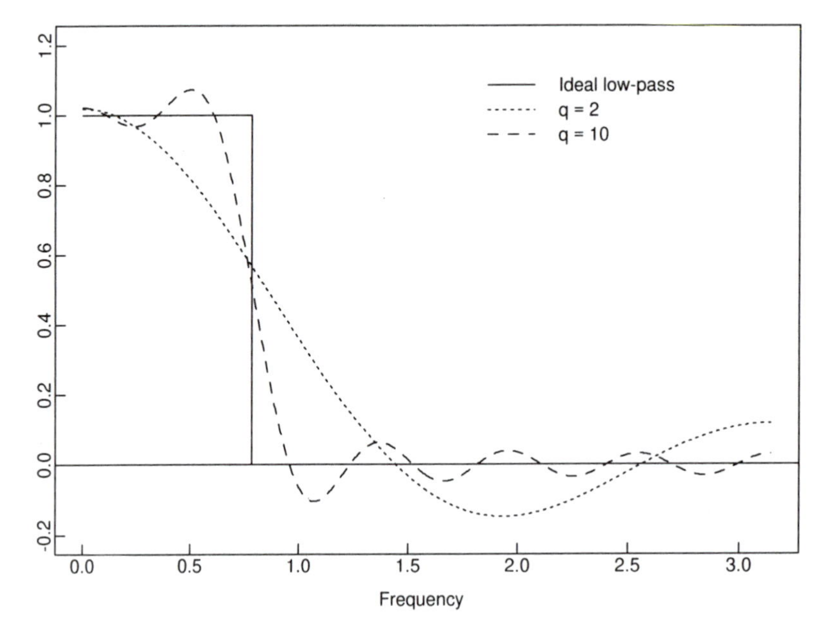

Figure 4-13 The transfer function for the ideal low-pass filter and truncated Fourier approximations $\Psi^{(q)}$ for $q = 2, 10$.

problem at ω_c and to improve the overall approximation of $\Psi^{(q)}(e^{-i\cdot})$ to $\Psi(e^{-i\cdot})$ (see Bloomfield, 1976).

4.4 The Spectral Density of an ARMA Process

In Section 4.1, the spectral density was computed for an MA(1) and for an AR(1) process. As an application of Proposition 4.3.1, we can now easily derive the spectral density of an arbitrary ARMA(p, q) process.

Spectral Density of an ARMA(*p*, *q*) Process:

If $\{X_t\}$ is a causal ARMA(p, q) process satisfying $\phi(B)X_t = \theta(B)Z_t$, then

$$f_X(\lambda) = \frac{\sigma^2}{2\pi}\frac{|\theta(e^{-i\lambda})|^2}{|\phi(e^{-i\lambda})|^2}, \qquad -\pi \le \lambda \le \pi. \tag{4.4.1}$$

Because the spectral density of an ARMA process is a ratio of trigonometric polynomials it is often called a **rational spectral density**.

Proof From (3.1.3), $\{X_t\}$ is obtained from $\{Z_t\}$ by application of the TLF with transfer function

$$\Psi(e^{-i\lambda}) = \frac{\theta(e^{-i\lambda})}{\phi(e^{-i\lambda})}.$$

Since $\{Z_t\}$ has spectral density $f_Z(\lambda) = \sigma^2/(2\pi)$, the result now follows from Proposition 4.3.1. ■

For any specified parameters $\phi_1, \ldots, \phi_p, \theta_1, \ldots, \theta_q$, and σ^2 the option *Spectral Density of Model* in the main menu of the program PEST can be used to plot the spectral density.

Example 4.4.1 The spectral density of an AR(2) process

For an AR(2) process (4.4.1) becomes

$$
\begin{aligned}
f_X(\lambda) &= \frac{\sigma^2}{2\pi(1 - \phi_1 e^{-i\lambda} - \phi_2 e^{-2i\lambda})(1 - \phi_1 e^{i\lambda} - \phi_2 e^{2i\lambda})} \\
&= \frac{\sigma^2}{2\pi(1 + \phi_1^2 + 2\phi_2 + \phi_2^2 + 2(\phi_1\phi_2 - \phi_1)\cos\lambda - 4\phi_2\cos^2\lambda)}
\end{aligned}
$$

Figure 4.14 shows the spectral density, found using the *Spectral Density of Model* option in PEST, for the model (3.2.20) fitted to the mean-corrected sunspot series. Notice the well-defined peak in the model spectral density. The frequency at which this peak occurs can be found by differentiating the denominator of the spectral density with respect to $\cos\lambda$ and setting the derivative equal to zero. This gives

$$\cos\lambda = \frac{\phi_1\phi_2 - \phi_1}{4\phi_2} = 0.849.$$

The corresponding frequency is $\lambda = 0.556$ radians per year or equivalently $c = \lambda/(2\pi) = 0.0885$ cycles per year, and the corresponding period is therefore

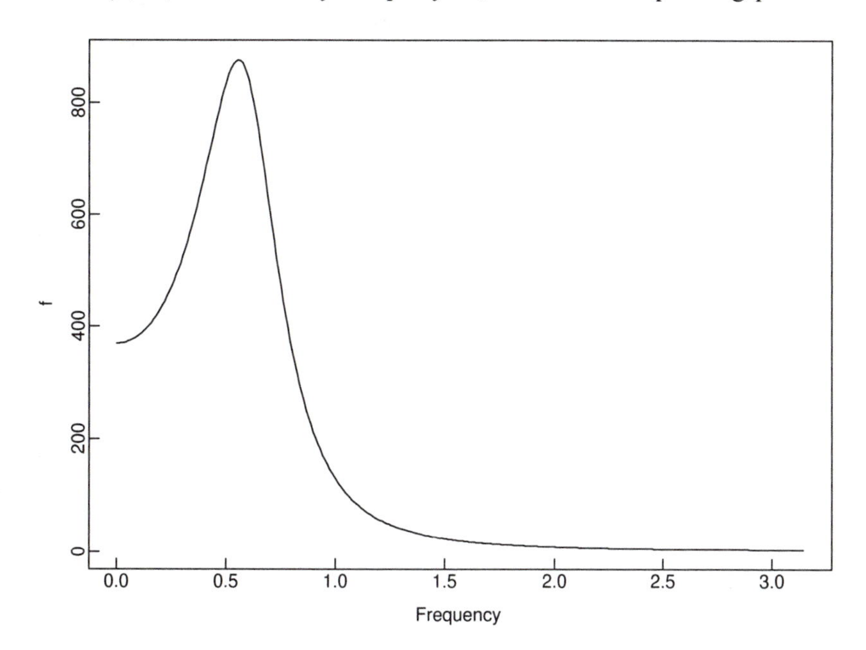

Figure 4-14 The spectral density $f_X(\lambda), 0 \leq \lambda \leq \pi$ of the AR(2) model (3.2.20) fitted to the mean-corrected sunspot series.

$1/0.0885 = 11.3$ years. The model thus reflects the approximate cyclic behavior of the data already pointed out in Example 4.2.2. The model spectral density in Figure 4.14 should be compared with the rescaled periodogram of the data and the nonparametric spectral density estimates of Figures 4.9–4.11. □

Example 4.4.2 The ARMA(1,1) process

In this case the expression (4.4.1) becomes

$$f_X(\lambda) = \frac{\sigma^2(1+\theta e^{i\lambda})(1+\theta e^{-i\lambda})}{2\pi(1-\phi e^{i\lambda})(1-\phi e^{-i\lambda})}$$

$$= \frac{\sigma^2(1+\theta^2+2\theta\cos\lambda)}{2\pi(1+\phi^2+2\phi\cos\lambda)}.$$

□

Rational Spectral Density Estimation

An alternative to the spectral density estimator of Definition 4.2.2 is the estimator obtained by fitting an ARMA model to the data and then computing the spectral density of the fitted model. The spectral density shown in Figure 4.14 can be regarded as such an estimate, obtained by fitting an AR(2) model to the mean-corrected sunspot data.

Provided there is an ARMA model that fits the data satisfactorily, this procedure has the advantage that it can be made systematic by selecting the model according (for example) to the AICC criterion (see Section 5.5.2). For further information see TSTM, Section 10.6.

Problems

4.1. Show that

$$\int_{-\pi}^{\pi} e^{i(k-h)\lambda}\, d\lambda = \begin{cases} 2\pi, & \text{if } k = h, \\ 0, & \text{otherwise.} \end{cases}$$

4.2. If $\{Z_t\} \sim \text{WN}(0, \sigma^2)$, apply Corollary 4.1.1 to compute the spectral density of $\{Z_t\}$.

4.3. Show that the vectors $\mathbf{e}_1, \ldots, \mathbf{e}_n$ are orthonormal in the sense of (4.2.3).

4.4. Use Corollary 4.1.1 to establish whether or not the following function is the autocovariance function of a stationary process $\{X_t\}$:

$$\gamma(h) = \begin{cases} 1 & \text{if } h = 0, \\ -0.5 & \text{if } h = \pm 2, \\ -0.25 & \text{if } h = \pm 3, \\ 0 & \text{otherwise.} \end{cases}$$

4.5. If $\{X_t\}$ and $\{Y_t\}$ are uncorrelated stationary processes with autocovariance functions $\gamma_X(\cdot)$ and $\gamma_Y(\cdot)$ and spectral distribution functions $F_X(\cdot)$ and $F_Y(\cdot)$, respectively, show that the process $\{Z_t = X_t + Y_t\}$ is stationary with autocovariance function $\gamma_Z = \gamma_X + \gamma_Y$ and spectral distribution function $F_Z = F_X + F_Y$.

4.6. Let $\{X_t\}$ be the process defined by

$$X_t = A\cos(\pi t/3) + B\sin(\pi t/3) + Y_t,$$

where $Y_t = Z_t + 2.5Z_{t-1}$, $\{Z_t\} \sim \text{WN}(0, \sigma^2)$, A and B are uncorrelated with mean 0 and variance ν^2, and Z_t is uncorrelated with A and B for each t. Find the autocovariance function and spectral distribution function of $\{X_t\}$.

4.7. Let $\{X_t\}$ denote the sunspot series filed as SUNSPOTS.DAT and let $\{Y_t\}$ denote the mean-corrected series $Y_t = X_t - 46.93$, $t = 1, \ldots, 100$. Use the program PEST to find the Yule-Walker AR(2) model

$$Y_t = \phi_1 Y_{t-1} + \phi_2 Y_{t-2} + Z_t, \quad \{Z_t\} \sim \text{WN}(0, \sigma^2),$$

i.e., find ϕ_1, ϕ_2 and σ^2. Use PEST to plot the spectral density of the fitted model and find the frequency at which it achieves its maximum value. What is the corresponding period?

4.8. a. Use PEST to compute and plot the spectral density of the stationary series $\{X_t\}$ satisfying

$$X_t - 0.99X_{t-3} = Z_t, \quad \{Z_t\} \sim \text{WN}(0, 1).$$

b. Does the spectral density suggest that the sample-paths of $\{X_t\}$ will exhibit approximately oscillatory behavior? If so, then with what period?

c. Use PEST to generate a realization of $X_1, \ldots, X_{60}$ and plot the realization. Save the generated series as X.DAT. Does the graph of the realization support the conclusion of part (b)?

d. Compute the spectral density of the filtered process

$$Y_t = \frac{1}{3}(X_{t-1} + X_t + X_{t+1})$$

and compare the numerical values of the spectral densities of $\{X_t\}$ and $\{Y_t\}$ at frequency $\omega = 2\pi/3$ radians per unit time. What effect would you expect the filter to have on the oscillations of $\{X_t\}$?

e. Use the program SMOOTH to apply the filter of part (d) to the realization generated in part (c). Comment on the result.

4.9. The spectral density of a real-valued time series $\{X_t\}$ is defined on $[0, \pi]$ by

$$f(\lambda) = \begin{cases} 100, & \text{if } \pi/6 - .01 < \lambda < \pi/6 + .01, \\ 0, & \text{otherwise,} \end{cases}$$

and on $[-\pi, 0]$ by $f(\lambda) = f(-\lambda)$.

a. Evaluate the ACVF of $\{X_t\}$ at lags 0 and 1.

b. Find the spectral density of the process $\{Y_t\}$ defined by

$$Y_t := \nabla_{12} X_t = X_t - X_{t-12}.$$

c. What is the variance of Y_t?

d. Sketch the power transfer function of the filter ∇_{12} and use the sketch to explain the effect of the filter on sinusoids with frequencies (i) near zero and (ii) near $\pi/6$.

4.10. Suppose that $\{X_t\}$ is the noncausal and noninvertible ARMA(1,1) process satisfying

$$X_t - \phi X_{t-1} = Z_t + \theta Z_{t-1}, \quad \{Z_t\} \sim \mathrm{WN}(0, \sigma^2),$$

where $|\phi| > 1$ and $|\theta| > 1$. Define $\tilde{\phi}(B) = 1 - \frac{1}{\phi} B$ and $\tilde{\theta}(B) = 1 + \frac{1}{\theta} B$ and let $\{W_t\}$ be the process given by

$$W_t := \tilde{\theta}^{-1}(B) \tilde{\phi}(B) X_t.$$

a. Show that $\{W_t\}$ has a constant spectral density function.

b. Conclude that $\{W_t\} \sim \mathrm{WN}(0, \sigma_w^2)$. Give an explicit formula for σ_w^2 in terms of ϕ, θ, and σ^2.

c. Deduce that $\tilde{\phi}(B) X_t = \tilde{\theta}(B) W_t$, so that $\{X_t\}$ is a causal and invertible ARMA(1,1) process relative to the white noise sequence $\{W_t\}$.

5 Modelling and Forecasting with ARMA Processes

The determination of an appropriate ARMA(p, q) model to represent an observed stationary time series involves a number of interrelated problems. These include the choice of p and q (order selection) and estimation of the mean, the coefficients $\{\phi_i, i = 1, \ldots, p\}$, $\{\theta_i, i = 1, \ldots, q\}$, and the white noise variance σ^2. Final selection of the model depends on a variety of goodness of fit tests, although it can be systematized to a large degree by use of criteria such as the AICC statistic discussed in Section 5.5.

This chapter is primarily devoted to the problem of estimating the parameters $\boldsymbol{\phi} = (\phi_1, \ldots, \phi_p)'$, $\boldsymbol{\theta} = (\theta_1, \ldots, \theta_q)'$, and σ^2 when p and q are assumed to be known, although reference will also be made to the closely related problem of order selection, i.e., the selection of appropriate values of p and q. It will be assumed throughout (unless the mean is believed a priori to be zero) that the data have been "mean-corrected" by subtraction of the sample mean, so that it is appropriate to fit a zero-mean ARMA model to the adjusted data, $x_1, \ldots, x_n$. If the model fitted to the mean-corrected data is

$$\phi(B)X_t = \theta(B)Z_t, \quad \{Z_t\} \sim \text{WN}(0, \sigma^2),$$

then the corresponding model for the original stationary series $\{Y_t\}$ is found on replacing X_t for each t by $Y_t - \bar{y}$, where $\bar{y} = n^{-1} \sum_{j=1}^{n} y_j$ is the sample mean of the original data, treated as a fixed constant.

When p and q are known, good estimators of ϕ and θ can be found by imagining the data to be observations of a stationary Gaussian time series and maximizing the likelihood with respect to the $p + q + 1$ parameters, $\phi_1, \dots, \phi_p, \theta_1, \dots, \theta_q$ and σ^2. The estimators obtained by this procedure are known as maximum likelihood (or maximum Gaussian likelihood) estimators. Maximum likelihood estimation is discussed in Section 5.2 and can be carried out in practice using the option *ARMA Estimation* in the program PEST.

The maximization of the likelihood described in the preceding paragraph is non-linear in the sense that the function to be maximized is not a quadratic function of the unknown parameters. Maximization is therefore carried out by searching numerically for the maximum. The algorithm used in PEST requires initial parameter values with which to begin the search. The closer these are to the maximum likelihood estimates, the faster the search will generally be.

To provide these initial values a number of alternative preliminary estimation procedures are described in Section 5.1 and implemented in the option *Preliminary estimation* of the program PEST. For pure autoregressive models a choice between Yule-Walker and Burg estimators is provided, while for models with $q > 0$ the choice is between preliminary estimation based on either the innovations algorithm or the Hannan-Rissanen algorithm. These four algorithms are discussed in Section 5.1. Once a preliminary model has been fitted, it is used to initiate the search for the model of the same order which maximizes the Gaussian likelihood. This is done in PEST by exiting from the *Preliminary estimation* option and then selecting the option *ARMA estimation* followed by the suboption *Optimize with current settings*.

Calculation of the exact Gaussian likelihood for an ARMA model (and in fact for *any* second-order model) is greatly simplified by use of the innovations algorithm. In Section 5.2 we take advantage of this simplification in discussing maximum likelihood estimation and consider also the construction of confidence intervals for the estimated coefficients.

Section 5.3 deals with goodness of fit tests for the chosen model and Section 5.4 with the use of the fitted model for forecasting. In Section 5.5 we discuss the theoretical basis for some of the criteria used for order selection.

For an overview of the general strategy for model-fitting see Section 6.2, p. 188.

5.1 Preliminary Estimation

In this section we shall consider four techniques for preliminary estimation of the parameters $\phi = (\phi_1, \dots, \phi_p)'$, $\theta = (\theta_1, \dots, \phi_p)'$, and σ^2 from observations $x_1, \dots, x_n$ of the causal ARMA(p, q) process defined by

$$\phi(B)X_t = \theta(B)Z_t, \quad \{Z_t\} \sim \mathrm{WN}(0, \sigma^2). \tag{5.1.1}$$

The Yule-Walker and Burg procedures apply to the fitting of pure autoregressive models. (Although the former can be adapted to models with $q > 0$, its performance is less efficient than when $q = 0$.) The innovation and Hannan-Rissanen algorithms are used in PEST to provide preliminary estimates of the ARMA parameters when $q > 0$.

For pure autoregressive models Burg's algorithm usually gives higher likelihoods than the Yule-Walker equations. For pure moving average models the innovations algorithm frequently gives slightly higher likelihoods than the Hannan-Rissanen algorithm (we use only the first two steps of the latter for preliminary estimation). For mixed models (i.e., those with $p > 0$ and $q > 0$) the Hannan-Rissanen algorithm is usually more successful in finding causal models (which are required for initialization of the likelihood maximization).

5.1.1 Yule-Walker Estimation

For a pure autoregressive model the moving average polynomial $\theta(z)$ is identically 1 and the causality assumption in (5.1.1) allows us to write X_t in the form

$$X_t = \sum_{j=0}^{\infty} \psi_j Z_{t-j}, \tag{5.1.2}$$

where, from Section 3.1, $\psi(z) = \sum_{j=0}^{\infty} \psi_j z^j = 1/\phi(z)$. Multiplying each side of (5.1.1) by X_{t-j}, $j = 0, 1, 2, \ldots, p$, taking expectations, and using (5.1.2) to evaluate the right-hand side of the first equation, we obtain the Yule-Walker equations

$$\Gamma_p \boldsymbol{\phi} = \boldsymbol{\gamma}_p \tag{5.1.3}$$

and

$$\sigma^2 = \gamma(0) - \boldsymbol{\phi}' \boldsymbol{\gamma}_p, \tag{5.1.4}$$

where Γ_p is the covariance matrix $[\gamma(i-j)]_{i,j=1}^{p}$ and $\boldsymbol{\gamma}_p = (\gamma(1), \ldots, \gamma(p))'$. These equations can be used to determine $\gamma(0), \ldots, \gamma(p)$ from σ^2 and $\boldsymbol{\phi}$.

On the other hand, if we replace the covariances $\gamma(j)$, $j = 0, \ldots, p$, appearing in (5.1.3) and (5.1.4) by the corresponding sample covariances $\hat{\gamma}(j)$, we obtain a set of equations for the so-called Yule-Walker estimators $\hat{\boldsymbol{\phi}}$ and $\hat{\sigma}^2$ of $\boldsymbol{\phi}$ and σ^2, namely

$$\hat{\Gamma}_p \hat{\boldsymbol{\phi}} = \hat{\boldsymbol{\gamma}}_p \tag{5.1.5}$$

and

$$\hat{\sigma}^2 = \hat{\gamma}(0) - \hat{\boldsymbol{\phi}}' \hat{\boldsymbol{\gamma}}_p, \tag{5.1.6}$$

where $\hat{\Gamma}_p = [\hat{\gamma}(i-j)]_{i,j=1}^{p}$ and $\hat{\boldsymbol{\gamma}}_p = (\hat{\gamma}(1), \ldots, \hat{\gamma}(p))'$.

If $\hat{\gamma}(0) > 0$ then $\hat{\Gamma}_m$ is nonsingular for every $m = 1, 2, \ldots$, (see TSTM, Problem 7.11) so we can rewrite equations (5.1.5) and (5.1.6) in the form

Sample Yule-Walker Equations:

$$\hat{\phi} = (\hat{\phi}_1, \ldots, \hat{\phi}_p)' = \hat{R}_p^{-1}\hat{\rho}_p \tag{5.1.7}$$

and

$$\hat{\sigma}^2 = \hat{\gamma}(0)[1 - \hat{\rho}_p'\hat{R}_p^{-1}\hat{\rho}_p], \tag{5.1.8}$$

where $\hat{\rho}_p = (\hat{\rho}(1), \ldots, \hat{\rho}(p))' = \hat{\gamma}_p/\hat{\gamma}(0)$.

With $\hat{\phi}$ as defined by (5.1.7), it can be shown that $1 - \hat{\phi}_1 z - \cdots - \hat{\phi}_p z^p \neq 0$ for $|z| \leq 1$ (see TSTM, Problem 8.3). Hence the fitted model,

$$X_t - \hat{\phi}_1 X_{t-1} - \cdots - \hat{\phi}_p X_{t-p} = Z_t, \quad \{Z_t\} \sim \mathrm{WN}(0, \hat{\sigma}^2),$$

is causal. The autocovariances $\gamma_F(h)$, $h = 0, \ldots, p$, of the fitted model therefore satisfy the $p + 1$ linear equations

$$\gamma_F(h) - \hat{\phi}_1\gamma_F(h-1) - \cdots - \hat{\phi}_p\gamma_F(h-p) = \begin{cases} 0, & h = 1, \ldots, p, \\ \hat{\sigma}^2, & h = 0. \end{cases}$$

However, from (5.1.5) and (5.1.6) we see that the solution of these equations is $\gamma_F(h) = \hat{\gamma}(h)$, $h = 0, \ldots, p$, so that the autocovariances of the fitted model at lags $0, 1, \ldots, p$ coincide with the corresponding sample autocovariances.

The argument of the preceding paragraph shows that for *every* nonsingular covariance matrix of the form $\Gamma_{p+1} = [\gamma(i-j)]_{i,j=1}^{p+1}$, there is an AR($p$) process whose autocovariances at lags $0, \ldots, p$ are $\gamma(0), \ldots, \gamma(p)$. (The required coefficients and white noise variance are found from (5.1.7) and (5.1.8) on replacing $\hat{\rho}(j)$ by $\gamma(j)/\gamma(0)$, $j = 0, \ldots, p$, and $\hat{\gamma}(0)$ by $\gamma(0)$.) There may not, however, be an MA(p) process with this property. For example, if $\gamma(0) = 1$ and $\gamma(1) = \gamma(-1) = \beta$, the matrix Γ_2 is a nonsingular covariance matrix for all $\beta \in (-1, 1)$. Consequently, there is an AR(1) process with autocovariances 1 and β at lags 0 and 1 for all $\beta \in (-1, 1)$. However, there is an MA(1) process with autocovariances 1 and β at lags 0 and 1 if and only if $|\beta| \leq 1/2$. (See Example 2.1.1.)

It is often the case that moment estimators, i.e., estimators that (like $\hat{\phi}$) are obtained by equating theoretical and sample moments, have much higher variances than estimators obtained by alternative methods such as maximum likelihood. However, the Yule-Walker estimators of the coefficients $\phi_1, \ldots, \phi_p$ of an AR(p) process have approximately the same distribution for large samples as the corresponding maximum likelihood estimators. For a precise statement of this result see TSTM, Section 8.10. For our purposes it suffices to note the following:

Large Sample Distribution of Yule-Walker Estimators:

For a large sample from an AR(p) process,

$$\hat{\phi} \approx N(\phi, n^{-1}\sigma^2\Gamma_p^{-1}).$$

If we replace σ^2 and Γ_p by their estimates $\hat{\sigma}^2$ and $\hat{\Gamma}_p$, we can use this result to find large-sample confidence regions for ϕ and each of its components as in (5.1.12) and (5.1.13) below.

Order Selection

In practice we do not know the true order of the model generating the data. In fact it will usually be the case that there is *no* true AR model, in which case our goal is simply to find one that represents the data optimally in some sense. Two useful techniques for selecting an appropriate AR model are given below. The second is more systematic and extends beyond the narrow class of pure autoregressive models.

- Some guidance in the choice of order is provided by a large-sample result (see TSTM, Section 8.10), which states that if $\{X_t\}$ is the causal AR(p) process defined by (5.1.1) with $\{Z_t\} \sim \text{IID}(0, \sigma^2)$ and if we fit a model with order $m > p$ using the Yule-Walker equations, i.e., if we fit a model with coefficient vector

  $$\hat{\phi}_m = \hat{R}_m^{-1}\hat{\rho}_m, \quad m > p,$$

  then the *last component*, $\hat{\phi}_{mm}$, of the vector ϕ_m is approximately normally distributed with mean 0 and variance $1/n$. Notice that $\hat{\phi}_{mm}$ is exactly the sample partial autocorrelation at lag m as defined in Section 3.2.3.

  Now we already know from Example 3.2.6 that for an AR(p) process the partial autocorrelations ϕ_{mm}, $m > p$ are zero. By the result of the previous paragraph, if an AR(p) model is appropriate for the data, then the values $\hat{\phi}_{kk}$, $k > p$, should be compatible with observations from the distribution N$(0, 1/n)$. In particular for $k > p$, $\hat{\phi}_{kk}$ will fall between the bounds $\pm 1.96n^{-1/2}$ with probability close to 0.95. This suggests using as a preliminary estimator of p the smallest value m such that $|\hat{\phi}_{kk}| < 1.96n^{-1/2}$ for $k > m$.

  The program PEST plots the sample PACF $\{\hat{\phi}_{mm}, m = 1, 2, \ldots\}$ together with the bounds $\pm 1.96/\sqrt{n}$. From this graph it is easy to read off the preliminary estimator of p defined above.

- A more systematic approach to order selection is to find the values of p and ϕ_p that minimize the AICC statistic (see Section 5.5.2 below),

  $$\text{AICC} = -2\ln L(\phi_p, S(\phi_p)/n) + 2(p+1)n/(n-p-2),$$

where L is the Gaussian likelihood defined in (5.2.9) and S is defined in (5.2.11). The *Preliminary estimation* option of PEST allows you to search for the minimum AICC Yule-Walker (or Burg) models by specifying the order of autoregression to be fitted as -1. This causes the program to fit autoregressions of orders $0, 1, \ldots, 26$ and to return the model with smallest AICC value.

Definition 5.1.1. The **fitted Yule-Walker AR(m) model** is

$$X_t - \hat{\phi}_{m1}X_{t-1} - \cdots - \hat{\phi}_{mm}X_{t-m} = Z_t, \quad \{Z_t\} \sim \mathrm{WN}(0, \hat{v}_m), \tag{5.1.9}$$

where

$$\hat{\boldsymbol{\phi}}_m = (\hat{\phi}_{m1}, \ldots, \hat{\phi}_{mm})' = \hat{R}_m^{-1}\hat{\boldsymbol{\rho}}_m \tag{5.1.10}$$

and

$$\hat{v}_m = \hat{\gamma}(0)[1 - \hat{\boldsymbol{\rho}}_m'\hat{R}_m^{-1}\hat{\boldsymbol{\rho}}_m]. \tag{5.1.11}$$

For both approaches to order selection, we need to fit AR models of gradually increasing order to our given data. The problem of solving the Yule-Walker equations with gradually increasing orders has already been encountered in a slightly different context in Section 2.5.1, where we derived a recursive scheme for solving the equations (5.1.3) and (5.1.4) with p successively taking the values $1, 2, \ldots$. Here we can use exactly the same scheme (the Durbin-Levinson algorithm) to solve the Yule-Walker equations (5.1.5) and (5.1.6), the only difference being that the covariances in (5.1.3) and (5.1.4) are replaced by their sample counterparts. This is the algorithm used by PEST to perform the necessary calculations.

Confidence Regions for the Coefficients

Under the assumption that the order p of the fitted model is the correct value, we can use the asymptotic distribution of $\hat{\boldsymbol{\phi}}_p$ to derive approximate large-sample confidence regions for the true coefficient vector $\boldsymbol{\phi}_p$ and for its individual components, ϕ_{pj}. Thus, if $\chi^2_{1-\alpha}(p)$ denotes the $(1-\alpha)$ quantile of the chi-squared distribution with p degrees of freedom, then for large sample-size n the region

$$\{\boldsymbol{\phi} \in \mathbf{R}^p : (\hat{\boldsymbol{\phi}}_p - \boldsymbol{\phi})'\hat{\Gamma}_p(\hat{\boldsymbol{\phi}}_p - \boldsymbol{\phi}) \le n^{-1}\hat{v}_p\chi^2_{1-\alpha}(p)\} \tag{5.1.12}$$

contains $\boldsymbol{\phi}_p$ with probability close to $(1-\alpha)$. (This follows from Problem A.7 and the fact that $\sqrt{n}(\hat{\boldsymbol{\phi}}_p - \boldsymbol{\phi}_p)$ is approximately normally distributed with mean $\mathbf{0}$ and covariance matrix $\hat{v}_p\hat{\Gamma}_p^{-1}$.) Similarly if $\Phi_{1-\alpha}$ denotes the $(1-\alpha)$ quantile of the standard normal distribution and $\hat{v}_{jj}$ is the jth diagonal element of $\hat{v}_p\hat{\Gamma}_p^{-1}$, then for large n the interval bounded by

$$\hat{\phi}_{pj} \pm \Phi_{1-\alpha/2}n^{-1/2}\hat{v}_{jj}^{1/2} \tag{5.1.13}$$

contains ϕ_{pj} with probability close to $(1-\alpha)$.

Example 5.1.1 The Dow-Jones Utilities Index, Aug. 28–Dec. 18, 1972; DOWJ.DAT

The very slowly decaying positive sample ACF of this time series suggests differencing at lag 1 before attempting to fit a stationary model. One application of the operator $(1 - B)$ produces a new series $\{Y_t\}$ with no obvious deviations from stationarity. We shall therefore try fitting an AR process to this new series,

$$Y_t = D_t - D_{t-1},$$

using the Yule-Walker equations. There are 77 values of Y_t which we shall denote as $Y_1, \ldots, Y_{77}$. (We ignore the unequal spacing of the original data resulting from the five-day working week.) The sample autocovariances of the series $y_1, \ldots, y_{77}$ are $\hat{\gamma}(0) = 0.17992$, $\hat{\gamma}(1) = 0.07590$, $\hat{\gamma}(2) = 0.04885$, etc.

Applying the Durbin-Levinson algorithm to fit successively higher order autoregressive processes to the data, we obtain

$$\hat{\phi}_{11} = \hat{\rho}(1) = 0.4219$$

$$\hat{v}_1 = \hat{\gamma}(0)[1 - \hat{\rho}^2(1)] = 0.1479,$$

$$\hat{\phi}_{22} = [\hat{\gamma}(2) - \hat{\phi}_{11}\hat{\gamma}(1)]/\hat{v}_1 = 0.1138,$$

$$\hat{\phi}_{21} = \hat{\phi}_{11} - \hat{\phi}_{11}\hat{\phi}_{22} = 0.3739,$$

$$\hat{v}_2 = \hat{v}_1[1 - \hat{\phi}_{22}^2] = 0.1460.$$

The sample ACF and PACF of the data can be computed using the *Data entry* option of the program PEST. These are shown in Figures 5.1 and 5.2, respectively. Also plotted are the bounds $\pm 1.96/\sqrt{77}$. Since the PACF values at lags greater than 1 all lie between the bounds, the order-selection criterion described above indicates that we should fit an AR(1) model to the data set $\{Y_t\}$. Unless we wish to assume that $\{Y_t\}$ is a zero-mean process, we should subtract the sample mean from the data before attempting to fit a (zero-mean) AR(1) model. This can be done using Option 9 of the *Data entry* menu. If you fail to do this before exiting from the *Data entry* menu, you will see a warning on the screen that allows you to subtract the sample mean before continuing. In this example we subtract the mean from the data giving the new series

$$X_t = Y_t - 0.1336.$$

We have already computed $\hat{\phi}_{11}$ and $\hat{v}_1$ above using the Durbin-Levinson algorithm. The fitted model for $\{X_t\}$ is therefore

$$X_t - 0.4219 X_{t-1} = Z_t, \quad \{Z_t\} \sim \text{WN}(0, 0.1479), \tag{5.1.14}$$

and the corresponding model for $\{Y_t\}$ is

$$Y_t - 0.1336 - 0.4219(Y_{t-1} - 0.1336) = Z_t, \quad \{Z_t\} \sim \text{WN}(0, 0.1479). \tag{5.1.15}$$

Assuming that our observed data really are a sample from an AR process with $p = 1$, (5.1.13) gives us approximate 95% confidence bounds for the autoregressive

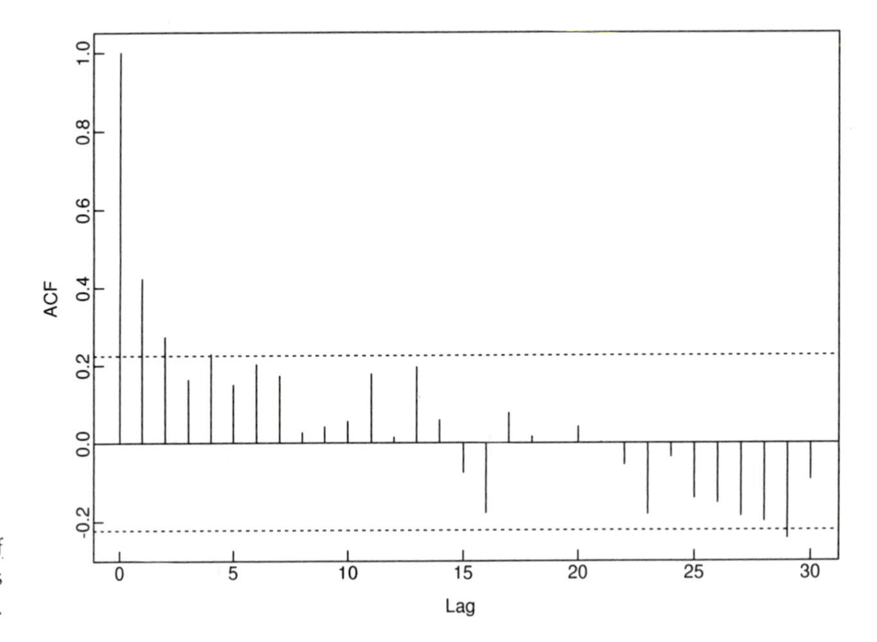

Figure 5-1 The sample ACF of the differenced series $\{Y_t\}$ in Example 5.1.1.

coefficient ϕ,

$$-0.4219 \pm \frac{(1.96)(.1479)}{(.17992)\sqrt{77}} = (0.2194, 0.6244).$$

The calculations required to solve the Yule-Walker equations and to find confidence intervals for the AR coefficients can easily be done using the option *Preliminary*

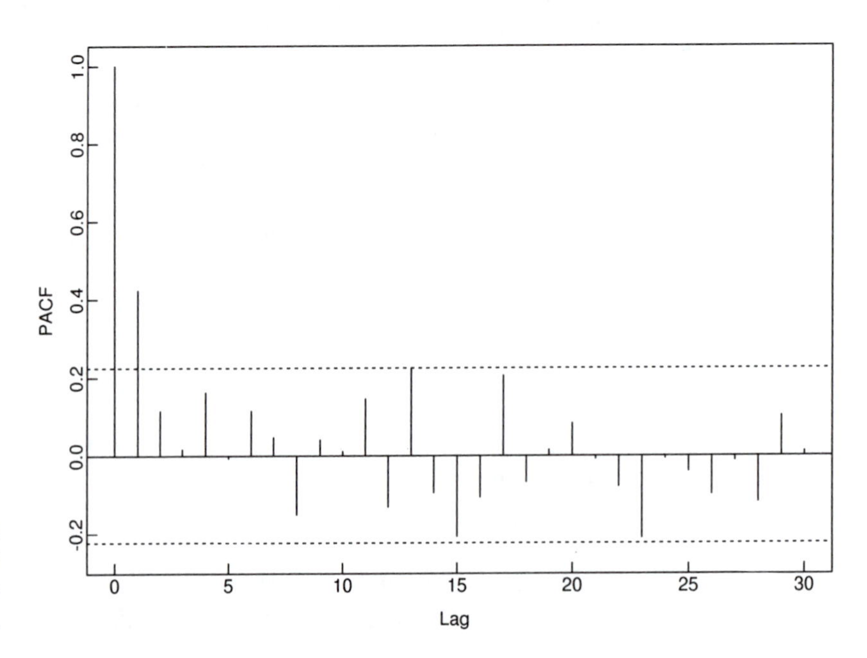

Figure 5-2 The sample PACF of the differenced series $\{Y_t\}$ in Example 5.1.1.

estimation of the program PEST. After entering the data file DOWJ.DAT, differencing once at lag 1 and subtracting the mean of the resulting series to get the observations of $\{X_t\}$, return to the main PEST menu and select the *Preliminary Estimation* option. Enter the order of the autoregression to be fitted (1 in this example) and 0 for the order of the moving average component. Then select the *Yule-Walker* option and each estimated coefficient will appear on the screen. You will also see the ratio of each coefficient to 1.96 times its estimated standard deviation. From these numbers large-sample 95% confidence intervals for each of the coefficients are easily obtained. For this particular example there is just one coefficient estimate $\hat{\phi}_1 = 0.4219$ with ratio of coefficient to 1.96×standard error equal to 2.0832. Hence the required 95% confidence bounds are $0.4219 \pm 0.4219/2.0832 = (0.2194, 0.6244)$ as found above.

A much simpler way to arrive at the AR(1) model for $\{X_t\}$ is to use the automatic AICC minimization option for autoregressions in the *Preliminary estimation* option of PEST. Instead of choosing order 1 as in the previous paragraph, choose -1 followed by *Yule-Walker* and the program will search through all the Yule-Walker AR(p) models, $p = 0, 1, \ldots, 26$, selecting the one with smallest AICC value. You will find that this is the same model obtained in the preceding paragraph (with $p = 1$ and $\hat{\phi} = 0.4219$). □

Yule-Walker Estimation with $q > 0$; Moment Estimators

The Yule-Walker estimates for the parameters in an AR(p) model are examples of moment estimators—the autocovariances at lags $0, 1, \ldots, p$ are replaced by the corresponding sample estimates in the Yule-Walker equations (5.1.3) which are then solved for the parameters $\phi = (\phi_1, \ldots, \phi_p)'$ and σ^2. The analogous procedure for ARMA(p, q) models with $q > 0$ is easily formulated, but the corresponding equations are nonlinear in the unknown coefficients, leading to possible nonexistence and nonuniqueness of solutions for the required estimators.

From (3.2.5), the equations to be solved for $\phi_1, \ldots, \phi_p, \theta_1, \ldots, \theta_q$ and σ^2 are

$$\hat{\gamma}(k) - \phi_1\hat{\gamma}(k-1) - \cdots - \phi_p\hat{\gamma}(k-p) = \sigma^2 \sum_{j=k}^{q} \theta_j \psi_{j-k}, \tag{5.1.16}$$

for $0 \le k \le p + q$, where ψ_j must first be expressed in terms of ϕ and $\boldsymbol{\theta}$ using the identity $\psi(z) = \theta(z)/\phi(z)$ ($\theta_0 := 1$ and $\theta_j = \psi_j = 0$ for $j < 0$).

Example 5.1.2 For the MA(1) model, the equations (5.1.16) are equivalent to

$$\hat{\gamma}(0) = \hat{\sigma}^2(1 + \hat{\theta}_1^2) \tag{5.1.17}$$

$$\hat{\rho}(1) = \frac{\hat{\theta}_1}{1 + \hat{\theta}_1^2}. \tag{5.1.18}$$

If $|\hat{\rho}(1)| > .5$, there is no real solution so we define $\hat{\theta}_1 = \hat{\rho}(1)/|\hat{\rho}(1)|$. If $|\hat{\rho}(1)| \le .5$, then the solution of (5.1.17)–(5.1.18) (with $|\hat{\theta}| \le 1$) is

$$\hat{\theta}_1 = (1 - (1 - 4\hat{\rho}^2(1))^{1/2})/(2\hat{\rho}(1))$$

$$\hat{\sigma}^2 = \hat{\gamma}(0)/(1 + \hat{\theta}_1^2).$$

For the overshort data of Example 3.2.8, $\hat{\rho}(1) = -0.5035$ and $\hat{\gamma}(0) = 3416$ so the fitted MA(1) model has parameters $\hat{\theta}_1 = -1.0$ and $\hat{\sigma}^2 = 1708$. □

Relative Efficiency of Estimators

The performance of two competing estimators is often measured by computing their asymptotic relative efficiency. In a general statistics estimation problem, suppose $\hat{\theta}_n^{(1)}$ and $\hat{\theta}_n^{(2)}$ are two estimates of the parameter θ in the parameter space Θ based on the observations $X_1, \ldots, X_n$. If $\hat{\theta}_n^{(i)}$ is approximately $\mathrm{N}(\theta, \sigma_i^2(\theta))$ for large n, $i = 1, 2$, then the **asymptotic efficiency** of $\hat{\theta}_n^{(1)}$ relative to $\hat{\theta}_n^{(2)}$ is defined to be

$$e(\theta, \hat{\theta}^{(1)}, \hat{\theta}^{(2)}) = \frac{\sigma_2^2(\theta)}{\sigma_1^2(\theta)}.$$

If $e(\theta, \hat{\theta}^{(1)}, \hat{\theta}^{(2)}) \le 1$ for all $\theta \in \Theta$, then we say that $\hat{\theta}_n^{(2)}$ is a more efficient estimator of θ than $\hat{\theta}_n^{(1)}$ (strictly more efficient if in addition $e(\theta, \hat{\theta}^{(1)}, \hat{\theta}^{(2)}) < 1$ for some $\theta \in \Theta$). For the MA(1) process, the moment estimator $\theta_n^{(1)}$ discussed in Example 5.1.2 is approximately $\mathrm{N}(\theta_1, \sigma_1^2(\theta_1)/n)$ with

$$\sigma_1^2(\theta_1) = (1 + \theta_1^2 + 4\theta_1^4 + \theta_1^6 + \theta_1^8)/(1 - \theta_1^2)^2$$

(see TSTM, p. 254). On the other hand, the innovations estimator $\hat{\theta}_n^{(2)}$ discussed in the next section is distributed approximately as $\mathrm{N}(\theta_1, n^{-1})$. Thus, $e(\theta_1, \hat{\theta}^{(1)}, \hat{\theta}^{(2)}) = \sigma_1^{-2}(\theta_1) \le 1$ for all $|\theta_1| < 1$ with strict inequality when $\theta \ne 1$. In particular,

$$e(\theta_1, \hat{\theta}^{(1)}, \hat{\theta}^{(2)}) = \begin{cases} .82, & \theta_1 = .25, \\ .37, & \theta_1 = .50, \\ .06, & \theta_1 = .75, \end{cases}$$

demonstrating the superiority, at least in terms of asymptotic relative efficiency, of $\hat{\theta}_n^{(2)}$ over $\hat{\theta}_n^{(1)}$. On the other hand (Section 5.2) the maximum likelihood estimator $\hat{\theta}_n^{(3)}$ of θ_1 is approximately $\mathrm{N}(\theta_1, (1 - \theta_1^2)/n)$. Hence,

$$e(\theta_1, \hat{\theta}^{(2)}, \hat{\theta}^{(3)}) = \begin{cases} .94, & \theta_1 = .25, \\ .75, & \theta_1 = .50, \\ .44, & \theta_1 = .75. \end{cases}$$

While $\hat{\theta}_n^{(3)}$ is more efficient, $\hat{\theta}_n^{(2)}$ has reasonably good efficiency, except when $|\theta_1|$ is close to 1, and can serve as initial value for the nonlinear optimization procedure in computing the maximum likelihood estimator.

While the method of moments is an effective procedure for fitting autoregressive models, it does not perform as well for ARMA models with $q > 0$. From a computational point of view, it requires as much computing time as the more efficient estimators based on either the innovations algorithm or the Hannan-Rissanen procedure and is therefore rarely used except when $q = 0$.

5.1.2 Burg's Algorithm

The Yule-Walker coefficients $\hat{\phi}_{p1}, \ldots, \hat{\phi}_{pp}$ are precisely the coefficients of the best linear predictor of X_{p+1} in terms of $\{X_p, \ldots, X_1\}$ under the assumption that the ACF of $\{X_t\}$ coincides with the sample ACF at lags $1, \ldots, p$.

Burg's algorithm estimates the PACF $\{\phi_{11}, \phi_{22}, \ldots\}$ by successively minimizing sums of squares of forward and backward one-step prediction errors with respect to the coefficients ϕ_{ii}. Given observations $\{x_1, \ldots, x_n\}$ of a stationary zero-mean time series $\{X_t\}$ we define $u_i(t), t = i+1, \ldots, n, 0 \le i < n$, to be the difference between $x_{n+1+i-t}$ and the best linear estimate of $x_{n+1+i-t}$ in terms of the preceding i observations. Similarly we define $v_i(t), t = i+1, \ldots, n, 0 \le i < n$, to be the difference between x_{n+1-t} and the best linear estimate of x_{n+1-t} in terms of the subsequent i observations. Then it can be shown (see Problem 5.6) that the **forward and backward prediction errors** $\{u_i(t)\}$ and $\{v_i(t)\}$ satisfy the recursions

$$u_0(t) = v_0(t) = x_{n+1-t},$$

$$u_i(t) = u_{i-1}(t-1) - \phi_{ii} v_{i-1}(t), \tag{5.1.19}$$

and

$$v_i(t) = v_{i-1}(t) - \phi_{ii} u_{i-1}(t-1). \tag{5.1.20}$$

Burg's estimate $\phi_{11}^{(B)}$ of ϕ_{11} is found by minimizing

$$\sigma_1^2 := \frac{1}{2(n-1)} \sum_{t=2}^{n} [u_1^2(t) + v_1^2(t)],$$

with respect to ϕ_{11}. This gives corresponding numerical values for $u_1(t)$ and $v_1(t)$ and σ_1^2 that can then be substituted into (5.1.19) and (5.1.20) with $i = 2$. Then we minimize

$$\sigma_2^2 := \frac{1}{2(n-2)} \sum_{t=3}^{n} [u_2^2(t) + v_2^2(t)],$$

with respect to ϕ_{22} to obtain the Burg estimate $\phi_{22}^{(B)}$ of ϕ_{22} and corresponding values of $u_2(t)$, $v_2(t)$, and σ_2^2. This process can clearly be continued to obtain estimates $\phi_{pp}^{(B)}$ and corresponding minimum values, $\sigma_p^{(B)2}$, $p \le n-1$. Estimates of the coefficients ϕ_{pj}, $1 \le j \le p-1$, in the best linear predictor,

$$P_p X_{p+1} = \phi_{p1} X_p + \cdots + \phi_{pp} X_1,$$

are then found by substituting the estimates $\phi_{ii}^{(B)}$, $i = 1, \ldots, p$, for ϕ_{ii} in the recursions (2.5.20)–(2.5.22). The resulting estimates of ϕ_{pj}, $j = 1, \ldots, p$ are the coefficient estimates of the Burg AR(p) model for the data $\{x_1, \ldots, x_n\}$. The Burg estimate of the white noise variance is the minimum value $\sigma_p^{(B)2}$ found in the determination of $\phi_{pp}^{(B)}$. The calculation of the estimates of ϕ_{pp} and σ_p^2 described above is equivalent (Problem 5.7) to solving the recursions,

Burg's Algorithm:

$$d(1) = \sum_{t=2}^{n} (u_0^2(t-1) + v_0^2(t)),$$

$$\phi_{ii}^{(B)} = \frac{2}{d(i)} \sum_{t=i+1}^{n} v_{i-1}(t) u_{i-1}(t-1),$$

$$d(i+1) = (1 - \phi_{ii}^{(B)2}) d(i) - v_i^2(i+1) - u_i^2(n),$$

$$\sigma_i^{(B)2} = [(1 - \phi_{ii}^{(B)2}) d(i)]/[2(n-i)].$$

The large-sample distribution of the estimated coefficients for the Burg estimators of the coefficients of an AR(p) process is the same as for the Yule-Walker estimators, namely $N(\phi, n^{-1}\sigma^2\Gamma_p^{-1})$. Approximate large-sample confidence intervals for the coefficients can be found as in Section 5.1.1 by substituting estimated values for σ^2 and Γ_p.

Example 5.1.3 The Dow-Jones Utilities Index

The fitting of AR models using Burg's algorithm in the program PEST is completely analogous to the use of the Yule-Walker equations. Applying the same transformations as in Example 5.1.1 to the Dow-Jones Utilities Index and selecting *Burg* instead of *Yule-Walker* in the *Preliminary Estimation* option, we obtain the minimum AICC Burg model,

$$X_t - 0.4371 X_{t-1} = Z_t, \quad \{Z_t\} \sim \text{WN}(0, 0.1423). \tag{5.1.21}$$

This is slightly different from the Yule-Walker AR(1) model fitted in Example 5.1.1, and has a larger likelihood L, i.e., a smaller value of $-2 \ln L$ (see Section 5.2). Although the two methods give estimators with the same *large-sample* distributions, for finite sample sizes the Burg model usually has smaller estimated white noise variance and larger Gaussian likelihood. From the ratio of the estimated coefficient to (1.96× standard error) displayed by PEST, we obtain the 95% confidence bounds for ϕ: $0.4371 \pm 0.4371/2.1668 = (0.2354, 0.6388)$. □

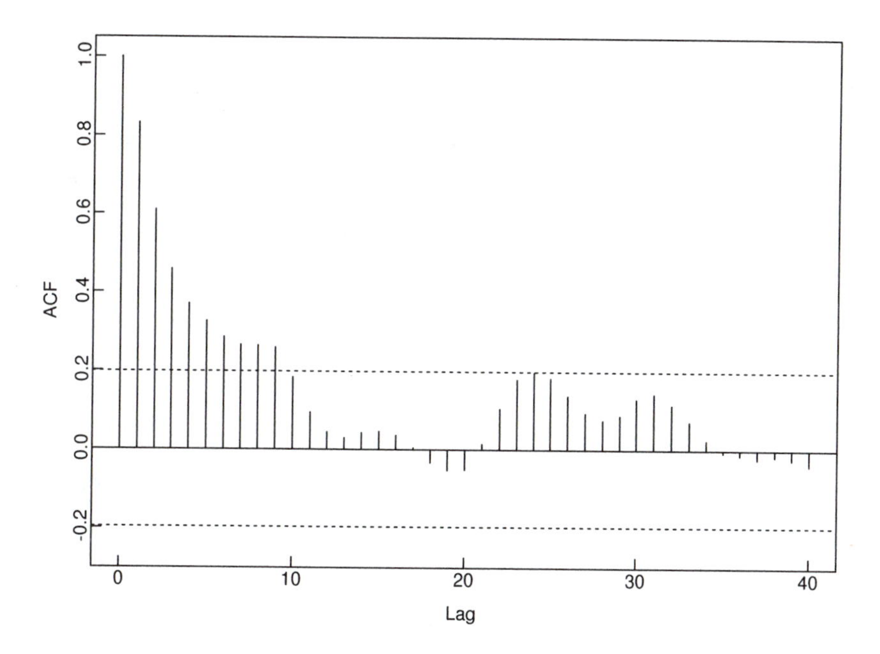

Figure 5-3 The sample ACF of the lake data in Example 5.1.4.

Example 5.1.4 The lake data

This series $\{Y_t, t = 1, \ldots, 98\}$ has already been studied in Example 1.3.5. In this example we shall consider the problem of fitting an AR process directly to the data without first removing any trend component. A graph of the data was displayed in Figure 1.9. The sample ACF and PACF are shown in Figures 5.3 and 5.4, respectively.

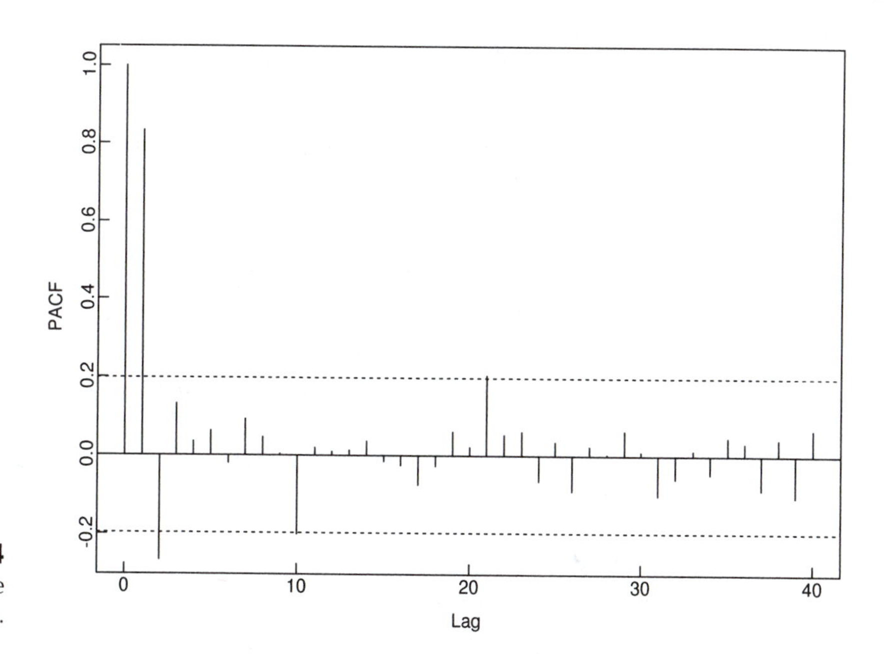

Figure 5-4 The sample PACF of the lake data in Example 5.1.4.

The sample PACF shown in Figure 5.4 strongly suggests fitting an AR(2) model to the mean-corrected data, $X_t = Y_t - 9.0041$. Before leaving the *Data entry* option of PEST, subtract the sample mean of $\{Y_t\}$ in order to fit a model to $\{X_t\}$. Then select the *Preliminary estimation* option with autoregression order 2 and moving average order 0. Selecting Burg's method for estimation we find the model

$$X_t - 1.0449X_{t-1} + 0.2456X_{t-2} = Z_t, \quad \{Z_t\} \sim \text{WN}(0, 0.4706),$$

with 95% confidence bounds

$$\phi_1 : 1.0449 \pm 1.0449/5.5295 = (0.8559, 1.2339),$$

$$\phi_2 : -0.2456 \pm 0.2456/1.2997 = (-0.4346, -0.0566).$$

Selecting the Yule-Walker method for estimation we find the model

$$X_t - 1.0538X_{t-1} + 0.2668X_{t-2} = Z_t, \quad \{Z_t\} \sim \text{WN}(0, 0.4920),$$

with 95% confidence bounds

$$\phi_1 : 1.0538 \pm 1.0538/5.5227 = (0.8630, 1.2446),$$

$$\phi_2 : -0.2668 \pm 0.2668/1.3980 = (-0.4576, -.0760).$$

We notice, as in Example 5.1.3, that the Burg model again has smaller white noise variance and larger Gaussian likelihood.

If we determine the minimum AICC Yule-Walker and Burg models (by choosing -1 as the autoregressive order) we find that they are precisely the second-order models found above. For this data set the order suggested by the sample PACF thus coincides again with the minimum AICC order for both estimation procedures. □

5.1.3 The Innovations Algorithm

Just as we can fit autoregressive models of orders 1, 2, ..., to the data $\{x_1, \ldots, x_n\}$ by applying the Durbin-Levinson algorithm to the sample autocovariances, we can also fit moving average models

$$X_t = Z_t + \hat{\theta}_{m1}Z_{t-1} + \cdots + \hat{\theta}_{mm}Z_{t-m}, \quad \{Z_t\} \sim \text{WN}(0, \hat{v}_m) \tag{5.1.22}$$

of orders $m = 1, 2, \ldots$, by means of the innovations algorithm (Section 2.5.2). The estimated coefficient vectors $\hat{\boldsymbol{\theta}}_m := (\hat{\theta}_{m1}, \ldots, \hat{\theta}_{mm})'$, and white noise variances $\hat{v}_m$, $m = 1, 2, \ldots$, are specified in the following definition. (The justification for using estimators defined in this way is contained in Remark 1 following the definition.)

Definition 5.1.2

The **fitted innovations MA(m) model** is

$$X_t = Z_t + \hat{\theta}_{m1}Z_{t-1} + \cdots + \hat{\theta}_{mm}Z_{t-m}, \quad \{Z_t\} \sim \text{WN}(0, \hat{v}_m),$$

where $\hat{\boldsymbol{\theta}}_m$ and $\hat{v}_m$ are found from the innovations algorithm with the ACVF replaced by the sample ACVF.

Remark 1. It can be shown (see Brockwell and Davis, 1988) that if $\{X_t\}$ is an invertible MA(q) process

$$X_t = Z_t + \theta_1 Z_{t-1} + \cdots + \theta_q Z_{t-q}, \quad \{Z_t\} \sim \text{IID}(0, \sigma^2),$$

with $EZ_t^4 < \infty$, and if we define $\theta_0 = 1$ and $\theta_j = 0$ for $j > q$, then the innovation estimates have the following large-sample properties. If $n \to \infty$ and $m(n)$ is any sequence of positive integers such that $m(n) \to \infty$ but $n^{-1/3}m(n) \to 0$, then for each positive integer k the joint distribution function of

$$n^{1/2}(\hat{\theta}_{m1} - \theta_1, \hat{\theta}_{m2} - \theta_2, \ldots, \hat{\theta}_{mk} - \theta_k)',$$

converges to that of the multivariate normal distribution with mean $\mathbf{0}$ and covariance matrix $A = [a_{ij}]_{i,j=1}^k$ where

$$a_{ij} = \sum_{r=1}^{\min(i,j)} \theta_{i-r}\theta_{j-r}. \tag{5.1.23}$$

This result enables us to find approximate large-sample confidence intervals for the moving average coefficients from the innovation estimates as described in the examples below. Moreover, the estimator $\hat{v}_m$ is **consistent** for σ^2 in the sense that for every $\epsilon > 0$, $P(|\hat{v}_m - \sigma^2| > \epsilon) \to 0$ as $m \to \infty$. □

Remark 2. Although the recursive fitting of moving average models using the innovations algorithm is closely analogous to the recursive fitting of autoregressive models using the Durbin-Levinson algorithm, there is one important distinction. For an AR(p) process the Yule-Walker and Burg estimators $\hat{\phi}_p$ are consistent estimators of $(\phi_1, \ldots, \phi_p)'$ as the sample size $n \to \infty$. However, for an MA(q) process the estimator $\hat{\theta}_q = (\theta_{q1}, \ldots, \theta_{qq})'$ is not consistent for $(\theta_1, \ldots, \theta_q)'$. For consistency it is necessary to use the estimators $(\theta_{m1}, \ldots, \theta_{mq})'$ with $m(n)$ satisfying the conditions of Remark 1. The choice of m for any fixed sample size can be made by increasing m until the vector $(\theta_{m1}, \ldots, \theta_{mq})'$ stabilizes. It is found in practice that there is a large range of values of m for which the fluctuations in θ_{mj} are small compared with the estimated asymptotic standard deviation $n^{-1/2}(\sum_{i=0}^{j-1} \hat{\theta}_{mi}^2)^{1/2}$ as found from (5.1.23) when the coefficients θ_j are replaced by their estimated values $\hat{\theta}_{mj}$. □

Order Selection

Three useful techniques for selecting an appropriate MA model are given below. The third is more systematic and extends beyond the narrow class of pure moving average models.

- We know from Section 3.2.2 that for an MA(q) process the autocorrelations $\rho(m)$, $m > q$ are zero. Moreover, we know from Bartlett's formula (Section 2.4) that the sample autocorrelation $\hat{\rho}(m)$, $m > q$, is approximately normally distributed with mean $\rho(m) = 0$ and variance $n^{-1}[1 + 2\rho^2(1) + \cdots + 2\rho^2(q)]$. This result enables us to use the graph of $\hat{\rho}(m)$, $m = 1, 2, \ldots$, both to decide whether or not a given data set can be plausibly modelled by a moving average process and also to obtain a preliminary estimate of the order q as the smallest value of m such that $\hat{\rho}(k)$ is not significantly different from zero for all $k > m$. For practical purposes "significantly different from zero" is often interpreted as "larger than $1.96/\sqrt{n}$ in absolute value" (cf. the corresponding approach to order selection for AR models based on the sample PACF and described in Section 5.1.1.)

- If, in addition to examining $\hat{\rho}(m)$, $m = 1, 2, \ldots$, we examine the coefficient vectors $\hat{\theta}_m$, $m = 1, 2, \ldots$, we are able not only to assess the appropriateness of a moving average model and estimate its order q, but at the same time to obtain preliminary estimates $\hat{\theta}_{m1}, \ldots, \hat{\theta}_{mq}$ of the coefficients. By inspecting the estimated coefficients $\hat{\theta}_{m1}, \ldots, \hat{\theta}_{mm}$ for $m = 1, 2, \ldots$, and the ratio of each coefficient estimate $\hat{\theta}_{mj}$ to 1.96 times its approximate standard deviation $\sigma_j = n^{-1/2}[\sum_{i=0}^{j-1} \hat{\theta}_{mi}^2]^{1/2}$, we can see which of the coefficient estimates are most significantly different from zero, estimate the order of the model to be fitted as the largest lag j for which the ratio is larger than 1 in absolute value, and at the same time read off estimated values for each of the coefficients. A default value of m is set by the program, but it may be altered manually. As m is increased the values $\hat{\theta}_{m1}, \ldots, \hat{\theta}_{mm}$ stabilize in the sense that the fluctuations in each component are of the order $n^{-1/2}$, the asymptotic standard deviation of θ_{m1}.
- As for autoregressive models, a more systematic approach to order selection for moving average models is to find the values of q and $\hat{\boldsymbol{\theta}}_q = (\hat{\theta}_{m1}, \ldots, \hat{\theta}_{mq})'$ that minimize the AICC statistic

$$\text{AICC} = -2 \ln L(\boldsymbol{\theta}_q, S(\boldsymbol{\theta}_q)/n) + 2(q+1)n/(n-q-2),$$

where L is the Gaussian likelihood defined in (5.2.9) and S is defined in (5.2.11). (See Section 5.5 for further details.)

Confidence Regions for the Coefficients

Asymptotic confidence regions for the coefficient vector $\boldsymbol{\theta}_q$ and for its individual components can be found with the aid of the large-sample distribution specified in Remark 1. For example, approximate 95% confidence bounds for θ_j are given by

$$\hat{\theta}_{mj} \pm 1.96 n^{-1/2} \Big(\sum_{i=0}^{j-1} \hat{\theta}_{mi}^2\Big)^{1/2}. \tag{5.1.24}$$

Example 5.1.5 The Dow-Jones Utilities Index

In Example 5.1.1 we fitted an AR(1) model to the differenced Dow-Jones Utilities Index. The sample ACF of the differenced data shown in Figure 5.1 suggests the possibility that an MA(2) model might also provide a good fit to the data. To apply the innovation technique for preliminary estimation, we proceed as in Example 5.1.1

to difference the series DOWJ.DAT to obtain observations of the differenced series $\{Y_t\}$ and then to subtract off the mean of the differences to obtain observations of the differenced and mean-corrected series $\{X_t\}$. We then choose the *Preliminary estimation* option and enter 0 for the order of autoregression and 2 for the order of moving average. Next we select *Innovations* as the estimation method. We must then specify a value of m, which is set by default in this case to 17. If we accept the default value the program will compute $\hat{\theta}_{17,1}, \ldots, \hat{\theta}_{17,17}$ and print out the first two values as the estimates of θ_1 and θ_2, together with the ratios of the estimated values to their estimated standard deviations. These are

```
MA COEFFICIENT
                    .4269      .2704
COEFFICIENT/(1.96*STANDARD ERROR)
                   1.9114     1.1133
```

The remaining parameter in the model is the white noise variance, for which two estimates are given.

```
WN VARIANCE ESTIMATE = (RESID SS)/N
                                .1470
INNOVATION WN VARIANCE ESTIMATE
                                .1122
```

The first of these is the average of the squares of the rescaled one-step prediction errors under the fitted MA(2) model, i.e., $\frac{1}{77}\sum_{j=1}^{77}(X_j - \hat{X}_j)^2/r_{j-1}$. The second value is the innovation estimate $\hat{v}_{17}$. When you exit from the *Preliminary estimation* option the program will retain the first value. If you wish instead to use the innovation estimate you must change the white noise variance by choosing the *Current Model and Data File Status* option in the main menu of PEST. The fitted model for $X_t (= Y_t - .1336)$ is thus

$$X_t = Z_t + 0.4269 Z_{t-1} + 0.2704 Z_{t-2}, \quad \{Z_t\} \sim \mathrm{WN}(0, 0.1470).$$

To see all 17 estimated coefficients $\hat{\theta}_{17,j}$, $j = 1, \ldots, 17$, we select the option to try another model, this time fitting an MA(17) model with $m = 17$. The coefficients and ratios for the resulting model are found to be as follows:

```
MA COEFFICIENT
   .4269    .2704   .1183    .1589   .1355   .1568    .1284   -.0060
   .0148   -.0017   .1974   -.0463   .2023   .1285   -.0213   -.2575
   .0760
COEFFICIENT/(1.96*STANDARD ERROR)
  1.9114   1.1133   .4727    .6314   .5331   .6127    .4969   -.0231
   .0568   -.0064   .7594   -.1757   .7667   .4801   -.0792   -.9563
   .2760
```

The ratios indicate that the estimated coefficients most significantly different from zero are the first and second, reinforcing our original intention of fitting an MA(2) model to the data. Estimated coefficients $\hat{\theta}_{mj}$ for other values of m can be examined in the same way and it is found that the values obtained for $m > 17$ change only slightly from the values tabulated above.

By fitting MA(q) models of orders 0, 1, 2, ..., 26 using the innovations algorithm with the default settings for m, we find that the minimum AICC model is the one with $q = 2$ found above. Thus the model suggested by the sample ACF again coincides with the more systematically chosen minimum AICC model. □

Innovations Algorithm Estimates when $p > 0$ and $q > 0$

The causality assumption (Section 3.1) ensures that

$$X_t = \sum_{j=0}^{\infty} \psi_j Z_{t-j},$$

where the coefficients ψ_j satisfy

$$\psi_j = \theta_j + \sum_{i=1}^{\min(j,p)} \phi_i \psi_{j-i}, \quad j = 0, 1, \ldots, \tag{5.1.25}$$

and we define $\theta_0 := 1$ and $\theta_j := 0$ for $j > q$. To estimate $\psi_1, \ldots, \psi_{p+q}$ we can use the innovation estimates $\hat{\theta}_{m1}, \ldots, \hat{\theta}_{m,p+q}$ whose large-sample behavior is specified in Remark 1. Replacing ψ_j by $\hat{\theta}_{mj}$ in (5.1.25) and solving the resulting equations,

$$\hat{\theta}_{mj} = \theta_j + \sum_{i=1}^{\min(j,p)} \phi_i \hat{\theta}_{m,j-i}, \quad j = 1, \ldots, p+q, \tag{5.1.26}$$

for $\boldsymbol{\phi}$ and $\boldsymbol{\theta}$, we obtain initial parameter estimates $\hat{\boldsymbol{\phi}}$ and $\hat{\boldsymbol{\theta}}$. To solve (5.1.26) we first find $\boldsymbol{\phi}$ from the last q equations,

$$\begin{bmatrix} \hat{\theta}_{m,q+1} \\ \hat{\theta}_{m,q+2} \\ \vdots \\ \hat{\theta}_{m,q+p} \end{bmatrix} = \begin{bmatrix} \hat{\theta}_{mq} & \hat{\theta}_{m,q-1} & \cdots & \hat{\theta}_{m,q+1-p} \\ \hat{\theta}_{m,q+1} & \hat{\theta}_{m,q} & \cdots & \hat{\theta}_{m,q+2-p} \\ \vdots & \vdots & & \vdots \\ \hat{\theta}_{m,q+p-1} & \hat{\theta}_{m,q+p-2} & \cdots & \hat{\theta}_{m,q} \end{bmatrix} \begin{bmatrix} \phi_1 \\ \phi_2 \\ \vdots \\ \phi_p \end{bmatrix}. \tag{5.1.27}$$

Having solved (5.1.27) for $\hat{\boldsymbol{\phi}}$ (which may not be causal), we can easily determine the estimate of $\boldsymbol{\theta}$ from

$$\hat{\theta}_j = \hat{\theta}_{mj} - \sum_{i=1}^{\min(j,p)} \hat{\phi}_i \hat{\theta}_{m,j-i}, \quad j = 1, \ldots, q.$$

Finally the white noise variance σ^2 is estimated by

$$\hat{\sigma}^2 = n^{-1} \sum_{t=1}^{n} (X_t - \hat{X}_t)^2 / r_{t-1},$$

where $\hat{X}_t$ is the one-step predictor of X_t computed from the fitted coefficient vectors $\hat{\phi}$ and $\hat{\theta}$, and r_{t-1} is defined in (3.3.8).

The above calculations are all carried out in the *Preliminary Estimation* option of the program PEST. In addition the ratio of each estimated coefficient to 1.96 times its estimated (large-sample) standard deviation is computed in the special case when $p = q$ so that approximate 95% confidence intervals can easily be obtained in this case. If the fitted model is noncausal it cannot be used to initialize the search for the maximum likelihood estimators and so the program PEST then automatically sets all the coefficients equal to .001 (giving a causal model for all values of $p \leq 26$). If this occurs it may still be the case that the Hannan-Rissanen algorithm (Section 5.1.4) will give a causal model. If not it will also automatically set all the coefficients equal to .001. If both the innovation and Hannan-Rissanen algorithms give noncausal models, it is an indication (but not a conclusive one) that the assumed values of p and q may not be appropriate for the data.

Order Selection for Mixed Models

For models with $p > 0$ and $q > 0$, the sample ACF and PACF are difficult to recognize and of far less value in order selection than in the special cases when $p = 0$ or $q = 0$. A systematic approach, however, is still available through minimization of the AICC statistic,

$$\text{AICC} = -2 \ln L(\phi_p, \theta_q, S(\phi_p, \theta_q)/n) + 2(p + q + 1)n/(n - p - q - 2),$$

which is discussed in more detail in Section 5.5. For fixed p and q it is clear from the definition that the AICC value is minimized by the parameter values that maximize the likelihood. Hence, final decisions regarding the orders p and q that minimize AICC must be based on maximum likelihood estimation as described in Section 5.2.

Example 5.1.6 The lake data

In Example 5.1.4 we fitted AR(2) models to the mean corrected lake data using the Yule-Walker equations and Burg's algorithm. If instead we fit an ARMA(1,1) model using the innovations method in the *Preliminary Estimation* option of PEST (with the default value $m = 17$), we obtain the model

$$X_t - 0.7234X_{t-1} = Z_t + 0.3596Z_{t-1}, \quad \{Z_t\} \sim \text{WN}(0, 0.4757),$$

for the mean-corrected series, $X_t = Y_t - 9.0041$. The ratio of the two coefficient estimates $\hat{\phi}$ and $\hat{\theta}$ to 1.96 times their estimated standard deviations are given by PEST as 3.2064 and 1.8513, respectively. The corresponding 95% confidence intervals are therefore

$$\phi : 0.7234 \pm 0.7234/3.2064 = (0.4978, 0.9490),$$

$$\theta : 0.3596 \pm 0.3596/1.8513 = (0.1654, 0.5538).$$

It is interesting to note that the value of AICC for this model is 212.89, which is smaller than the corresponding values for the Burg and Yule-Walker AR(2) models in Example 5.1.4. This suggests that an ARMA(1,1) model may be superior to a pure autoregressive model for these data. Preliminary estimation of a variety of ARMA(p, q) models shows that the minimum AICC value does in fact occur when $p = q = 1$. (Before committing ourselves to this model, however, we need to compare AICC values for the corresponding *maximum likelihood* models. We shall do this in Section 5.2.) □

5.1.4 The Hannan-Rissanen Algorithm

The defining equations for a causal AR(p) model have the form of a linear regression model with coefficient vector $\phi = (\phi_1, \ldots, \phi_p)'$. This suggests the use of simple least squares regression for obtaining preliminary parameter estimates when $q = 0$. Application of this technique when $q > 0$ is complicated by the fact that in the general ARMA(p, q) equations X_t is regressed not only on $X_{t-1}, \ldots, X_{t-p}$, but also on the unobserved quantities $Z_{t-1}, \ldots, Z_{t-q}$. Nevertheless, it is still possible to apply least squares regression to the estimation of ϕ and θ by first replacing the unobserved quantities $Z_{t-1}, \ldots, Z_{t-q}$ in (5.1.1) by estimated values $\hat{Z}_{t-1}, \ldots, \hat{Z}_{t-q}$. The parameters ϕ and θ are then estimated by regressing X_t onto $X_{t-1}, \ldots, X_{t-p}, \hat{Z}_{t-1}, \ldots, \hat{Z}_{t-q}$. These are the main steps in the Hannan-Rissanen estimation procedure that we now describe in more detail.

Step 1. A high order AR(m) model (with $m > \max(p, q)$) is fitted to the data using the Yule-Walker estimates of Section 5.1.1. If $(\hat{\phi}_{m1}, \ldots, \hat{\phi}_{mm})'$, is the vector of estimated coefficients, then the estimated residuals are computed from the equations

$$\hat{Z}_t = X_t - \hat{\phi}_{m1} X_{t-1} - \cdots - \hat{\phi}_{mm} X_{t-m},$$

$t = m + 1, \ldots, n$.

Step 2. Once the estimated residuals $\hat{Z}_t$, $t = m + 1, \ldots, n$ have been computed as in Step 1, the vector of parameters, $\beta = (\phi', \theta')'$ is estimated by least squares linear regression of X_t onto $(X_{t-1}, \ldots, X_{t-p}, \hat{Z}_{t-1}, \ldots, \hat{Z}_{t-q})$, $t = m + 1, \ldots, n$, i.e., by minimizing the sum of squares

$$S(\beta) = \sum_{t=m+1+q}^{n} (X_t - \phi_1 X_{t-1} - \cdots - \phi_p X_{t-p} - \theta_1 \hat{Z}_{t-1} - \cdots - \theta_q \hat{Z}_{t-q})^2$$

with respect to β. This gives the Hannan-Rissanen estimator

$$\hat{\beta} = (Z'Z)^{-1} Z' \mathbf{X}_n,$$

where $\mathbf{X}_n = (X_{m+1+q}, \ldots, X_n)'$ and Z is the $(n-m-q) \times (p+q)$ matrix

$$Z = \begin{bmatrix} X_{m+q} & X_{m+q-1} & \cdots & X_{m+q+1-p} & \hat{Z}_{m+q} & \hat{Z}_{m+q-1} & \cdots & \hat{Z}_{m+1} \\ X_{m+q+1} & X_{m+q} & \cdots & X_{m+q+2-p} & \hat{Z}_{m+q+1} & \hat{Z}_{m+q} & \cdots & \hat{Z}_{m+2} \\ \vdots & \vdots & \cdots & \vdots & \vdots & \vdots & \cdots & \vdots \\ X_{n-1} & X_{n-2} & \cdots & X_{n-p} & \hat{Z}_{n-1} & \hat{Z}_{n-2} & \cdots & \hat{Z}_{n-q} \end{bmatrix}.$$

The Hannan-Rissanen estimate of the white noise variance is

$$\hat{\sigma}^2_{HR} = \frac{S(\hat{\beta})}{n-m-q}.$$

Example 5.1.7 The lake data

In Example 5.1.6, an ARMA(1,1) model was fitted to the mean corrected lake data using the innovations algorithm. We can fit an ARMA(1,1) model to this data using the Hannan-Rissanen estimates by selecting *Hannan-Rissanen* in the *Preliminary estimation* option of PEST. The fitted model is

$$X_t - 0.6961X_{t-1} = Z_t + 0.3788Z_{t-1}, \quad \{Z_t\} \sim \mathrm{WN}(0, 0.4774),$$

for the mean-corrected series, $X_t = Y_t - 9.0041$. (Two estimates of the white noise variance are computed in PEST for the Hannan-Rissanen procedure, $\hat{\sigma}^2_{HR}$ and $\sum_{j=1}^{n}(X_t - \hat{X}_{t-1})^2/n$, with the latter being the one that is generally preferred.) The ratios of the two coefficient estimates to 1.96 times their standard deviation are 4.5289 and 1.3120, respectively. The corresponding 95% confidence bounds for ϕ and θ are

$$\phi : 0.6961 \pm 0.6961/4.5289 = (0.5424, 0.8498),$$

$$\theta : 0.3788 \pm 0.3788/1.3120 = (0.0901, 0.6675).$$

Clearly, there is little difference between this model and the one fitted using the innovations method in Example 5.1.6. In fact, the value of $-2\ln(L)$ for the two models (206.93 for the current model and 206.64 for the model fitted in Example 5.1.6) are virtually the same. □

Hannan and Rissanen include a third step in their procedure to improve the estimates.

Step 3. Using the estimate $\hat{\beta} = (\hat{\phi}_1, \ldots, \hat{\phi}_p, \hat{\theta}_1, \ldots, \hat{\theta}_q)'$ from Step 2, set

$$\tilde{Z}_t = \begin{cases} 0, & \text{if } t \le \max(p,q), \\ X_t - \sum_{j=1}^{p} \hat{\phi}_j X_{t-j} - \sum_{j=1}^{q} \hat{\theta}_j \tilde{Z}_{t-j}, & \text{if } t > \max(p,q). \end{cases}$$

Now for $t = 1, \ldots, n$, put

$$V_t = \begin{cases} 0, & \text{if } t \le \max(p, q), \\ \sum_{j=1}^{p} \hat{\phi}_j V_{t-j} + \tilde{Z}_t, & \text{if } t > \max(p, q) \end{cases}$$

and

$$W_t = \begin{cases} 0, & \text{if } t \le \max(p, q), \\ -\sum_{j=1}^{p} \hat{\theta}_j W_{t-j} + \tilde{Z}_t, & \text{if } t > \max(p, q). \end{cases}$$

(Observe that both V_t and W_t satisfy the AR recursions $\hat{\phi}(B)V_t = \tilde{Z}_t$ and $\hat{\theta}(B)W_t = \tilde{Z}_t$ for $t = 1, \ldots, n$.) If $\hat{\beta}^\dagger$ is the regression estimate of β found by regressing $\tilde{Z}_t$ on $(V_{t-1}, \ldots, V_{t-p}, W_{t-1}, \ldots, W_{t-q})$, i.e., if $\hat{\beta}^\dagger$ minimizes

$$S^\dagger(\beta) = \sum_{t=\max(p,q)+1}^{n} \left(\tilde{Z}_t - \sum_{j=1}^{p} \beta_j V_{t-j} - \sum_{k=1}^{q} \beta_{k+p} W_{t-k}\right)^2,$$

then the improved estimate of β is $\tilde{\beta} = \hat{\beta}^\dagger + \hat{\beta}$. The new estimator $\tilde{\beta}$ then has the same asymptotic efficiency as the maximum likelihood estimator. In PEST, however, we eliminate Step 3, using the model produced by Step 2 as the initial model for the calculation (by numerical maximization) of the maximum likelihood estimator itself.

5.2 Maximum Likelihood Estimation

Suppose that $\{X_t\}$ is a Gaussian time series with mean zero and autocovariance function $\kappa(i, j) = E(X_i X_j)$. Let $\mathbf{X}_n = (X_1, \ldots, X_n)'$ and let $\hat{\mathbf{X}}_n = (\hat{X}_1, \ldots, \hat{X}_n)'$ where $\hat{X}_1 = 0$ and $\hat{X}_j = E(X_j|X_1, \ldots, X_{j-1}) = P_{j-1}X_j$, $j \ge 2$. Let Γ_n denote the covariance matrix, $\Gamma_n = E(\mathbf{X}_n\mathbf{X}_n')$, and assume that Γ_n is nonsingular.

The likelihood of $\mathbf{X}_n$ is

$$L(\Gamma_n) = (2\pi)^{-n/2}(\det\Gamma_n)^{-1/2}\exp\left(-\frac{1}{2}\mathbf{X}_n'\Gamma_n^{-1}\mathbf{X}_n\right). \tag{5.2.1}$$

As we shall now show, the direct calculation of $\det\Gamma_n$ and Γ_n^{-1} can be avoided by expressing this in terms of the one-step prediction errors $X_j - \hat{X}_j$ and their variances v_{j-1}, $j = 1, \ldots, n$, both of which are easily calculated recursively from the innovations algorithm (Section 2.5.2).

Let θ_{ij}, $j = 1, \ldots, i$; $i = 1, 2, \ldots$, denote the coefficients obtained when the innovations algorithm is applied to the autocovariance function κ of $\{X_t\}$, and let C_n be the $n \times n$ lower triangular matrix defined in Section 2.5.2. From (2.5.27) we have the identity

$$\mathbf{X}_n = C_n(\mathbf{X}_n - \hat{\mathbf{X}}_n). \tag{5.2.2}$$

We also know from Remark 5 of Section 2.5.2 that the components of $\mathbf{X}_n - \hat{\mathbf{X}}_n$ are uncorrelated. Consequently, by the definition of v_j, $\mathbf{X}_n - \hat{\mathbf{X}}_n$ has the diagonal covariance matrix

$$D_n = \operatorname{diag}\{v_0, \ldots, v_{n-1}\}.$$

From (5.2.2) and (A.2.5) we conclude that

$$\Gamma_n = C_n D_n C_n'. \tag{5.2.3}$$

From (5.2.2) and (5.2.3) we see that

$$\mathbf{X}_n' \Gamma_n^{-1} \mathbf{X}_n = (\mathbf{X}_n - \hat{\mathbf{X}}_n)' D_n^{-1} (\mathbf{X}_n - \hat{\mathbf{X}}_n) = \sum_{j=1}^{n} (X_j - \hat{X}_j)^2 / v_{j-1} \tag{5.2.4}$$

and

$$\det \Gamma_n = (\det C_n)^2 (\det D_n) = v_0 v_1 \cdots v_{n-1}. \tag{5.2.5}$$

The likelihood (5.2.1) of the vector $\mathbf{X}_n$ therefore reduces to

$$L(\Gamma_n) = \frac{1}{\sqrt{(2\pi)^n v_0 \cdots v_{n-1}}} \exp\left\{ -\frac{1}{2} \sum_{j=1}^{n} (X_j - \hat{X}_j)^2 / v_{j-1} \right\}. \tag{5.2.6}$$

If Γ_n is expressible in terms of a finite number of unknown parameters $\beta_1, \ldots, \beta_r$ (as is the case when $\{X_t\}$ is an ARMA(p, q) process) the **maximum likelihood estimators** of the parameters are those values that maximize L for the given data set. When $X_1, X_2, \ldots, X_n$ are iid, it is known, under mild assumptions and for n large, that maximum likelihood estimators are approximately normally distributed with variances that are at least as small as those of other asymptotically normally distributed estimators (see e.g., Lehmann, 1983).

Even if $\{X_t\}$ is not Gaussian, it still makes sense to regard (5.2.6) as a measure of goodness of fit of the model to the data, and to choose the parameters $\beta_1, \ldots, \beta_r$ in such a way as to maximize (5.2.6). We shall always refer to the estimators, $\hat{\beta}_1, \ldots, \hat{\beta}_r$, so obtained as "maximum likelihood" estimators, even when $\{X_t\}$ is not Gaussian. Regardless of the joint distribution of $X_1, \ldots, X_n$, we shall refer to (5.2.1) and its algebraic equivalent (5.2.6) as the "likelihood" (or "Gaussian likelihood") of $X_1, \ldots, X_n$. A justification for using maximum Gaussian likelihood estimators of ARMA coefficients is that the large-sample distribution of the estimators is the same for $\{Z_t\} \sim \text{IID}(0, \sigma^2)$, regardless of whether or not $\{Z_t\}$ is Gaussian (see TSTM, Section 10.8).

The likelihood for data from an ARMA(p, q) process is easily computed from the innovations form of the likelihood (5.2.6) by evaluating the one-step predictors $\hat{X}_{i+1}$ and the corresponding mean squared errors v_i. These can be found from the recursions (Section 3.3)

$$\hat{X}_{n+1} = \begin{cases} \sum_{j=1}^{n} \theta_{nj}(X_{n+1-j} - \hat{X}_{n+1-j}), & 1 \le n < m, \\ \phi_1 X_n + \cdots + \phi_p X_{n+1-p} + \sum_{j=1}^{q} \theta_{nj}(X_{n+1-j} - \hat{X}_{n+1-j}), & n \ge m, \end{cases} \tag{5.2.7}$$

and

$$E(X_{n+1} - \hat{X}_{n+1})^2 = \sigma^2 E(W_{n+1} - \hat{W}_{n+1})^2 = \sigma^2 r_n, \tag{5.2.8}$$

where θ_{nj} and r_n are determined by the innovations algorithm with κ as in (3.3.3). Substituting in the general expression (5.2.6), we obtain the following.

The Gaussian Likelihood for an ARMA Process:

$$L(\boldsymbol{\phi}, \boldsymbol{\theta}, \sigma^2) = \frac{1}{\sqrt{(2\pi\sigma^2)^n r_0 \cdots r_{n-1}}} \exp\left\{ -\frac{1}{2\sigma^2} \sum_{j=1}^{n} \frac{(X_j - \hat{X}_j)^2}{r_{j-1}} \right\}. \tag{5.2.9}$$

Differentiating $\ln L(\boldsymbol{\phi}, \boldsymbol{\theta}, \sigma^2)$ partially with respect to σ^2 and noting that $\hat{X}_j$ and r_j are independent of σ^2, we find that the maximum likelihood estimators $\hat{\boldsymbol{\phi}}$, $\hat{\boldsymbol{\theta}}$ and $\hat{\sigma}^2$ satisfy the equations (Problem 5.8)

Maximum Likelihood Estimators:

$$\hat{\sigma}^2 = n^{-1} S(\hat{\boldsymbol{\phi}}, \hat{\boldsymbol{\theta}}), \tag{5.2.10}$$

where

$$S(\hat{\boldsymbol{\phi}}, \hat{\boldsymbol{\theta}}) = \sum_{j=1}^{n} (X_j - \hat{X}_j)^2 / r_{j-1}, \tag{5.2.11}$$

and $\hat{\boldsymbol{\phi}}$, $\hat{\boldsymbol{\theta}}$ are the values of $\boldsymbol{\phi}$, $\boldsymbol{\theta}$ that minimize

$$\ell(\boldsymbol{\phi}, \boldsymbol{\theta}) = \ln(n^{-1} S(\boldsymbol{\phi}, \boldsymbol{\theta})) + n^{-1} \sum_{j=1}^{n} \ln r_{j-1}. \tag{5.2.12}$$

Minimization of $\ell(\boldsymbol{\phi}, \boldsymbol{\theta})$ must be done numerically. Initial values for $\boldsymbol{\phi}$ and $\boldsymbol{\theta}$ can be obtained from PEST using the methods described in Section 5.1. The program then searches systematically for the values of $\boldsymbol{\phi}$ and $\boldsymbol{\theta}$ that minimize the reduced likelihood (5.2.12) and computes the corresponding maximum likelihood estimate of σ^2 from (5.2.10).

Least Squares Estimation for Mixed Models

The least squares estimates $\tilde{\phi}$ and $\tilde{\theta}$ of ϕ and θ are obtained by minimizing the function S as defined in (5.2.11) rather than ℓ as defined in (5.2.12), subject to the constraints that the model be causal and invertible. The least squares estimate of σ^2 is

$$\tilde{\sigma}^2 = \frac{S(\tilde{\phi}, \tilde{\theta})}{n - p - q}.$$

Order Selection

In Section 5.1 we introduced minimization of the AICC value as a major criterion for the selection of the orders p and q. To apply this criterion:

AICC Criterion:

Choose p, q, ϕ_p, and θ_q, to minimize

$$\text{AICC} = -2 \ln L(\phi_p, \theta_q, S(\phi_p, \theta_q)/n) + 2(p + q + 1)n/(n - p - q - 2).$$

For any fixed p and q it is clear that the estimates of the remaining parameters are those that minimize $-2 \ln L(\phi_p, \theta_q, \sigma^2)$, i.e., the maximum likelihood estimators. Final decisions with respect to order selection should therefore be made on the basis of maximum likelihood estimators (rather than the preliminary estimators of Section 5.1, which serve primarily as a guide). The AICC statistic and its justification are discussed in detail in Section 5.5.

Confidence Regions for the Coefficients

For large sample size the maximum likelihood estimator $\hat{\beta}$ of $\beta := (\phi_1, \ldots, \phi_p, \theta_1, \ldots, \theta_q)'$ is approximately normally distributed with mean β and covariance matrix which can be approximated by $2H^{-1}(\beta)$ where H is the Hessian matrix, $[\partial^2 \ell(\beta)/\partial\beta_i \partial\beta_j]_{i,j=1}^{p+q}$. PEST prints out the approximate standard deviations and correlations of the coefficient estimators based on the Hessian matrix evaluated numerically at $\hat{\beta}$ unless this matrix is not positive definite in which case PEST instead computes the theoretical asymptotic covariance matrix in Section 8.8 of TSTM. The resulting covariances can be used to compute confidence bounds for the parameters.

Large Sample Distribution of Maximum Likelihood Estimators:

For a large sample from an ARMA(p, q) process,

$$\hat{\beta} \approx N(\beta, n^{-1} V(\beta)).$$

The general form of $V(\beta)$ can be found in TSTM, Section 8.8. The following are several special cases.

Example 5.2.1 An AR(p) model

The asymptotic covariance matrix in this case is the same as that for the Yule-Walker estimates given by

$$V(\boldsymbol{\phi}) = \sigma^2 \Gamma_p^{-1}.$$

In the special cases $p = 1$ and $p = 2$, we have

$$\text{AR}(1): V(\boldsymbol{\phi}) = (1 - \phi_1^2),$$

$$\text{AR}(2): V(\boldsymbol{\phi}) = \begin{bmatrix} 1 - \phi_2^2 & -\phi_1(1+\phi_2) \\ -\phi_1(1+\phi_2) & 1 - \phi_2^2 \end{bmatrix}. \quad \square$$

Example 5.2.2 An MA(q) model

Let Γ_q^* be the covariance matrix of $Y_1, \ldots, Y_q$ where $\{Y_t\}$ is the autoregressive process with autoregressive polynomial $\theta(z)$, i.e.,

$$Y_t + \theta_1 Y_{t-1} + \cdots + \theta_q Y_{t-q} = Z_t, \quad \{Z_t\} \sim \text{WN}(0, 1).$$

Then

$$V(\boldsymbol{\theta}) = \Gamma_q^*.$$

Inspection of the results of Example 5.2.1 (replace ϕ_i with $-\theta_i$), yields

$$\text{MA}(1): V(\boldsymbol{\theta}) = (1 - \theta_1^2),$$

$$\text{MA}(2): V(\boldsymbol{\theta}) = \begin{bmatrix} 1 - \theta_2^2 & \theta_1(1-\theta_2) \\ \theta_1(1-\theta_2) & 1 - \theta_2^2 \end{bmatrix}. \quad \square$$

Example 5.2.3 An ARMA(1, 1) model

In this case,

$$V(\phi, \theta) = \frac{1+\phi\theta}{(\phi+\theta)^2} \begin{bmatrix} (1-\phi^2)(1+\phi\theta) & -(1-\theta^2)(1-\phi^2) \\ -(1-\theta^2)(1-\phi^2) & (1-\theta^2)(1+\phi\theta) \end{bmatrix}. \quad \square$$

Example 5.2.4 The Dow-Jones Utilities Index

For the Burg and Yule-Walker AR(1) models derived for the differenced and mean-corrected series in Examples 5.1.1 and 5.1.3, it is easy to check in the *Preliminary estimation* option of PEST that $-2\ln(L) = 70.330$ for the Burg model and $-2\ln(L) = 70.378$ for the Yule-Walker model. Since maximum likelihood estimation attempts to minimize $-2\ln L$, the Burg estimate appears to be a slightly better initial estimate of ϕ. We therefore exit from the *Preliminary estimation* section of PEST retaining the Burg AR(1) model and then select *ARMA Estimation* followed by *Optimize with current settings*. The Burg coefficient estimates provide initial parameter values to start the search for the minimizing values. The model found on completion of the minimization is

$$Y_t - 0.4471 Y_{t-1} = Z_t, \quad \{Z_t\} \sim \mathrm{WN}(0, 0.02117).$$

This model is different again from the Burg and Yule-Walker models. It has $-2\ln(L) = 70.321$, corresponding to a slightly higher likelihood. The standard error (or estimated standard deviation) of the estimator $\hat{\phi}$ is found from the program to be 0.1050. This is in good agreement with the estimated standard deviation, $\sqrt{(1-(.4471)^2)/77} = .1019$, based on the large sample approximation given in Example 5.2.1. Using the value computed from PEST, approximate 95% confidence bounds for ϕ are $0.4471 \pm 1.96 \times 0.1050 = (0.2413, 0.6529)$. These are quite close to the bounds found earlier based on the Yule-Walker and Burg estimates. □

Example 5.2.5 The lake data

We use the preliminary ARMA(1,1) model obtained using the innovations algorithm in Example 5.1.6 as the initial model for fitting an ARMA(1,1) model by maximum likelihood. To do this we proceed as in Example 5.1.6, mean-correcting the data in LAKE.DAT and fitting an ARMA(1,1) using the *Preliminary estimation* option. We then return to the main PEST menu, retaining the innovation model, and select the *ARMA parameter estimation* option, followed by *Optimize with current settings*. This gives the maximum likelihood model

$$X_t - 0.7453 X_{t-1} = Z_t + 0.3208 Z_{t-1}, \quad \{Z_t\} \sim \mathrm{WN}(0, 0.4750),$$

for the mean-corrected series, $X_t = Y_t - 9.0041$. The estimated standard deviations of the two coefficient estimates $\hat{\phi}$ and $\hat{\theta}$ are found from PEST to be 0.0771 and 0.1116, respectively. (The respective estimated standard deviations based on the large sample approximation given in Example 5.2.3 are .0718 and .1020.) The corresponding 95% confidence bounds are therefore

$$\phi : 0.7453 \pm 1.96 \times 0.0771 = (0.5942, 0.8964),$$

$$\theta : 0.3208 \pm 1.96 \times 0.1116 = (0.1020, 0.5395).$$

The value of AICC for this model is 212.77.

If we repeat the above steps, substituting preliminary AR(2) estimation for preliminary ARMA(1,1) estimation and then going to the *ARMA Estimation* option to find the maximum likelihood AR(2) model for the mean-corrected series $\{X_t = Y_t - 9.0041\}$, we obtain,

$$X_t - 1.0415 X_{t-1} + 0.2494 X_{t-2} = Z_t, \quad \{Z_t\} \sim \mathrm{WN}(0, 0.4790).$$

with AICC $= 213.54$. On the basis of the AICC criterion, the ARMA(1,1) model of the preceding paragraph is thus slightly superior to this AR(2) model. □

5.3 Diagnostic Checking

Typically the goodness of fit of a statistical model to a set of data is judged by comparing the observed values with the corresponding predicted values obtained from the fitted model. If the fitted model is appropriate, then the residuals should behave in a manner that is consistent with the model.

When we fit an ARMA(p, q) model to a given series we determine the maximum likelihood estimators $\hat{\phi}$, $\hat{\theta}$, and $\hat{\sigma}^2$ of the parameters ϕ, θ, and σ^2. In the course of this procedure the predicted values $\hat{X}_t(\hat{\phi}, \hat{\theta})$ of X_t based on $X_1, \ldots, X_{t-1}$ are computed for the fitted model. The **residuals** are then defined, in the notation of Section 3.3, by

$$\hat{W}_t = (X_t - \hat{X}_t(\hat{\phi}, \hat{\theta}))/(r_{t-1}(\hat{\phi}, \hat{\theta}))^{1/2}, \quad t = 1, \ldots, n. \tag{5.3.1}$$

If we were to assume that the maximum likelihood ARMA(p, q) model is the *true* process generating $\{X_t\}$, then we could say that $\{\hat{W}_t\} \sim \mathrm{WN}(0, \hat{\sigma}^2)$. However, to check the appropriateness of an ARMA(p, q) model for the data we should assume only that $X_1, \ldots, X_n$ is generated by an ARMA(p, q) process with unknown parameters ϕ, θ, and σ^2, whose maximum likelihood *estimators* are $\hat{\phi}$, $\hat{\theta}$, and $\hat{\sigma}^2$, respectively. Then it is not true that $\{\hat{W}_t\}$ is white noise. Nonetheless $\hat{W}_t$, $t = 1, \ldots, n$ should have properties that are similar to those of the white noise sequence

$$W_t(\phi, \theta) = (X_t - \hat{X}_t(\phi, \theta))/(r_{t-1}(\phi, \theta))^{1/2}, \quad t = 1, \ldots, n.$$

Moreover $W_t(\phi, \theta)$ approximates the white noise term in the defining equation (5.1.1) in the sense that $E(W_t(\phi, \theta) - Z_t)^2 \to 0$ as $t \to \infty$ (TSTM, Section 8.11). Consequently the properties of the residuals $\{\hat{W}_t\}$ should reflect those of the white noise sequence $\{Z_t\}$ generating the underlying ARMA(p, q) process. In particular the sequence $\{\hat{W}_t\}$ should be approximately (i) uncorrelated if $\{Z_t\} \sim \mathrm{WN}(0, \sigma^2)$, (ii) independent if $\{Z_t\} \sim \mathrm{IID}(0, \sigma^2)$, and (iii) normally distributed if $Z_t \sim \mathrm{N}(0, \sigma^2)$.

The **rescaled residuals** $\hat{R}_t$ are obtained by dividing $\hat{W}_t$ by the estimated white noise standard deviation. Thus,

$$\hat{R}_t = \hat{W}_t / \hat{\sigma}. \tag{5.3.2}$$

If the fitted model is appropriate, the rescaled residuals should have properties similar to those of a WN(0, 1) sequence or of an IID(0,1) sequence if we make the stronger assumption that the white noise $\{Z_t\}$ driving the ARMA process is independent white noise.

The following diagnostic checks are all based on the expected properties of the residuals or rescaled residuals under the assumption that the fitted model is correct and that $\{Z_t\} \sim \mathrm{IID}(0, \sigma^2)$. They are the same tests introduced in Section 1.6.

5.3.1 The Graph of $\{\hat{R}_t, t = 1, \ldots, n\}$

If the fitted model is appropriate, then the graph of the rescaled residuals $\{\hat{R}_t, t = 1, \ldots, n\}$ should resemble that of a white noise sequence with variance one. While it is difficult to identify the correlation structure of $\{\hat{R}_t\}$ (or any time series for that matter) from its graph, deviations of the mean from zero are sometimes clearly indicated by a trend or cyclic component and nonconstancy of the variance by fluctuations in $\hat{R}_t$, whose magnitude depends strongly on t.

The rescaled residuals obtained from the ARMA(1,1) model fitted to the mean-corrected lake data in Example 5.2.5 are displayed in Figure 5.5. The graph gives no indication of a nonzero mean or nonconstant variance, so on this basis there is no reason to doubt the compatibility of $\hat{R}_1, \ldots, \hat{R}_n$ with unit variance white noise.

The next step is to check that the sample autocorrelation function of $\{\hat{W}_t\}$ (or equivalently of $\{\hat{R}_t\}$) behaves as it should under the assumption that the fitted model is appropriate.

5.3.2 The Sample ACF of the Residuals

We know from Section 1.6 that for large n the sample autocorrelations of an iid sequence $Y_1, \ldots, Y_n$ with finite variance are approximately iid with distribution $N(0, 1/n)$. We can therefore test whether or not the observed residuals are consistent with iid noise by examining the sample autocorrelations of the residuals and rejecting the iid noise hypothesis if more than two or three out of 40 fall outside the bounds $\pm 1.96/\sqrt{n}$ or if one falls far outside the bounds. (As indicated above our *estimated* residuals will not be precisely iid even if the true model generating the data is as assumed. To correct for this the bounds $\pm 1.96/\sqrt{n}$ should be modified to give a more precise test as in Box and Pierce (1970) and TSTM, Section 9.4.) The bounds $1.96/\sqrt{n}$ are automatically plotted with the sample ACF of the residuals after model fitting in the option *ARMA Estimation* of PEST. Figure 5.6 shows the sample ACF

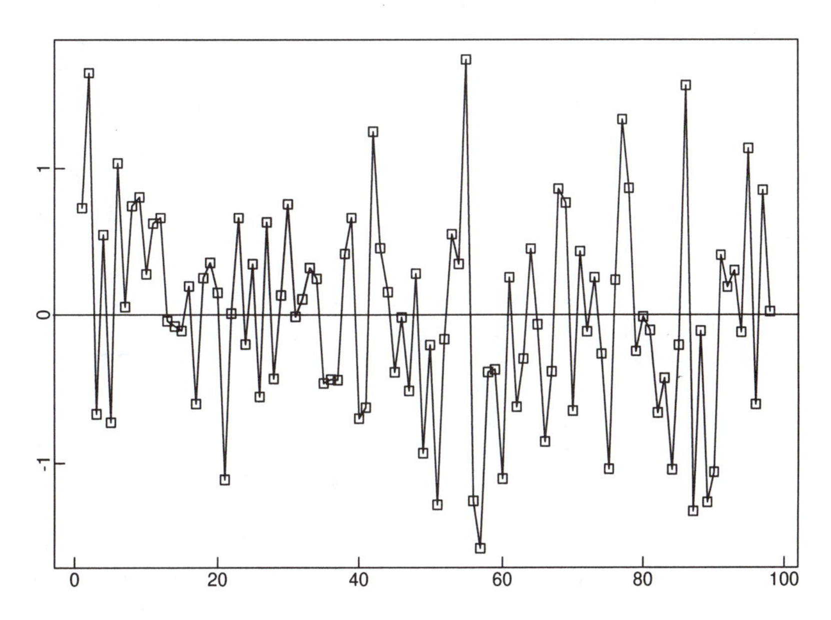

Figure 5-5 The rescaled residuals after fitting the ARMA(1,1) model of Example 5.2.5 to the lake data.

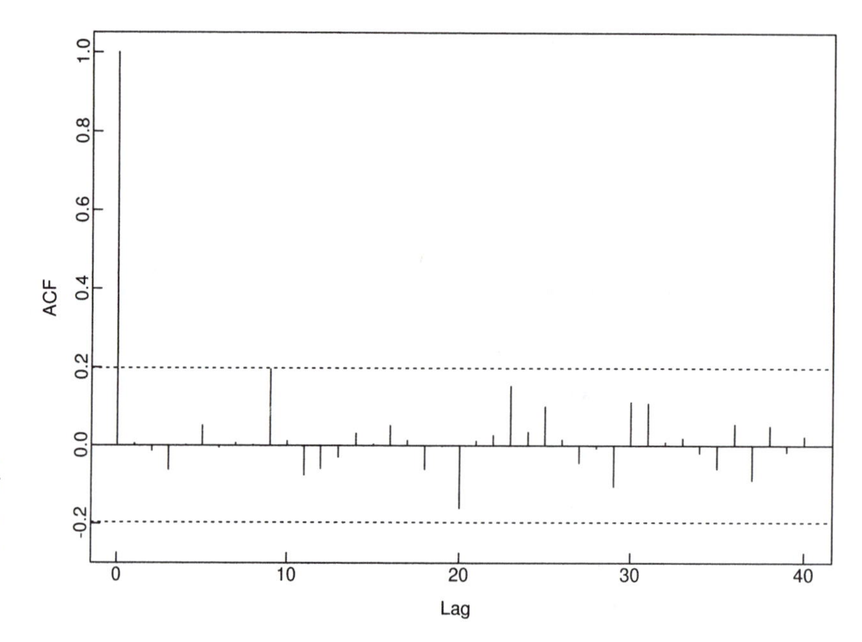

Figure 5-6 The sample ACF of the residuals after fitting the ARMA(1,1) model of Example 5.2.5 to the lake data.

of the residuals after fitting the ARMA(1,1) of Example 5.2.5 to the lake data. As can be seen from the graph, there is no cause to reject the fitted model on the basis of these autocorrelations.

5.3.3 Tests for Randomness of the Residuals

A simple test for whiteness of the residuals is to fit autoregressive models of orders $p = 0, 1, \ldots, 26$ and to record the value of p for which the AICC value is minimum. Compatibility of these residuals with white noise is indicated by selection of the value 0. This test, in conjunction with those of Section 1.6 (b), (c), (d), and (e), can be carried out with PEST by selecting *File and analyze residuals* then *Tests for randomness* after model-fitting under the option *ARMA estimation* of the main menu. Following this procedure for the ARMA(1,1) model fitted to the mean-corrected lake data and using the default value $h = 22$ as suggested for the portmanteau tests, we obtain the following results.

```
RANDOMNESS TEST STATISTICS
LJUNG-BOX PORTM.   =   10.23    CHISQUR(20)
MCLEOD-LI PORTM.   =   16.54    CHISQUR(22)
TURNING POINTS     =      69   ANORMAL(64.0, 4.14**2)
DIFFERENCE-SIGN    =      50   ANORMAL(48.5, 2.87**2)
RANK TEST          =    2088   ANORMAL(2376, 162.9**2)
ORDER OF MIN AICC YW MODEL FOR RESIDUALS = 0
```

This table shows the observed values of the statistics defined in Section 1.6, with each followed by its large sample distribution under the null hypothesis that the residuals are iid. The observed values can thus be checked easily for compatibility with their distributions under the null hypothesis. In this particular example the portmanteau statistics are both clearly less than the 95 percentile of the chi-squared distribution with 20 and 22 degrees of freedom and the other three test statistics are all within 1.96 standard deviations of the means of their asymptotically normal distributions. None of the test statistics therefore lead us to reject the null hypothesis at level .05. The order of the minimum AICC autoregressive model for the residuals also suggests the compatibility of the residuals with white noise.

The McLeod-Li test is a test for iid normal residuals. Another check on normality is provided by visual inspection of the histogram of the rescaled residuals obtained by selecting the option *Plot rescaled residuals* after conducting the tests for randomness. No obvious deviation from normality is apparent in the histogram for the residuals after fitting the ARMA(1,1) model to the lake data.

5.4 Forecasting

Once a model has been fitted to the data, forecasting future values of the time series can be carried out using the method described in Section 3.3. We illustrate this method with one of the examples from Section 3.2.

Example 5.4.1 For the overshort data $\{X_t\}$ of Example 3.2.8, use of the PEST options *Preliminary estimation* and *ARMA Estimation* leads to the following maximum likelihood MA(1) model for $\{X_t\}$:

$$X_t + 4.035 = Z_t - .818Z_{t-1}, \quad \{Z_t\} \sim \text{WN}(0, 2040.75). \tag{5.4.1}$$

To predict the next 7 days of overshorts, we treat (5.4.1) as the true model for the data, and use the results of Example 3.3.3 with $\phi = 0$. From (3.3.11), the predictors are given by

$$\begin{aligned} P_{57}X_{57+h} &= -4.035 + \sum_{j=h}^{1} \theta_{57+h-1,j}(X_{57+h-j} - \hat{X}_{57+h-j}) \\ &= \begin{cases} -4.035 + \theta_{57,1}(X_{57} - \hat{X}_{57}), & \text{if } h = 1, \\ -4.035, & \text{if } h > 1, \end{cases} \end{aligned}$$

with mean squared error

$$E(X_{57+h} - P_{57}X_{57+h})^2 = \begin{cases} 2040.75r_{57}, & \text{if } h = 1, \\ 2040.75(1 + (-.818)^2), & \text{if } h > 1, \end{cases}$$

where $\theta_{57,1}$ and r_{57} are computed recursively from (3.3.9) with $\theta = -.818$.

These calculations are performed in PEST and the results can be obtained, after fitting the maximum likelihood model (5.4.1), by selecting the *Forecasting* option from the main menu. This option produces the 1-step, 2-step, ... and 7-step forecasts of X_t shown in Table 5.1. Notice that the predictor of X_t for $t \geq 59$ is equal to the sample mean since under the MA(1) model $\{X_t, t \geq 59\}$ is uncorrelated with $\{X_t, t \leq 57\}$. □

Assuming that the innovations $\{Z_t\}$ are normally distributed, an approximate 95% prediction interval for X_{64} is given by

$$-4.0351 \pm 1.96 \times 58.3602 = (-118.42, 110.35).$$

The mean squared errors of prediction, as computed in Section 3.3 and the example above, are based on the assumption that the fitted model is in fact the true model for the data. As a result, they do not reflect the variability in the estimation of the model parameters. To illustrate this point, suppose the data $X_1, \ldots, X_n$ are generated from the causal AR(1) model

$$X_t = \phi X_{t-1} + Z_t, \quad \{Z_t\} \sim \text{IID}(0, \sigma^2).$$

If $\hat{\phi}$ is the maximum likelihood estimate of ϕ, based on $X_1, \ldots, X_n$, then the one-step ahead forecast of X_{n+1} is $\hat{\phi} X_n$, which has mean squared error

$$\begin{aligned} E(X_{n+1} - \hat{\phi} X_n)^2 &= E((\phi - \hat{\phi}) X_n + Z_{n+1})^2 \\ &= E((\phi - \hat{\phi}) X_n)^2 + \sigma^2. \end{aligned} \tag{5.4.2}$$

The second equality follows from the independence of Z_{n+1} and $(\hat{\phi}, X_n)'$. To evaluate the first term in (5.4.2), first condition on X_n and then use the approximations

$$E\left((\phi - \hat{\phi})^2 | X_n\right) \approx E(\phi - \hat{\phi})^2 \approx (1 - \phi^2)/n,$$

where the second relation comes from the formula for the asymptotic variance of $\hat{\phi}$ given by $\sigma^2 \Gamma_1^{-1} = (1 - \phi^2)$ (see Example 5.2.1). The one-step mean squared error is

Table 5.1 Forecasts of the next 7 observations of the overshort data of Example 3.2.8 using model (5.4.1).

#	XHAT	SQRT(MSE)	XHAT + MEAN
58	1.0097	45.1753	−3.0254
59	0.0000	58.3602	−4.0351
60	0.0000	58.3602	−4.0351
61	0.0000	58.3602	−4.0351
62	0.0000	58.3602	−4.0351
63	0.0000	58.3602	−4.0351
64	0.0000	58.3602	−4.0351

then approximated by

$$E(\phi - \hat{\phi})^2 E X_n^2 + \sigma^2 \approx n^{-1}(1-\phi^2)(1-\phi^2)^{-1}\sigma^2 + \sigma^2$$
$$= \frac{n+1}{n}\sigma^2.$$

Thus, the error in parameter estimation contributes the term σ^2/n to the mean squared error of prediction. If the sample size is large, this factor is negligible and so, for the purpose of mean squared error computation, the estimated parameters can be treated as the true model parameters. On the other hand, for small sample sizes, ignoring parameter variability can lead to a severe underestimate of the actual mean squared error of the forecast.

5.5 Order Selection

Once the data have been transformed (e.g., by some combination of Box-Cox and differencing transformations or by removal of trend and seasonal components) to the point where the transformed series $\{X_t\}$ can potentially be fitted by a zero-mean ARMA model, we are faced with the problem of selecting appropriate values for the orders p and q.

It is not advantageous from a forecasting point of view to choose p and q arbitrarily large. Fitting a very high order model will generally result in a small estimated white noise variance, but when using the fitted model for forecasting, the mean squared error of the forecasts will depend not only on the white noise variance of the fitted model but also on errors arising from estimation of the parameters of the model (see the paragraphs following Example 5.4.1). These will be larger for higher order models. For this reason we need to introduce a "penalty factor" to discourage the fitting of models with too many parameters.

Many criteria based on such penalty factors have been proposed in the literature since the problem of model selection arises frequently in statistics, particularly in regression analysis. We shall restrict attention here to a brief discussion of the FPE, AIC, and BIC criteria of Akaike and a bias-corrected version of the AIC known as the AICC.

5.5.1 The FPE Criterion

The FPE criterion was developed by Akaike (1969) to select the appropriate order of an AR process to fit to a time series $\{X_1, \ldots, X_n\}$. Instead of trying to choose the order p to make the estimated white noise variance as small as possible, the idea is to choose the model for $\{X_t\}$ in such a way as to minimize the one-step mean squared error when the model fitted to $\{X_t\}$ is used to predict an independent realization $\{Y_t\}$ of the same process that generated $\{X_t\}$.

Suppose then that $\{X_1, \ldots, X_n\}$ is a realization of an AR(p) process with coefficients $\phi_1, \ldots, \phi_p$, $p < n$, and that $\{Y_1, \ldots, Y_n\}$ is an independent realization of the same process. If $\hat{\phi}_1, \ldots, \hat{\phi}_p$, are the maximum likelihood estimators of the coefficients based on $\{X_1, \ldots, X_n\}$ and if we use these to compute the one-step predictor $\hat{\phi}_1 Y_n + \cdots + \hat{\phi}_p Y_{n+1-p}$ of Y_{n+1}, then the mean-square prediction error is

$$\begin{aligned} &E(Y_{n+1} - \hat{\phi}_1 Y_n - \cdots - \hat{\phi}_p Y_{n+1-p})^2 \\ &\quad = E[Y_{n+1} - \phi_1 Y_n - \cdots - \phi_p Y_{n+1-p} - (\hat{\phi}_1 - \phi_1) Y_n - \cdots - (\hat{\phi}_p - \phi_p) Y_{n+1-p}]^2 \\ &\quad = \sigma^2 + E[(\hat{\boldsymbol{\phi}}_p - \boldsymbol{\phi}_p)'[Y_{n+1-i} Y_{n+1-j}]_{i,j=1}^p (\hat{\boldsymbol{\phi}}_p - \boldsymbol{\phi})], \end{aligned}$$

where $\boldsymbol{\phi}_p' = (\phi_1, \ldots, \phi_p)'$, $\hat{\boldsymbol{\phi}}_p' = (\hat{\phi}_1, \ldots, \hat{\phi}_p)'$, and σ^2 is the white noise variance of the AR(p) model. Writing the last term in the preceding equation as the expectation of the conditional expectation given $X_1, \ldots, X_n$, and using the independence of $\{X_1, \ldots, X_n\}$ and $\{Y_1, \ldots, Y_n\}$, we obtain

$$E(Y_{n+1} - \hat{\phi}_1 Y_n - \cdots - \hat{\phi}_p Y_{n+1-p})^2 = \sigma^2 + E[(\hat{\boldsymbol{\phi}}_p - \boldsymbol{\phi}_p)' \Gamma_p (\hat{\boldsymbol{\phi}}_p - \boldsymbol{\phi})],$$

where $\Gamma_p = E[Y_i Y_j]_{i,j=1}^p$. We can approximate the last term by assuming that $n^{-1/2}(\hat{\boldsymbol{\phi}}_p - \boldsymbol{\phi}_p)$ has its large-sample distribution $N(\mathbf{0}, \sigma^2 \Gamma_p^{-1})$ from Example 5.21. Using Problem 5.13, this gives

$$E(Y_{n+1} - \hat{\phi}_1 Y_n - \cdots - \hat{\phi}_p Y_{n+1-p})^2 \approx \sigma^2 \left(1 + \frac{p}{n}\right). \tag{5.5.1}$$

If $\hat{\sigma}^2$ is the maximum likelihood estimator of σ^2, then for large n, $n\hat{\sigma}^2/\sigma^2$ is distributed approximately as chi-squared with $(n - p)$ degrees of freedom (see TSTM, Section 8.9). We therefore replace σ^2 in (5.5.1) by the estimator $n\hat{\sigma}^2/(n - p)$ to get the estimated mean square prediction error of Y_{n+1},

$$\text{FPE}_p = \hat{\sigma}^2 \frac{n + p}{n - p}. \tag{5.5.2}$$

To apply the FPE criterion for autoregressive order selection we therefore choose the value of p that minimizes FPE_p as defined in (5.5.2).

Example 5.5.1 Selection of an AR model for the lake data

In Example 5.2.5 we used maximum likelihood to fit an AR(2) model to the lake data. The choice $p = 2$ was motivated by the sample partial autocorrelation function shown in Figure 5.4. To use the FPE criterion to select p, we have shown in Table 5.2 the values of FPE for values of p from 0 to 10. These values are found by fitting maximum likelihood AR models using the *ARMA estimation* option from the main menu of PEST. Also shown in the table are the values of the maximum likelihood estimates of σ^2 for the same values of p. Whereas $\hat{\sigma}_p^2$ decreases steadily with p, the

Table 5.2 $\hat{\sigma}_p^2$ and FPE_p for AR(p) models fitted to the lake data.

p	σ_p^2	FPE_p
0	1.7203	1.7203
1	0.5097	0.5202
2	0.4790	0.4989
3	0.4728	0.5027
4	0.4708	0.5109
5	0.4705	0.5211
6	0.4705	0.5318
7	0.4679	0.5399
8	0.4664	0.5493
9	0.4664	0.5607
10	0.4453	0.5465

values of FPE_p have a clear minimum at $p = 2$, confirming our earlier choice of $p = 2$ as the most appropriate for this data set. □

5.5.2 The AICC Criterion

A more generally applicable criterion for model selection than the FPE is the information criterion of Akaike (1973), known as the AIC. This was designed to be an approximately unbiased estimate of the Kullback-Leibler index of the fitted model relative to the true model (defined below). Here we use a bias-corrected version of the AIC, referred to as the AICC, suggested by Hurvich and Tsai (1989).

If $\mathbf{X}$ is an n-dimensional random vector whose probability density belongs to the family $\{f(\cdot;\psi), \psi \in \Psi\}$, the Kullback-Leibler discrepancy between $f(\cdot;\psi)$ and $f(\cdot;\theta)$ is defined as

$$d(\psi|\theta) = \Delta(\psi|\theta) - \Delta(\theta|\theta),$$

where

$$\begin{aligned}\Delta(\psi|\theta) &= E_\theta(-2\ln f(\mathbf{X};\psi)) \\ &= \int_{\mathbb{R}^n} -2\ln(f(\mathbf{x};\psi))f(\mathbf{x};\theta)\,d\mathbf{x},\end{aligned}$$

is the Kullback-Leibler index of $f(\cdot;\psi)$ relative to $f(\cdot;\theta)$. (Note that in general $\Delta(\psi|\theta) \neq \Delta(\theta|\psi)$.) By Jensen's inequality (see e.g., Mood et al., 1974),

$$\begin{aligned} d(\psi|\theta) &= \int_{\mathbb{R}^n} -2\ln\left(\frac{f(\mathbf{x};\psi)}{f(\mathbf{x};\theta)}\right) f(\mathbf{x};\theta)\,d\mathbf{x} \\ &\geq -2\ln\left(\int_{\mathbb{R}^n} \frac{f(\mathbf{x};\psi)}{f(\mathbf{x};\theta)} f(\mathbf{x};\theta)\,d\mathbf{x}\right) \\ &= -2\ln\left(\int_{\mathbb{R}^n} f(\mathbf{x};\psi)\,d\mathbf{x}\right) \\ &= 0 \end{aligned}$$

with equality holding if and only if $f(\mathbf{x};\psi) = f(\mathbf{x};\theta)$.

Given observations $X_1, \ldots, X_n$ of an ARMA process with unknown parameters $\theta = (\boldsymbol{\beta}, \sigma^2)$, the true model could be identified if it were possible to compute the Kullback-Leibler discrepancy between all candidate models and the true model. Since this is not possible we *estimate* the Kullback-Leibler discrepancies and choose the model whose estimated discrepancy (or index) is minimum. In order to do this, we assume that the true model and the alternatives are all Gaussian. Then for any given $\theta = (\boldsymbol{\beta}, \sigma^2)$, $f(\cdot;\theta)$ is the probability density of $(Y_1, \ldots, Y_n)'$, where $\{Y_t\}$ is a Gaussian ARMA(p,q) process with coefficient vector $\boldsymbol{\beta}$ and white noise variance σ^2. (The dependence of θ on p and q is through the dimension of the autoregressive and moving average coefficients in $\boldsymbol{\beta}$.)

Suppose therefore that our observations $X_1, \ldots, X_n$ are from a Gaussian ARMA process with parameter vector $\theta = (\boldsymbol{\beta}, \sigma^2)$ and assume for the moment that the true order is (p,q). Let $\hat{\theta} = (\hat{\boldsymbol{\beta}}, \hat{\sigma}^2)$ be the maximum likelihood estimator of θ based on $X_1, \ldots, X_n$ and let $Y_1, \ldots, Y_n$ be an independent realization of the true process (with parameter θ). Then

$$-2\ln L_Y(\hat{\boldsymbol{\beta}}, \hat{\sigma}^2) = -2\ln L_X(\hat{\boldsymbol{\beta}}, \hat{\sigma}^2) + \hat{\sigma}^{-2} S_Y(\hat{\boldsymbol{\beta}}) - n,$$

where L_X, L_Y, S_X, and S_Y are defined as in (5.2.9) and (5.2.11). Hence,

$$\begin{aligned} E_\theta(\Delta(\hat{\theta}|\theta)) &= E_{\boldsymbol{\beta},\sigma^2}(-2\ln L_Y(\hat{\boldsymbol{\beta}}, \hat{\sigma}^2)) \\ &= E_{\boldsymbol{\beta},\sigma^2}(-2\ln L_X(\hat{\boldsymbol{\beta}}, \hat{\sigma}^2)) + E_{\boldsymbol{\beta},\sigma^2}\left(\frac{S_Y(\hat{\boldsymbol{\beta}})}{\hat{\sigma}^2}\right) - n. \end{aligned} \tag{5.5.3}$$

It can be shown using large-sample approximations (see TSTM, Section 9.3 for details) that

$$E_{\boldsymbol{\beta},\sigma^2}\left(\frac{S_Y(\hat{\boldsymbol{\beta}})}{\hat{\sigma}^2}\right) \approx \frac{2(p+q+1)n}{n-p-q-2},$$

from which we see that $-2\ln L_X(\hat{\boldsymbol{\beta}}, \hat{\sigma}^2) + 2(p+q+1)n/(n-p-q-2)$ is an approximately unbiased estimator of the expected Kullback-Leibler index $E_\theta(\Delta(\hat{\theta}|\theta))$

in (5.5.3). Since the preceding calculations (and the maximum likelihood estimators $\hat{\beta}$ and $\hat{\sigma}^2$) are based on the assumption that the true order is (p, q), we therefore select the values of p and q for our fitted model to be those that minimize AICC($\hat{\beta}$), where

$$\text{AICC}(\boldsymbol{\beta}) := -2 \ln L_X(\boldsymbol{\beta}, S_X(\boldsymbol{\beta})/n) + 2(p+q+1)n/(n-p-q-2). \quad (5.5.4)$$

The AIC statistic, defined as

$$\text{AIC}(\boldsymbol{\beta}) := -2 \ln L_X(\boldsymbol{\beta}, S_X(\boldsymbol{\beta})/n) + 2(p+q+1),$$

can be used in the same way. Both AICC($\boldsymbol{\beta}, \sigma^2$) and AIC($\boldsymbol{\beta}, \sigma^2$) can be defined for arbitrary σ^2 by replacing $S_X(\boldsymbol{\beta})/n$ in the preceding definitions by σ^2. The value $S_X(\boldsymbol{\beta})/n$ is used in (5.5.4) since AICC($\boldsymbol{\beta}, \sigma^2$) (like AIC($\boldsymbol{\beta}, \sigma^2$)) is minimized for any given $\boldsymbol{\beta}$ by setting $\sigma^2 = S_X(\boldsymbol{\beta})/n$.

For fitting autoregressive models, Monte Carlo studies (Jones, 1975, Shibata, 1976) suggest that the AIC has a tendency to overestimate p. The penalty factors $2(p+q+1)n/(n-p-q-2)$ and $2(p+q+1)$, for the AICC and AIC statistics are asymptotically equivalent as $n \to \infty$. The AICC statistic, however, has a more extreme penalty for large-order models, which counteracts the overfitting tendency of the AIC. The BIC is another criterion that attempts to correct the overfitting nature of the AIC. For a zero-mean causal invertible ARMA(p, q) process, it is defined (Akaike, 1978) to be

$$\begin{aligned} \text{BIC} = \ & (n-p-q) \ln[n\hat{\sigma}^2/(n-p-q)] + n(1 + \ln \sqrt{2\pi}) \\ & + (p+q) \ln \left[\left(\sum_{t=1}^{n} X_t^2 - n\hat{\sigma}^2\right)/(p+q) \right], \end{aligned} \quad (5.5.5)$$

where $\hat{\sigma}^2$ is the maximum likelihood estimate of the white noise variance.

The BIC is a consistent order selection criterion in the sense that if the data $\{X_1, \ldots, X_n\}$ *are* in fact observations of an ARMA(p, q) process, and if $\hat{p}$ and $\hat{q}$ are the estimated orders found by minimizing the BIC, then $\hat{p} \to p$ and $\hat{q} \to q$ with probability 1 as $n \to \infty$ (Hannan, 1980). This property is not shared by the AICC or AIC. On the other hand, order selection by minimization of the AICC, AIC, or FPE is asymptotically efficient for autoregressive processes while order selection by BIC minimization is not (Shibata, 1980; Hurvich and Tsai, 1989). Efficiency is a desirable property defined in terms of the one-step mean square prediction error achieved by the fitted model. For more details see TSTM, Section 9.3.

In the modelling of real data there is rarely such a thing as the "true order." For the process $X_t = \sum_{j=0}^{\infty} \psi_j Z_{t-j}$ there may be many polynomials $\theta(z), \phi(z)$ such that the coefficients of z^j in $\theta(z)/\phi(z)$ closely approximate ψ_j for moderately small values of j. Correspondingly there may be many ARMA processes with properties similar to $\{X_t\}$. This problem of identifiability becomes much more serious for multivariate processes. The AICC criterion does, however, provide us with a rational criterion for choosing between competing models. It has been suggested (Duong, 1984) that models with AIC values within c of the minimum value should be considered competitive (with $c = 2$ as a typical value). Selection from among the competitive models can then be based on such factors as whiteness of the residuals (Section 5.3) and model simplicity.

We frequently have occasion, particularly in analyzing seasonal data, to fit ARMA(p, q) models in which all except $m(\leq p + q)$ of the coefficients are constrained to be zero. In such cases the definition (5.5.4) is replaced by

$$\text{AICC}(\beta) := -2 \ln L_X(\beta, S_X(\beta)/n) + 2(m + 1)n/(n - m - 2). \qquad (5.5.6)$$

Example 5.5.1 The lake data

In Example 5.2.5 we fitted maximum likelihood AR(2) and ARMA(1,1) models to the mean-corrected lake data. From the *ARMA estimation* option of PEST we obtain values of AICC and BIC for each of these models. In the same way we can fit ARMA(p, q) models for all values of p and q up to some reasonable limit, checking the values of the order-selection criteria for each of the fitted models. We easily find in this way that for $p+q \leq 6$ both the AICC and BIC are minimized by the ARMA(1,1) model for which

$$\text{AICC} = 212.77, \quad \text{BIC} = 216.85.$$

The next smallest values are achieved by the AR(2) model with

$$\text{AICC} = 213.54, \quad \text{BIC} = 217.64.$$

Thus the AICC and BIC both select the ARMA(1,1) model,

$$X_t - 0.7453X_{t-1} = Z_t + 0.3208Z_{t-1}, \quad \{Z_t\} \sim \text{WN}(0, 0.4750),$$

for the mean-corrected series, $X_t = Y_t - 9.0041$. Both models pass the diagnostic checks of Section 5.3, and in view of the small difference between the AICC values there is no strong reason to prefer one model or the other. □

Problems

5.1. The sunspot numbers $\{X_t, t = 1, \ldots, 100\}$, filed as SUNSPOTS.DAT, have sample autocovariances $\hat{\gamma}(0) = 1382.2$, $\hat{\gamma}(1) = 1114.4$, $\hat{\gamma}(2) = 591.73$ and $\hat{\gamma}(3) = 96.216$. Use these values to find the Yule-Walker estimates of ϕ_1, ϕ_2 and σ^2 in the model

$$Y_t = \phi_1 Y_{t-1} + \phi_2 Y_{t-2} + Z_t, \quad \{Z_t\} \sim \text{WN}(0, \sigma^2),$$

for the mean-corrected series $Y_t = X_t - 46.93$, $t = 1, \ldots, 100$. Assuming the data really are a realization of an AR(2) process, find 95% confidence intervals for ϕ_1 and ϕ_2.

5.2. From the information given in the previous problem, use the Durbin-Levinson algorithm to compute the sample partial autocorrelations $\hat{\phi}_{11}$, $\hat{\phi}_{22}$, and $\hat{\phi}_{33}$ of the sunspot series. Is the value of $\hat{\phi}_{33}$ compatible with the hypothesis that the data are generated by an AR(2) process? (Use significance level .05.)

5.3. Consider the AR(2) process $\{X_t\}$ satisfying

$$X_t - \phi X_{t-1} - \phi^2 X_{t-2} = Z_t, \quad \{Z_t\} \sim \mathrm{WN}(0, \sigma^2).$$

a. For what values of ϕ is this a causal process?

b. The following sample moments were computed after observing $X_1, \ldots, X_{200}$:

$$\hat{\gamma}(0) = 6.06, \quad \hat{\rho}(1) = .687.$$

Find estimates of ϕ and σ^2 by solving the Yule-Walker equations. (If you find more than one solution, choose the one that is causal.)

5.4. Two hundred observations of a time series, $X_1, \ldots, X_{200}$, gave the following sample statistics:

sample mean:	$\bar{x}_{200} = 3.82$
sample variance:	$\hat{\gamma}(0) = 1.15$
sample ACF:	$\hat{\rho}(1) = .427$
	$\hat{\rho}(2) = .475$
	$\hat{\rho}(3) = .169$

a. Based on these sample statistics, is it reasonable to suppose that $\{X_t - \mu\}$ is white noise?

b. Assuming that $\{X_t - \mu\}$ can be modelled as the AR(2) process,

$$X_t - \mu - \phi_1(X_{t-1} - \mu) - \phi_2(X_{t-2} - \mu) = Z_t,$$

where $\{Z_t\} \sim \mathrm{IID}(0, \sigma^2)$, find estimates of μ, ϕ_1, ϕ_2, and σ^2.

c. Would you conclude that $\mu = 0$?

d. Construct 95% confidence intervals for ϕ_1 and ϕ_2.

e. Assuming that the data was generated from an AR(2) model, derive estimates of the PACF for all lags $h \geq 1$.

5.5. Use the program PEST to simulate and file 20 realizations of length 200 of the Gaussian MA(1) process,

$$X_t = Z_t + \theta Z_{t-1}, \quad \{Z_t\} \sim \mathrm{WN}(0, 1),$$

with $\theta = 0.6$.

a. For each series find the moment estimate of θ as defined in Example 5.1.2.

b. For each series use the innovations algorithm in the *Preliminary estimation* option of PEST to find an estimate of θ. (Use the default value of the parameter m.) As soon as you have found this preliminary estimate for a particular series, return to the main menu of PEST, then choose the option *ARMA parameter estimation* and suboption *Optimize with current settings* to find the maximum likelihood estimate of θ for the series.

c. Compute the sample means and sample variances of your three sets of estimates.

d. Use the asymptotic formulae given at the end of Section 5.1.1 (with $n = 200$) to compute the variances of the moment, innovation and maximum likelihood estimators of θ. Compare with the corresponding sample variances found in (c).

e. What do the results of (c) suggest concerning the relative merits of the three estimators?

5.6. Establish the recursions (5.1.19) and (5.1.20) for the forward and backward prediction errors $u_i(t)$ and $v_i(t)$ in Burg's algorithm.

5.7. Derive the recursions for the Burg estimates $\phi_{ii}^{(B)}$ and $\sigma_i^{(B)2}$.

5.8. From the innovation form of the likelihood (5.2.9) derive the equations (5.2.10), (5.2.11), and (5.2.12) for the maximum likelihood estimators of the parameters of an ARMA process.

5.9. Use equation (5.2.9) to show that for $n > p$, the likelihood of the observations $\{X_1, \ldots, X_n\}$ of the causal AR(p) process defined by

$$X_t = \phi_1 X_{t-1} + \cdots + \phi_p X_{t-p} + Z_t, \quad \{Z_t\} \sim \mathrm{WN}(0, \sigma^2),$$

is

$$L(\phi, \sigma^2) = (2\pi\sigma^2)^{-n/2} (\det G_p)^{-1/2}$$

$$\times \exp\left\{ -\frac{1}{2\sigma^2} \left[\mathbf{X}_p' G_p^{-1} \mathbf{X}_p + \sum_{t=p+1}^{n} (X_t - \phi_1 X_{t-1} - \cdots - \phi_p X_{t-p})^2 \right] \right\}$$

where $\mathbf{X}_p = (X_1, \ldots, X_p)'$ and $G_p = \sigma^{-2}\Gamma_p = \sigma^{-2} E(\mathbf{X}_p \mathbf{X}_p')$.

5.10. Use the result of Problem 5.9 to derive a pair of linear equations for the least squares estimates of ϕ_1 and ϕ_2 for a causal AR(2) process (with mean zero). Compare your equations with those for the Yule-Walker estimates. (Assume that the mean is known to be zero in writing down the latter equations, so that the sample autocovariances are $\hat{\gamma}(h) = \frac{1}{n} \sum_{t=1}^{n-h} X_{t+h} X_t$ for $h \geq 0$.)

5.11. Given two observations x_1 and x_2 from the causal AR(1) process satisfying

$$X_t = \phi X_{t-1} + Z_t, \quad \{Z_t\} \sim \mathrm{WN}(0, \sigma^2),$$

and assuming that $|x_1| \neq |x_2|$, find the maximum likelihood estimates of ϕ and σ^2.

5.12. Derive a cubic equation for the maximum likelihood estimate of the coefficient ϕ of a causal AR(1) process based on the observations $X_1, \ldots, X_n$.

5.13. Use the result of Problem A.7 and the approximate large-sample normal distribution of the maximum likelihood estimator $\hat{\phi}_p$ to establish the approximation (5.5.1).

6 Nonstationary and Seasonal Time Series Models

In this chapter we shall examine the problem of finding an appropriate model for a given set of observations $\{x_1, \ldots, x_n\}$ that are not necessarily generated by a stationary time series. If the data (a) exhibit no apparent deviations from stationarity and (b) have a rapidly decreasing autocovariance function, we attempt to fit an ARMA model to the mean-corrected data using the techniques developed in Chapter 5. Otherwise we look first for a transformation of the data that generates a new series with the properties (a) and (b). This can frequently be achieved by differencing, leading us to consider the class of ARIMA (autoregressive integrated moving average) models, defined in Section 6.1. We have in fact already encountered ARIMA processes. The model fitted in Example 5.1.1 to the Dow-Jones Utilities Index was obtained by fitting an AR model to the differenced data, thereby effectively fitting an ARIMA model to the original series. In Section 6.1 we shall give a more systematic account of such models.

In Section 6.2, we discuss the problem of finding an appropriate transformation for the data and identifying a satisfactory ARMA(p, q) model for the transformed data. The latter can be handled using the techniques developed in Chapter 5. The sample ACF and PACF and the preliminary estimators $\hat{\phi}_m$ and $\hat{\theta}_m$ of Section 5.1 can provide useful guidance in this choice. However, our prime criterion for model selection will be the AICC statistic discussed in Section 5.5.2. To apply this criterion we compute maximum likelihood estimators of ϕ, θ, and σ^2 for a variety of competing p

and q values and choose the fitted model with smallest AICC value. Other techniques, in particular those that use the R and S arrays of Gray et al. (1978), are discussed in the survey of model identification by de Gooijer et al. (1985). If the fitted model is satisfactory, the residuals (see Section 5.3) should resemble white noise. Tests for this were described in Section 5.3 and should be applied to the minimum AICC model to make sure that the residuals are consistent with their expected behavior under the model. If they are not, then competing models (models with AICC value close to the minimum) should be checked until we find one that passes the goodness of fit tests. In some cases a small difference in AICC value (say less than 2) between two satisfactory models may be ignored in the interest of model simplicity. In Section 6.3 we consider the problem of testing for a unit root of either the autoregressive or moving average polynomial. An autoregressive unit root suggests that the data require differencing and a moving average unit root suggests that they have been overdifferenced. Section 6.4 considers the prediction of ARIMA processes, which can be carried out using an extension of the techniques developed for ARMA processes in Sections 3.3 and 5.4. In Section 6.5 we examine the fitting and prediction of seasonal ARIMA (SARIMA) models, whose analysis, except for certain aspects of model identification, is quite analogous to that of ARIMA processes. Finally we consider the problem of regression allowing for dependence between successive residuals from the regression. Such models are known as regression models with time series residuals and often occur in practice as natural representations for data containing both trend and serially dependent errors.

6.1 ARIMA Models for Nonstationary Time Series

We have already discussed the importance of the class of ARMA models for representing stationary series. A generalization of this class, which incorporates a wide range of nonstationary series, is provided by the ARIMA processes, i.e., processes which, after differencing finitely many times, reduce to ARMA processes.

Definition 6.1.1

If d is a nonnegative integer, then $\{X_t\}$ is an **ARIMA(p,d,q) process** if $Y_t := (1-B)^d X_t$ is a causal ARMA(p,q) process.

This definition means that $\{X_t\}$ satisfies a difference equation of the form

$$\phi^*(B)X_t \equiv \phi(B)(1-B)^d X_t = \theta(B)Z_t, \quad \{Z_t\} \sim \mathrm{WN}(0,\sigma^2), \tag{6.1.1}$$

where $\phi(z)$ and $\theta(z)$ are polynomials of degrees p and q, respectively, and $\phi(z) \neq 0$ for $|z| \le 1$. The polynomial $\phi^*(z)$ has a zero of order d at $z = 1$. The process $\{X_t\}$ is stationary if and only if $d = 0$, in which case it reduces to an ARMA(p,q) process.

Notice that if $d \ge 1$ we can add an arbitrary polynomial trend of degree $(d-1)$ to $\{X_t\}$ without violating the difference equation (6.1.1). ARIMA models are therefore useful for representing data with trend (see Sections 1.5 and 6.2). It should be noted, however, that ARIMA processes can also be appropriate for modelling series with no trend. Except when $d = 0$, the mean of $\{X_t\}$ is not determined by equation (6.1.1) and it can in particular be zero (as in Example 1.3.3). Since for $d \geq 1$, equation (6.1.1) determines the second-order properties of $\{(1-B)^d X_t\}$ but not those of $\{X_t\}$ (Problem 6.1), estimation of $\boldsymbol{\phi}$, $\boldsymbol{\theta}$, and σ^2 will be based on the observed differences $(1-B)^d X_t$. Additional assumptions are needed for prediction (see Section 6.4).

Example 6.1.1 $\{X_t\}$ is an ARIMA(1,1,0) process if for some $\phi \in (-1, 1)$

$$(1-\phi B)(1-B)X_t = Z_t, \quad \{Z_t\} \sim \mathrm{WN}(0, \sigma^2).$$

We can then write

$$X_t = X_0 + \sum_{j=1}^{t} Y_j, t \geq 1,$$

where

$$Y_t = (1-B)X_t = \sum_{j=0}^{\infty} \phi^j Z_{t-j}.$$

A realization of $\{X_1, \ldots, X_{200}\}$ with $X_0 = 0$, $\phi = 0.8$, and $\sigma^2 = 1$ is shown in Figure 6.1, with the corresponding sample autocorrelation and partial autocorrelation functions in Figures 6.2 and 6.3, respectively. □

A distinctive feature of the data which suggests the appropriateness of an ARIMA model is the slowly decaying positive sample autocorrelation function in Figure 6.2.

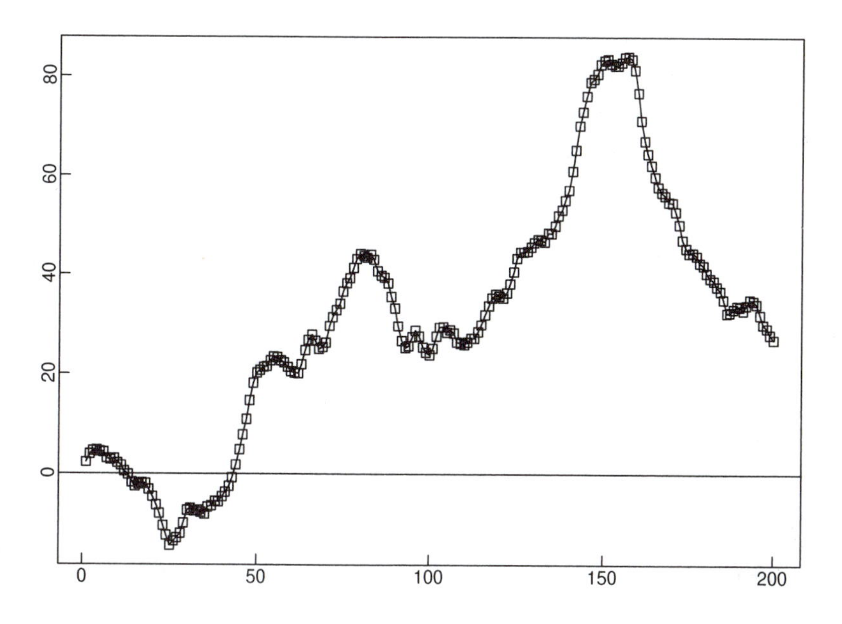

Figure 6-1 200 observations of the ARIMA(1,1,0) series X_t of Example 6.1.1.

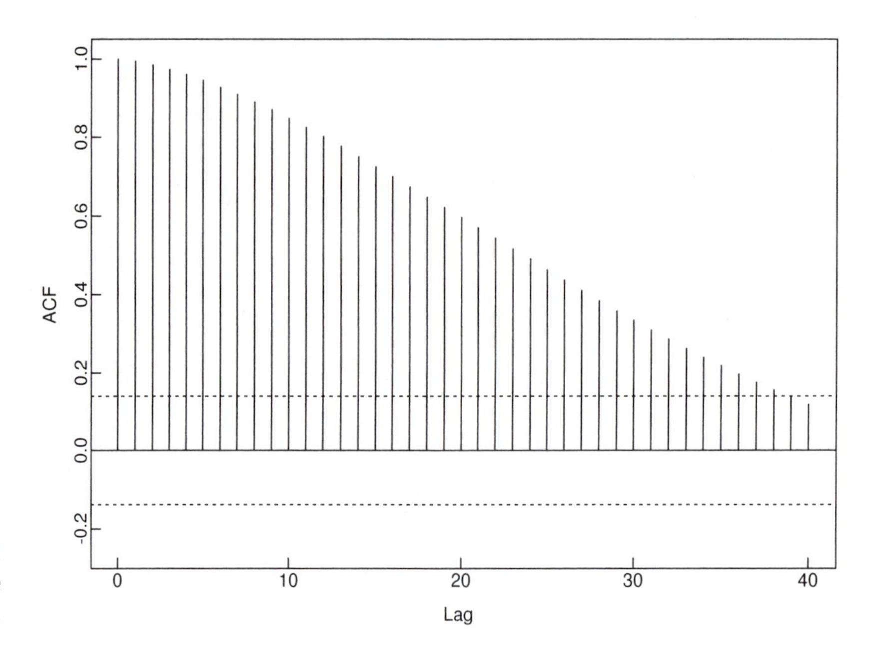

Figure 6-2
The sample ACF of the data in Figure 6.1.

If, therefore, we were given only the data and wished to find an appropriate model, it would be natural to apply the operator $\nabla = 1 - B$ repeatedly in the hope that for some j, $\{\nabla^j X_t\}$ will have a rapidly decaying sample autocorrelation function compatible with that of an ARMA process with no zeroes of the autoregressive polynomial near the unit circle. For this particular time series, one application of the operator ∇ produces the realization shown in Figure 6.4, whose sample ACF and PACF (Figures

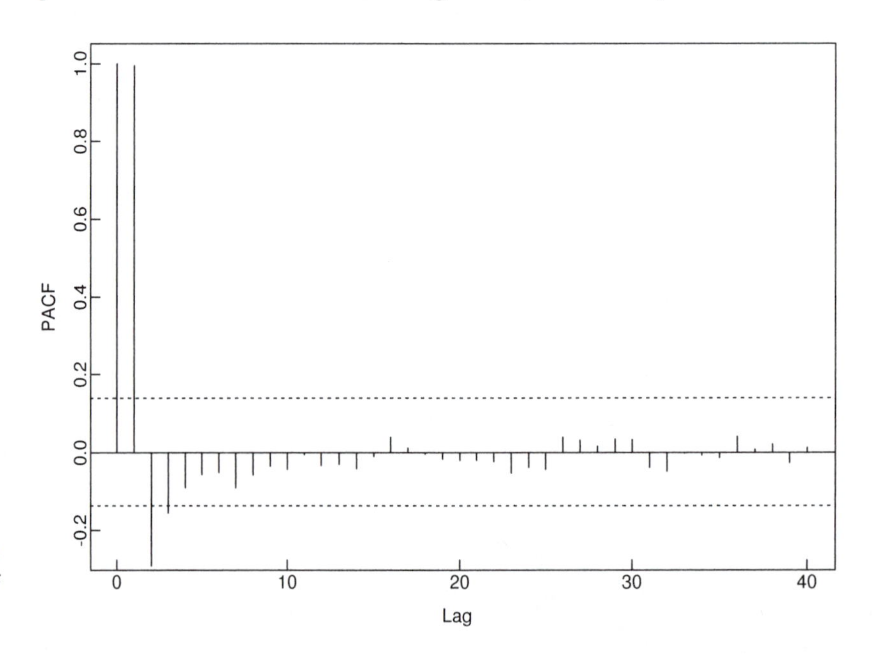

Figure 6-3
The sample PACF of the data in Figure 6.1.

6.5 and 6.6) suggest an AR(1) (or possibly AR(2)) model for $\{\nabla X_t\}$. The maximum likelihood estimates of ϕ and σ^2 obtained from PEST under the assumption that $E(\nabla X_t) = 0$ (found by *not* subtracting the mean after differencing the data) are .808 and .978, respectively, giving the model

$$(1 - 0.808B)(1 - B)X_t = Z_t, \quad \{Z_t\} \sim \mathrm{WN}(0, 0.978), \tag{6.1.2}$$

which bears a close resemblance to the true underlying process,

$$(1 - 0.8B)(1 - B)X_t = Z_t, \quad \{Z_t\} \sim \mathrm{WN}(0, 1). \tag{6.1.3}$$

Instead of differencing the series in Figure 6.1 we could proceed more directly by attempting to fit an AR(2) process as suggested by the sample PACF of the original series in Figure 6.3. Maximum likelihood estimation, carried out using PEST after fitting a preliminary model with Burg's algorithm and assuming that $EX_t = 0$, gives the model

$$(1 - 1.808B + 0.811B^2)X_t = (1 - 0.825B)(1 - .983B)X_t = Z_t,$$

$$\{Z_t\} \sim \mathrm{WN}(0, 0.970), \tag{6.1.4}$$

which, although stationary, has coefficients closely resembling those of the true non-stationary process (6.1.3). (To obtain the model (6.1.4), two optimizations were carried out in the *ARMA estimation* option of PEST, the first with the default settings and the second after setting the accuracy parameter to 0.00001.)

From a sample of finite length it will be extremely difficult to distinguish between a nonstationary process such as (6.1.3) for which $\phi^*(1) = 0$, and a process such as (6.1.4), which has very similar coefficients but for which ϕ^* has all of its zeroes outside

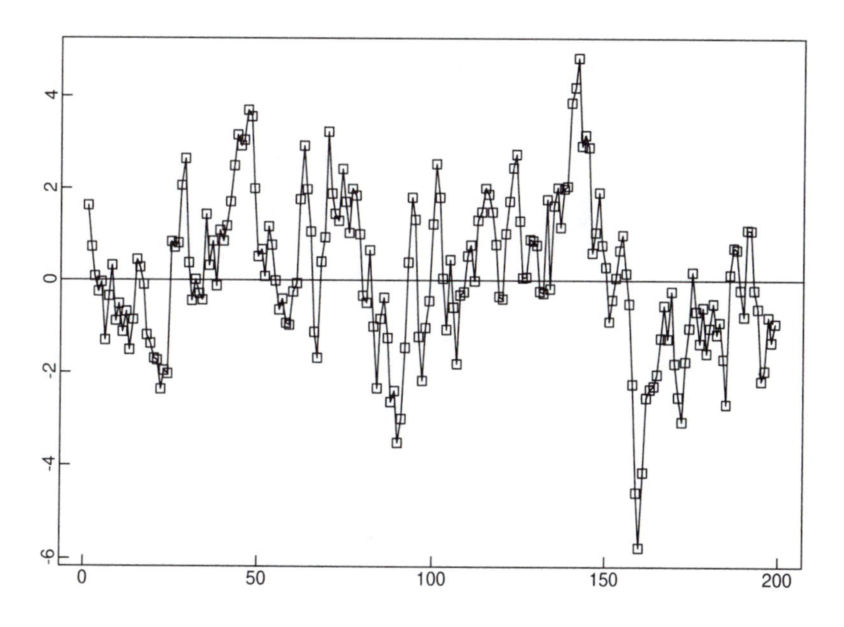

Figure 6-4 199 observations of the series $Y_t = \nabla X_t$ with $\{X_t\}$ as in Figure 6.1.

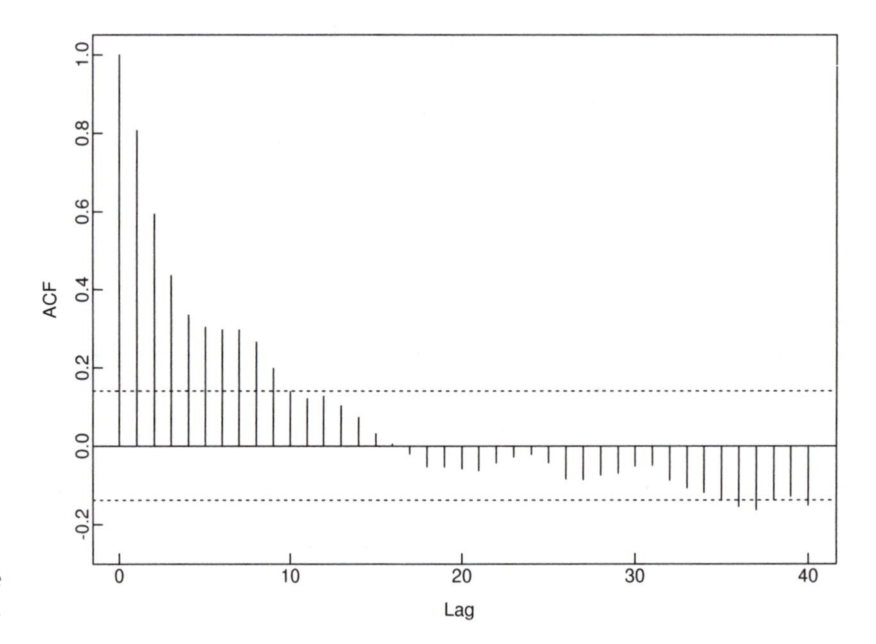

Figure 6-5
The sample ACF of the series $\{Y_t\}$ in Figure 6.4.

the unit circle. In either case, however, if it is possible by differencing to generate a series with rapidly decaying sample ACF, then the differenced data set can be fitted by a low-order ARMA process whose autoregressive polynomial ϕ^* has zeroes that are comfortably outside the unit circle. This means that the fitted parameters will be well away from the boundary of the allowable parameter set. This is desirable for numerical computation of parameter estimates and can be quite critical for some

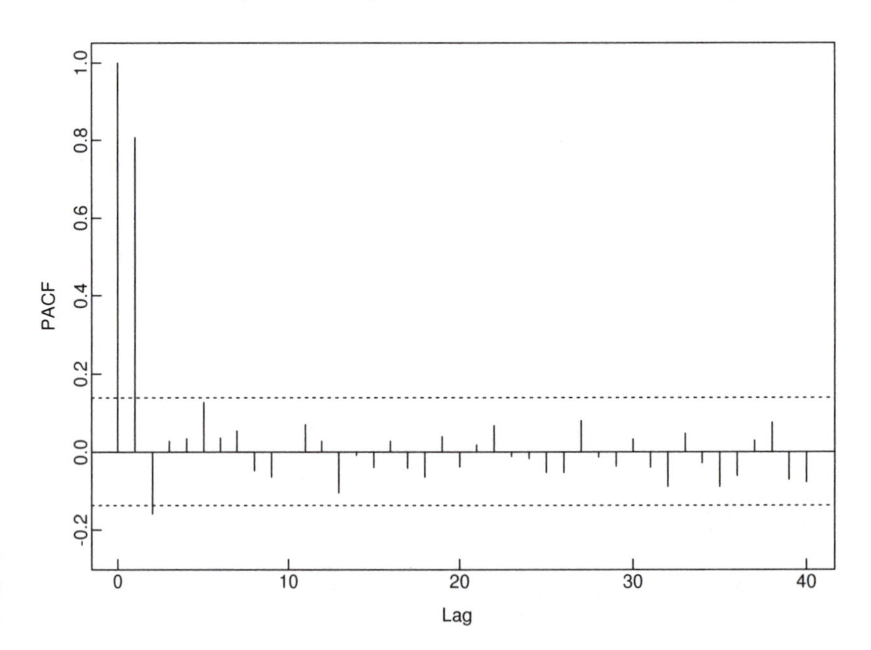

Figure 6-6
The sample PACF of the series $\{Y_t\}$ in Figure 6.4.

methods of estimation. For example, if we apply the Yule-Walker equations to fit an AR(2) model to the data in Figure 6.1, we obtain the model

$$(1 - 1.282B + 0.290B^2)X_t = Z_t, \quad \{Z_t\} \sim \text{WN}(0, 6.435), \tag{6.1.5}$$

which bears little resemblance to either the maximum likelihood model (6.1.4) or the true model (6.1.3). In this case the matrix $\hat{R}_2$ appearing in (5.1.7) is nearly singular.

An obvious limitation in fitting an ARIMA(p, d, q) process $\{X_t\}$ to data is that $\{X_t\}$ is permitted to be nonstationary only in a very special way, i.e., by allowing the polynomial $\phi^*(B)$ in the representation $\phi^*(B)X_t = Z_t$ to have a zero of multiplicity d at the point 1 on the unit circle. Such models are appropriate when the sample ACF is a slowly decaying positive function as in Figure 6.2, since sample autocorrelation functions of this form are associated with models $\phi^*(B)X_t = \theta(B)Z_t$ in which ϕ^* has a zero either at or close to 1.

Sample autocorrelations with slowly decaying oscillatory behavior as in Figure 6.8 are associated with models $\phi^*(B)X_t = \theta(B)Z_t$ in which ϕ^* has a zero close to $e^{i\omega}$ for some $\omega \in (-\pi, \pi]$ other than 0. Figure 6.8 is the sample ACF of the series of 200 observations in Figure 6.7, obtained from PEST by simulating the AR(2) process

$$X_t - (2r^{-1}\cos\omega)X_{t-1} + r^{-2}X_{t-2} = Z_t, \quad \{Z_t\} \sim \text{WN}(0, 1), \tag{6.1.6}$$

with $r = 1.005$ and $\omega = \pi/3$, i.e.,

$$X_t - 0.9950X_{t-1} + 0.9901X_{t-2} = Z_t, \quad \{Z_t\} \sim \text{WN}(0, 1).$$

The autocorrelation function of the model (6.1.6) can be derived by noting that

$$1 - (2r^{-1}\cos\omega)B + r^{-2}B^2 = (1 - r^{-1}e^{i\omega}B)(1 - r^{-1}e^{-i\omega}B) \tag{6.1.7}$$

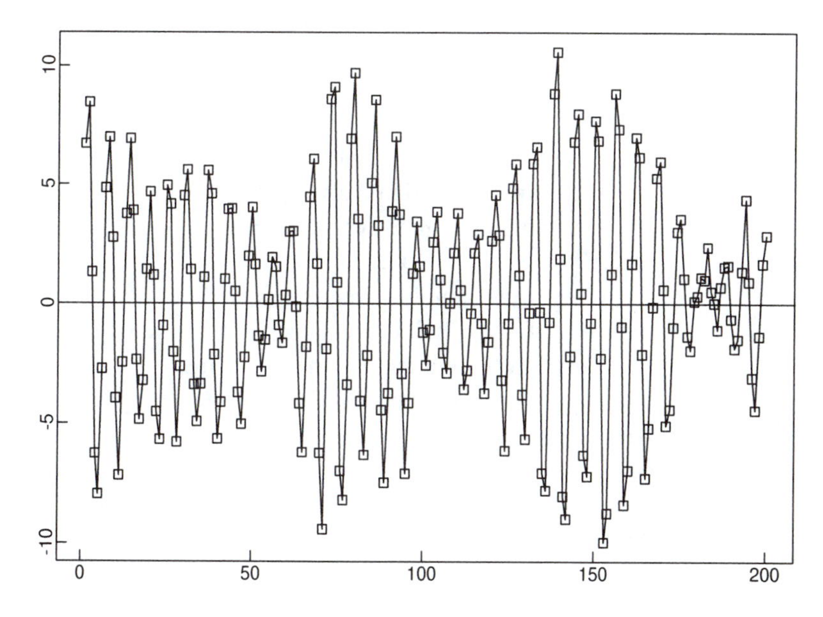

Figure 6-7
200 observations of the AR(2) process defined by (6.1.6) with $r = 1.005$ and $\omega = \pi/3$.

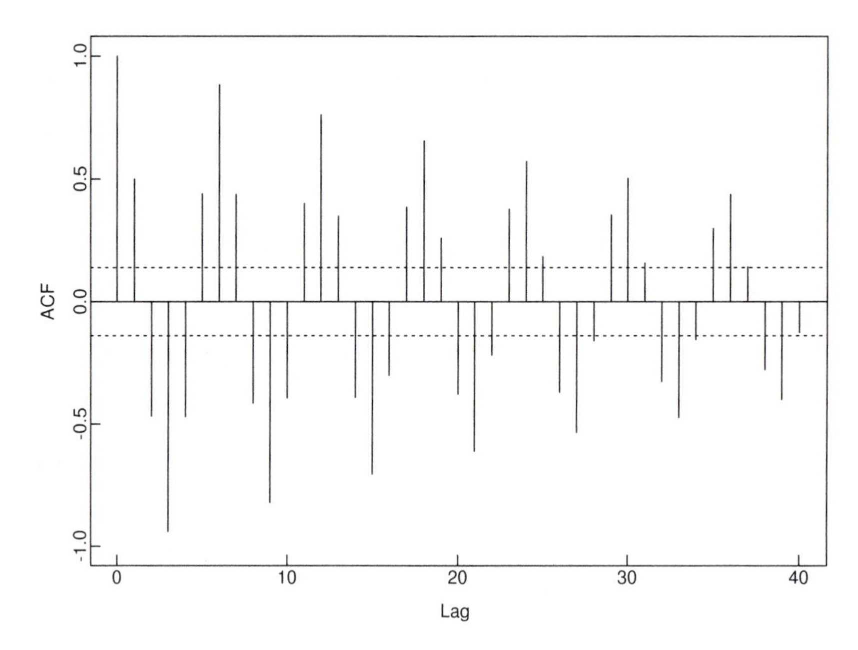

Figure 6-8 The sample ACF of the data in Figure 6.7.

and using (3.2.12). This gives

$$\rho(h) = r^{-h}\frac{\sin(h\omega + \psi)}{\sin\psi}, h \geq 0, \tag{6.1.8}$$

where

$$\tan\psi = \frac{r^2+1}{r^2-1}\tan\omega. \tag{6.1.9}$$

It is clear from these equations that

$$\rho(h) \to \cos(h\omega) \text{ as } r \downarrow 1. \tag{6.1.10}$$

With $r = 1.005$ and $\omega = \pi/3$ as in the model generating Figure 6.7, the model ACF (6.1.8) is a damped sine wave with damping ratio $1/1.005$ and period 6. These properties are reflected in the sample ACF shown in Figure 6.8. For values of r closer to 1, the damping will be even slower, as the model ACF approaches its limiting form (6.1.10).

If we were simply given the data shown in Figure 6.7, with no indication of the model from which it was generated, the slowly damped sinusoidal sample ACF with period 6 would suggest trying to make the sample ACF decay more rapidly by applying the operator (6.1.7) with $r = 1$ and $\omega = \pi/3$, i.e., $(1 - B + B^2)$. If it happens, as in this case, that the period $2\pi/\omega$ is close to some integer s (in this case 6), then the operator $1 - B^s$ can also be applied to produce a series with more rapidly decaying autocorrelation function (see also Section 6.5). Figures 6.9 and 6.10 show the sample autocorrelation functions obtained after applying the operators $1 - B + B^2$ and $1 - B^6$, respectively, to the data shown in Figure 6.7. For either one of these two

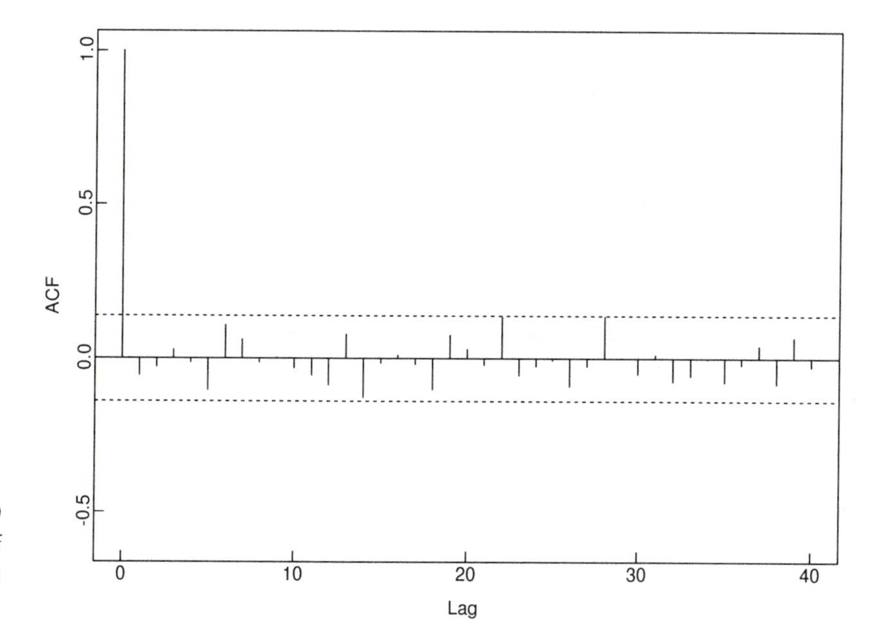

Figure 6-9 The sample ACF of $(1 - B + B^2)X_t$ with $\{X_t\}$ as in Figure 6.7.

differenced series, it is then not difficult to fit an ARMA model $\phi(B)X_t = \theta(B)Z_t$, for which the zeroes of ϕ are well outside the unit circle. Techniques for identifying and determining such ARMA models have already been introduced in Chapter 5. For convenience we shall collect these together in the following sections with a number of illustrative examples.

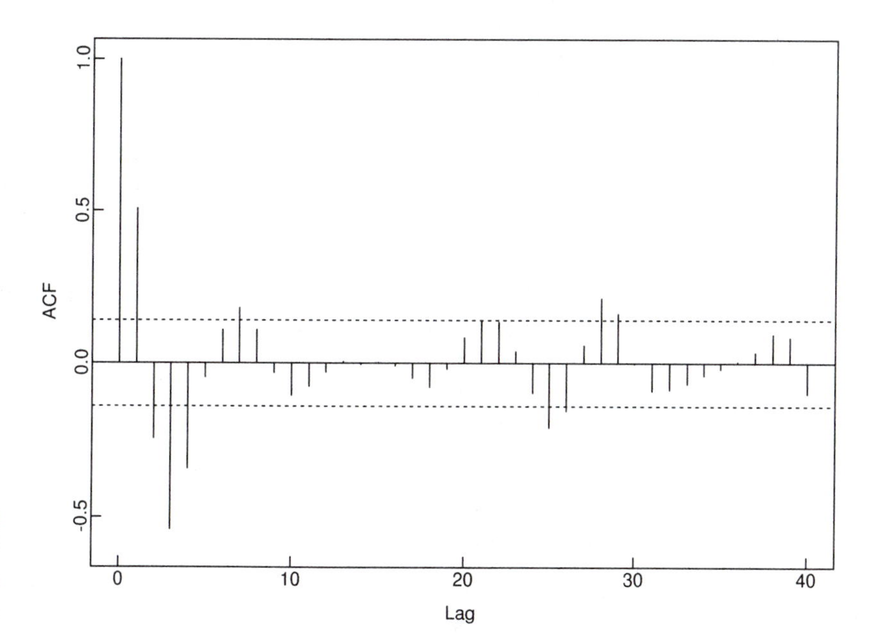

Figure 6-10 The sample ACF of $(1 - B^6)X_t$ with $\{X_t\}$ as in Figure 6.7.

6.2 Identification Techniques

(a) *Preliminary Transformations.* The estimation methods of Chapter 5 enable us to find, for given values of p and q, an ARMA(p, q) model to fit a given series of data. For this procedure to be meaningful it must be at least plausible that the data are in fact a realization of an ARMA process and in particular a realization of a stationary process. If the data display characteristics suggesting nonstationarity (e.g., trend and seasonality), then it may be necessary to make a transformation so as to produce a new series that is more compatible with the assumption of stationarity.

Deviations from stationarity may be suggested by the graph of the series itself or by the sample autocorrelation function or both.

Inspection of the graph of the series will occasionally reveal a strong dependence of variability on the level of the series, in which case the data should first be transformed to reduce or eliminate this dependence. For example, Figure 1.1 shows the Australian monthly red wine sales from January 1980, through October 1991, and Figure 1.17 shows how the increasing variability with sales level is reduced by taking natural logarithms of the original series. The logarithmic transformation $V_t = \ln U_t$ used here is in fact appropriate whenever $\{U_t\}$ is a series whose standard deviation increases linearly with the mean. For a systematic account of a general class of variance-stabilizing transformations, we refer the reader to Box and Cox (1964). The defining equation for the general Box-Cox transformation f_λ is

$$f_\lambda(U_t) = \begin{cases} \lambda^{-1}(U_t^\lambda - 1), & U_t \ge 0, \lambda > 0, \\ \ln U_t, & U_t > 0, \lambda = 0, \end{cases}$$

and the program PEST provides the option of applying f_λ (with $0 \le \lambda \le 1.5$) prior to the elimination of trend and/or seasonality from the data. In practice, if a Box-Cox transformation is necessary, it is often the case that either f_0 or $f_{0.5}$ is adequate.

Trend and seasonality are usually detected by inspecting the graph of the (possibly transformed) series. However, they are also characterized by autocorrelation functions that are slowly decaying and nearly periodic, respectively. The elimination of trend and seasonality was discussed in Section 1.5, where we described two methods:

i. "classical decomposition" of the series into a trend component, a seasonal component, and a random residual component, and
ii. differencing.

The program PEST (in the *Data entry* option) offers a choice between these techniques. The results of applying methods (i) and (ii) to the transformed red wine data $V_t = \ln U_t$ in Figure 1.17 are shown in Figures 6.11 and 6.12, respectively. Figure 6.11 was obtained from PEST by estimating and removing from $\{V_t\}$ a linear trend component and a seasonal component with period 12. Figure 6.12 was obtained by applying the operator $(1 - B^{12})$ to $\{V_t\}$. Neither of the two resulting series displays any

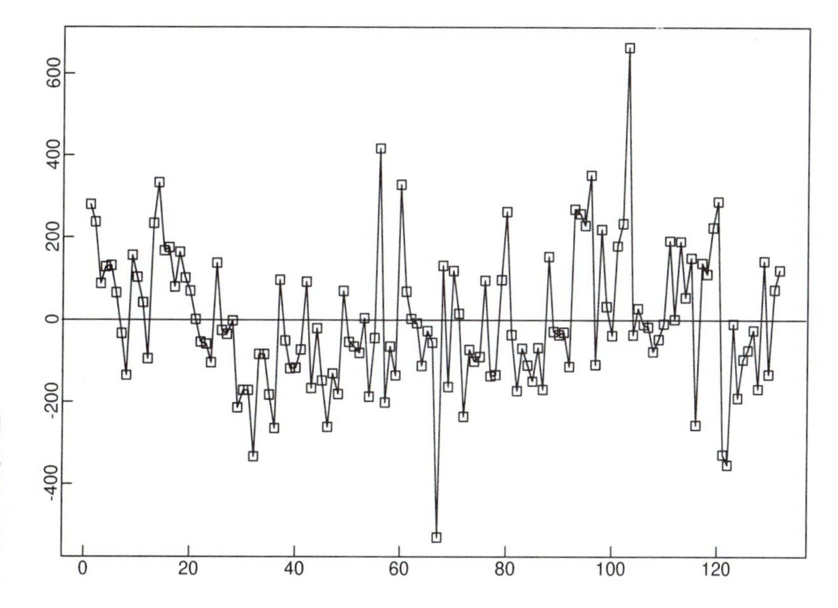

Figure 6-11 The Australian red wine data after taking natural logarithms and removing a seasonal component of period 12 and a linear trend.

apparent deviations from stationarity, nor do their sample autocorrelation functions. The sample ACF and PACF of $\{(1 - B^{12})V_t\}$ are shown in Figures 6.13 and 6.14, respectively.

After the elimination of trend and seasonality, it is still possible that the sample autocorrelation function may appear to be that of a nonstationary (or nearly nonstationary) process, in which case further differencing may be carried out.

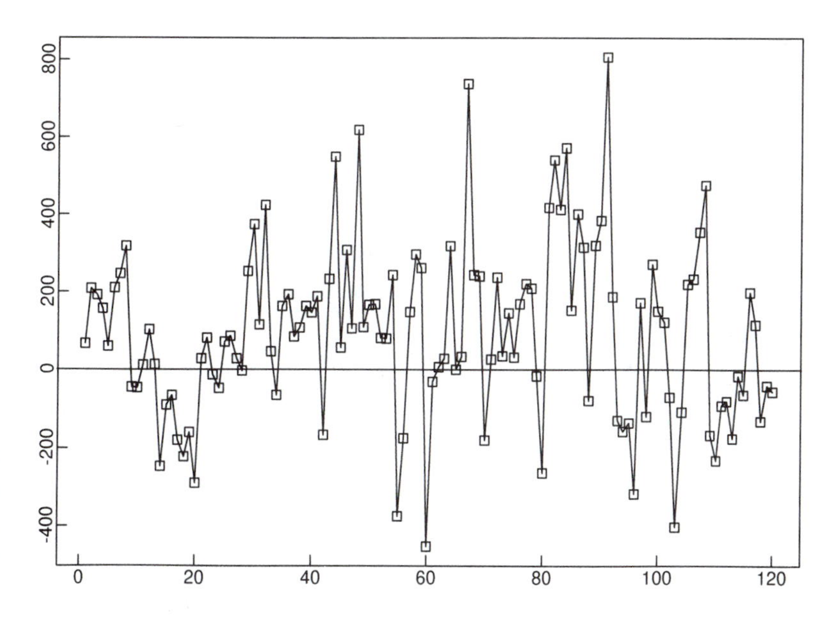

Figure 6-12 The Australian red wine data after taking natural logarithms and differencing at lag 12.

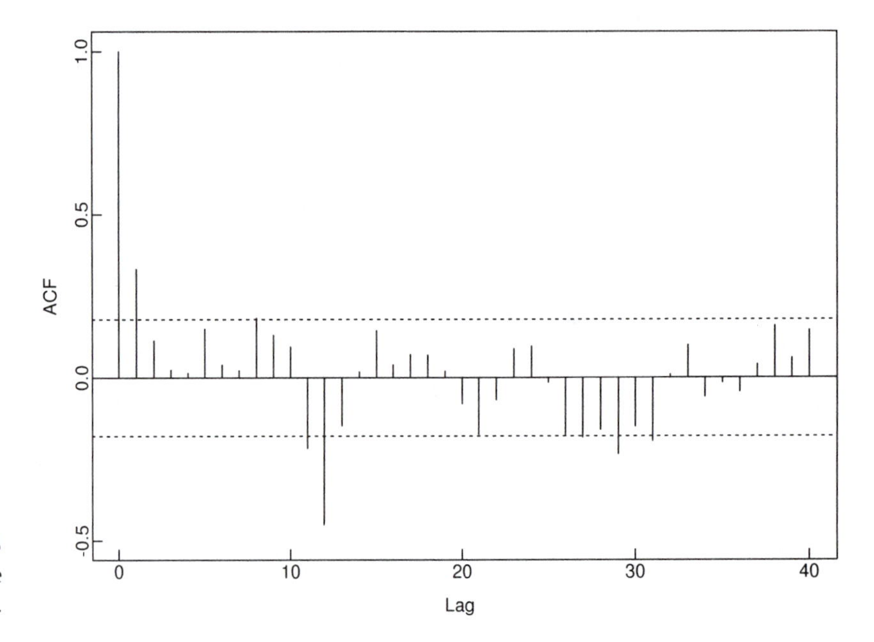

Figure 6-13 The sample ACF of the data in Figure 6.12.

(b) *Identification and Estimation.* Let $\{X_t\}$ be the mean-corrected transformed series found as described in (a). The problem now is to find the most satisfactory ARMA(p, q) model to represent $\{X_t\}$. If p and q were known in advance this would be a straightforward application of the estimation techniques described in Chapter 5. However, this is usually not the case, so it becomes necessary also to identify appropriate values for p and q.

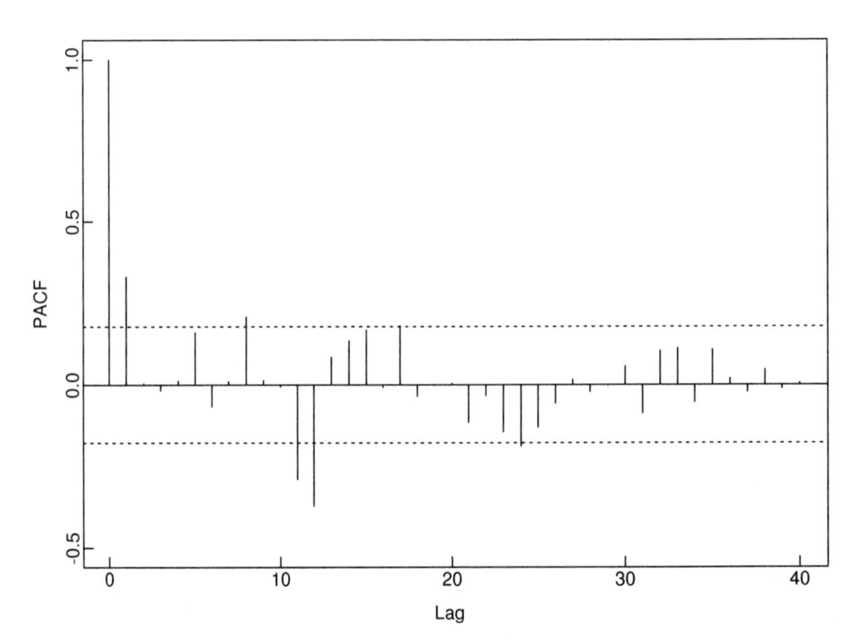

Figure 6-14 The sample PACF of the data in Figure 6.12.

It might appear at first sight that the higher the values chosen for p and q, the better the resulting fitted model will be. However, as pointed out in Section 5.5, estimation of too large a number of parameters introduces estimation errors that adversely affect the use of the fitted model for prediction as illustrated in Section 5.4. We therefore minimize one of the model selection criteria discussed in Section 5.5 in order to choose the values of p and q. Each of these criteria includes a penalty term to discourage the fitting of too many parameters. We shall base our choice of p and q primarily on the minimization of the AICC statistic, defined as

$$\text{AICC}(\boldsymbol{\phi}, \boldsymbol{\theta}) = -2 \ln L(\boldsymbol{\phi}, \boldsymbol{\theta}, S(\boldsymbol{\phi}, \boldsymbol{\theta})/n) + 2(p+q+1)n/(n-p-q-2), \quad (6.2.1)$$

where $L(\boldsymbol{\phi}, \boldsymbol{\theta}, \sigma^2)$ is the likelihood of the data under the Gaussian ARMA model with parameters $(\boldsymbol{\phi}, \boldsymbol{\theta}, \sigma^2)$ and $S(\boldsymbol{\phi}, \boldsymbol{\theta})$ is the residual sum of squares defined in (5.2.11). Once a model has been found that minimizes the AICC value, it is then necessary to check the model for goodness of fit (essentially by checking that the residuals are like white noise) as discussed in Section 5.3.

For any fixed values of p and q, the maximum likelihood estimates of $\boldsymbol{\phi}$ and $\boldsymbol{\theta}$ are the values that minimize the AICC. Hence, the minimum AICC model (over any given range of p and q values) can be found by computing the maximum likelihood estimators for each fixed p and q and choosing from these the maximum likelihood model with the smallest value of AICC. However, exact maximum likelihood estimation is relatively slow, so it is useful to get a preliminary idea of the optimum orders by finding *preliminary* estimators of $\boldsymbol{\phi}$ and $\boldsymbol{\theta}$ and computing the corresponding values of AICC from (6.2.1). Preliminary estimates of p and q are the values for which the AICC of the preliminary coefficient estimates is minimum.

The program PEST automates the determination of the preliminary estimate of p when a purely autoregressive process is being fitted. In the option *Preliminary estimation*, if you choose -1 for the order of autoregression, the program will fit autoregressions from orders 0 to 26 and select from these the model with smallest AICC value. This optimization can be carried out using either Burg or Yule-Walker estimation. (The AICC values from Burg's algorithm are usually lower than those from the Yule-Walker equations and hence closer to the values for the corresponding maximum likelihood models.) For fitting more general ARMA(p, q) models, minimization of the AICC value must be done by checking one model at a time. The search can be done systematically, starting with values (p, q) such that $p+q=0$ and working up through values such that $p+q=1$, $p+q=2, \ldots$, etc. For models with $q > 0$ the difference between the AICC for the preliminary and the maximum likelihood estimates is generally greater than for models with $q=0$. When $q > 0$, therefore, the estimates of p and q obtained by checking the AICC of preliminary models are less reliable indicators than in the case $q=0$. In fitting models with $q > 0$, minimization of the AICC is best left until maximum likelihood estimation has been performed, with preliminary estimation used only to provide a model with which to initialize the likelihood maximization.

The steps in model identification and estimation can be summarized as follows:

- After transforming the data (if necessary) to make a stationary model reasonable, examine the sample ACF and PACF to get some idea of potential p and q values.
- Using the option *Preliminary estimation* of PEST make preliminary estimates of the coefficients for selected p and q values. (If you are checking purely autoregressive models, enter -1 for the selected autoregressive order and select Burg's algorithm. The program will then give you the minimum AICC Burg model with $p \leq 26$.)
- Starting from any promising model found in the *Preliminary estimation* stage, go to the *ARMA estimation* option and find the maximum likelihood model with the same orders as the preliminary model.
- From the fitted maximum likelihood models choose the one with smallest AICC value, making a note of other candidate models whose AICC value is close to the minimum.
- Examination of the fitted coefficients and their standard errors may suggest that some of them can be set to zero. If this is the case then a **subset model** can be fitted by selecting the option *Constrain optimized coefficients* and setting the selected coefficients to zero. Optimization will then give the maximum likelihood model with the chosen coefficients constrained to be zero. The constrained model is assessed by comparing its AICC value with those of the other candidate models.
- Check the candidate model(s) for goodness of fit as described in Section 5.3. These tests can be performed in the *ARMA estimation* option by selecting *File and analyze residuals* immediately after maximizing the likelihood.

Example 6.2.1 The Australian red wine data

Let $\{X_1, \ldots, X_{130}\}$ denote the series obtained from the red wine data of Example 1.1.1 after taking natural logarithms, differencing at lag 12, and subtracting the mean (0.0681) of the differences. The data prior to mean correction are shown in Figure 6.12. The sample PACF of $\{X_t\}$, shown in Figure 6.14, suggests that an AR(12) model might be appropriate for this series. To explore this possibility we use the PEST option *Preliminary estimation* with autoregressive order set equal to -1 and Burg's algorithm as the method of estimation. As anticipated, the fitted Burg models do indeed have minimum AICC when $p = 12$. The fitted model is

$$(1 - .245B - .069B^2 - .012B^3 - .021B^4 - .200B^5 + .025B^6 + .004B^7$$
$$-.133B^8 + .010B^9 - .095B^{10} + .118B^{11} + .384B^{12})X_t = Z_t,$$

with $\{Z_t\} \sim \text{WN}(0, 0.0135)$ and AICC value -158.77. Returning to the main menu of PEST and selecting *ARMA estimation* followed by *Optimize with current settings* gives the maximum likelihood AR(12) model, which is very similar to the Burg model and has AICC value -158.87. Inspection of the standard errors of the coefficient estimators suggests the possibility of setting those at lags 2,3,4,6,7,9,10, and 11 equal to zero. If we do this in the *Constrain optimized coefficients* option in the Estimation Menu and then reoptimize, we obtain the model

$$(1 - .270B - .224B^5 - .149B^8 + .099B^{11} + .353B^{12})X_t = Z_t,$$

with $\{Z_t\} \sim \mathrm{WN}(0, 0.0138)$ and AICC value -172.49. The sample ACF suggests that an MA model might also be appropriate for this data set. If we check preliminary MA(q) models of orders 1 through 15 using either the innovations or Hannan-Rissanen algorithm we find that the minimum AICC model is the one obtained from the innovations algorithm with $q = 13$. Switching to *ARMA estimation*, retaining the preliminary MA(13) model and maximizing the likelihood, we obtain an MA(13) model with AICC value equal to -172.53 (somewhat better than the unconstrained maximum likelihood AR(12) model found above). Inspecting the standard errors of the coefficient estimators suggests the possibility of setting the coefficients at lags 3,4,6,7,9, and 11 equal to zero. Doing this and reoptimizing, we arrive at the subset MA(13) model,

$$X_t = (1 + .248B + .205B^2 + .230B^5 + .275B^8 + .266B^{10} - .608B^{12} - .263B^{13})Z_t,$$

with $\{Z_t\} \sim \mathrm{WN}(0, 0.0110)$ and AICC value -181.52. In order to check ARMA(p, q) models with both p and q greater than zero we return to the *Preliminary estimation* option of PEST and enter selected p and q values. A convenient initial search procedure is to set $p = q$ and gradually increase from $p = 0$ upward using the Hannan-Rissanen algorithm (the innovations algorithm for mixed models has a tendency to return noncausal models that cannot be used to initialize the maximum likelihood estimation). It soon becomes clear that the AICC of the preliminary mixed models is substantially worse than that of the pure AR and MA models fitted previously. At this stage the most promising model is the subset MA(13) candidate. To confirm this we need to check goodness of fit using the *File and analyze residuals* option of the estimation menu in PEST. We find that all the tests are passed and that the sample ACF and PACF of the residuals from the subset MA(13) model are more compatible with white noise than those from the subset AR(12). Since it also has substantially smaller AICC value, we therefore select the former. Although this model is not invertible, the program PEST does not require this of fitted models since the algorithms it uses can compute exact predictors and mean squared errors for either invertible or noninvertible models. We note, however, that the white noise variance and the one-step mean square prediction error are not the same for noninvertible models. □

Example 6.2.2 The lake data

Let $\{X_t, t = 1, \ldots, 98\}$ denote the mean-corrected lake data considered earlier in Examples 5.1.4, 5.1.6, 5.1.7, and 5.2.5. Conducting a systematic search for the minimum AICC model, we begin with preliminary AR estimation, selecting order -1 and Burg's algorithm. This gives $p = 2$ for the order of the preliminary Burg model with smallest AICC value. Similarly the preliminary Yule-Walker model with smallest AICC has $p = 2$ (but slightly larger AICC than the Burg model). Checking preliminary MA models fitted by the innovations algorithm, we find that the minimum AICC model has $q = 7$. Similarly the preliminary Hannan-Rissanen MA model with smallest AICC has $q = 7$ (but larger AICC than the innovations model). Checking the preliminary mixed models (i.e., those with both p and q greater than 0, starting from $p = q = 0$ and working up), we find that both the Hannan-Rissanen and innovation algorithms give minimum AICC at $p = q = 1$, with the latter giving a slightly lower AICC model. Notice also that for higher order models the innovations method has more tendency to give noncausal models (which are of no further value) than the Hannan-Rissanen algorithm. Preliminary estimation thus suggests three candidate models, AR(2), MA(7), and ARMA(1,1). To finalize our selection we need to carry out maximum likelihood estimation and search for the minimum AICC model, guided by these preliminary results and using the preliminary models as initial models for likelihood maximization. It is found that the maximum likelihood ARMA(1,1) model gives the smallest AICC value, with the maximum likelihood AR(2) model a close second. The results of tests for goodness of fit of the ARMA(1,1) model were given in Section 5.3 and were all found to be satisfactory. □

6.3 Unit Roots in Time Series Models

The unit root problem in time series arises when either the autoregressive or moving average polynomial of an ARMA model has a root on or near the unit circle. A unit root in either of these polynomials has important implications for modelling. For example, a root near 1 of the autoregressive polynomial suggests that the data should be differenced before fitting an ARMA model, whereas a root near 1 of the moving average polynomial indicates that the data were overdifferenced. In this section, we consider inference procedures for detecting the presence of a unit root in the autoregressive and moving average polynomials.

6.3.1 Unit Roots in Autoregressions

In Section 6.1, we discussed the use of differencing to transform a nonstationary time series with a slowly decaying sample ACF and values near 1 at small lags into one with a rapidly decreasing sample ACF. The degree of differencing of a time series $\{X_t\}$ was largely determined by applying the difference operator repeatedly until the sample ACF of $\{\nabla^d X_t\}$ decays quickly. The differenced time series could then be modelled by a low order ARMA(p,q) process, and hence the resulting ARIMA(p,d,q) model for the original data has an autoregressive polynomial, $(1-\phi_1 z-\cdots-\phi_p z^p)(1-z)^d$ (see (6.1.1)), with d roots on the unit circle. In this subsection we discuss a more systematic approach to testing for the presence of a unit root of the autoregressive polynomial in order to decide whether or not a time series should be differenced. This approach was pioneered by Dickey and Fuller (1979).

Let $X_1,\ldots,X_n$ be observations from the AR(1) model

$$X_t-\mu=\phi_1(X_{t-1}-\mu)+Z_t, \qquad \{Z_t\}\sim \mathrm{WN}(0,\sigma^2), \tag{6.3.1}$$

where $|\phi_1|<1$ and $\mu=EX_t$. For large n, the maximum likelihood estimator $\hat\phi_1$ of ϕ_1 is approximately $\mathrm{N}(\phi_1,(1-\phi_1^2)/n)$. For the unit root case, this normal approximation is no longer applicable, even asymptotically, which precludes its use for testing the unit root hypothesis $H_0:\phi_1=1$ vs. $H_1:\phi_1<1$. To construct a test of H_0, write the model (6.3.1) as

$$\nabla X_t=X_t-X_{t-1}=\phi_0^*+\phi_1^*X_{t-1}+Z_t, \qquad \{Z_t\}\sim \mathrm{WN}(0,\sigma^2), \tag{6.3.2}$$

where $\phi_0^*=\mu(1-\phi_1)$ and $\phi_1^*=\phi_1-1$. Now let $\hat\phi_1^*$ be the ordinary least squares (OLS) estimator of ϕ_1^* found by regressing ∇X_t on 1 and X_{t-1}. The estimated standard error of $\hat\phi_1^*$ is

$$\widehat{SE}(\hat\phi_1^*)=S\Big/\left(\sum_{t=2}^{n}(X_{t-1}-\bar X)^2\right)^{1/2},$$

where $S^2=\sum_{t=2}^{n}(\nabla X_t-\hat\phi_0^*-\hat\phi_1^*X_{t-1})^2/(n-3)$ and $\bar X$ is the sample mean of $X_1,\ldots,X_{n-1}$. Dickey and Fuller derived the limit distribution as $n\to\infty$ of the t-ratio

$$\hat\tau_\mu:=\hat\phi_1^*/\widehat{SE}(\hat\phi_1^*) \tag{6.3.3}$$

under the unit root assumption $\phi_1^*=0$, from which a test of the null hypothesis $H_0:\phi_1=1$ can be constructed. The .01, .05, and .10 quantiles of the limit distribution of $\hat\tau_\mu$ (see Table 8.5.2 of Fuller, 1976) are -3.43, -2.86, and -2.57, respectively. The augmented Dickey-Fuller test then rejects the null hypothesis of a unit root at say level .05 if $\hat\tau_\mu<-2.86$. Notice that the cutoff value for this test statistic is much smaller than the standard cutoff value of -1.645 obtained from the normal approximation to the t-distribution so that the unit root hypothesis is less likely to be rejected using the correct limit distribution.

The above procedure can be extended to the case when $\{X_t\}$ follows the AR(p) model with mean μ given by

$$X_t-\mu=\phi_1(X_{t-1}-\mu)+\cdots+\phi_p(X_{t-p}-\mu)+Z_t, \qquad \{Z_t\}\sim \mathrm{WN}(0,\sigma^2).$$

This model can be rewritten as (see Problem 6.2)

$$\nabla X_t = \phi_0^* + \phi_1^* X_{t-1} + \phi_2^* \nabla X_{t-1} + \cdots + \phi_p^* \nabla X_{t-p+1} + Z_t, \tag{6.3.4}$$

where $\phi_0 = \mu(1-\phi_1-\cdots-\phi_p)$, $\phi_1^* = \sum_{i=1}^p \phi_i - 1$ and $\phi_j^* = -\sum_{i=j}^p \phi_i$, $j = 2, \ldots, p$. If the autoregressive polynomial has a unit root at 1, then $0 = \phi(1) = -\phi_1^*$, and the differenced series, $\{\nabla X_t\}$, is an AR$(p-1)$ process. Consequently, testing the hypothesis of a unit root at 1 of the autoregressive polynomial is equivalent to testing $\phi_1^* = 0$. As in the AR(1) example, ϕ_1^* can be estimated as the coefficient of X_{t-1} in the OLS regression of ∇X_t onto $1, X_{t-1}, \nabla X_{t-1}, \ldots, \nabla X_{t-p+1}$. For large n the t-ratio,

$$\hat{\tau}_\mu := \hat{\phi}_1^* / \widehat{SE}(\hat{\phi}_1^*), \tag{6.3.5}$$

where $\widehat{SE}(\hat{\phi}_1^*)$ is the estimated standard error of $\hat{\phi}_1^*$, has the same limit distribution as the test statistic in (6.3.3). The augmented Dickey-Fuller test in this case is applied in exactly the same manner as for the AR(1) case using the test statistic (6.3.5) and the cutoff values given above.

Example 6.3.1 Consider testing the time series of Example 6.1.1 (see Figure 6.1) for the presence of a unit root in the autoregressive operator. The sample PACF in Figure 6.3 suggests fitting an AR(2) or possibly an AR(3) model to the data. Regressing ∇X_t on $1, X_{t-1}, \nabla X_{t-1}, \nabla X_{t-2}$ for $t = 4, \ldots, 200$ using OLS gives

$$\nabla X_t = \underset{(.1135)}{.1503} - \underset{(.0028)}{.0041} X_{t-1} + \underset{(.0707)}{.9335} \nabla X_{t-1} - \underset{(.0708)}{.1548} \nabla X_{t-2} + Z_t,$$

where $\{Z_t\} \sim \mathrm{WN}(0, .9639)$. The test statistic for testing the presence of a unit root is

$$\hat{\tau}_\mu = \frac{-.0041}{.0028} = -1.464.$$

Since $-1.464 > -2.57$, the unit root hypothesis is not rejected at level .10. In contrast, if we had mistakenly used the t-distribution with 193 degrees of freedom as an approximation to $\hat{\tau}_\mu$, then we would have rejected the unit root hypothesis at the .10 level (p-value is .074). The t-ratios for the other coefficients, ϕ_0^*, ϕ_2^*, and ϕ_3^*, have an approximate t-distribution with 193 degrees of freedom. Based on these t-ratios, the intercept should be 0 while the coefficient of ∇X_{t-2} is barely significant. The evidence is much stronger in favor of a unit root if the analysis is repeated without a mean term. The fitted model without a mean term is

$$\nabla X_t = \underset{(.0018)}{.0012} X_{t-1} + \underset{(.0707)}{.9395} \nabla X_{t-1} - \underset{(.0709)}{.1585} \nabla X_{t-2} + Z_t,$$

where $\{Z_t\} \sim \mathrm{WN}(0, .9677)$. The .01, .05, and .10 cutoff values for the corresponding test statistic when a mean term is excluded from the model are -2.58, -1.95, and

-1.62 (see Table 8.5.2 of Fuller, 1976). In this example, the test statistic is

$$\hat{\tau} = \frac{-.0012}{.0018} = -.667,$$

which is substantially larger than the .10 cutoff value of -1.62. □

Further extensions of the above test to AR models with $p = O(n^{1/3})$ and to ARMA(p, q) models can be found in Said and Dickey (1984). However, as reported in Schwert (1987) and Pantula (1991), this test must be used with caution if the underlying model orders are not correctly specified.

6.3.2 Unit Roots in Moving Averages

A unit root in the moving average polynomial can have a number of interpretations depending on the modelling application. For example, let $\{X_t\}$ be a causal-invertible ARMA(p, q) process satisfying the equations

$$\phi(B)X_t = \theta(B)Z_t, \qquad \{Z_t\} \sim \text{WN}(0, \sigma^2).$$

Then the differenced series $Y_t := \nabla X_t$ is a noninvertible ARMA$(p, q+1)$ process with moving average polynomial $\theta(z)(1-z)$. Consequently, testing for a unit root in the moving average polynomial is equivalent to testing that the time series has been overdifferenced.

As a second application, it is possible to distinguish between the competing models

$$\nabla^k X_t = a + V_t$$

and

$$X_t = c_0 + c_1 t + \cdots + c_k t^k + W_t,$$

where $\{V_t\}$ and $\{W_t\}$ are invertible ARMA processes. For the former model the differenced series $\{\nabla^k X_t\}$ has no moving average unit roots, while for the latter model $\{\nabla^k X_t\}$ has a multiple moving average unit root of order k. We can therefore distinguish between the two models by using the observed values of $\{\nabla^k X_t\}$ to test for the presence of a moving average unit root.

We confine our discussion of unit root tests to first-order moving-average models, the general case being considerably more complicated and not fully resolved. Let $X_1, \ldots, X_n$ be observations from the MA(1) model

$$X_t = Z_t + \theta Z_{t-1}, \qquad \{Z_t\} \sim \text{IID}(0, \sigma^2).$$

Davis and Dunsmuir (1996) showed that under the assumption $\theta = -1$, $n(\hat{\theta}+1)$ ($\hat{\theta}$ is the maximum likelihood estimator) converges in distribution. A test of $H_0 : \theta = -1$ vs. $H_1 : \theta > -1$ can be fashioned on this limiting result by rejecting H_0 when

$$\hat{\theta} > -1 + c_\alpha / n,$$

where c_α is the $(1-\alpha)$-quantile of the limit distribution of $n(\hat{\theta}+1)$. (From Table 3.2 of Davis, Chen, and Dunsmuir (1995), $c_{.01} = 11.93$, $c_{.05} = 6.80$, and $c_{.10} = 4.90$.) In particular, if $n = 50$, then the null hypothesis is rejected at level .05 if $\hat{\theta} > -1 + 6.80/50 = -.864$.

The likelihood ratio test can also be used for testing the unit root hypothesis. The likelihood ratio for this problem is $L(-1, S(-1)/n)/L(\hat{\theta}, \hat{\sigma}^2)$, where $L(\theta, \sigma^2)$ is the Gaussian likelihood of the data based on an MA(1) model, $S(-1)$ is the sum of squares given by (5.2.11) when $\theta = -1$ and $\hat{\theta}$ and $\hat{\sigma}^2$ are the maximum likelihood estimators of θ and σ^2. The null hypothesis is rejected at level α if

$$\lambda_n := -2\ln\left(\frac{L(-1, S(-1)/n)}{L(\hat{\theta}, \hat{\sigma}^2)}\right) > c_{LR,\alpha}$$

where the cutoff value is chosen such that $P_{\theta=-1}[\lambda_n > c_{LR,\alpha}] = \alpha$. The limit distribution of λ_n was derived by Davis et al. (1995), who also gave selected quantiles of the limit. It was found that these quantiles provide a good approximation to their finite sample counterparts for time series of length $n \geq 50$. The limiting quantiles for λ_n under H_0 are $c_{LR,.01} = 4.41$, $c_{LR,.05} = 1.94$, and $c_{LR,.10} = 1.00$.

Example 6.3.2 For the overshort data $\{X_t\}$ of Example 3.2.8, the maximum likelihood MA(1) model for the mean corrected data $\{Y_t = X_t + 4.035\}$ was (see Example 5.4.1)

$$Y_t = Z_t - 0.818Z_{t-1}, \quad \{Z_t\} \sim \text{WN}(0, 2040.75).$$

In the structural formulation of this model given in Example 3.2.8, the moving average parameter θ was related to the measurement error variances σ_U^2 and σ_V^2 through the equation

$$\frac{\theta}{1+\theta^2} = \frac{-\sigma_U^2}{2\sigma_U^2 + \sigma_V^2}.$$

(These error variances correspond to the daily measured amounts of fuel in the tank and the daily measured adjustments due to sales and deliveries.) A value of $\theta = -1$ indicates that there is no appreciable measurement error due to sales and deliveries (i.e., $\sigma_V^2 = 0$) and hence testing for a unit root in this case is equivalent to testing that $\sigma_U^2 = 0$. Assuming the mean is known, the unit root hypothesis is rejected at $\alpha = .05$ since $-.818 > -1 + 6.80/57 = -.881$. The evidence against H_0 is stronger using the likelihood ratio statistic. Using PEST and entering the MA(1) model $\theta = -1$ and $\sigma^2 = 2203.12$, we find that $-2\ln L(-1, 2203.12) = 604.584$ while $-2\ln L(\hat{\theta}, \hat{\sigma}^2) = 597.267$. Comparing the likelihood ratio statistic $\lambda_n = 604.584 - 597.267 = 7.317$ with the cutoff value $c_{LR,.01}$, we reject H_0 at level $\alpha = .01$ and conclude that the measurement error associated with sales and deliveries is nonzero.

In the above example, it was assumed that the mean was known. In practice, these tests should be adjusted for the fact that the mean is also being estimated.

Tanaka (1990) proposed a locally best invariant unbiased (LBIU) test for the unit root hypothesis. It was found that the LBIU test has slightly larger power than the likelihood ratio test for alternatives close to $\theta = -1$ but has less power for alternatives further away from -1 (see Davis et al., 1995). The LBIU test has been extended to cover more general models by Tanaka (1990) and Tam and Reinsel (1995). Similar extensions to tests based on the maximum likelihood estimator and the likelihood ratio statistic have been explored in Davis, Chen, and Dunsmuir (1996). □

6.4 Forecasting ARIMA Models

In this section we demonstrate how the methods of Section 3.3 and 5.4 can be adapted to forecast the future values of an ARIMA(p, d, q) process $\{X_t\}$. (The required numerical calculations can all be carried out using the program PEST.)

If $d \geq 1$ the first and second moments EX_t and $E(X_{t+h}X_t)$ are not determined by the difference equations (6.1.1). We cannot expect therefore to determine best linear predictors for $\{X_t\}$ without further assumptions.

For example, suppose that $\{Y_t\}$ is a causal ARMA(p, q) process and that X_0 is any random variable. Define

$$X_t = X_0 + \sum_{j=1}^{t} Y_j, t = 1, 2, \ldots.$$

Then $\{X_t, t \geq 0\}$ is an ARIMA$(p, 1, q)$ process with mean $EX_t = EX_0$ and autocovariances $E(X_{t+h}X_t) - (EX_0)^2$ that depend on $\text{Var}(X_0)$ and $\text{Cov}(X_0, Y_j), j = 1, 2, \ldots$. The best linear predictor of X_{n+1} based on $\{1, X_0, X_1, \ldots, X_n\}$ is the same as the best linear predictor in terms of the set $\{1, X_0, Y_1, \ldots, Y_n\}$ since each linear combination of the latter is a linear combination of the former and vice versa. Hence, using P_n to denote best linear predictor in terms of either set and using the linearity of P_n, we can write

$$P_n X_{n+1} = P_n(X_0 + Y_1 + \cdots + Y_{n+1}) = P_n(X_n + Y_{n+1}) = X_n + P_n Y_{n+1}.$$

To evaluate $P_n Y_{n+1}$ it is necessary (see Section 2.5) to know $E(X_0 Y_j), j = 1, \ldots, n+1$, and EX_0^2. However, if we assume that X_0 is uncorrelated with $\{Y_t, t \geq 1\}$ then $P_n Y_{n+1}$ is the same (Problem 6.5) as the best linear predictor $\hat{Y}_{n+1}$ of Y_{n+1} in terms of $\{1, Y_1, \ldots, Y_n\}$, which can be calculated as described in Section 3.3. The assumption that X_0 is uncorrelated with $Y_1, Y_2, \ldots$, therefore suffices to determine the best linear predictor $P_n X_{n+1}$ in this case.

Turning now to the general case, we shall assume that our observed process $\{X_t\}$ satisfies the difference equations

$$(1 - B)^d X_t = Y_t, t = 1, 2, \ldots,$$

where $\{Y_t\}$ is a causal ARMA(p, q) process, and that the random vector $(X_{1-d}, \ldots, X_0)$ is uncorrelated with $Y_t, t > 0$. The difference equations can be rewritten in the form

$$X_t = Y_t - \sum_{j=1}^{d} \binom{d}{j} (-1)^j X_{t-j}, t = 1, 2, \ldots. \tag{6.4.1}$$

It is convenient, by relabelling the time axis if necessary, to assume that we observe $X_{1-d}, X_{2-d}, \ldots, X_n$. (The observed values of $\{Y_t\}$ are then $Y_1, \ldots, Y_n$.) As usual we shall use P_n to denote best linear prediction in terms of the observations up to time n (in this case $1, X_{1-d}, \ldots, X_n$ or equivalently $1, X_{1-d}, \ldots, X_0, Y_1, \ldots, Y_n$).

Our goal is to compute the best linear predictors $P_n X_{n+h}$. This can be done by applying the operator P_n to each side of (6.4.1) (with $t = n+h$) and using the linearity of P_n to obtain

$$P_n X_{n+h} = P_n Y_{n+h} - \sum_{j=1}^{d} \binom{d}{j} (-1)^j P_n X_{n+h-j}. \tag{6.4.2}$$

Now the assumption that $(X_{1-d}, \ldots, X_0)$ is uncorrelated with $Y_t, t > 0$ enables us to identify $P_n Y_{n+h}$ with the best linear predictor of Y_{n+h} in terms of $\{1, Y_1, \ldots, Y_n\}$ and this can be calculated as described in Section 3.3. The predictor $P_n X_{n+1}$ is obtained directly from (6.4.2) by noting that $P_n X_{n+1-j} = X_{n+1-j}$ for each $j \geq 1$. The predictor $P_n X_{n+2}$ can then be found from (6.4.2) using the previously calculated value of $P_n X_{n+1}$. The predictors $P_n X_{n+3}$, $P_n X_{n+4}$, ..., can be computed recursively in the same way.

To find the mean squared error of prediction it is convenient to express $P_n Y_{n+h}$ in terms of $\{X_j\}$. For $n \geq 0$ we denote the one-step predictors by $\hat{Y}_{n+1} = P_n Y_{n+1}$ and $\hat{X}_{n+1} = P_n X_{n+1}$. Then from (6.4.1) and (6.4.2) we have

$$X_{n+1} - \hat{X}_{n+1} = Y_{n+1} - \hat{Y}_{n+1}, n \geq 1,$$

and hence from (3.3.12), if $n > m = \max(p, q)$ and $h \geq 1$, we can write

$$P_n Y_{n+h} = \sum_{i=1}^{p} \phi_i P_n Y_{n+h-i} + \sum_{j=h}^{q} \theta_{n+h-1,j} (X_{n+h-j} - \hat{X}_{n+h-j}). \tag{6.4.3}$$

Setting $\phi^*(z) = (1 - z)^d \phi(z) = 1 - \phi_1^* z - \cdots - \phi_{p+d}^* z^{p+d}$, we find from (6.4.2) and (6.4.3) that

$$P_n X_{n+h} = \sum_{j=1}^{p+d} \phi_j^* P_n X_{n+h-j} + \sum_{j=h}^{q} \theta_{n+h-1,j} (X_{n+h-j} - \hat{X}_{n+h-j}), \tag{6.4.4}$$

which is analogous to the h-step prediction formula (3.3.12) for an ARMA process. As in (3.3.13), the mean squared error of the h-step predictor is

$$\sigma_n^2(h) = E(X_{n+h} - P_n X_{n+h})^2 = \sum_{j=0}^{h-1} \left(\sum_{r=0}^{j} \chi_r \theta_{n+h-r-1,j-r} \right)^2 v_{n+h-j-1}, \tag{6.4.5}$$

where $\theta_{n0} = 1$,

$$\chi(z) = \sum_{r=0}^{\infty} \chi_r z^r = (1 - \phi_1^* z - \cdots - \phi_{p+d}^* z^{p+d})^{-1},$$

and

$$v_{n+h-j-1} = E(X_{n+h-j} - \hat{X}_{n+h-j})^2 = E(Y_{n+h-j} - \hat{Y}_{n+h-j})^2.$$

The coefficients χ_j can be found from the recursions (3.3.14) with ϕ_j^* replacing ϕ_j. For large n we can approximate (6.4.5), provided $\theta(\cdot)$ is invertible, by

$$\sigma_n^2(h) = \sum_{j=0}^{h-1} \psi_j^2 \sigma^2, \tag{6.4.6}$$

where

$$\psi(z) = \sum_{j=0}^{\infty} \psi_j z^j = (\phi^*(z))^{-1}\theta(z).$$

6.4.1 The Forecast Function

Inspection of equation (6.4.4) shows that for fixed $n > m = \max(p, q)$ and for all $h > q$, the h step predictors,

$$g(h) := P_n X_{n+h}, h = q+1, q+2, \ldots,$$

satisfy the homogeneous linear difference equations with constant coefficients,

$$g(h) - \phi_1^* g(h-1) - \cdots - \phi_{p+d}^* g(h-p-d) = 0, h > q, \tag{6.4.7}$$

where $\phi_1^*, \ldots, \phi_{p+d}^*$ are the coefficients of $z, \ldots, z^{p+d}$ in

$$\phi^*(z) = (1-z)^d \phi(z).$$

The solution of (6.4.7) is well known from the theory of linear difference equations (see TSTM, Section 3.6). If we assume that the zeroes of $\phi(z)$ (denoted by $\xi_1, \ldots, \xi_p$) are all distinct, then the solution is

$$g(h) = a_0 + a_1 h + \cdots + a_d h^{d-1} + b_1 \xi_1^{-h} + \cdots + b_p \xi_p^{-h}, h > q-p-d, \tag{6.4.8}$$

where the coefficients $a_1, \ldots, a_d$ and $b_1, \ldots, b_p$ can be determined from the $p+d$ equations obtained by equating the right-hand side of (6.4.8) for $q-p-d < h \le q$ to the corresponding values of $g(h)$ computed numerically (for $h \le 0$, $P_n X_{n+h} = X_{n+h}$ and for $1 \le h \le q$, $P_n X_{n+h}$ can be computed from (6.4.4) as already described). Once the constants a_i and b_i have been evaluated, the algebraic expression (6.4.8) gives the predictors for all values of $h > q$.

Example 6.4.1 An ARIMA(1,1,0) model

In Example 5.2.4 we found the maximum likelihood AR(1) model for the mean-corrected differences X_t of the Dow-Jones Utilities Index (Aug. 28 – Dec. 18, 1972). The model was

$$X_t - 0.4471X_{t-1} = Z_t, \quad \{Z_t\} \sim \mathrm{WN}(0, 0.1455), \tag{6.4.9}$$

where $X_t = D_t - D_{t-1} - 0.1336$, $t = 1, \ldots, 78$ and $\{D_t, t = 0, 1, 2, \ldots, 77\}$ is the original series. The model for $\{D_t\}$ is thus

$$(1 - 0.4471B)(1 - B)D_t = Z_t + 0.0739, \{Z_t\} \sim \mathrm{WN}(0, 0.1455).$$

The recursions (6.4.2) take the form

$$\begin{aligned} P_n D_{n+h} &= P_n X_{n+h} + 0.1336 + P_n D_{n+h-1} \\ &= (0.4471)^h X_n + 0.1336 + P_n D_{n+h-1} \end{aligned}$$

since the predictors $P_n X_{n+h}$ are the best linear predictors of the zero-mean AR(1) process defined by (6.4.9). To compute the one- and two-step predictors $P_{77}D_{78}$ and $P_{77}D_{79}$ we note from the data that $D_{77} = 121.23$ and $X_{77} = -0.9036$. Applying the recursions we therefore find that

$$P_{77}D_{78} = (0.4471)(-0.9036) + 0.1336 + 121.23 = 120.960$$

and

$$P_{77}D_{79} = (0.4471)^2(-0.9036) + 0.1336 + 120.96 = 120.913.$$

From (6.4.5) we find that the corresponding mean squared errors are

$$\sigma_{77}^2(1) = v_{77} = \sigma^2 = .1455$$

and

$$\sigma_{77}^2(2) = v_{78} + \phi_1^{*2} v_{77} = \sigma^2(1 + 1.4471^2) = .4502.$$

(Notice that the approximation (6.4.6) is exact in this case.) The predictors and their mean squared errors are easily obtained from the program PEST by entering the data DOWJ.DAT, differencing at lag 1, subtracting the mean, carrying out preliminary estimation with Burg's algorithm, retaining the preliminary AR(1) model and carrying out maximum likelihood estimation using the *ARMA estimation* option. Once the maximum likelihood model for the differenced and mean-corrected data has been found, a menu will appear on the screen from which you should select *Predict future data*. □

6.5 Seasonal ARIMA Models

We have already seen how differencing the series $\{X_t\}$ at lag s is a convenient way of eliminating a seasonal component of period s. If we fit an ARMA(p, q) model $\phi(B)Y_t = \theta(B)Z_t$ to the differenced series $Y_t = (1 - B^s)X_t$, then the model for the original series is $\phi(B)(1 - B^s)X_t = \theta(B)Z_t$. This is a special case of the general seasonal ARIMA (SARIMA) model defined as follows.

Definition 6.5.1

If d and D are nonnegative integers, then $\{X_t\}$ is a **seasonal ARIMA$(p,d,q) \times (P,D,Q)_s$ process with period s** if the differenced series $Y_t = (1-B)^d(1-B^s)^D X_t$ is a causal ARMA process defined by

$$\phi(B)\Phi(B^s)Y_t = \theta(B)\Theta(B^s)Z_t, \qquad \{Z_t\} \sim \text{WN}(0, \sigma^2), \tag{6.5.1}$$

where $\phi(z) = 1 - \phi_1 z - \cdots - \phi_p z^p$, $\Phi(z) = 1 - \Phi_1 z - \cdots - \Phi_P z^P$, $\theta(z) = 1 + \theta_1 z + \cdots + \theta_q z^q$ and $\Theta(z) = 1 + \Theta_1 z + \cdots + \Theta_Q z^Q$.

Remark 1. Note that the process $\{Y_t\}$ is causal if and only if $\phi(z) \neq 0$ and $\Phi(z) \neq 0$ for $|z| \leq 1$. In applications D is rarely more than one and P and Q are typically less than three. □

Remark 2. The equation (6.5.1) satisfied by the differenced process $\{Y_t\}$ can be rewritten in the equivalent form

$$\phi^*(B)Y_t = \theta^*(B)Z_t, \tag{6.5.2}$$

where $\phi^*(\cdot), \theta^*(\cdot)$ are polynomials of degree $p + sP$ and $q + sQ$, respectively, whose coefficients can all be expressed in terms of $\phi_1, \ldots, \phi_p$, $\Phi_1, \ldots, \Phi_P$, $\theta_1, \ldots, \theta_q$, and $\Theta_1, \ldots, \Theta_Q$. Provided $p < s$ and $q < s$ the constraints on the coefficients of $\phi^*(\cdot)$ and $\theta^*(\cdot)$ can all be expressed as multiplicative relations

$$\phi^*_{is+j} = \phi^*_{is}\phi^*_j, \; i = 1, 2, \ldots; \; j = 1, \ldots, s - 1$$

and

$$\theta^*_{is+j} = \theta^*_{is}\theta^*_j, \; i = 1, 2, \ldots; \; j = 1, \ldots, s - 1.$$

In Section 1.5 we discussed the classical decomposition model incorporating trend, seasonality, and random noise, namely $X_t = m_t + s_t + Y_t$. In modelling real data it might not be reasonable to assume, as in the classical decomposition model, that the seasonal component s_t repeats itself precisely in the same way cycle after cycle. Seasonal ARIMA models allow for randomness in the seasonal pattern from one cycle to the next. □

Example 6.5.1 Suppose we have r years of monthly data which we tabulate as follows:

Year/Month	1	2	...	12
1	Y_1	Y_2	...	Y_{12}
2	Y_{13}	Y_{14}	...	Y_{24}
3	Y_{25}	Y_{26}	...	Y_{36}
⋮	⋮	⋮		⋮
r	$Y_{1+12(r-1)}$	$Y_{2+12(r-1)}$	...	$Y_{12+12(r-1)}$

Each column in this table may itself be viewed as a realization of a time series. Suppose that each one of these twelve time series is generated by the same ARMA(P, Q) model, or more specifically that the series corresponding to the jth month, Y_{j+12t}, $t = 0, \ldots, r-1$, satisfies a difference equation of the form

$$\begin{aligned} Y_{j+12t} = {} & \Phi_1 Y_{j+12(t-1)} + \cdots + \Phi_P Y_{j+12(t-P)} + U_{j+12t} \\ & + \Theta_1 U_{j+12(t-1)} + \cdots + \Theta_Q U_{j+12(t-Q)}, \end{aligned} \tag{6.5.3}$$

where

$$\{U_{j+12t}, t = \ldots, -1, 0, 1, \ldots\} \sim \mathrm{WN}(0, \sigma_U^2). \tag{6.5.4}$$

Then since the same ARMA(P, Q) model is assumed to apply to each month, (6.5.3) holds for each $j = 1, \ldots, 12$. (Notice, however, that $E(U_t U_{t+h})$ is not necessarily zero except when h is an integer multiple of 12.) We can thus write (6.5.3) in the compact form

$$\Phi(B^{12})Y_t = \Theta(B^{12})U_t, \tag{6.5.5}$$

where $\Phi(z) = 1 - \Phi_1 z - \cdots - \Phi_P z^P$, $\Theta(z) = 1 + \Theta_1 z + \cdots + \Theta_Q z^Q$, and $\{U_{j+12t}, t = \ldots, -1, 0, 1, \ldots\} \sim \mathrm{WN}(0, \sigma_U^2)$ for each j. We refer to the model (6.5.5) as the between-year model. □

Example 6.5.2 Suppose $P = 0$, $Q = 1$ and $\Theta_1 = -0.4$ in (6.5.5). Then the series for any particular month is a moving average of order 1. If $E(U_t U_{t+h}) = 0$ for all h, i.e., if the white noise sequences for different months are uncorrelated with each other, then the columns themselves are uncorrelated. The correlation function for such a process is shown in Figure 6.15. □

Example 6.5.3 Suppose $P = 1$, $Q = 0$ and $\Phi_1 = 0.7$ in (6.5.5). In this case the 12 series (one for each month) are AR(1) processes that are uncorrelated if the white noise sequences for different months are uncorrelated. A graph of the autocorrelation function of this process is shown in Figure 6.16. □

In each of the Examples 6.5.1, 6.5.2, and 6.5.3, the 12 series corresponding to the different months are uncorrelated. To incorporate dependence between these series

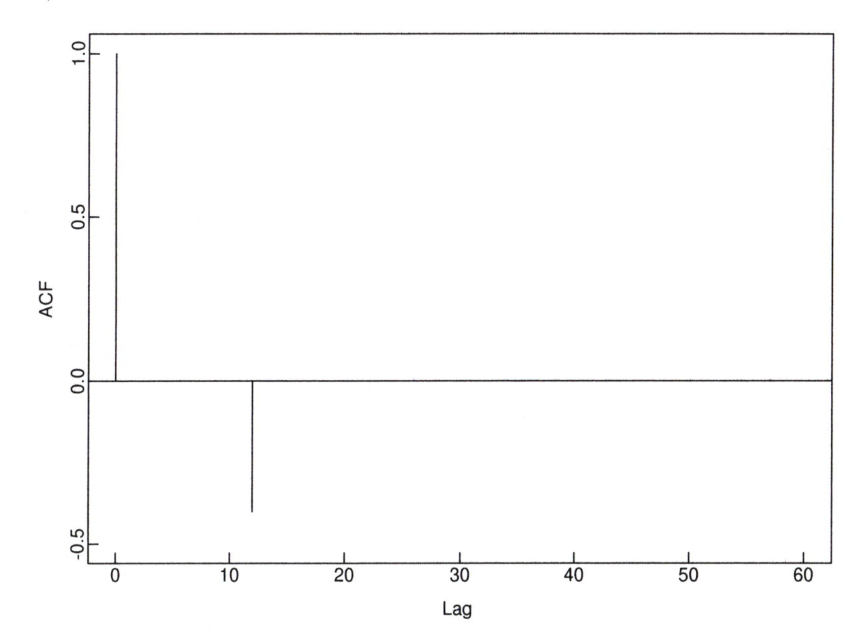

Figure 6-15 The ACF of the model $X_t = U_t - 0.4U_{t-12}$ of Example 6.5.2.

we allow the process $\{U_t\}$ in (6.5.5) to follow an ARMA(p, q) model,

$$\phi(B)U_t = \theta(B)Z_t, \qquad \{Z_t\} \sim \text{WN}(0, \sigma^2). \tag{6.5.6}$$

This assumption not only implies possible nonzero correlation between consecutive values of U_t, but also within the twelve sequences $\{U_{j+12t}, t = \ldots, -1, 0, 1, \ldots\}$, each of which was assumed to be uncorrelated in the preceding examples. In this

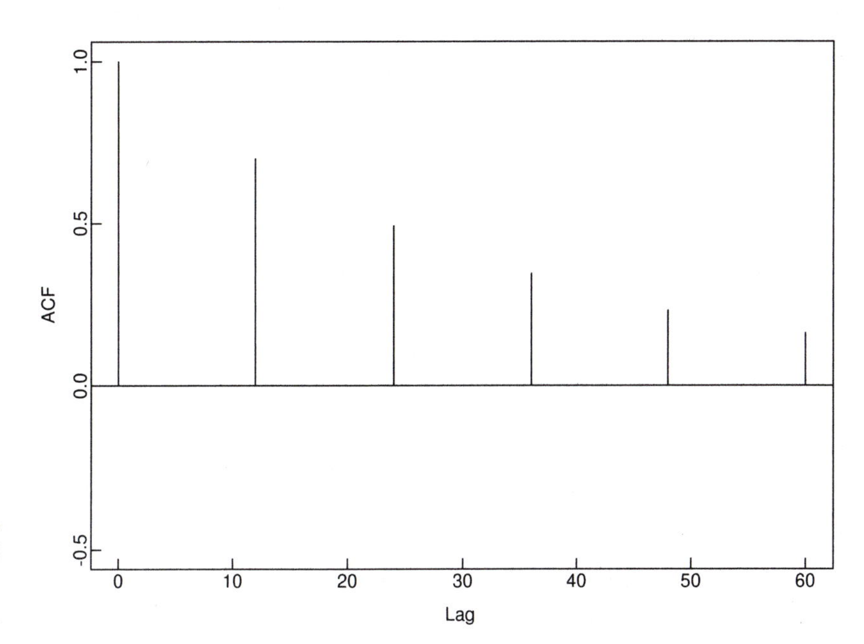

Figure 6-16 The ACF of the model $X_t - 0.7X_{t-12} = U_t$ of Example 6.5.3.

case (6.5.4) may no longer hold; however, the coefficients in (6.5.6) will frequently have values such that $E(U_t U_{t+12j})$ is small for $j = \pm 1, \pm 2, \ldots$. Combining the two models (6.5.5) and (6.5.6) and allowing for possible differencing leads directly to Definition 6.5.1 of the general SARIMA model as given above.

The first steps in identifying SARIMA models for a (possibly transformed) data set are to find d and D so as to make the differenced observations

$$Y_t = (1 - B)^d (1 - B^s)^D X_t$$

stationary in appearance (see Sections 6.1–6.3). Next we examine the sample ACF and PACF of $\{Y_t\}$ at lags that are multiples of s for an indication of the orders P and Q in the model (6.5.5). If $\hat{\rho}(\cdot)$ is the sample ACF of $\{Y_t\}$, then P and Q should be chosen so that $\hat{\rho}(ks)$, $k = 1, 2, \ldots$, is compatible with the ACF of an ARMA(P, Q) process. The orders p and q are then selected by trying to match $\hat{\rho}(1), \ldots, \hat{\rho}(s-1)$ with the ACF of an ARMA(p, q) process. Ultimately the AICC criterion (Section 5.5) and the goodness of fit tests (Section 5.3) are used to select the best SARIMA model from competing alternatives.

For given values of p, d, q, P, D, and Q, the parameters $\boldsymbol{\phi}$, $\boldsymbol{\theta}$, Φ, Θ, and σ^2 can be found using the maximum likelihood procedure of Section 5.2. The differences $Y_t = (1 - B)^d (1 - B^s)^D X_t$ constitute an ARMA($p + sP, q + sQ$) process in which some of the coefficients are zero and the rest are functions of the $(p + P + q + Q)$-dimensional vector $\boldsymbol{\beta}' = (\boldsymbol{\phi}', \Phi', \boldsymbol{\theta}', \Theta')$. For any fixed $\boldsymbol{\beta}$ the reduced likelihood $\ell(\boldsymbol{\beta})$ of the differences $Y_{t+d+sD}, \ldots, Y_n$, is easily computed as described in Section 5.2. The maximum likelihood estimator of $\boldsymbol{\beta}$ is the value that minimizes $\ell(\boldsymbol{\beta})$ and the maximum likelihood estimate of σ^2 is given by (5.2.10). The estimates can be found using the program PEST by specifying the required multiplicative relationships between the coefficients as given in Remark 2 above.

A more direct approach to modelling the differenced series $\{Y_t\}$ is simply to fit a subset ARMA model of the form (6.5.2) without making use of the multiplicative form of $\phi^*(\cdot)$ and $\theta^*(\cdot)$ in (6.5.1).

Example 6.5.4 Monthly accidental deaths

In Figure 1.27 we showed the series $\{Y_t = (1 - B^{12})(1 - B)X_t\}$ obtained by differencing the accidental deaths series $\{X_t\}$ once at lag 12 and once at lag 1. The sample ACF of $\{Y_t\}$ is shown in Figure 6.17. □

The values $\hat{\rho}(12) = -0.333$, $\hat{\rho}(24) = -0.099$ and $\hat{\rho}(36) = 0.013$ suggest a moving average of order 1 for the between-year model (i.e., $P = 0$ and $Q = 1$). Moreover, inspection of $\hat{\rho}(1), \ldots, \hat{\rho}(11)$ suggests that $\rho(1)$ is the only short-term correlation different from zero, so we also choose a moving average of order 1 for the between-month model (i.e., $p = 0$ and $q = 1$). Taking into account the sample

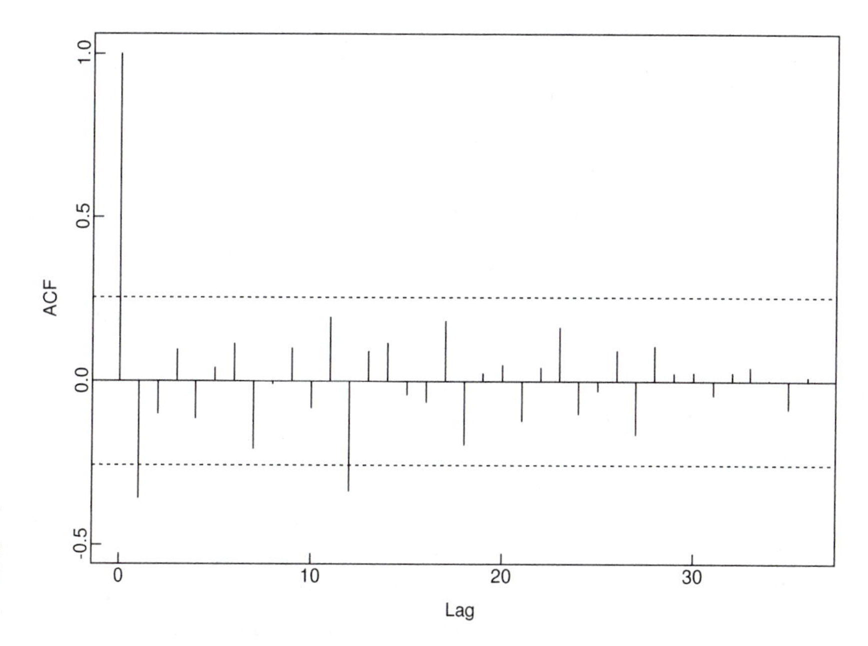

Figure 6-17 The sample ACF of the differenced accidental deaths $\{\nabla\nabla_{12}X_t\}$.

mean (28.831) of the differences $\{Y_t\}$, we therefore arrive at the model

$$Y_t = 28.831 + (1 + \theta_1 B)(1 + \Theta_1 B^{12})Z_t, \{Z_t\} \sim \text{WN}(0, \sigma^2), \tag{6.5.7}$$

for the series $\{Y_t\}$. The maximum likelihood estimates of the parameters are obtain from PEST as follows. After entering, differencing and mean-correcting the data, choose the *Entry of an ARMA model* option in the main menu. Enter an MA(13) model with $\theta_1 = -0.3$, $\theta_{12} = -0.3$, and $\theta_{13} = 0.09$. (This corresponds to the initial guess $Y_t = (1-0.3B)(1-0.3B^{12})Z_t$.) Then choose the *ARMA estimation* option followed by *Constrain optimized coefficients*. Select *Define multiplicative relationships*, specify that one relationship is to be imposed, and define the relationship by entering 1, 12, 13 to indicate that $\theta_1 \times \theta_{12} = \theta_{13}$. From the menu, choose *Return to optimization menu* and *Optimize with current settings*. Optimization will then be carried out and you will obtain the fitted model (6.5.7) with

$$\hat{\theta}_1 = -0.478,$$

$$\hat{\Theta}_1 = -0.588,$$

and

$$\hat{\sigma}^2 = 94390,$$

with AICC value 855.53. The corresponding fitted model for $\{X_t\}$ is thus the SARIMA $(0, 1, 1) \times (0, 1, 1)_{12}$ process,

$$\nabla\nabla_{12}X_t = 28.831 + (1 - 0.478B)(1 - 0.588B^{12})Z_t, \tag{6.5.8}$$

where $\{Z_t\} \sim \text{WN}(0, 94390)$.

If we adopt the alternative approach of fitting a subset ARMA model to $\{Y_t\}$ without seeking a multiplicative structure for the operators $\phi^*(B)$ and $\theta^*(B)$ in (6.5.2), we begin by fitting a preliminary MA(13) model (as suggested in Figure 6.17) to the series $\{Y_t\}$. Returning to the main menu we then fit a maximum likelihood MA(13) model and examine the standard errors of the coefficient estimators. This suggests setting the coefficients at lags 2,3,8,10, and 11 equal to zero as these are all less than one standard error from zero. To do this select the option *Continue with parameter optimization* followed by *Constrain optimized coefficients* and *Set nonzero coefficient to zero*. After setting each of the coefficients listed above to zero, choose *Return to optimization menu* and *Optimize with current settings*. The coefficients that have been set to zero will be held at that value and the optimization will be with respect to the remaining coefficients. Carrying out the optimization gives a model with substantially smaller AICC than the unconstrained MA(13) model. Examining the standard errors again we see that the coefficients at lags 4, 5, and 7 are promising candidates to be set to zero as each of them is less than one standard error from zero. Selecting *Continue with parameter optimization*, setting these coefficients to zero, and reoptimizing gives a further reduction in AICC. Setting the coefficient at lag 9 to zero and reoptimizing gives a further reduction in AICC (to 855.61) and the fitted model

$$\begin{aligned}\nabla\nabla_{12}X_t &= 28.831 + Z_t - 0.596Z_{t-1} - 0.407Z_{t-6} - 0.685Z_{t-12}\\ &\quad + 0.460Z_{t-13}, \quad \{Z_t\} \sim \mathrm{WN}(0, 71240).\end{aligned} \tag{6.5.9}$$

The AICC value 855.61 is quite close to the value 855.53 for the model (6.5.8).

As soon as the maximum likelihood model has been fitted you will see the Results Menu, which includes the option *File and analyze residuals*. This option allows you to carry out the goodness of fit tests described in Section 5.3. The models (6.5.8) and (6.5.9) both satisfy the tests.

6.5.1 Forecasting SARIMA Processes

Forecasting SARIMA processes is completely analogous to the forecasting of ARIMA processes discussed in Section 6.4. Expanding out the operator $(1-B)^d(1-B^s)^D$ in powers of B, rearranging the equation

$$(1-B)^d(1-B^s)^D X_t = Y_t$$

and setting $t = n+h$ gives the analogue

$$X_{n+h} = Y_{n+h} + \sum_{j=1}^{d+Ds} a_j X_{n+h-j} \tag{6.5.10}$$

of equation (6.4.2). Under the assumption that the first $d+Ds$ observations $X_{-d-Ds+1}$, $\ldots, X_0$ are uncorrelated with $\{Y_t, t \geq 1\}$, we can determine the best linear predictors $P_n X_{n+h}$ of X_{n+h} based on $\{1, X_{-d-Ds+1}, \ldots, X_n\}$ by applying P_n to each side of

(6.5.10) to obtain

$$P_n X_{n+h} = P_n Y_{n+h} + \sum_{j=1}^{d+Ds} a_j P_n X_{n+h-j}. \tag{6.5.11}$$

The first term on the right is just the best linear predictor of the (possibly nonzero-mean) ARMA process $\{Y_t\}$ in terms of $\{1, Y_1, \ldots, Y_n\}$, which can be calculated as described in Section 3.3. The predictors $P_n X_{n+h}$ can then be computed recursively for $h = 1, 2, \ldots$ from (6.5.11), noting that $P_n X_{n+1-j} = X_{n+1-j}$ for each $j \geq 1$.

For large n the prediction mean squared error can be approximated by

$$\sigma_n^2(h) = E(X_{n+h} - P_n X_{n+h})^2 = \sum_{j=0}^{h-1} \psi_j^2 \sigma^2, \tag{6.5.12}$$

where

$$\psi(z) = \sum_{j=0}^{\infty} \psi_j z^j = \frac{\theta(z)\Theta(z^s)}{\phi(z)\Phi(z^s)(1-z)^d(1-z^s)^D}, \quad |z| < 1.$$

The required calculations can all be carried out with the aid of the program PEST. The mean squared errors are computed from the large-sample approximation (6.5.12) if the fitted model is invertible. If the fitted model is not invertible, PEST computes the mean squared errors by converting the model to the equivalent (in terms of Gaussian likelihood) invertible model and then uses (6.5.12).

Example 6.5.5 Monthly accidental deaths

Continuing with Example 6.5.4 we shall now use PEST to predict six future values of the Accidental Deaths series using the fitted models (6.5.8) and (6.5.9). (Notice that the latter is not invertible.) Immediately after fitting the chosen model and carrying out goodness of fit tests you will again see the Results Menu, which contains the

Table 6.1 Predicted values of the Accidental Deaths series for $t = 73, \ldots, 78$, the standard deviations σ_t of the prediction errors, and the corresponding observed values of X_t for the same period.

t	73	74	75	76	77	78
Model (6.5.8)						
Predictors	8441	7704	8549	8885	9843	10279
σ_t	308	348	383	415	445	474
Model (6.5.9)						
Predictors	8345	7619	8356	8742	9795	10179
σ_t	292	329	366	403	442	486
Observed values						
X_t	7798	7406	8363	8460	9217	9316

option *Predict future data*. (This option is also available for *any* specified model from the main menu of PEST.) After selecting *Predict future data* you will be asked to specify the number of values required, in this case 6. You will then be asked if you wish to change the white noise variance from the maximum likelihood value. (This has the effect of changing the estimated mean squared errors but not the predicted values.) Answer *No* to this question. You will then see the predicted values of the next six $\{Y_t\}$ values in the right column and the square roots of their mean squared errors. The next step is to select *Undo differencing*, after which you will see the six predicted values of the original series $\{X_t\}$ and the square roots of the corresponding mean squared errors. These are the values shown in Table 6.1. The original series plus six predictors will then be plotted on the screen. If a preliminary Box-Cox transformation had been performed prior to differencing the data, you would also have the option of inverting the transformation and plotting the untransformed original data with the corresponding predicted values. Since this is not relevant to the present example, you will be returned to the main menu after the graph is plotted. □

6.6 Regression with ARMA Errors

In the standard linear regression model, the errors are assumed to be independent and identically distributed. We now consider generalizations of this basic model to the case when the errors follow an ARMA(p, q) model. The general model takes the form

$$Y_t = x_{t1}\beta_1 + \cdots + x_{tk}\beta_k + W_t, \quad t = 1, \ldots, n.$$

Using matrix notation, this can be expressed as

$$\mathbf{Y} = X\boldsymbol{\beta} + \mathbf{W}, \tag{6.6.1}$$

where $\mathbf{Y} = (Y_1, \ldots, Y_n)'$ is the response vector (or vector of time series observations), X is the design matrix consisting of the n vectors of explanatory variables $\mathbf{x}_t = (x_{t1}, \ldots, x_{tk})'$, $\boldsymbol{\beta} = (\beta_1, \ldots, \beta_k)'$ is the vector of regression parameters, and $\mathbf{W} = (W_1, \ldots, W_n)'$ are observations from the zero-mean ARMA(p, q) process satisfying

$$\phi(B)W_t = \theta(B)Z_t, \qquad \{Z_t\} \sim \mathrm{WN}(0, \sigma^2).$$

We have already seen one application of this model for estimating the trend of a time series in Section 1.3.2. In this case, each x_{tj} is a function of t only. For example, a model with quadratic trend would correspond to $x_{t1} = 1, x_{t2} = t$, and $x_{t3} = t^2$, while a model with a sinusoidal trend at frequency ω would correspond to $x_{t1} = 1, x_{t2} = \cos(\omega t)$ and $x_{t3} = \sin(\omega t)$. It is not necessary that x_{tj} be a function only of t. In the overshort data of Example 3.2.8, one might wish to include a time series of temperature readings or other available meteorological variables in the model.

The **ordinary least squares** (OLS) estimate of $\boldsymbol{\beta}$ minimizes the sum of squares

$$\sum_{t=1}^{n}(Y_t - \boldsymbol{\beta}'\mathbf{x}_t)^2$$

with respect to $\boldsymbol{\beta}$. The solution is given by

$$\hat{\boldsymbol{\beta}}_{OLS} = (X'X)^{-1}X'\mathbf{Y} \tag{6.6.2}$$

($(X'X)^{-1}$ is a generalized inverse if the design matrix is not of full rank). The OLS estimate is the maximum likelihood estimate when $\{W_t\}$ is IID N$(0, \sigma^2)$ noise. Even in the case of nonnormal and nonindependent W_t, the OLS estimate still has a number of desirable properties. In particular if the vector $\mathbf{x}_t$ is nonrandom, then the OLS estimate is unbiased (i.e., $E(\hat{\boldsymbol{\beta}}_{OLS}) = \boldsymbol{\beta}$) with covariance matrix

$$\text{Cov}(\hat{\boldsymbol{\beta}}_{OLS}) = (X'X)^{-1}X'\Gamma_n X(X'X)^{-1}, \tag{6.6.3}$$

where $\Gamma_n = E(\mathbf{W}\mathbf{W}')$ is the covariance matrix of $\mathbf{W}$.

A second estimator, called the **generalized least squares** (GLS) or **best linear unbiased estimator**, minimizes the weighted sum of squares

$$(\mathbf{Y} - X\boldsymbol{\beta})'\Gamma_n^{-1}(\mathbf{Y} - X\boldsymbol{\beta})$$

with respect to $\boldsymbol{\beta}$. The minimizer, obtained by setting the partial derivative of the quadratic form with respect to $\boldsymbol{\beta}$ equal to 0 and solving for $\boldsymbol{\beta}$, is given by

$$\hat{\boldsymbol{\beta}}_{GLS} = (X'\Gamma_n^{-1}X)^{-1}X'\Gamma_n^{-1}\mathbf{Y}. \tag{6.6.4}$$

Of course to compute this estimate, the parameters $\boldsymbol{\phi}$ and $\boldsymbol{\theta}$ of the ARMA model for $\{W_t\}$ must be known.

Again assuming that each x_{tj} is nonrandom, the GLS estimator is unbiased with covariance matrix

$$\text{Cov}(\hat{\boldsymbol{\beta}}_{GLS}) = (X'\Gamma_n^{-1}X)^{-1}. \tag{6.6.5}$$

Since the OLS estimate is also a linear unbiased estimator and the GLS estimate is the best linear unbiased estimator, the matrix $\text{Cov}(\hat{\boldsymbol{\beta}}_{OLS}) - \text{Cov}(\hat{\boldsymbol{\beta}}_{GLS})$ must be nonnegative definite and hence

$$\text{Var}(\mathbf{c}'\hat{\boldsymbol{\beta}}_{GLS}) \le \text{Var}(\mathbf{c}'\hat{\boldsymbol{\beta}}_{OLS})$$

for any k-dimensional vector $\mathbf{c}$. While the GLS estimator is superior to the OLS, the GLS is only computable when $\boldsymbol{\phi}$ and $\boldsymbol{\theta}$ are known.

A third alternative is to maximize the Gaussian likelihood with respect to all the model parameters. The likelihood based on model (6.6.1) is given by

$$L(\boldsymbol{\beta}, \Gamma_n) = (2\pi)^{-n/2}(\det \Gamma_n)^{-1/2}\exp\{-\frac{1}{2}(\mathbf{Y} - X\boldsymbol{\beta})'\Gamma_n^{-1}(\mathbf{Y} - X\boldsymbol{\beta})\}.$$

This function can be computed easily by applying the innovations algorithm to the mean-corrected data, $Y_t^* := Y_t - \boldsymbol{\beta}'\mathbf{x}_t$. For fixed $\boldsymbol{\beta}$, Y_t^* is a zero-mean ARMA process so that

$$\begin{aligned} S(\beta, \phi, \theta) &:= \sigma^2(\mathbf{Y} - X\beta)'\Gamma_n^{-1}(\mathbf{Y} - X\beta) \\ &= \sigma^2 \mathbf{Y}^{*\prime}\Gamma_n^{-1}\mathbf{Y}^* \\ &= \sum_{t=1}^{n} (Y_t^* - \hat{Y}_t^*)^2 / r_{t-1}. \end{aligned}$$

Following the derivation given in Section 5.2, we find that the maximum likelihood estimates, $\hat{\beta}$, $\hat{\phi}$, and $\hat{\theta}$, minimize

$$\ell(\beta, \phi, \theta) = \ln(n^{-1} S(\beta, \phi, \theta)) + n^{-1} \sum_{t=1}^{n} \ln r_{t-1}.$$

The maximum likelihood estimator of σ^2 is then given by

$$\hat{\sigma}^2 = n^{-1} S(\hat{\beta}, \hat{\phi}, \hat{\theta}).$$

Notice that if $\hat{\beta}_{GLS}(\phi, \theta)$ is the GLS estimate of β corresponding to a fixed ϕ and θ, then $\hat{\beta} = \hat{\beta}_{GLS}(\hat{\phi}, \hat{\theta})$. This observation suggests an iterative estimation procedure of the parameters which may be computationally preferable, especially if the number of explanatory variables is large. The steps, which are analogous to those of the procedure of Cochran and Orcutt (1949), are as follows:

Step 0. Set $\hat{\beta}^{(0)} = \hat{\beta}_{OLS}$ and define $\tilde{W}_t^{(0)} = Y_t - \hat{\beta}^{(0)\prime}\mathbf{x}_t$, $t = 1, \ldots, n$. Let $\hat{\phi}^{(0)}$ and $\hat{\theta}^{(0)}$ be the maximum likelihood estimates of ϕ and θ in fitting an ARMA(p, q) model to $\tilde{W}_1^{(0)}, \ldots, \tilde{W}_n^{(0)}$.

Step i. Calculate $\hat{\beta}^{(i)} = \hat{\beta}_{GLS}(\hat{\phi}^{(i-1)}, \hat{\theta}^{(i-1)})$ and define $\tilde{W}_t^{(i)} = Y_t - \hat{\beta}^{(i)\prime}\mathbf{x}_t$, $t = 1, \ldots, n$. Let $\hat{\phi}^{(i)}$ and $\hat{\theta}^{(i)}$ be the maximum likelihood estimates of ϕ and θ in fitting an ARMA(p, q) model to $\tilde{W}_1^{(i)}, \ldots, \tilde{W}_n^{(i)}$.

Steps $0, 1, \ldots,$ are followed until no further change occurs in the parameter estimates. Depending on the underlying ARMA model, the calculation of the GLS estimate may be the most difficult part of each step since it potentially involves a number of large matrix operations. This computation can be greatly accelerated if the factorization, $\Gamma_n^{-1} = \sigma^{-2} T'T$ with T a lower triangular matrix, is easily computable. Then by premultiplying both sides of (6.6.1) by T, we obtain the transformed model

$$\tilde{\mathbf{Y}} = \tilde{X}\beta + \tilde{\mathbf{W}}, \tag{6.6.6}$$

where $\tilde{\mathbf{Y}} = T\mathbf{Y}$, $\tilde{X} = TX$, and $\tilde{\mathbf{W}} = T\mathbf{W}$. Since $\tilde{W}_1, \ldots, \tilde{W}_n$ are uncorrelated with mean 0 and variance σ^2, the GLS estimate of β for the model (6.6.1) is the same as the OLS estimate for the transformed model (6.6.6), namely

$$\hat{\beta}_{GLS} = (\tilde{X}'\tilde{X})^{-1}\tilde{X}'\tilde{\mathbf{Y}}.$$

For autoregressive processes, the matrix T is band limited, i.e., the (i, j) element of T is zero for $|i - j| > p$ (see Problem 6.12). For an AR(1) process,

$$T = \begin{bmatrix} (1-\phi_1^2)^{1/2} & 0 & 0 & \cdots & 0 & 0 \\ -\phi_1 & 1 & 0 & \cdots & 0 & 0 \\ 0 & -\phi_1 & 1 & \cdots & 0 & 0 \\ \vdots & \vdots & \ddots & \ddots & \vdots & \vdots \\ 0 & 0 & 0 & \ddots & 1 & 0 \\ 0 & 0 & 0 & \cdots & -\phi_1 & 1 \end{bmatrix} \tag{6.6.7}$$

(To verify this choice of T, observe that $T\mathbf{W} = (W_1(1-\phi_1^2)^{1/2}, Z_2, \ldots, Z_n)'$ so that $T\Gamma_n T' = \sigma^2 I_n$ and hence $\Gamma_n^{-1} = \sigma^{-2}T'T$.)

If $\{W_t\}$ is a causal and invertible ARMA process, then under mild conditions on the explanatory variables $\mathbf{x}_t$, the maximum likelihood estimates are asymptotically multivariate normal (see Fuller, 1976). In addition, the estimated regression coefficients are asymptotically independent of the estimated ARMA parameters. The large-sample covariance matrix of the regression estimates, suitably normalized, has a complicated form that involves both the regression variables $\mathbf{x}_t$ and the covariance function of $\{W_t\}$. It is therefore convenient to estimate the covariance matrix as H^{-1} where H is the Hessian matrix of the observed log(likelihood).

The OLS, GLS, and maximum likelihood estimates of the regression parameter all have the same limiting covariance matrix. Nevertheless, the formulas (6.6.3) and (6.6.5) for computing the covariance matrix of the OLS and GLS estimates can be vastly different even for moderate sample sizes. This point is illustrated in the following examples.

Example 6.6.1 The overshort data

The analysis of the overshort data in Example 3.2.8 suggested the model

$$Y_t = \beta + W_t,$$

where β is interpreted as the daily leak rate of the underground storage tank and $\{W_t\}$ is the MA(1) process

$$W_t = Z_t + \theta Z_{t-1}, \qquad \{Z_t\} \sim \mathrm{WN}(0, \sigma^2).$$

(Here $k = 1$ and $x_{t1} = 1$.) The OLS estimate of β is simply the sample mean $\hat{\beta}_{OLS} = \bar{Y}_n = -4.035$. Under the assumption that $\{W_t\}$ is IID noise, the variance of the OLS is estimated as $\hat{\gamma}_Y(0)/57 = 59.92$. However, since this estimate of the variance ignores the dependence structure in the model, it is not reliable. Using PEST, the fitted MA(1) model for $\{W_t\}$ is given by

$$W_t = Z_t - .818 Z_{t-1}, \qquad \{Z_t\} \sim \mathrm{WN}(0, 2040.75). \tag{6.6.8}$$

Table 6.2 Estimates of β and θ_1 for the overshort data of Example 6.6.1.

Iteration i	$\hat{\beta}^{(i)}$	$\hat{\theta}_1^{(i)}$
0	−4.035	−.818
1	−4.745	−.848
2	−4.781	−.848

The estimated variance of the OLS estimator, using (6.6.3) is now 2.214. We see that the large negative lag one correlation of $\{W_t\}$ has substantially reduced the variance of the estimate. (For a positively correlated time series, the variance would increase.)

The GLS estimate under model (6.6.8) is $\hat{\beta}_{GLS} = -4.75$ with an estimated variance (calculated from (6.6.5)) of 1.408 (see Problem 6.13). This amounts to a 36% reduction in estimated variance of the leak rate. Estimated 95% confidence bounds for β using the GLS estimate are $-4.75 \pm 1.96(1.408)^{1/2} = (-7.07, -2.43)$, strongly suggesting that the storage tank has a leak. Such a conclusion would not have been reached without taking into account the dependence in the data.

The results from estimating β and θ using the iterative procedure described above are summarized in Table 6.2. Notice the rapid convergence of the estimates. □

Example 6.6.2 The lake data

In Example 5.2.5 we fitted maximum likelihood AR(2) and ARMA(1,1) models to the mean-corrected lake data. Now let us consider fitting a linear trend to the data with AR(2) noise. The choice of an AR(2) model was suggested by an analysis of the residuals obtained after removing a linear trend from the data using OLS. Our model now takes the form

$$Y_t = \beta_0 + \beta_1 t + W_t,$$

where $\{W_t\}$ is the AR(2) process

$$W_t = \phi_1 W_{t-1} + \phi_2 W_{t-2} + Z_t, \qquad \{Z_t\} \sim \mathrm{WN}(0, \sigma^2).$$

From Example 1.3.5, we find that the OLS estimate of β is $\hat{\beta}_{OLS} = (10.202, -.0242)'$. Ignoring the correlation structure in the noise, the covariance matrix of $\hat{\beta}_{OLS}$ is estimated by

$$\hat{\gamma}_Y(0)(X'X)^{-1} = \hat{\gamma}_Y(0)\begin{bmatrix} n & \sum_{t=1}^n t \\ \sum_{t=1}^n t & \sum_{t=1}^n t^2 \end{bmatrix}^{-1}$$
$$= \begin{bmatrix} .07203 & -.00110 \\ -.00110 & .00002 \end{bmatrix}. \tag{6.6.9}$$

Table 6.3 Estimates of β and ϕ for the lake data after 3 iterations.

Iteration i	$\hat{\beta}_1^{(i)}$	$\hat{\beta}_2^{(i)}$	$\hat{\phi}_1^{(i)}$	$\hat{\phi}_2^{(i)}$
0	10.20	−.0242	1.008	−.295
1	10.09	−.0216	1.005	−.291
2	10.09	−.0216	1.005	−.291

The estimated model for the noise process, found by fitting an AR(2) model to the residuals $Y_t - \hat{\beta}'_{OLS}\mathbf{x}_t$, is

$$W_t = 1.008W_{t-1} - .295W_{t-2} + Z_t, \quad \{Z_t\} \sim \text{WN}(0, .4571).$$

Assuming this is the true model for $\{W_t\}$, the GLS estimate is found to be $(10.091, -.0216)'$, which is in close agreement with the OLS estimate. The estimated covariance matrices for the OLS and GLS estimates are given by

$$\text{Cov}(\hat{\beta}_{OLS}) = \begin{bmatrix} .22177 & -.00335 \\ -.00335 & .00007 \end{bmatrix}$$

and

$$\text{Cov}(\hat{\beta}_{GLS}) = \begin{bmatrix} .21392 & -.00321 \\ -.00321 & .00006 \end{bmatrix}.$$

Notice now that the variance of the OLS and GLS estimates are nearly three times the magnitude of the corresponding variance estimates of the OLS calculated under the independence assumption (see (6.6.9)). Estimated 95% confidence bounds for the slope β_1 using the GLS estimate are $-.0216 \pm 1.96(.00006)^{1/2} = -.0216 \pm .0048$, indicating a significant decreasing trend in the level of Lake Huron during the years 1875–1972.

The iterative procedure described above was also used to produce estimates of the parameters. The results from each iteration are summarized in Table 6.3. As in Example 6.6.1, the convergence of the estimates is very rapid. □

Problems

6.1. Suppose that $\{X_t\}$ is an ARIMA(p, d, q) process, satisfying the difference equations

$$\phi(B)(1 - B)^d X_t = \theta(B)Z_t, \{Z_t\} \sim \text{WN}(0, \sigma^2).$$

Show that these difference equations are also satisfied by the process $W_t = X_t + A_0 + A_1 t + \cdots + A_{d-1}t^{d-1}$, where $A_0, \ldots, A_{d-1}$ are arbitrary random variables.

6.2. Verify the representation given in (6.3.4).

6.3. Test the data in Example 6.3.1 for the presence of a unit root in an AR(2) model using the augmented Dickey-Fuller test.

6.4. Apply the augmented Dickey-Fuller test to the levels of Lake Huron data (LAKE.DAT). Perform two analyses assuming AR(1) and AR(2) models.

6.5. If $\{Y_t\}$ is a causal ARMA process (with zero mean) and if X_0 is a random variable with finite second moment such that X_0 is uncorrelated with Y_t for each $t = 1, 2, \ldots$, show that the best linear predictor of Y_{n+1} in terms of $1, X_0, Y_1, \ldots, Y_n$ is the same as the best linear predictor of Y_{n+1} in terms of $1, Y_1, \ldots, Y_n$.

6.6. Let $\{X_t\}$ be the ARIMA(2,1,0) process satisfying

$$(1 - 0.8B + 0.25B^2)\nabla X_t = Z_t, \{Z_t\} \sim \text{WN}(0, 1).$$

a. Determine the forecast function $g(h) = P_n X_{n+h}$ for $h > 0$.

b. Assuming that n is large, compute $\sigma_n^2(h)$ for $h = 1, \ldots, 5$.

6.7. Use a text editor or WORD6 to create a new data set ASHORT.DAT that consists of the data in AIRPASS.DAT with the last twelve values deleted. Use PEST to find an ARIMA model for the logarithms of the data in ASHORT.DAT. Your analysis should include

a. a logical explanation of the steps taken to find the chosen model,

b. approximate 95% bounds for the components of $\boldsymbol{\phi}$ and $\boldsymbol{\theta}$,

c. an examination of the residuals to check for whiteness as described in Section 1.6,

d. a graph of the series ASHORT.DAT with forecasts of the next 12 values from the end of the data set,

e. a numerical value for the 12-step ahead forecast from the end of the data set together with corresponding 95% prediction bounds (PEST will give you an estimated mean squared error for the logged data from which, assuming approximate normality, you can obtain 95% prediction bounds for the logged series. To convert these to prediction bounds for the original series you will need to invert the log transformation manually),

f. a table of the actual forecast errors, i.e., the true value (deleted from AIRPASS.DAT) minus the forecast value, for each of the twelve forecasts. Does the last value of AIRPASS.DAT lie within the corresponding 95% prediction bounds?

6.8. Repeat Problem 6.7 but, instead of differencing, apply the classical decomposition method to the logarithms of the data in ASHORT.DAT by deseasonalizing, subtracting a quadratic trend, and then finding an appropriate ARMA model for the residuals. Compare the twelve forecast errors found from this approach with those found in Problem 6.7.

6.9. Repeat Problem 6.7 for the series BEER.DAT, deleting the last twelve values to create a file named BSHORT.DAT.

6.10. Repeat Problem 6.8 for the series BEER.DAT and the shortened series BSHORT.DAT.

6.11. A time series $\{X_t\}$ is differenced at lag 12, then at lag 1 to produce a zero-mean series $\{Y_t\}$ with the following sample ACF:

$$\hat{\rho}(12j) \approx (.8)^j, \qquad j = 0, \pm 1, \pm 2, \ldots,$$

$$\hat{\rho}(12j \pm 1) \approx (.4)(.8)^j, \quad j = 0, \pm 1, \pm 2, \ldots,$$

$$\hat{\rho}(h) \approx 0, \qquad \text{otherwise},$$

and $\hat{\gamma}(0) = 25$.

a. Suggest a SARIMA model for $\{X_t\}$ specifying all parameters.

b. For large n, express the one- and twelve-step linear predictors, $P_n X_{n+1}$ and $P_n X_{n+12}$ in terms of X_t, $t = -12, -11, \ldots, n$ and $Y_t - \hat{Y}_t$, $t = 1, \ldots, n$.

c. Find the mean squared errors of the predictors in (b).

6.12. a. If T_n is the covariance matrix of $(X_1, \ldots, X_n)'$ where $\{X_t\}$ is an AR(1) process with coefficient ϕ_1 and white noise variance σ^2, show that the matrix T defined by (6.6.7) satisfies $T_n^{-1} = \sigma^{-2} T'T$.

b. If in (a) we replace $\{X_t\}$ by an AR(2) process with coefficients ϕ_1 and ϕ_2, find a lower-triangular matrix T such that $T_n^{-1} = \sigma^{-2} T'T$.

c. How would you generalize the arguments in (a) and (b) to derive a lower triangular matrix T satisfying $T_n^{-1} = \sigma^{-2} T'T$ when $\{X_t\}$ is an AR(p) process?

6.13. Verify the computation of the GLS estimator and its variance given in Example 6.6.1.

7 Multivariate Time Series

Many time series arising in practice are best considered as components of some vector-valued (multivariate) time series $\{\mathbf{X}_t\}$ having not only serial dependence within each component series $\{X_{ti}\}$ but also interdependence between the different component series $\{X_{ti}\}$ and $\{X_{tj}\}$, $i \neq j$. Much of the theory of univariate time series extends in a natural way to the multivariate case; however, new problems arise. In this chapter we introduce the basic properties of multivariate series and consider the multivariate extensions of some of the techniques developed earlier. In Section 7.1 we introduce two sets of bivariate time series data for which we develop multivariate models later in the chapter. In Section 7.2 we discuss the basic properties of stationary multivariate time series, namely the mean vector $\boldsymbol{\mu} = E\mathbf{X}_t$ and the covariance matrices $\Gamma(h) = E(\mathbf{X}_{t+h}\mathbf{X}_t') - \boldsymbol{\mu}\boldsymbol{\mu}'$, $h = 0, \pm 1, \pm 2, \ldots$, with reference to some simple examples, including multivariate white noise. Section 7.3 deals with estimation of $\boldsymbol{\mu}$ and $\Gamma(\cdot)$ and the question of testing for serial independence on the basis of observations of $\mathbf{X}_1, \ldots, \mathbf{X}_n$. In Section 7.4 we introduce multivariate ARMA processes. and illustrate the problem of multivariate model identification with an example of a multivariate AR(1) process that also has an MA(1) representation. (Such examples do not exist in the univariate case.) The identification problem can be avoided by confining attention to multivariate autoregressive (or VAR) models. Forecasting multivariate time series with known second-order properties is discussed in Section 7.5, and in Section 7.6 we consider the modelling and forecasting of multivariate time series using the multivariate Yule-Walker equations and Whittle's generalization of the Durbin-Levinson algorithm. Section 7.7 contains a brief introduction to the notion of cointegrated time series.

7.1 Examples

In this section we introduce two examples of bivariate time series. A bivariate time series is a series of two-dimensional vectors $(X_{t1}, X_{t2})'$ observed at times t (usually $t = 1, 2, 3, \ldots$). The two component series $\{X_{t1}\}$and $\{X_{t2}\}$ could be studied independently as univariate time series, each characterized, from a second-order point of view, by its own mean and autocovariance function. Such an approach, however, fails to take into account possible dependence *between* the two component series and such cross-dependence may be of great importance, for example in predicting future values of the two component series.

We therefore consider the series of random vectors, $\mathbf{X}_t = (X_{t1}, X_{t2})'$ and define the mean vector

$$\boldsymbol{\mu}_t := E\mathbf{X}_t = \begin{bmatrix} EX_{t1} \\ EX_{t2} \end{bmatrix}$$

and covariance matrices

$$\Gamma(t+h, t) := \operatorname{Cov}(\mathbf{X}_{t+h}, \mathbf{X}_t) = \begin{bmatrix} \operatorname{cov}(X_{t+h,1}, X_{t1}) & \operatorname{cov}(X_{t+h,1}, X_{t2}) \\ \operatorname{cov}(X_{t+h,2}, X_{t1}) & \operatorname{cov}(X_{t+h,2}, X_{t2}) \end{bmatrix}.$$

The bivariate series $\{\mathbf{X}_t\}$ is said to be **(weakly) stationary** if the moments $\boldsymbol{\mu}_t$ and $\Gamma(t+h, t)$ are both independent of t, in which case we use the notation

$$\boldsymbol{\mu} = E\mathbf{X}_t = \begin{bmatrix} EX_{t1} \\ EX_{t2} \end{bmatrix}$$

and

$$\Gamma(h) = \operatorname{Cov}(\mathbf{X}_{t+h}, \mathbf{X}_t) = \begin{bmatrix} \gamma_{11}(h) & \gamma_{12}(h) \\ \gamma_{21}(h) & \gamma_{22}(h) \end{bmatrix}.$$

The diagonal elements are the autocovariance functions of the univariate series $\{X_{t1}\}$ and $\{X_{t2}\}$ as defined in Chapter 2, while the off-diagonal elements are the covariances between $X_{t+h,i}$ and X_{tj}, $i \neq j$. Notice that $\gamma_{12}(h) = \gamma_{21}(-h)$.

A natural estimator of the mean vector $\boldsymbol{\mu}$ in terms of the observations $\mathbf{X}_1, \ldots, \mathbf{X}_n$, is the vector of sample means,

$$\overline{\mathbf{X}}_n = \frac{1}{n}\sum_{t=1}^{n} \mathbf{X}_t,$$

and a natural estimator of $\Gamma(h)$ is

$$\hat{\Gamma}(h) = \begin{cases} n^{-1} \sum_{t=1}^{n-h} (\mathbf{X}_{t+h} - \overline{\mathbf{X}}_n)(\mathbf{X}_t - \overline{\mathbf{X}}_n)' & \text{for } 0 \le h \le n-1, \\ \hat{\Gamma}(-h)' & \text{for } -n+1 \le h < 0. \end{cases}$$

The correlation $\rho_{ij}(h)$ between $X_{t+h,i}$ and $X_{t,j}$ is estimated by

$$\hat{\rho}_{ij}(h) = \hat{\gamma}_{ij}(h)(\hat{\gamma}_{ii}(0)\hat{\gamma}_{jj}(0))^{-1/2}.$$

If $i = j$, $\hat{\rho}_{ij}$ reduces to the sample autocorrelation function of the ith series. These estimators will be discussed in more detail in Section 7.2.

Example 7.1.1 Dow-Jones–All-ordinaries Indices; DJAO2.DAT

Figure 7.1 shows the closing values $D_0, \ldots, D_{250}$ of the Dow-Jones Index of stocks on the New York Stock Exchange and the closing values $A_0, \ldots, A_{250}$ of the Australian All-ordinaries Index of Share Prices, recorded at the termination of trading on 251 successive trading days up to August 26th, 1994. (Because of the time difference between Sydney and New York, the markets do not close simultaneously in both places; however, the time difference is such that the most recent values of the indices at the beginning of trading in either place are the closing prices for the preceding trading day.) The efficient market hypothesis suggests that these processes should resemble random walks with uncorrelated increments. In order to model the data as a stationary bivariate time series we first reexpress them as percentage relative price

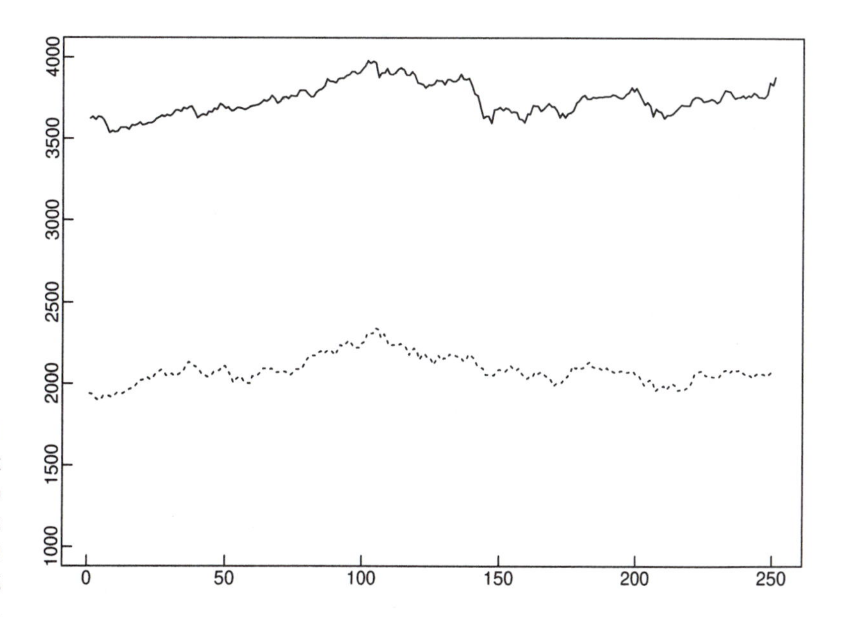

Figure 7-1 The Dow-Jones Index (top) and Australian All-ordinaries Index (bottom) at closing on 251 trading days ending August 26th, 1994.

changes (filed as DJAOPC2.DAT),

$$X_{t1} = 100\frac{(D_t - D_{t-1})}{D_{t-1}}, \quad t = 1, \ldots, 250,$$

and

$$X_{t2} = 100\frac{(A_t - A_{t-1})}{A_{t-1}}, \quad t = 1, \ldots, 250.$$

The estimators $\hat{\rho}_{11}(h)$, $\hat{\rho}_{22}(h)$, and $\hat{\rho}_{21}(h)$ are shown in Figures 7.2–7.4, respectively. We see from these figures that the correlations $\hat{\rho}_{ii}(h)$, $i = 1, 2$, are all small, while there is a much larger correlation between $X_{t-1,1}$ and $X_{t,2}$. This indicates the importance of considering *both* series as components of a bivariate time series. It also suggests that the value of $X_{t-1,1}$ may be of assistance in predicting the value of $X_{t,2}$. This last observation is supported by the scatterplot of the points $(x_{t-1,1}, x_{t,2})$, $t = 2, \ldots, 250$ shown in Figure 7.5. □

Example 7.1.2 Sales with a leading indicator; LS2.DAT

In this example we consider the sales data $\{Y_{t2}, t = 1, \ldots, 150\}$ with leading indicator $\{Y_{t1}, t = 1, \ldots, 150\}$ given by Box and Jenkins (1976), p. 537. The two series are stored in the ITSM data files SALES.DAT and LEAD.DAT, respectively, and in bivariate format as LS2.DAT. The graphs of the two series and their sample autocorrelation functions strongly suggest that both series are nonstationary. Application of the operator $(1 - B)$ yields the two differenced series $\{D_{t1}\}$ and $\{D_{t2}\}$ whose properties are compatible with those of low-order ARMA processes. Using the program PEST,

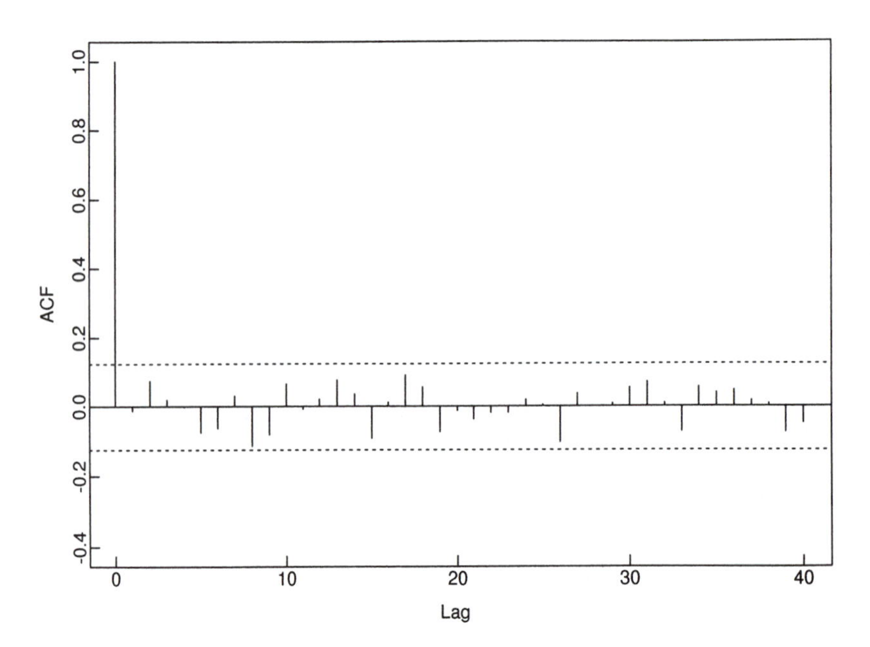

Figure 7-2 The sample ACF $\hat{\rho}_{11}$ of the observed values of $\{X_{t1}\}$ in Example 7.1.1, showing the bounds $\pm 1.96n^{-1/2}$

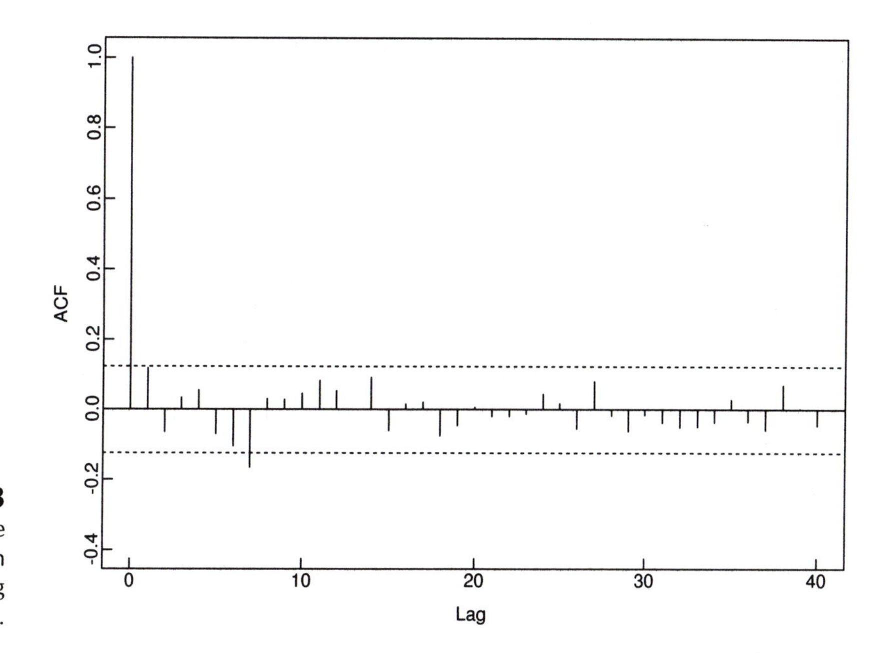

Figure 7-3 The sample ACF $\hat{\rho}_{22}$ of the observed values of $\{X_{t2}\}$ in Example 7.1.1, showing the bounds $\pm 1.96n^{-1/2}$.

we find that the models

$$D_{t1} - .0228 = Z_{t1} - .474Z_{t-1,1}, \quad \{Z_{t1}\} \sim \text{WN}(0, .0779), \tag{7.1.1}$$

$$D_{t2} - .838D_{t-1,2} - .0676 = Z_{t2} - .610Z_{t-1,2}, \quad \{Z_{t2}\} \sim \text{WN}(0, 1.754), \tag{7.1.2}$$

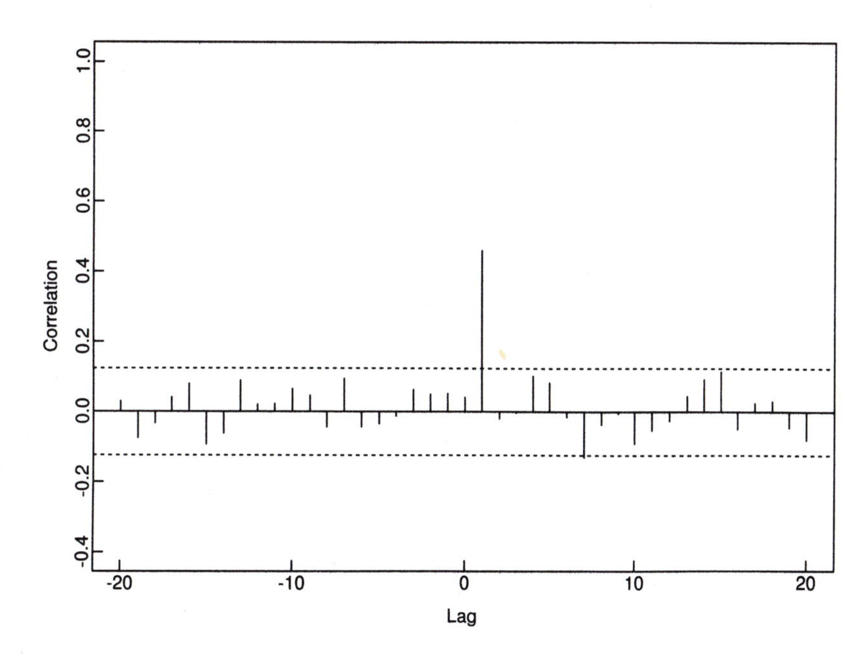

Figure 7-4 The sample cross-correlations $\hat{\rho}_{21}(h)$ between $X_{t-h,1}$ and $X_{t,2}$ for Example 7.1.1, showing the bounds $\pm 1.96n^{-1/2}$

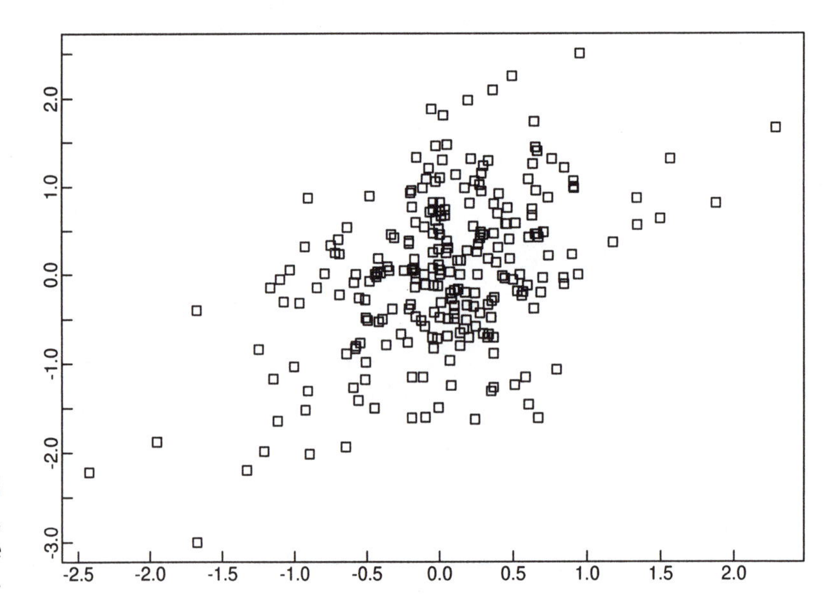

Figure 7-5
Scatterplot of $(x_{t-1,1}, x_{t,2})$, $t = 2, \ldots, 250$, for the data in Example 7.1.1.

provide good fits to the series $\{D_{t1}\}$ and $\{D_{t2}\}$.

The sample cross-correlation function of $\{D_{t1}\}$ and $\{D_{t2}\}$, computed using the program TRANS, is shown in Figure 7.6. As we shall see in Section 7.3, care must be taken in interpreting this function without first taking into account the autocorrelation structures of $\{D_{t1}\}$ and $\{D_{t2}\}$. □

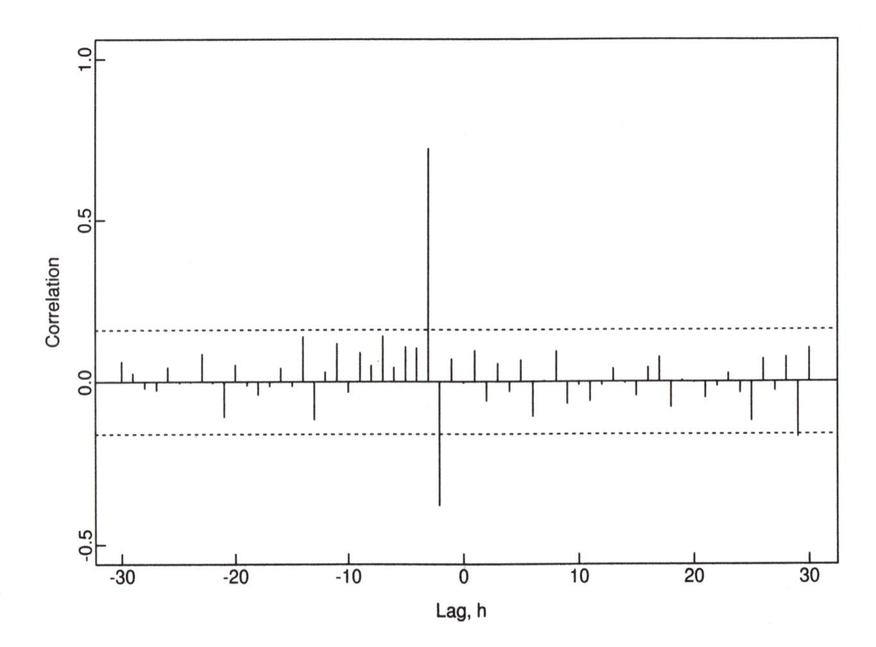

Figure 7-6
The sample cross-correlations function between $D_{t+h,1}$ and D_{t2}, Example 7.1.2, showing the bounds $\pm 1.96 n^{-1/2}$.

7.2 Second-Order Properties of Multivariate Time Series

Consider m time series $\{X_{ti}, t = 0, \pm 1, \ldots, \}, i = 1, \ldots, m$, with $EX_{ti}^2 < \infty$ for all t and i. If all the finite dimensional distributions of the random variables $\{X_{ti}\}$ were multivariate normal, then the distributional properties of $\{X_{ti}\}$ would be completely determined by the means

$$\mu_{ti} := EX_{ti} \tag{7.2.1}$$

and the covariances

$$\gamma_{ij}(t+h, t) := E[(X_{t+h,i} - \mu_{ti})(X_{tj} - \mu_{tj})]. \tag{7.2.2}$$

Even when the observations $\{X_{ti}\}$ do not have joint normal distributions, the quantities μ_{ti} and $\gamma_{ij}(t+h, t)$ specify the second-order properties, the covariances providing us with a measure of the dependence, not only between observations in the same series, but also between the observations in different series.

It is more convenient when dealing with m interrelated series to use vector notation. Thus we define

$$\mathbf{X}_t := \begin{bmatrix} X_{t1} \\ \vdots \\ X_{tm} \end{bmatrix}, \quad t = 0, \pm 1, \ldots. \tag{7.2.3}$$

The second-order properties of the multivariate time series $\{\mathbf{X}_t\}$ are then specified by the mean vectors

$$\boldsymbol{\mu}_t := E\mathbf{X}_t = \begin{bmatrix} \mu_{t1} \\ \vdots \\ \mu_{tm} \end{bmatrix} \tag{7.2.4}$$

and covariance matrices

$$\Gamma(t+h, t) := \begin{bmatrix} \gamma_{11}(t+h, t) & \cdots & \gamma_{1m}(t+h, t) \\ \vdots & \ddots & \vdots \\ \gamma_{m1}(t+h, t) & \cdots & \gamma_{mm}(t+h, t) \end{bmatrix}, \tag{7.2.5}$$

where

$$\gamma_{ij}(t+h, t) := \mathrm{Cov}(X_{t+h,i}, X_{t,j}).$$

Remark 1. The matrix $\Gamma(t+h, t)$ can also be expressed as

$$\Gamma(t+h, t) := E[(\mathbf{X}_{t+h} - \boldsymbol{\mu}_{t+h})(\mathbf{X}_t - \boldsymbol{\mu}_t)'].$$

where, as usual, the expected value of a random matrix A is the matrix whose components are the expected values of the components of A. □

As in the univariate case, a particularly important role is played by the class of **multivariate stationary time series**, defined as follows.

Definition 7.2.1

The m-variate series $\{\mathbf{X}_t\}$ is **(weakly) stationary** if

(i) $\mu_X(t)$ is independent of t

and

(ii) $\Gamma_X(t+h, t)$ is independent of t for each h.

For a stationary time series we shall use the notation

$$\boldsymbol{\mu} := E\mathbf{X}_t = \begin{bmatrix} \mu_1 \\ \vdots \\ \mu_m \end{bmatrix} \tag{7.2.6}$$

and

$$\Gamma(h) := E[(\mathbf{X}_{t+h} - \boldsymbol{\mu})(\mathbf{X}_t - \boldsymbol{\mu})'] = \begin{bmatrix} \gamma_{11}(h) & \cdots & \gamma_{1m}(h) \\ \vdots & \ddots & \vdots \\ \gamma_{m1}(h) & \cdots & \gamma_{mm}(h) \end{bmatrix}. \tag{7.2.7}$$

We shall refer to $\boldsymbol{\mu}$ as the mean of the series and to $\Gamma(h)$ as the covariance matrix at lag h. Notice that if $\{\mathbf{X}_t\}$ is stationary with covariance matrix function $\Gamma(\cdot)$, then for each i, $\{X_{ti}\}$ is stationary with covariance function $\gamma_{ii}(\cdot)$. The function $\gamma_{ij}(\cdot)$, $i \neq j$, is called the cross-covariance function of the two series $\{X_{ti}\}$ and $\{X_{tj}\}$. It should be noted that $\gamma_{ij}(\cdot)$ is not in general the same as $\gamma_{ji}(\cdot)$. The correlation matrix function $R(\cdot)$ is defined by

$$R(h) := \begin{bmatrix} \rho_{11}(h) & \cdots & \rho_{1m}(h) \\ \vdots & \ddots & \vdots \\ \rho_{m1}(h) & \cdots & \rho_{mm}(h) \end{bmatrix}, \tag{7.2.8}$$

where $\rho_{ij}(h) = \gamma_{ij}(h)/[\gamma_{ii}(0)\gamma_{jj}(0)]^{1/2}$. The function $R(\cdot)$ is the covariance matrix function of the normalized series obtained by subtracting $\boldsymbol{\mu}$ from $\mathbf{X}_t$ and then dividing each component by its standard deviation.

Example 7.2.1 Consider the bivariate stationary process $\{\mathbf{X}_t\}$ defined by

$$X_{t1} = Z_t,$$

$$X_{t2} = Z_t + 0.75Z_{t-10},$$

where $\{Z_t\} \sim \mathrm{WN}(0, 1)$. Elementary calculations yield $\boldsymbol{\mu} = \mathbf{0}$,

$$\Gamma(-10) = \begin{bmatrix} 0 & 0.75 \\ 0 & 0.75 \end{bmatrix}, \Gamma(0) = \begin{bmatrix} 1 & 1 \\ 1 & 1.5625 \end{bmatrix}, \Gamma(10) = \begin{bmatrix} 0 & 0 \\ 0.75 & 0.75 \end{bmatrix},$$

and $\Gamma(j) = 0$ otherwise. The correlation matrix function is given by

$$R(-10) = \begin{bmatrix} 0 & 0.60 \\ 0 & 0.48 \end{bmatrix}, R(0) = \begin{bmatrix} 1 & 0.8 \\ 0.8 & 1 \end{bmatrix}, R(10) = \begin{bmatrix} 0 & 0 \\ 0.60 & 0.48 \end{bmatrix},$$

and $R(j) = 0$ otherwise. □

Basic Properties of $\Gamma(\cdot)$:

1. $\Gamma(h) = \Gamma'(-h)$,
2. $|\gamma_{ij}(h)| \le [\gamma_{ii}(0)\gamma_{jj}(0)]^{1/2}, i, j, = 1, \ldots, m$,
3. $\gamma_{ii}(\cdot)$ is an autocovariance function, $i = 1, \ldots, m$, and
4. $\sum_{j,k=1}^{n} \mathbf{a}_j'\Gamma(j-k)\mathbf{a}_k \ge 0$ for all $n \in \{1, 2, \ldots\}$ and $\mathbf{a}_1, \ldots, \mathbf{a}_n \in \mathbb{R}^m$.

Proof The first property follows at once from the definition, the second from the fact that correlations cannot be greater than one in absolute value, and the third from the observation that $\gamma_{ii}(\cdot)$ is the autocovariance function of the stationary series $\{X_{ti}, t = 0, \pm 1, \ldots\}$. Property 4 is a statement of the obvious fact that

$$E\left(\sum_{j=1}^{n} \mathbf{a}_j'(\mathbf{X}_j - \boldsymbol{\mu})\right)^2 \ge 0.$$

■

Remark 2. The basic properties of the matrices $\Gamma(h)$ are shared also by the corresponding matrices of correlations, $R(h) = [\rho_{ij}(h)]_{i,j=1}^{m}$, which have the additional property

$$\rho_{ii}(0) = 1 \quad \text{for all } i.$$

The correlation $\rho_{ij}(0)$ is the correlation between X_{ti} and X_{tj}, which is generally not equal to 1 if $i \neq j$ (see Example 7.2.1). It is also possible that $|\gamma_{ij}(h)| > |\gamma_{ij}(0)|$ if $i \neq j$ (see Problem 7.1). □

The simplest multivariate time series is multivariate white noise, the definition of which is quite analogous to that of univariate white noise.

Definition 7.2.2 The m-variate series $\{\mathbf{Z}_t\}$ is called **white noise with mean 0 and covariance matrix Σ**, written

$$\{\mathbf{Z}_t\} \sim \mathrm{WN}(\mathbf{0}, \Sigma), \tag{7.2.9}$$

if and only if $\{\mathbf{Z}_t\}$ is stationary with mean vector $\mathbf{0}$ and covariance matrix function

$$\Gamma(h) = \begin{cases} \Sigma, & \text{if } h = 0, \\ 0, & \text{otherwise.} \end{cases} \tag{7.2.10}$$

Definition 7.2.3

The m-variate series $\{\mathbf{Z}_t\}$ is called **IID noise with mean 0 and covariance matrix** Σ, written

$$\{\mathbf{Z}_t\} \sim \text{IID}(\mathbf{0}, \Sigma), \tag{7.2.11}$$

if the random vectors $\{\mathbf{Z}_t\}$ are independent and identically distributed with mean $\mathbf{0}$ and covariance matrix Σ.

Multivariate white noise $\{\mathbf{Z}_t\}$ is used as a building block from which can be constructed an enormous variety of multivariate time series. The linear processes are generated as follows.

Definition 7.2.4

The m-variate series $\{\mathbf{X}_t\}$ is a **linear process** if it has the representation

$$\mathbf{X}_t = \sum_{j=-\infty}^{\infty} C_j \mathbf{Z}_{t-j}, \quad \{\mathbf{Z}_t\} \sim \text{WN}(\mathbf{0}, \Sigma), \tag{7.2.12}$$

where $\{C_j\}$ is a sequence of $m \times m$ matrices whose components are absolutely summable.

The linear process (7.2.12) is stationary (Problem 7.2) with mean $\mathbf{0}$ and covariance function

$$\Gamma(h) = \sum_{j=-\infty}^{\infty} C_{j+h} \Sigma C_j', \quad h = 0, \pm 1, \ldots. \tag{7.2.13}$$

An **MA(∞) process** is a linear process with $C_j = 0$ for $j < 0$. Thus $\{\mathbf{X}_t\}$ is a linear process if and only if there exists a white noise sequence $\{\mathbf{Z}_t\}$ and a sequence of matrices C_j with absolutely summable components such that

$$\mathbf{X}_t = \sum_{j=0}^{\infty} C_j \mathbf{Z}_{t-j}.$$

Multivariate ARMA processes will be discussed in Section 7.4, where it will be shown in particular that any causal ARMA(p, q) process can be expressed as an MA(∞) process, while any invertible ARMA(p, q) process can be expressed as an **AR(∞) process**, i.e., a process satisfying equations of the form

$$\mathbf{X}_t + \sum_{j=1}^{\infty} A_j \mathbf{X}_{t-j} = \mathbf{Z}_t,$$

in which the matrices A_j have absolutely summable components.

Second-Order Properties in the Frequency Domain

Provided the components of the covariance matrix function $\Gamma(\cdot)$ have the property $\sum_{h=-\infty}^{\infty} |\gamma_{ij}(h)| < \infty, i, j = 1, \ldots, m$, then Γ has a **matrix-valued spectral density function**,

$$f(\lambda) = \frac{1}{2\pi} \sum_{h=-\infty}^{\infty} e^{-i\lambda h} \Gamma(h), \quad -\pi \le \lambda \le \pi,$$

and Γ can be expressed in terms of f as

$$\Gamma(h) = \int_{-\pi}^{\pi} e^{i\lambda h} f(\lambda) d\lambda.$$

The second-order properties of the stationary process $\{\mathbf{X}_t\}$ can therefore be described equivalently in terms of $f(\cdot)$ rather than $\Gamma(\cdot)$. Similarly $\{\mathbf{X}_t\}$ has a **spectral representation**

$$\mathbf{X}_t = \int_{-\pi}^{\pi} e^{i\lambda t} d\mathbf{Z}(\lambda),$$

where $\{\mathbf{Z}(\lambda), -\pi \le \lambda \le \pi\}$ is a process whose components are complex-valued processes satisfying

$$E(dZ_j(\lambda) d\overline{Z}_k(\mu)) = \begin{cases} f_{jk}(\lambda) d\lambda & \text{if } \lambda = \mu, \\ 0 & \text{if } \lambda \ne \mu, \end{cases}$$

and $\overline{Z}_k$ denotes the complex conjugate of Z_k. We shall not go into the spectral representation in this book. For details see TSTM.

7.3 Estimation of the Mean and Covariance Function

As in the univariate case, the estimation of the mean vector and covariances of a stationary multivariate time series plays an important role in describing and modelling the dependence structure of the component series. In this section we introduce estimators, for a stationary m-variate time series $\{\mathbf{X}_t\}$, of the components μ_j, $\gamma_{ij}(h)$, and $\rho_{ij}(h)$ of $\boldsymbol{\mu}$, $\Gamma(h)$ and $R(h)$, respectively. We also examine the large-sample properties of these estimators.

7.3.1 Estimation of $\boldsymbol{\mu}$

A natural unbiased estimator of the mean vector $\boldsymbol{\mu}$ based on the observations $\mathbf{X}_1, \ldots, \mathbf{X}_n$, is the vector of sample means

$$\overline{\mathbf{X}}_n = \frac{1}{n} \sum_{t=1}^{n} \mathbf{X}_t.$$

The resulting estimate of the mean of the jth time series is then the univariate sample mean $(1/n)\sum_{t=1}^{n} X_{tj}$. If each of the univariate autocovariance functions $\gamma_{ii}(\cdot)$, $i = 1, \ldots, m$, satisfies the conditions of Proposition 2.4.1, then the consistency of the estimator $\overline{\mathbf{X}}_n$ can be established by applying the proposition to each of the component time series $\{X_{ti}\}$. This immediately gives the following result.

Proposition 7.3.1 *If $\{\mathbf{X}_t\}$ is a stationary multivariate time series with mean $\boldsymbol{\mu}$ and covariance function $\Gamma(\cdot)$, then as $n \to \infty$,*

$$E(\overline{\mathbf{X}}_n - \boldsymbol{\mu})'(\overline{\mathbf{X}}_n - \boldsymbol{\mu}) \to 0 \quad \text{if } \gamma_{ii}(n) \to 0, \quad 1 \le i \le m,$$

and

$$nE(\overline{\mathbf{X}}_n - \boldsymbol{\mu})'(\overline{\mathbf{X}}_n - \boldsymbol{\mu}) \to \sum_{i=1}^{m} \sum_{h=-\infty}^{\infty} \gamma_{ii}(h) \quad \text{if } \sum_{h=-\infty}^{\infty} |\gamma_{ii}(h)| < \infty, \quad 1 \le i \le m.$$

Under more restrictive assumptions on the process $\{\mathbf{X}_t\}$ it can also be shown that $\overline{\mathbf{X}}_n$ is approximately normally distributed for large n. Determination of the covariance matrix of this distribution would allow us to obtain confidence regions for $\boldsymbol{\mu}$. However, this is quite complicated and the following simple approximation is useful in practice.

For each i we construct a confidence interval for μ_i based on the sample mean $\overline{X}_i$ of the univariate series $X_{1i}, \ldots, X_{ti}$ and combine these to form a confidence region for $\boldsymbol{\mu}$. If $f_i(\omega)$ is the spectral density of the ith process $\{X_{ti}\}$ and if the sample size n is large, then we know, under the same conditions as in Section 2.4, that $\sqrt{n}(\overline{X}_i - \mu_i)$ is approximately normally distributed with mean zero and variance

$$2\pi f_i(0) = \sum_{k=-\infty}^{\infty} \gamma_{ii}(k).$$

It can also be shown (see e.g., Anderson, 1971) that

$$2\pi \hat{f}_i(0) := \sum_{|h| \le r} \left(1 - \frac{|h|}{r}\right) \hat{\gamma}(h)$$

is a consistent estimator of $2\pi f_i(0)$ provided $r = r_n$ is a sequence of numbers depending on n in such a way that $r_n \to \infty$ and $r_n/n \to 0$ as $n \to \infty$. Thus if $\overline{X}_i$ denotes the sample mean of the ith process and Φ_α is the α-quantile of the standard normal distribution, then the bounds

$$\overline{X}_i \pm \Phi_{1-\alpha/2}(2\pi \hat{f}_i(0)/n)^{1/2}$$

are asymptotic $(1 - \alpha)$ confidence bounds for μ_i. Hence

$$\begin{aligned} P(|\mu_i - \overline{X}_i| &\le \Phi_{1-\alpha/2}(2\pi \hat{f}_i(0)/n)^{1/2}, i = 1, \ldots, m) \\ &\ge 1 - \sum_{i=1}^{m} P(|\mu_i - \overline{X}_i| > \Phi_{1-\alpha/2}(2\pi \hat{f}_i(0)/n)^{1/2}) \end{aligned}$$

where the right-hand side converges to $1 - m\alpha$ as $n \to \infty$. Consequently as $n \to \infty$, the set of m-dimensional vectors bounded by

$$\{x_i = \overline{X}_i \pm \Phi_{1-(\alpha/(2m))}(2\pi \hat{f}_i(0)/n)^{1/2}, i = 1, \ldots, m\} \tag{7.3.1}$$

has a confidence coefficient which converges to a value *greater than or equal to* $1-\alpha$ (and substantially greater if m is large). Nevertheless the region defined by (7.3.1) is easy to determine and is of reasonable size provided m is not too large.

7.3.2 Estimation of $\Gamma(h)$

As in the univariate case, a natural estimator of the covariance $\Gamma(h) = E[(\mathbf{X}_{t+h} - \mu)(\mathbf{X}_t - \mu)']$ is

$$\hat{\Gamma}(h) = \begin{cases} n^{-1} \sum_{t=1}^{n-h} (\mathbf{X}_{t+h} - \overline{\mathbf{X}}_n)(\mathbf{X}_t - \overline{\mathbf{X}}_n)' & \text{for } 0 \le h \le n-1, \\ \hat{\Gamma}'(-h) & \text{for } -n+1 \le h < 0. \end{cases}$$

Writing $\hat{\gamma}_{ij}(h)$ for the (i, j)-component of $\hat{\Gamma}(h)$, $i, j = 1, 2, \ldots$, we estimate the cross-correlations by

$$\hat{\rho}_{ij}(h) = \hat{\gamma}_{ij}(h)(\hat{\gamma}_{ii}(0)\hat{\gamma}_{jj}(0))^{-1/2}.$$

If $i = j$, $\hat{\rho}_{ij}$ reduces to the sample autocorrelation function of the ith series.

Derivation of the large sample properties of $\hat{\gamma}_{ij}$ and $\hat{\rho}_{ij}$ is quite complicated in general. Here we shall simply note one result that is of particular importance for testing the independence of two component series. For details of the proof of this and related results, see TSTM.

Theorem 7.3.1 *Let $\{\mathbf{X}_t\}$ be the bivariate time series whose components are defined by*

$$X_{t1} = \sum_{k=-\infty}^{\infty} \alpha_k Z_{t-k,1}, \quad \{Z_{t1}\} \sim \text{IID}(0, \sigma_1^2),$$

and

$$X_{t2} = \sum_{k=-\infty}^{\infty} \beta_k Z_{t-k,2}, \quad \{Z_{t2}\} \sim \text{IID}(0, \sigma_2^2),$$

where the two sequences $\{Z_{t1}\}$ and $\{Z_{t2}\}$ are independent, $\sum_k |\alpha_k| < \infty$ and $\sum_k |\beta_k| < \infty$.

Then for all integers h and k with $h \neq k$, the random variables $n^{1/2}\hat{\rho}_{12}(h)$ and $n^{1/2}\hat{\rho}_{12}(k)$ are approximately bivariate normal with mean $\mathbf{0}$, variance $\sum_{j=-\infty}^{\infty} \rho_{11}(j)\rho_{22}(j)$ and covariance $\sum_{j=-\infty}^{\infty} \rho_{11}(j)\rho_{22}(j+k-h)$, for n large.

[For a related result that does not require the independence of the two series $\{X_{t1}\}$ and $\{X_{t2}\}$ see Theorem 7.3.2 below.]

Theorem 7.3.1 is useful in testing for correlation between two time series. If one of the two processes in the theorem is white noise, then it follows at once from the theorem that $\hat{\rho}_{12}(h)$ is approximately normally distributed with mean 0 and variance $1/n$, in which case it is straightforward to test the hypothesis that $\rho_{12}(h) = 0$. However, if neither process is white noise, then a value of $\hat{\rho}_{12}(h)$ that is large relative to $n^{-1/2}$ does not necessarily indicate that $\rho_{12}(h)$ is different from zero. For example, suppose that $\{X_{t1}\}$ and $\{X_{t2}\}$ are two independent AR(1) processes with $\rho_{11}(h) = \rho_{22}(h) = .8^{|h|}$. Then the large-sample variance of $\hat{\rho}_{12}(h)$ is $n^{-1}(1 + 2\sum_{k=1}^{\infty}(.64)^k) = 4.556n^{-1}$. It would therefore not be surprising to observe a value of $\hat{\rho}_{12}(h)$ as large as $3n^{-1/2}$ even though $\{X_{t1}\}$ and $\{X_{t2}\}$ are independent. If on the other hand $\rho_{11}(h) = .8^{|h|}$ and $\rho_{22}(h) = (-.8)^{|h|}$, then the large-sample variance of $\hat{\rho}_{12}(h)$ is $.2195n^{-1}$ and an observed value of $3n^{-1/2}$ for $\hat{\rho}_{12}(h)$ would be very unlikely.

7.3.3 Testing for Independence of Two Stationary Time Series

Since by Theorem 7.3.1 the large-sample distribution of $\hat{\rho}_{12}(h)$ depends on both $\rho_{11}(\cdot)$ and $\rho_{22}(\cdot)$, any test for independence of the two component series cannot be based solely on estimated values of $\rho_{12}(h)$, $h = 0, \pm 1, \ldots$, without taking into account the nature of the two component series.

This difficulty can be circumvented by "prewhitening" the two series before computing the cross-correlations $\hat{\rho}_{12}(h)$, i.e., by transforming the two series to white noise by application of suitable filters. If $\{X_{t1}\}$ and $\{X_{t2}\}$ are invertible ARMA(p, q) processes, this can be achieved by the transformations

$$Z_{ti} = \sum_{j=0}^{\infty} \pi_j^{(i)} X_{t-j,i}$$

where $\sum_{j=0}^{\infty} \pi_j^{(i)} z^j = \phi^{(i)}(z)/\theta^{(i)}(z)$ and $\phi^{(i)}$, $\theta^{(i)}$ are the autoregressive and moving average polynomials of the ith series, $i = 1, 2$.

Since in practice the true model is nearly always unknown and since the data X_{tj}, $t \le 0$, are not available, it is convenient to replace the sequences $\{Z_{ti}\}$ by the residuals $\{\hat{W}_{ti}\}$ after fitting a maximum likelihood ARMA model to each of the component series (see (5.3.1)). If the fitted ARMA models were in fact the true models, the series $\{\hat{W}_{ti}\}$ would be white noise sequences for $i = 1, 2$.

To test the hypothesis H_0 that $\{X_{t1}\}$ and $\{X_{t2}\}$ are independent series, we observe that under H_0, the corresponding two prewhitened series $\{Z_{t1}\}$ and $\{Z_{t2}\}$ are also independent. Theorem 7.3.1 then implies that the sample autocorrelations $\hat{\rho}_{12}(h)$, $\hat{\rho}_{12}(k)$, $h \neq k$, of $\{Z_{t1}\}$ and $\{Z_{t2}\}$ are for large n approximately independent and normally distributed with means 0 and variances n^{-1}. An approximate test for independence can therefore be obtained by comparing the values of $|\hat{\rho}_{12}(h)|$ with $1.96n^{-1/2}$, exactly as in Section 5.3.2. If we prewhiten only one of the two original series, say $\{X_{t1}\}$, then under H_0 Theorem 7.3.1 implies that the sample autocorrelations $\tilde{\rho}_{12}(h)$, $\tilde{\rho}_{12}(k)$, $h \neq k$, of $\{Z_{t1}\}$ and $\{X_{t2}\}$ are for large n approximately normal with means 0, vari-

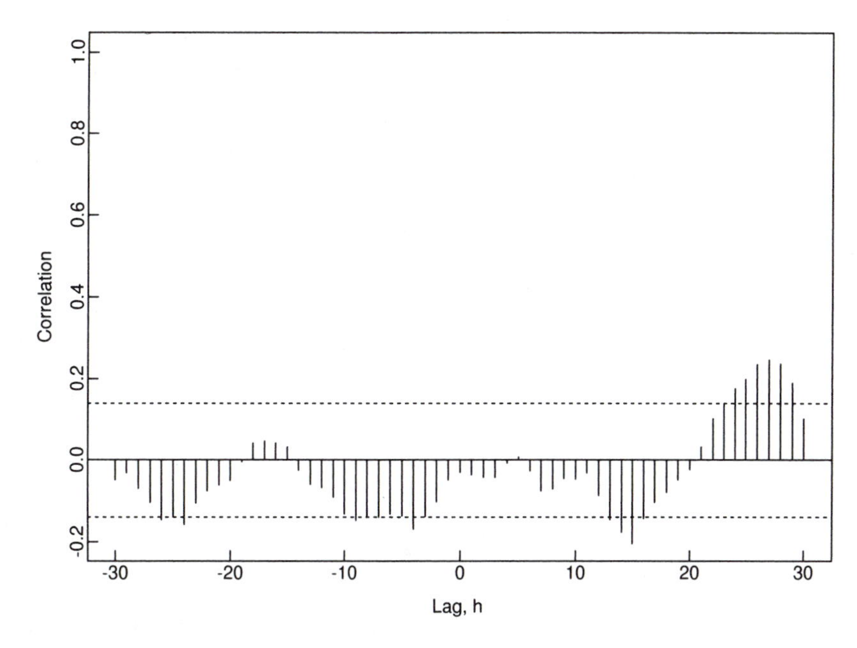

Figure 7-7 The sample cross-correlations between $X_{t+h,1}$ and X_{t2}, Example 7.3.1, showing the bounds $\pm 1.96n^{-1/2}$.

ances n^{-1} and covariance $n^{-1}\rho_{22}(k-h)$, where $\rho_{22}(\cdot)$ is the autocorrelation function of $\{X_{t2}\}$. Hence, for any fixed h, $\tilde{\rho}_{12}(h)$ also falls (under H_0) between the bounds $\pm 1.96n^{-1/2}$ with a probability of approximately 0.95.

Example 7.3.1 The sample cross-correlation function $\hat{\rho}_{12}(\cdot)$ of a bivariate time series of length $n = 200$ is shown in Figure 7.7. Without knowing the correlation function of each

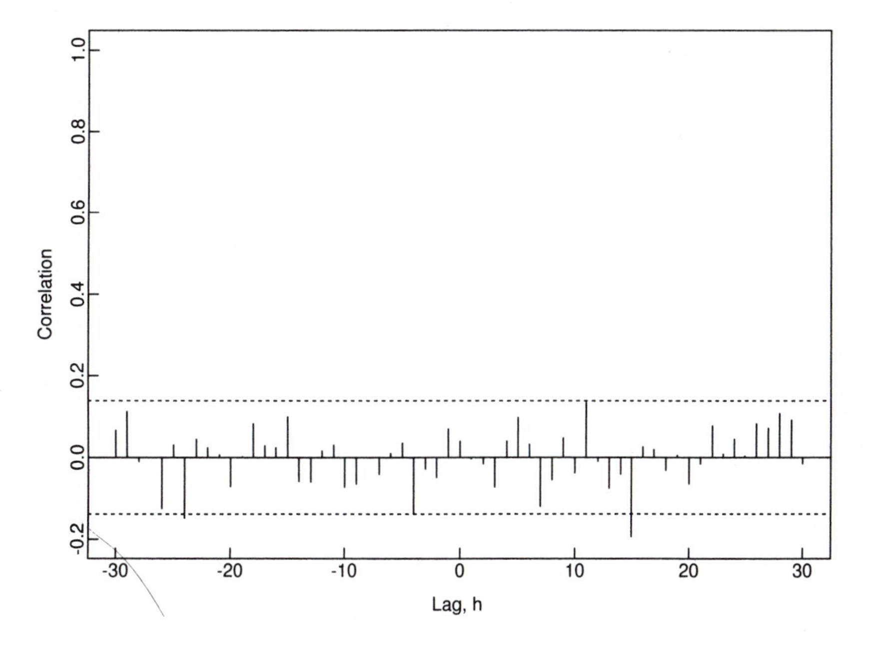

Figure 7-8 The sample cross-correlations between the residuals $\hat{W}_{t+h,1}$ and $\hat{W}_{t2}$, Example 7.3.1, showing the bounds $\pm 1.96n^{-1/2}$.

component process, it is impossible to decide if the two processes are uncorrelated with one another. Notice that many of the values of $\hat{\rho}_{12}(h)$ lie outside the bounds $\pm 1.96n^{-1/2} = \pm .139$. After inspection of the sample ACF and PACF of the two component series, they were both modelled as (zero-mean) AR(1) processes, fitted using the *ARMA estimation* option of PEST. The two residual series were then stored and their cross-correlation function computed with the program TRANS. This function is shown in Figure 7.8. All except two of the correlations are between the bounds $\pm .139$, suggesting by Theorem 7.3.1 that the two residual series (and hence the two original series) are uncorrelated. The data for this example were in fact generated as two independent AR(1) series with $\phi = 0.8$ and $\sigma^2 = 1$. □

7.3.4 Bartlett's Formula

In Section 2.4 we gave Bartlett's formula for the large-sample distribution of the sample autocorrelation vector $\hat{\rho} = (\hat{\rho}(1), \ldots, \hat{\rho}(k))'$ of a univariate time series. The following theorem gives a large sample approximation to the covariances of the sample cross-correlations $\hat{\rho}_{12}(h)$ and $\hat{\rho}_{12}(k)$ of the bivariate time series $\{\mathbf{X}_t\}$ under the assumption that $\{\mathbf{X}_t\}$ is Gaussian. However, it is not assumed (as in Theorem 7.3.1) that $\{X_{t1}\}$ is independent of $\{X_{t2}\}$.

Bartlett's Formula:

If $\{\mathbf{X}_t\}$ is a bivariate Gaussian time series with covariances satisfying $\sum_{h=-\infty}^{\infty} |\gamma_{ij}(h)| < \infty$, $i, j = 1, 2$, then

$$\lim_{n\to\infty} n\mathrm{Cov}(\hat{\rho}_{12}(h), \hat{\rho}_{12}(k)) = \sum_{j=-\infty}^{\infty} [\rho_{11}(j)\rho_{22}(j+k-h) + \rho_{12}(j+k)\rho_{21}(j-h)$$
$$- \rho_{12}(h)\{\rho_{11}(j)\rho_{12}(j+k) + \rho_{22}(j)\rho_{21}(j-k)\}$$
$$- \rho_{12}(k)\{\rho_{11}(j)\rho_{12}(j+h) + \rho_{22}(j)\rho_{21}(j-h)\}$$
$$+ \rho_{12}(h)\rho_{12}(k)\{\frac{1}{2}\rho_{11}^2(j) + \rho_{12}^2(j) + \frac{1}{2}\rho_{22}^2(j)\}]$$

Corollary 7.3.1 *If $\{\mathbf{X}_t\}$ satisfies the conditions for Bartlett's formula, if either $\{X_{t1}\}$ or $\{X_{t2}\}$ is white noise, and if*

$$\rho_{12}(h) = 0, \quad h \notin [a, b],$$

then

$$\lim_{n\to\infty} n\mathrm{Var}(\hat{\rho}_{12}(h)) = 1, \quad h \notin [a, b].$$

Example 7.3.2 Sales with a leading indicator

We consider again the differenced series $\{D_{t1}\}$ and $\{D_{t2}\}$ of Example 7.1.2, for which we found the maximum likelihood models (7.1.1) and (7.1.2) using PEST. The residuals from the two models (which can be filed by PEST) are the two "whitened" series $\{\hat{W}_{t1}\}$ and $\{\hat{W}_{t2}\}$ with sample variances .0779 and 1.754, respectively.

The sample cross-correlation function of $\{D_{t1}\}$ and $\{D_{t2}\}$ was shown in Figure 7.6. Without taking into account the autocorrelation structures of $\{D_{t1}\}$ and $\{D_{t2}\}$, however, it is not possible to draw any conclusions from this function.

Examination of the sample cross-correlation function of the whitened series $\{\hat{W}_{t1}\}$ and $\{\hat{W}_{t2}\}$ on the other hand is much more informative. From Figure 7.9 it is apparent that there is one large sample cross-correlation (between $\hat{W}_{t-3,1}$ and $\hat{W}_{t,2}$) while the others are all between $\pm 1.96n^{-1/2}$. □

If $\{\hat{W}_{t1}\}$ and $\{\hat{W}_{t2}\}$ are assumed to be jointly Gaussian, Corollary 7.3.1 indicates the compatibility of the cross-correlations with a model for which

$$\rho_{12}(-3) \neq 0$$

and

$$\rho_{12}(h) = 0, \quad h \neq -3.$$

The value $\hat{\rho}_{12}(-3) = .969$ suggests the model

$$\hat{W}_{t2} = 4.74\hat{W}_{t-3,1} + N_t, \tag{7.3.2}$$

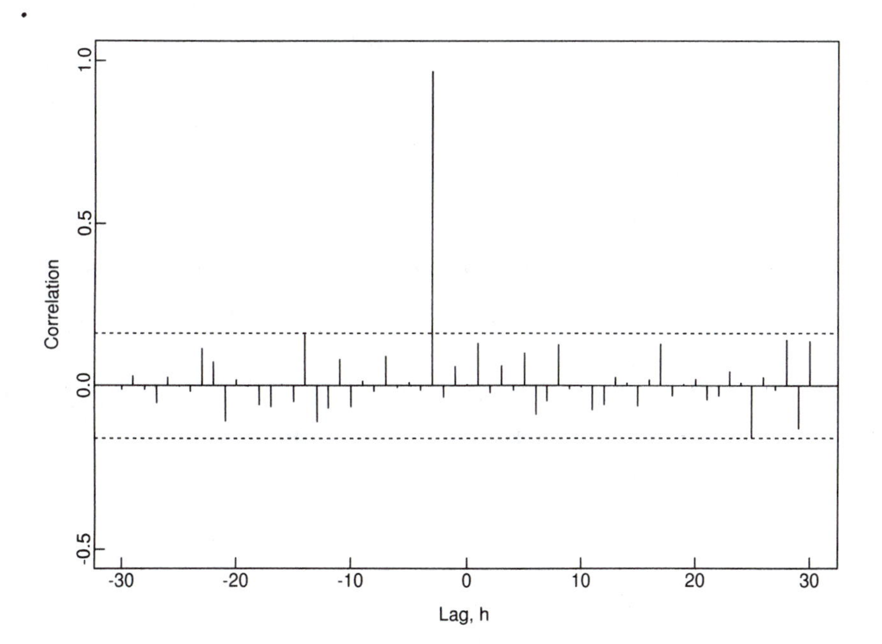

Figure 7-9 The sample cross-correlations between $\hat{W}_{t+h,1}$ and $\hat{W}_{t2}$, Example 7.3.2, showing the bounds $\pm 1.96n^{-1/2}$.

where the stationary noise $\{N_t\}$ has small variance compared with $\{\hat{W}_{t2}\}$ and $\{\hat{W}_{t1}\}$ and the coefficient 4.74 is the square root of the ratio of sample variances of $\{\hat{W}_{t2}\}$ and $\{\hat{W}_{t1}\}$. A study of the sample values of $\{\hat{W}_{t2} - 4.74\hat{W}_{t-3,1}\}$ suggests the model for $\{N_t\}$,

$$(1 + .345B)N_t = U_t, \quad \{U_t\} \sim \mathrm{WN}(0, .0782). \tag{7.3.3}$$

Finally, replacing $\hat{W}_{t2}$ and $\hat{W}_{t-3,1}$ in (7.3.2) by Z_{t2} and $Z_{t-3,1}$, respectively, and then using (7.1.1) and (7.1.2) to express Z_{t2} and $Z_{t-3,1}$ in terms of $\{D_{t2}\}$ and $\{D_{t1}\}$, we obtain a model relating $\{D_{t1}\}$, $\{D_{t2}\}$, and $\{U_{t1}\}$, namely

$$D_{t2} + .0773 = (1 - .610B)(1 - .838B)^{-1}[4.74(1 - .474B)^{-1}D_{t-3,1} + (1 + .345B)^{-1}U_t].$$

This model should be compared with the one derived later in Section 10.1 by the more systematic technique of transfer function modelling.

7.4 Multivariate ARMA Processes

As in the univariate case, we can define an extremely useful class of multivariate stationary processes $\{\mathbf{X}_t\}$ by requiring that $\{\mathbf{X}_t\}$ should satisfy a set of linear difference equations with constant coefficients. Multivariate white noise $\{\mathbf{Z}_t\}$ (see Definition 7.2.2) is a fundamental building block from which these ARMA processes are constructed.

Definition 7.4.1

$\{\mathbf{X}_t\}$ is an **ARMA(p, q) process** if $\{\mathbf{X}_t\}$ is stationary and if for every t,

$$\mathbf{X}_t - \Phi_1\mathbf{X}_{t-1} - \cdots - \Phi_p\mathbf{X}_{t-p} = \mathbf{Z}_t + \Theta_1\mathbf{Z}_{t-1} + \cdots + \Theta_q\mathbf{Z}_{t-q}, \tag{7.4.1}$$

where $\{\mathbf{Z}_t\} \sim \mathrm{WN}(\mathbf{0}, \Sigma)$. ($\{\mathbf{X}_t\}$ is an **ARMA(p, q) process with mean μ**, if $\{\mathbf{X}_t - \mu\}$ is an ARMA(p, q) process.)

The equations (7.4.1) can be written in the more compact form

$$\Phi(B)\mathbf{X}_t = \Theta(B)\mathbf{Z}_t, \quad \{\mathbf{Z}_t\} \sim \mathrm{WN}(\mathbf{0}, \Sigma), \tag{7.4.2}$$

where $\Phi(z) := I - \Phi_1 z - \cdots - \Phi_p z^p$ and $\Theta(z) := I + \Theta_1 z + \cdots + \Theta_q z^q$ are matrix-valued polynomials, I is the $m \times m$ identity matrix and B as usual denotes the backward shift operator. (Each component of the matrices $\Phi(z)$, $\Theta(z)$ is a polynomial with real coefficients and degree less than or equal to p, q, respectively.)

Example 7.4.1 The multivariate AR(1) process

Setting $p = 1$ and $q = 0$ in (7.4.1) gives the defining equations

$$\mathbf{X}_t = \Phi\mathbf{X}_{t-1} + \mathbf{Z}_t, \quad \{\mathbf{Z}_t\} \sim \mathrm{WN}(\mathbf{0}, \Sigma), \tag{7.4.3}$$

for the multivariate AR(1) series $\{\mathbf{X}_t\}$. By exactly the same argument as used in Example 2.2.1, we can express $\mathbf{X}_t$ as

$$\mathbf{X}_t = \sum_{j=0}^{\infty} \Phi^j \mathbf{Z}_{t-j}, \tag{7.4.4}$$

provided all the eigenvalues of Φ are less than 1 in absolute value, i.e., provided

$$\det(I - z\Phi) \neq 0 \quad \text{for all } z \in \mathbb{C} \text{ such that } |z| \leq 1. \tag{7.4.5}$$

If this condition is satisfied, then the coefficients Φ^j are absolutely summable and hence the series in (7.4.4) converges, i.e., each component of the matrix $\sum_{j=0}^{n} \Phi^j \mathbf{Z}_{t-j}$ converges (see Remark 1 of Section 2.2). The same argument as in Example 2.2.1 also shows that (7.4.4) is the unique stationary solution of (7.4.3). The condition that all the eigenvalues of Φ should be less than 1 in absolute value (or equivalently (7.4.5)) is just the multivariate analogue of the condition $|\phi| < 1$ required for the existence of a causal stationary solution of the univariate AR(1) equations (2.2.8). □

Causality and invertibility of a multivariate ARMA(p, q) process are defined precisely as in Section 3.1, except that the coefficients ψ_j, π_j in the representations $X_t = \sum_{j=0}^{\infty} \psi_j Z_{t-j}$ and $Z_t = \sum_{j=0}^{\infty} \pi_j X_{t-j}$ are replaced by $m \times m$ matrices Ψ_j and Π_j whose components are required to be absolutely summable. The following two theorems (proofs of which can be found in TSTM) provide us with criteria for causality and invertibility analogous to those of Section 3.1.

Causality:

An ARMA(p, q) process $\{\mathbf{X}_t\}$ is **causal**, or a **causal function of $\{\mathbf{Z}_t\}$**, if there exist matrices $\{\Psi_j\}$ with absolutely summable components such that

$$\mathbf{X}_t = \sum_{j=0}^{\infty} \Psi_j \mathbf{Z}_{t-j} \quad \text{for all } t. \tag{7.4.6}$$

Causality is equivalent to the condition

$$\det \Phi(z) \neq 0 \text{ for all } z \in \mathbb{C} \text{ such that } |z| \leq 1. \tag{7.4.7}$$

The matrices Ψ_j are found recursively from the equations

$$\Psi_j = \Theta_j + \sum_{k=1}^{\infty} \Phi_k \Psi_{j-k}, \quad j = 0, 1, \ldots, \tag{7.4.8}$$

where we define $\Theta_0 = I$, $\Theta_j = 0$ for $j > q$, $\Phi_j = 0$ for $j > p$, and $\Psi_j = 0$ for $j < 0$.

Invertibility:

An ARMA(p, q) process $\{\mathbf{X}_t\}$ is **invertible** if there exist matrices $\{\Pi_j\}$ with absolutely summable components such that

$$\mathbf{Z}_t = \sum_{j=0}^{\infty} \Pi_j \mathbf{X}_{t-j} \text{ for all } t. \tag{7.4.9}$$

Invertibility is equivalent to the condition

$$\det \Theta(z) \neq 0 \text{ for all } z \in \mathbb{C} \text{ such that } |z| \leq 1. \tag{7.4.10}$$

The matrices Π_j are found recursively from the equations

$$\Pi_j = -\Phi_j - \sum_{k=1}^{\infty} \Theta_k \Pi_{j-k}, \quad j = 0, 1, \ldots, \tag{7.4.11}$$

where we define $\Phi_0 = -I$, $\Phi_j = 0$ for $j > p$, $\Theta_j = 0$ for $j > q$, and $\Pi_j = 0$ for $j < 0$.

Example 7.4.2 For the multivariate AR(1) process defined by (7.4.3), the recursions (7.4.8) give

$$\Psi_0 = I,$$

$$\Psi_1 = \Phi\Psi_0 = \Phi,$$

$$\Psi_2 = \Phi\Psi_1 = \Phi^2,$$

$$\vdots$$

$$\Psi_j = \Phi\Psi_{j-1} = \Phi^j, \quad j \geq 3,$$

as already found in Example 7.4.1. □

Remark 3. For the bivariate AR(1) process (7.4.3) with

$$\Phi = \begin{bmatrix} 0 & 0.5 \\ 0 & 0 \end{bmatrix}$$

it is easy to check that $\Psi_j = \Phi^j = 0$ for $j > 1$ and hence that $\{\mathbf{X}_t\}$ has the alternative representation

$$\mathbf{X}_t = \mathbf{Z}_t + \Phi \mathbf{Z}_{t-1},$$

as an MA(1) process. This example shows that it is not always possible to distinguish between multivariate ARMA models of different orders without imposing further restrictions. If, for example, attention is restricted to pure AR processes, the problem does not arise. For detailed accounts of the identification problem for general ARMA(p, q) models see Hannan and Deistler (1988) and Lütkepohl (1993). □

7.4.1 The Covariance Matrix Function of a Causal ARMA Process

From (7.2.13) we can express the covariance matrix $\Gamma(h) = E(\mathbf{X}_{t+h}\mathbf{X}_t')$ of the causal process (7.4.6) as

$$\Gamma(h) = \sum_{j=0}^{\infty} \Psi_{h+j} \Sigma \Psi_j', \quad h = 0, \pm 1, \ldots, \tag{7.4.12}$$

where the matrices Ψ_j are found from (7.4.8) and $\Psi_j := 0$ for $j < 0$.

The covariance matrices $\Gamma(h)$, $h = 0, \pm 1, \ldots$, can also be found by solving the Yule-Walker equations

$$\Gamma(j) - \sum_{r=1}^{p} \Phi_r \Gamma(j-r) = \sum_{j \le r \le q} \Theta_r \Sigma \Psi_{r-j}, \quad j = 0, 1, 2, \ldots, \tag{7.4.13}$$

obtained by post-multiplying (7.4.1) by $\mathbf{X}_{t-j}'$ and taking expectations. The first $p+1$ of the equations (7.4.13) can be solved for the components of $\Gamma(0), \ldots, \Gamma(p)$ using the fact that $\Gamma(-h) = \Gamma'(h)$. The remaining equations then give $\Gamma(p+1)$, $\Gamma(p+2)$, $\ldots$ recursively. An explicit form of the solution of these equations can be written down by making use of Kronecker products and vec notation (see e.g., Lütkepohl, 1993).

Remark 4. If z_0 is the root of $\det \Phi(z) = 0$ with smallest absolute value, then it can be shown from the recursions (7.4.8) that $\Psi_j / r^j \to 0$ as $j \to \infty$ for all r such that $|z_0|^{-1} < r < 1$. Hence, there is a constant C such that each component of Ψ_j is smaller in absolute value than Cr^j. This implies in turn that there is a constant K such that each component of the matrix $\Psi_{h+j} \Sigma \Psi_j'$ on the right of (7.4.12) is bounded in absolute value by Kr^{2j}. Provided $|z_0|$ is not very close to 1, this means that the series (7.4.12) converges rapidly and the error incurred in each component by truncating the series after the term with $j = k-1$ is smaller in absolute value than $\sum_{j=k}^{\infty} Kr^{2j} = Kr^{2k}/(1-r^2)$. □

7.5 Best Linear Predictors of Second-Order Random Vectors

Let $\{\mathbf{X}_t = (X_{t1}, \ldots, X_{tm})'\}$ be an m-variate time series with means $E\mathbf{X}_t = \boldsymbol{\mu}_t$ and covariance function given by the $m \times m$ matrices

$$K(i, j) = E(\mathbf{X}_i \mathbf{X}_j') - \boldsymbol{\mu}_i \boldsymbol{\mu}_j'.$$

If $\mathbf{Y} = (Y_1, \ldots, Y_m)'$ is a random vector with finite second moments and $E\mathbf{Y} = \boldsymbol{\mu}$, we define

$$P_n(\mathbf{Y}) = (P_n Y_1, \ldots, P_n Y_m)', \tag{7.5.1}$$

where $P_n Y_j$ is the best linear predictor of the component Y_j of $\mathbf{Y}$ in terms of all of the components of the vectors $\mathbf{X}_t$, $t = 1, \dots, n$ and the constant 1. It follows immediately from the properties of the prediction operator (Section 2.5) that

$$P_n(\mathbf{Y}) = \boldsymbol{\mu} + A_1(\mathbf{X}_n - \boldsymbol{\mu}_n) + \cdots + A_n(\mathbf{X}_1 - \boldsymbol{\mu}_1), \tag{7.5.2}$$

for some matrices $A_1, \dots, A_n$, and that

$$\mathbf{Y} - P_n(\mathbf{Y}) \perp \mathbf{X}_{n+1-i}, \quad i = 1, \dots, n, \tag{7.5.3}$$

where we say that two m-dimensional random vectors $\mathbf{X}$ and $\mathbf{Y}$ are orthogonal (written $\mathbf{X} \perp \mathbf{Y}$) if $E(\mathbf{X}\mathbf{Y}')$ is a matrix of zeroes. The vector of best predictors (7.5.1) is uniquely determined by (7.5.2) and (7.5.3), although it is possible that there may be more than one possible choice for $A_1, \dots, A_n$.

As a special case of the above, if $\{\mathbf{X}_t\}$ is a zero-mean time series, the best linear predictor $\hat{\mathbf{X}}_{n+1}$ of $\mathbf{X}_{n+1}$ in terms of $\mathbf{X}_1, \dots, \mathbf{X}_n$ is obtained on replacing $\mathbf{Y}$ by $\mathbf{X}_{n+1}$ in (7.5.1). Thus

$$\hat{\mathbf{X}}_{n+1} = \begin{cases} \mathbf{0}, & \text{if } n = 0, \\ P_n(\mathbf{X}_{n+1}), & \text{if } n \geq 1. \end{cases}$$

Hence, we can write

$$\hat{\mathbf{X}}_{n+1} = \Phi_{n1}\mathbf{X}_n + \cdots + \Phi_{nn}\mathbf{X}_1, \quad n = 1, 2, \dots, \tag{7.5.4}$$

where, from (7.5.3), the coefficients Φ_{nj}, $j = 1, \dots, n$ are such that

$$E(\hat{\mathbf{X}}_{n+1}\mathbf{X}'_{n+1-i}) = E(\mathbf{X}_{n+1}\mathbf{X}'_{n+1-i}), \quad i = 1, \dots, n, \tag{7.5.5}$$

i.e.,

$$\sum_{j=1}^{n} \Phi_{nj} K(n+1-j, n+1-i) = K(n+1, n+1-i), \quad i = 1, \dots, n.$$

In the case when $\{\mathbf{X}_t\}$ is stationary with $K(i, j) = \Gamma(i - j)$, the prediction equations simplify to the m-dimensional analogues of (2.5.7), i.e.,

$$\sum_{j=1}^{n} \Phi_{nj}\Gamma(i - j) = \Gamma(i), \quad i = 1, \dots, n. \tag{7.5.6}$$

Provided the covariance matrix of the nm components of $\mathbf{X}_1, \dots, \mathbf{X}_n$ is nonsingular for every $n \geq 1$, the coefficients $\{\Phi_{nj}\}$ can be determined recursively using a multivariate version of the Durbin-Levinson algorithm given by Whittle (1963) (for details see TSTM, Proposition 11.4.1). Whittle's recursions also determine the covariance matrices of the one-step prediction errors, namely $V_0 = \Gamma(0)$ and, for $n \geq 1$,

$$\begin{aligned} V_n &= E(\mathbf{X}_{n+1} - \hat{\mathbf{X}}_{n+1})(\mathbf{X}_{n+1} - \hat{\mathbf{X}}_{n+1})' \\ &= \Gamma(0) - \Phi_{n1}\Gamma(-1) - \cdots - \Phi_{nn}\Gamma(-n). \end{aligned} \tag{7.5.7}$$

Remark 5. The innovations algorithm also has a multivariate version that can be used for prediction in much the same way as the univariate version described in Section 2.5.2 (for details see TSTM, Proposition 11.4.2). □

7.6 Modelling and Forecasting with Multivariate AR Processes

If $\{\mathbf{X}_t\}$ is any zero-mean second-order multivariate time series, it is easy to show from the results of Section 7.5 (Problem 7.4) that the one-step prediction errors $\mathbf{X}_j - \hat{\mathbf{X}}_j$, $j = 1, \ldots, n$, have the property

$$E(\mathbf{X}_j - \hat{\mathbf{X}}_j)(\mathbf{X}_k - \hat{\mathbf{X}}_k)' = 0 \text{ for } j \neq k. \tag{7.6.1}$$

Moreover, the matrix M, such that

$$\begin{bmatrix} \mathbf{X}_1 - \hat{\mathbf{X}}_1 \\ \mathbf{X}_2 - \hat{\mathbf{X}}_2 \\ \mathbf{X}_3 - \hat{\mathbf{X}}_3 \\ \vdots \\ \mathbf{X}_n - \hat{\mathbf{X}}_n \end{bmatrix} = M \begin{bmatrix} \mathbf{X}_1 \\ \mathbf{X}_2 \\ \mathbf{X}_3 \\ \vdots \\ \mathbf{X}_n \end{bmatrix} \tag{7.6.2}$$

is lower triangular with ones on the diagonal and therefore has determinant equal to 1.

If the series $\{\mathbf{X}_t\}$ is also Gaussian, then (7.6.1) implies that the prediction errors $\mathbf{U}_j = \mathbf{X}_j - \hat{\mathbf{X}}_j$, $j = 1, \ldots, n$, are independent with covariance matrices $V_0, \ldots, V_{n-1}$, respectively (as specified in (7.5.7)). Consequently the joint density of the prediction errors is the product

$$f(\mathbf{u}_1, \ldots, \mathbf{u}_n) = (2\pi)^{-nm/2} \left(\prod_{j=1}^{n} \det V_{j-1} \right)^{-1/2} \exp\left[-\frac{1}{2} \sum_{j=1}^{n} \mathbf{u}_j' V_{j-1}^{-1} \mathbf{u}_j \right].$$

Since the determinant of the matrix M in (7.6.2) is equal to 1, the joint density of the observations $\mathbf{X}_1, \ldots, \mathbf{X}_n$ at $\mathbf{x}_1, \ldots, \mathbf{x}_n$ is obtained on replacing $\mathbf{u}_1, \ldots, \mathbf{u}_n$ in the last expression by the values of $\mathbf{X}_j - \hat{\mathbf{X}}_j$ corresponding to the observations $\mathbf{x}_1, \ldots, \mathbf{x}_n$.

If we suppose that $\{\mathbf{X}_t\}$ is a zero-mean m-variate AR(p) process with coefficient matrices $\mathbf{\Phi} = \{\Phi_1, \ldots, \Phi_p\}$ and white noise covariance matrix Σ, we can therefore express the likelihood of the observations $\mathbf{X}_1, \ldots, \mathbf{X}_n$ as

$$L(\mathbf{\Phi}, \Sigma) = (2\pi)^{-nm/2} \left(\prod_{j=1}^{n} \det V_{j-1} \right)^{-1/2} \exp\left[-\frac{1}{2} \sum_{j=1}^{n} \mathbf{U}_j' V_{j-1}^{-1} \mathbf{U}_j \right],$$

where $\mathbf{U}_j = \mathbf{X}_j - \hat{\mathbf{X}}_j$, $j = 1, \ldots, n$, and $\hat{\mathbf{X}}_j$ and V_j are found from (7.5.4), (7.5.6), and (7.5.7).

Maximization of the Gaussian likelihood is much more difficult in the multivariate than in the univariate case because of the potentially large number of parameters involved and the fact that it is not possible to compute the maximum likelihood estimator of $\mathbf{\Phi}$ independently of Σ as in the univariate case. In principle maximum likelihood estimators can be computed with the aid of efficient nonlinear optimization algorithms, but it is important to begin the search with preliminary estimates that are reasonably close to the maximum. For pure AR processes good preliminary estimates can be obtained using Whittle's algorithm or a multivariate version of Burg's algorithm given by Jones (1978). We shall restrict our discussion here to the use of Whittle's algorithm but Jones's multivariate version of Burg's algorithm is available in ITSM as the program BURG. Other useful algorithms can be found in Lütkepohl (1993), in particular the method of conditional least squares and the method of Hannan and Rissanen (1982), the latter being useful also for preliminary estimation in the more difficult problem of fitting ARMA(p, q) models with $q > 0$. Spectral methods of estimation for multivariate ARMA processes are also frequently used. A discussion of these (as well as some time domain methods) is given in Anderson (1980).

Order selection for multivariate autoregressive models can be made by minimizing a multivariate analogue of the univariate AICC statistic,

$$\text{AICC} = -2\ln L(\Phi_1, \ldots, \Phi_p, \Sigma) + \frac{2(pm^2+1)nm}{nm - pm^2 - 2}. \tag{7.6.3}$$

7.6.1 Estimation for Autoregressive Processes Using Whittle's Algorithm

If $\{\mathbf{X}_t\}$ is the (causal) multivariate AR(p) process defined by the difference equations

$$\mathbf{X}_t = \Phi_1 \mathbf{X}_{t-1} + \cdots + \Phi_p \mathbf{X}_{t-p} + \mathbf{Z}_t, \quad \{\mathbf{Z}_t\} \sim \text{WN}(\mathbf{0}, \Sigma), \tag{7.6.4}$$

then post-multiplying by $\mathbf{X}'_{t-j}$, $j = 0, \ldots, p$ and taking expectations gives the equations

$$\Sigma = \Gamma(0) - \sum_{j=1}^{p} \Phi_j \Gamma(-j) \tag{7.6.5}$$

and

$$\Gamma(i) = \sum_{j=1}^{n} \Phi_j \Gamma(i-j), \quad i = 1, \ldots, p. \tag{7.6.6}$$

Given the matrices $\Gamma(0), \ldots, \Gamma(p)$, equations (7.6.6) can be used to determine the coefficient matrices $\Phi_1, \ldots, \Phi_p$. The white noise covariance matrix Σ can then be found from (7.6.5). The solution of these equations for $\Phi_1, \ldots, \Phi_p$, and Σ is identical to the solution of (7.5.6) and (7.5.7) for the prediction coefficient matrices $\Phi_{p1}, \ldots, \Phi_{pp}$ and the corresponding prediction error covariance matrix V_p. Consequently Whittle's algorithm can be used to carry out the algebra.

The Yule-Walker estimators $\hat{\Phi}_1, \ldots, \hat{\Phi}_p$ and $\hat{\Sigma}$ for the model (7.6.4) fitted to the data $\mathbf{X}_1, \ldots, \mathbf{X}_n$ are obtained by replacing $\Gamma(j)$ in (7.6.5) and (7.6.6) by $\hat{\Gamma}(j)$, $j = 0, \ldots, p$, and solving the resulting equations for $\Phi_1, \ldots, \Phi_p$ and Σ. The solution of these equations is computed by the program ARVEC making use of Whittle's algorithm. Note that ARVEC automatically subtracts the sample mean from the data, thus fitting an AR(p) model to the *mean-corrected* data.

In fitting multivariate autoregressive models using ARVEC, selection of the order $p = -1$ will cause the program to fit autoregressions of all orders from 0 to 20 and to select the model giving the smallest value of the AICC statistic (7.6.3).

Analogous calculations using Jones's multivariate version of Burg's algorithm for parameter estimation can be carried out using the program BURG.

Example 7.6.1 The Dow-Jones and All-ordinaries Indices

Returning now to Example 7.1.1, we can use the program ARVEC to find the minimum AICC Yule-Walker model for the bivariate series $\{(X_{t1}, X_{t2})', t = 1, \ldots, 250\}$ of Example 7.1.1. This series is contained in the ITSM data file under the name DJAOPC2.DAT. To run the program ARVEC, double click on the ARVEC icon, specify 2 for the dimension of the data vectors, and then read in the data file. No differencing is required (recalling from Example 7.1.1 that $\{X_{t1}\}$ and $\{X_{t2}\}$ were obtained from the original Dow-Jones and All-ordinaries indices by transforming to percent relative price changes). Select the option *Find minimum AICC Model* and the program selects the Yule-Walker AR(p) model, $p \le 20$, with minimum AICC value. The model found in this way is

$$\begin{bmatrix} X_{t1} \\ X_{t2} \end{bmatrix} - \begin{bmatrix} .0295 \\ .0309 \end{bmatrix} = \begin{bmatrix} -.0148 & .0357 \\ .6584 & .0998 \end{bmatrix} \left(\begin{bmatrix} X_{t-1,1} \\ X_{t-1,2} \end{bmatrix} - \begin{bmatrix} .0295 \\ .0309 \end{bmatrix} \right) + \begin{bmatrix} Z_{t1} \\ Z_{t2} \end{bmatrix},$$

where

$$\begin{bmatrix} Z_{t1} \\ Z_{t2} \end{bmatrix} \sim \text{WN}\left(\begin{bmatrix} 0 \\ 0 \end{bmatrix}, \begin{bmatrix} .3653 & .0224 \\ .0224 & .6016 \end{bmatrix} \right).$$

□

Example 7.6.2 Sales with a leading indicator

The series $\{Y_{t1}\}$ (leading indicator) and $\{Y_{t2}\}$ (sales) are stored in bivariate form (Y_{t1} in column 1 and Y_{t2} in column 2) in the file LS2.DAT. After reading the data into ARVEC you will be given the option of plotting the components. Inspection of the graphs immediately suggests, as in Example 7.1.2, that the differencing operator $\nabla = 1 - B$ should be applied to the data before a stationary AR model is fitted. After the data has been plotted you will be asked to specify if preliminary differencing is required. Answer yes and specify the differencing lag to be 1. The program will then carry out the specified differencing on both components and subtract the mean of the resulting differenced series. You will then be given the option of plotting the transformed data to check whether or not further differencing is required (up to two differencing operations are permitted in the programs ARVEC and BURG). Answer no this time. As a result of the preliminary transformations the program is now storing the transformed data

$$X_{t1} = (1 - B)Y_{t1} - .0228, \quad t = 1, \ldots, 149,$$

and

$$X_{t2} = (1 - B)Y_{t2} - .420, \quad t = 1, \ldots, 149.$$

The next step is to select the option *Find minimum AICC model*, which will cause the program to fit models of all orders from 0 to 20, and to display the one with smallest AICC value.

$$\hat{\Phi}_1 = \begin{bmatrix} -.517 & .024 \\ -.019 & -.051 \end{bmatrix}, \hat{\Phi}_2 = \begin{bmatrix} -.192 & -.018 \\ .047 & .250 \end{bmatrix}, \hat{\Phi}_3 = \begin{bmatrix} -.073 & .010 \\ 4.678 & .207 \end{bmatrix},$$

$$\hat{\Phi}_4 = \begin{bmatrix} -.032 & -.009 \\ 3.664 & .004 \end{bmatrix}, \hat{\Phi}_5 = \begin{bmatrix} .022 & .011 \\ 1.300 & .029 \end{bmatrix}, \hat{\Sigma} = \begin{bmatrix} .076 & -.003 \\ -.003 & .095 \end{bmatrix},$$

with AICC=109.49. (Analogous calculations with BURG leads to an AR(8) model for the differenced series.) The fitted model has the interesting property that the upper right component of each of the coefficient matrices is close to zero. This suggests that $\{X_{t1}\}$ can be effectively modelled independently of $\{X_{t2}\}$. In fact the MA(1) model

$$X_{t1} = (1 - .474B)U_t, \quad \{U_t\} \sim \text{WN}(0, .0779) \tag{7.6.7}$$

provides an adequate fit to the univariate series $\{X_{t1}\}$. Inspecting the bottom rows of the coefficient matrices and deleting small entries, we find that the relation between $\{X_{t1}\}$ and $\{X_{t2}\}$ can be expressed approximately as

$$X_{t2} = .250X_{t-2,2} + .207X_{t-3,2} + 4.678X_{t-3,1} + 3.664X_{t-4,1}$$
$$+ 1.300X_{t-5,1} + W_t,$$

or equivalently

$$X_{t2} = \frac{4.678B^3(1 + .783B + .278B^2)}{1 - .250B^2 - .207B^3} X_{t1} + \frac{W_t}{1 - .250B^2 - .207B^3}, \tag{7.6.8}$$

where $\{W_t\} \sim \text{WN}(0, .095)$. Moreover, since the estimated noise covariance matrix is essentially diagonal, it follows that the two sequences $\{X_{t1}\}$ and $\{W_t\}$ are uncorrelated. This reduced model defined by (7.6.7) and (7.6.8) is an example of a transfer function model that expresses the "output" series $\{X_{t2}\}$ as the output of a linear filter with "input" $\{X_{t1}\}$ plus added noise. A more direct approach to the fitting of transfer function models is given in Section 10.1 and applied to this same data set. □

7.6.2 Forecasting Multivariate Autoregressive Processes

The technique developed in Section 7.5 allows us to compute the minimum mean squared error one-step linear predictors $\hat{\mathbf{X}}_{n+1}$ for any multivariate stationary time series from the autocovariance matrices $\Gamma(h)$ by recursively determining the coefficients

$\Phi_{ni}, i = 1, \ldots, n$ and evaluating

$$\hat{\mathbf{X}}_{n+1} = \Phi_{n1}\mathbf{X}_n + \cdots + \Phi_{nn}\mathbf{X}_1. \tag{7.6.9}$$

The situation is simplified when $\{\mathbf{X}_t\}$ is the AR(p) process defined by (7.6.4) since for $n \geq p$ (as is almost always the case in practice)

$$\hat{\mathbf{X}}_{n+1} = \Phi_1\mathbf{X}_n + \cdots + \Phi_p\mathbf{X}_{n+1-p}. \tag{7.6.10}$$

To verify (7.6.10) it suffices to observe that the right-hand side has the required form (7.5.2) and that the prediction error,

$$\mathbf{X}_{n+1} - \Phi_1\mathbf{X}_n - \cdots - \Phi_p\mathbf{X}_{n+1-p} = \mathbf{Z}_{n+1},$$

is orthogonal to $\mathbf{X}_1, \ldots, \mathbf{X}_n$ in the sense of (7.5.3). (In fact the prediction error is orthogonal to *all* $\mathbf{X}_j, -\infty < j \leq n$ showing that if $n \geq p$ then (7.6.10) is also the best linear predictor of $\mathbf{X}_{n+1}$ in terms of all components of $\mathbf{X}_j, -\infty < j \leq n$.) The covariance matrix of the one-step prediction error is clearly $E(\mathbf{Z}_{n+1}\mathbf{Z}'_{n+1}) = \Sigma$.

To compute the best h-step linear predictor $P_n\mathbf{X}_{n+h}$ based on all the components of $\mathbf{X}_1, \ldots, \mathbf{X}_n$ we apply the linear operator P_n to (7.6.4) to obtain the recursions

$$P_n\mathbf{X}_{n+h} = \Phi_1 P_n\mathbf{X}_{n+h-1} + \cdots + \Phi_p P_n\mathbf{X}_{n+h-p}. \tag{7.6.11}$$

These equations are easily solved recursively, first for $P_n\mathbf{X}_{n+1}$, then for $P_n\mathbf{X}_{n+2}$, $P_n\mathbf{X}_{n+3}$, ..., etc. If $n \geq p$ then the h-step predictors based on all components of $\mathbf{X}_j, -\infty < j \leq n$ also satisfy (7.6.11) and are therefore the same as the h-step predictors based on $\mathbf{X}_1, \ldots, \mathbf{X}_n$.

To compute the h-step error covariance matrices, recall from (7.4.6) that

$$\mathbf{X}_{n+h} = \sum_{j=0}^{\infty} \Psi_j \mathbf{Z}_{n+h-j}, \tag{7.6.12}$$

where the coefficient matrices Ψ_j are found from the recursions (7.4.8) with $q = 0$. From (7.6.12) we find that for $n \geq p$,

$$P_n\mathbf{X}_{n+h} = \sum_{j=h}^{\infty} \Psi_j \mathbf{Z}_{n+h-j}. \tag{7.6.13}$$

Subtracting (7.6.13) from (7.6.12) gives the h-step prediction error,

$$\mathbf{X}_{n+h} - P_n\mathbf{X}_{n+h} = \sum_{j=0}^{h-1} \Psi_j \mathbf{Z}_{n+h-j}, \tag{7.6.14}$$

with covariance matrix

$$E[(\mathbf{X}_{n+h} - P_n\mathbf{X}_{n+h})(\mathbf{X}_{n+h} - P_n\mathbf{X}_{n+h})'] = \sum_{j=0}^{h-1} \Psi_j \Sigma \Psi'_j, \quad n \geq p. \tag{7.6.15}$$

The above calculations are all based on the assumption that the AR(p) model for the series is known. However, in practice the parameters of the model are usually estimated from the data and the uncertainty in the predicted values of the series will be larger than indicated by (7.6.15) because of parameter estimation errors. See Lütkepohl (1993).

Example 7.6.3 The Dow-Jones and All-ordinaries Indices

The VAR(1) model fitted to the series $\{\mathbf{X}_t, t = 1, \ldots, 250\}$ in Example 7.6.1 was

$$\begin{bmatrix} X_{t1} \\ X_{t2} \end{bmatrix} - \begin{bmatrix} .0295 \\ .0309 \end{bmatrix} = \begin{bmatrix} -.0148 & .0357 \\ .6584 & .0998 \end{bmatrix} \left(\begin{bmatrix} X_{t-1,1} \\ X_{t-1,2} \end{bmatrix} - \begin{bmatrix} .0295 \\ .0309 \end{bmatrix} \right) + \begin{bmatrix} Z_{t1} \\ Z_{t2} \end{bmatrix},$$

where

$$\begin{bmatrix} Z_{t1} \\ Z_{t2} \end{bmatrix} \sim \text{WN}\left(\begin{bmatrix} 0 \\ 0 \end{bmatrix}, \begin{bmatrix} .3653 & .0224 \\ .0224 & .6016 \end{bmatrix} \right).$$

The one-step mean squared error for prediction of X_{t2}, assuming the validity of this model, is thus 0.6016. This is a substantial reduction from the estimated mean squared error $\hat{\gamma}_{22}(0) = .7712$ when the sample mean $\hat{\mu}_2 = .0309$ is used as the one-step predictor.

If we fit a univariate model to the series $\{X_{t2}\}$ using PEST, we find that the autoregression with minimum AICC value (645.0) is

$$X_{t2} = .0273 + .1180 X_{t-1,2} + Z_t, \quad \{Z_t\} \sim \text{WN}(0, .7604).$$

Assuming the validity of this model, we thus obtain a mean squared error for one-step prediction of .7604, which is slightly less than the estimated mean squared error (.7911) incurred when the sample mean is used for one-step prediction.

The preceding calculations suggest that there is little to be gained from the point of view of one-step prediction by fitting a univariate model to $\{X_{t2}\}$, while there is a substantial reduction achieved by the bivariate AR(1) model for $\{\mathbf{X}_t = (X_{t1}, X_{t2})'\}$.

To test the models fitted above, we consider the next forty values $\{\mathbf{X}_t, t = 251, \ldots, 290\}$, which are stored as an ITSM data file under the name DJAOPCF.DAT. We can use these values, in conjunction with the bivariate and univariate models fitted to the data for $t = 1, \ldots, 250$, to compute one-step predictors of X_{t2}, $t = 251, \ldots, 290$. The results are as follows:

Predictor	Average Squared Error
$\hat{\mu} = 0.0309$	.4706
AR(1)	.4591
VAR(1)	.3962

It is clear from these results that the sample variance of the series $\{X_{t2}, t = 251, \ldots, 290\}$ is rather less than that of the series $\{X_{t2}, t = 1, \ldots, 250\}$ and consequently the average squared errors of all three predictors are substantially less than expected from the models fitted to the latter series. There is an improvement in one-step average squared error from both the univariate and bivariate autoregressive models for the data, but the improvement from the bivariate model is much more pronounced. □

The calculation of predictors and their error covariance matrices for multivariate ARIMA and SARIMA processes is analogous to the corresponding univariate calculation, so we shall simply state the pertinent results. Suppose that $\{\mathbf{Y}_t\}$ is a nonstationary process satisfying $D(B)\mathbf{Y}_t = \mathbf{U}_t$ where $D(z) = 1 - d_1 z - \cdots - d_r z^r$ is a polynomial with $D(1) = 0$ and $\{\mathbf{U}_t\}$ is a causal invertible ARMA process with mean $\boldsymbol{\mu}$. Then $\mathbf{X}_t = \mathbf{U}_t - \boldsymbol{\mu}$ satisfies

$$\Phi(B)\mathbf{X}_t = \Theta(B)\mathbf{Z}_t, \quad \{\mathbf{Z}_t\} \sim \mathrm{WN}(\mathbf{0}, \Sigma). \tag{7.6.16}$$

Under the assumption that the random vectors $\mathbf{Y}_{-r+1}, \ldots, \mathbf{Y}_0$ are uncorrelated with the sequence $\{\mathbf{Z}_t\}$, the best linear predictors $\tilde{P}_n\mathbf{Y}_j$ of $\mathbf{Y}_j$, $j > n > 0$, based on 1 and the components of $\mathbf{Y}_j$, $-r+1, \le j \le n$, are found as follows. Compute the observed values of $\mathbf{U}_t = D(B)\mathbf{Y}_t$, $t = 1, \ldots, n$, and use the ARMA model for $\mathbf{X}_t = \mathbf{U}_t - \boldsymbol{\mu}$ to compute predictors $P_n\mathbf{U}_{n+h}$. Then use the recursions

$$\tilde{P}_n\mathbf{Y}_{n+h} = P_n\mathbf{U}_{n+h} + \sum_{j=1}^{r} d_j \tilde{P}_n\mathbf{Y}_{n+h-j}, \tag{7.6.17}$$

to compute successively $\tilde{P}_n\mathbf{Y}_{n+1}$, $\tilde{P}_n\mathbf{Y}_{n+2}$, $\tilde{P}_n\mathbf{Y}_{n+3}$, etc. The error covariance matrices are approximately (for large n)

$$E[(\mathbf{Y}_{n+h} - \tilde{P}_n\mathbf{Y}_{n+h})(\mathbf{Y}_{n+h} - \tilde{P}_n\mathbf{Y}_{n+h})'] = \sum_{j=0}^{h-1} \Psi_j^* \Sigma \Psi_j^{*\prime}, \tag{7.6.18}$$

where Ψ_j^* is the coefficient of z^j in the power series expansion

$$\sum_{j=0}^{\infty} \Psi_j^* z^j = D(z)^{-1}\Phi^{-1}(z)\Theta(z) \quad |z| < 1.$$

The matrices Ψ_j^* are most readily found from the recursions (7.4.8) after replacing Φ_j, $j = 1, \ldots, p$, by Φ_j^*, $j = 1, \ldots, p+r$, where Φ_j^* is the coefficient of z^j in $D(z)\Phi(z)$.

Remark 6. In the special case when $\Theta(z) = I$ (i.e., in the purely autoregressive case) the expression (7.6.18) for the h-step error covariance matrix is exact for all $n \ge p$. The programs ARVEC and BURG allow up to two simple differencing operations to be applied to the data before a multivariate autoregression is fitted to the mean-corrected differences. Predicted values for the original series are then computed automatically by the programs and the standard deviation of the predictors (assuming that the fitted model is correct) are displayed together with the predicted values. □

Remark 7. In the multivariate case, simple differencing of the type discussed in this section where the same operator $D(B)$ is applied to all components of the random vectors is rather restrictive. It is useful to consider more general linear transformations of the data for the purpose of generating a stationary series. Such considerations lead to the class of **cointegrated models** discussed briefly in Section 7.7 below. □

Example 7.6.4 Sales with a leading indicator

Assume that the model fitted to the bivariate series $\{\mathbf{Y}_t, t = 0, \ldots, 149\}$ in Example 7.6.2 is correct, i.e., that

$$\Phi(B)\mathbf{X}_t = \mathbf{Z}_t, \quad \{\mathbf{Z}_t\} \sim \mathrm{WN}(\mathbf{0}, \hat{\Sigma}),$$

where

$$\mathbf{X}_t = (1 - B)\mathbf{Y}_t - (.0228, .420)', \quad t = 1, \ldots, 149,$$

$\Phi(B) = I - \hat{\Phi}_1 B - \cdots - \hat{\Phi}_5 B^5$ and $\hat{\Phi}_1, \ldots, \hat{\Phi}_5, \hat{\Sigma}$ are the matrices found in Example 7.6.2. Then the one- and two-step predictors of $\mathbf{X}_{150}$ and $\mathbf{X}_{151}$ are obtained from (7.6.11) as

$$P_{149}\mathbf{X}_{150} = \hat{\Phi}_1\mathbf{X}_{149} + \cdots + \hat{\Phi}_5\mathbf{X}_{145} = \begin{bmatrix} .163 \\ -.217 \end{bmatrix}$$

and

$$P_{149}\mathbf{X}_{151} = \hat{\Phi}_1 P_{149}\mathbf{X}_{150} + \hat{\Phi}_2\mathbf{X}_{149} + \cdots + \hat{\Phi}_5\mathbf{X}_{146} = \begin{bmatrix} -.027 \\ .816 \end{bmatrix}$$

with error covariance matrices, from (7.6.15),

$$\hat{\Sigma} = \begin{bmatrix} .076 & -.003 \\ -.003 & .095 \end{bmatrix}$$

and

$$\hat{\Sigma} + \hat{\Phi}_1 \hat{\Sigma} \hat{\Phi}_1' = \begin{bmatrix} .096 & -.002 \\ -.002 & .095 \end{bmatrix},$$

respectively.

Similarly the one- and two-step predictors of $\mathbf{Y}_{150}$ and $\mathbf{Y}_{151}$ are obtained from (7.6.17) as

$$\tilde{P}_{149}\mathbf{Y}_{150} = \begin{bmatrix} .0228 \\ .420 \end{bmatrix} + P_{149}\mathbf{X}_{150} + \mathbf{Y}_{149} = \begin{bmatrix} 13.59 \\ 262.90 \end{bmatrix}$$

and

$$\tilde{P}_{149}\mathbf{Y}_{151} = \begin{bmatrix} .0228 \\ .420 \end{bmatrix} + P_{149}\mathbf{X}_{151} + \tilde{P}_{149}\mathbf{Y}_{150} = \begin{bmatrix} 13.59 \\ 264.14 \end{bmatrix}$$

with error covariance matrices, from (7.6.18),

$$\hat{\Sigma} = \begin{bmatrix} .076 & -.003 \\ -.003 & .095 \end{bmatrix}$$

and

$$\hat{\Sigma} + (I + \hat{\Phi}_1)\hat{\Sigma}(I + \hat{\Phi}_1)' = \begin{bmatrix} .094 & -.003 \\ -.003 & .181 \end{bmatrix},$$

respectively. The predicted values and the standard deviations of the predictors can easily be verified with the aid of the program ARVEC. It is also of interest to compare the results with those obtained by fitting a transfer function model to the data as described in Section 10.1 below. □

7.7 Cointegration

We have seen that nonstationary univariate time series can frequently be made stationary by applying the differencing operator $\nabla = 1 - B$ repeatedly. If $\{\nabla^d X_t\}$ is stationary for some positive integer d but $\{\nabla^{d-1} X_t\}$ is nonstationary, we say that $\{X_t\}$ is **integrated of order d**, or more concisely, $\{X_t\} \sim I(d)$. Many macroeconomic time series are found to be integrated of order 1.

If $\{\mathbf{X}_t\}$ is a k-variate time series, we define $\{\nabla^d \mathbf{X}_t\}$ to be the series whose j^{th} component is obtained by applying the operator $(1 - B)^d$ to the jth component of $\{\mathbf{X}_t\}$, $j = 1, \ldots, k$. The idea of a cointegrated multivariate time series was introduced by Granger (1981) and developed by Engle and Granger (1987). Here we use the slightly different definition of Lütkepohl (1993). We say that the k-dimensional time series $\{\mathbf{X}_t\}$ is integrated of order d (or $\{\mathbf{X}_t\} \sim I(d)$) if d is a positive integer, $\{\nabla^d X_t\}$ is stationary, and $\{\nabla^{d-1} X_t\}$ is nonstationary. The $I(d)$ process $\{\mathbf{X}_t\}$ is said to be **cointegrated with cointegration vector α** if α is a $k \times 1$ vector such that $\{\alpha' \mathbf{X}_t\}$ is of order less than d.

Example 7.7.1 A simple example is provided by the bivariate process whose first component is the random walk,

$$X_t = \sum_{j=1}^{t} Z_j, \quad t = 1, 2, \ldots, \quad \{Z_t\} \sim \text{IID}(0, \sigma^2),$$

and whose second component consists of noisy observations of the same random walk,

$$Y_t = X_t + W_t, \quad t = 1, 2, \ldots, \quad \{W_t\} \sim \text{IID}(0, \tau^2),$$

where $\{W_t\}$ is independent of $\{Z_t\}$. Then $\{(X_t, Y_t)'\}$ is integrated of order 1 and cointegrated with cointegration vector $\boldsymbol{\alpha} = (1, -1)'$.

The notion of cointegration captures the idea of univariate nonstationary time series "moving together." Thus, even though $\{X_t\}$ and $\{Y_t\}$ in Example 7.7.1 are both nonstationary, they are linked in the sense that they differ only by the stationary sequence $\{W_t\}$. Series that behave in a cointegrated manner are often encountered in economics. Engle and Granger (1991) give as an illustratory example the prices of tomatoes U_t and V_t in Northern and Southern California. These are linked by the fact that if one were to increase sufficiently relative to the other, the profitability of buying in one market and selling for a profit in the other would tend to push the prices $(U_t, V_t)'$ toward the straight line $v = u$ in $\mathbb{R}^2$. This line is said to be an **attractor** for $(U_t, V_t)'$, since although U_t and V_t may both vary in a nonstationary manner as t increases, the points $(U_t, V_t)'$ will exhibit relatively small random deviations from the line $v = u$. □

Example 7.7.2 If we apply the operator $\nabla = 1 - B$ to the bivariate process defined in Example 7.7.1 in order to render it stationary, we obtain the series $(U_t, V_t)'$, where

$$U_t = Z_t$$

and

$$V_t = Z_t + W_t - W_{t-1}.$$

The series $\{(U_t, V_t)'\}$ is clearly a stationary multivariate MA(1) process

$$\begin{bmatrix} U_t \\ V_t \end{bmatrix} = \begin{bmatrix} 1 & 0 \\ 0 & 1 \end{bmatrix} \begin{bmatrix} Z_t \\ Z_t + W_t \end{bmatrix} - \begin{bmatrix} 0 & 0 \\ -1 & 1 \end{bmatrix} \begin{bmatrix} Z_{t-1} \\ Z_{t-1} + W_{t-1} \end{bmatrix}.$$

However, the process $\{(U_t, V_t)'\}$ cannot be represented as an AR(∞) process since the matrix $\left[\begin{smallmatrix} 1 & 0 \\ 0 & 1 \end{smallmatrix}\right] - z\left[\begin{smallmatrix} 0 & 0 \\ -1 & 1 \end{smallmatrix}\right]$ has zero determinant when $z = 1$, thus violating condition (7.4.10). Care is therefore needed in the estimation of parameters for such models (and the closely related error-correction models). We shall not go into the details here but refer the reader to Engle and Granger (1987) and Lütkepohl (1993). □

Problems

7.1. Let $\{Y_t\}$ be a stationary process and define the bivariate process, $X_{t1} = Y_t$, $X_{t2} = Y_{t-d}$, where $d \neq 0$. Show that $\{(X_{t1}, X_{t2})'\}$ is stationary and express its cross-correlation function in terms of the autocorrelation function of $\{Y_t\}$. If $\rho_Y(h) \to 0$ as $h \to \infty$, show that there exists a lag k for which $\rho_{12}(k) > \rho_{12}(0)$.

7.2. Show that the covariance matrix function of the multivariate linear process defined by (7.2.12) is as specified in (7.2.13).

7.3. Let $\{\mathbf{X}_t\}$ be the bivariate time series whose components are the MA(1) processes defined by

$$X_{t1} = Z_{t,1} + .8Z_{t-1,1}, \quad \{Z_{t1}\} \sim \text{IID}(0, \sigma_1^2),$$

and

$$X_{t2} = Z_{t,2} - .6Z_{t-1,2}, \quad \{Z_{t2}\} \sim \text{IID}(0, \sigma_2^2),$$

where the two sequences $\{Z_{t1}\}$ and $\{Z_{t2}\}$ are independent.

a. Find a large sample approximation to the variance of $n^{1/2}\hat{\rho}_{12}(h)$.

b. Find a large sample approximation to the covariance of $n^{1/2}\hat{\rho}_{12}(h)$ and $n^{1/2}\hat{\rho}_{12}(k)$ for $h \neq k$.

7.4. Use the characterization (7.5.3) of the multivariate best linear predictor of $\mathbf{Y}$ in terms of $\{\mathbf{X}_1, \ldots \mathbf{X}_n\}$ to establish the orthogonality of the one-step prediction errors $\mathbf{X}_j - \hat{\mathbf{X}}_j$ and $\mathbf{X}_k - \hat{\mathbf{X}}_k$, $j \neq k$, as asserted in (7.6.1).

7.5. Determine the covariance matrix function of the ARMA(1,1) process satisfying

$$\mathbf{X}_t - \Phi\mathbf{X}_{t-1} = \mathbf{Z}_t + \Theta\mathbf{Z}_{t-1}, \quad \{\mathbf{Z}_t\} \sim \text{WN}(\mathbf{0}, I_2),$$

where I_2 is the 2×2 identity matrix and $\Phi = \Theta' = \begin{bmatrix} 0.5 & 0.5 \\ 0 & 0.5 \end{bmatrix}$.

7.6. a. Let $\{\mathbf{X}_t\}$ be a causal AR(p) process satisfying the recursions

$$\mathbf{X}_t = \Phi_1\mathbf{X}_{t-1} + \cdots + \Phi_p\mathbf{X}_{t-p} + \mathbf{Z}_t, \quad \{\mathbf{Z}_t\} \sim \text{WN}(\mathbf{0}, \Sigma).$$

For $n \geq p$ write down recursions for the predictors $P_n\mathbf{X}_{n+h}$, $h \geq 0$, and find explicit expressions for the error covariance matrices in terms of the AR coefficients and Σ when $h = 1, 2$, and 3.

b. Suppose now that $\{\mathbf{Y}_t\}$ is the multivariate ARIMA($p, 1, 0$) process satisfying $\nabla\mathbf{Y}_t = \mathbf{X}_t$, where $\{\mathbf{X}_t\}$ is the AR process in (a). Assuming that $E(\mathbf{Y}_0\mathbf{X}_t') = 0$, for $t \geq 1$, show (using (7.6.17) with $r = 1$ and $d = 1$) that

$$\tilde{P}_n(\mathbf{Y}_{n+h}) = \mathbf{Y}_n + \sum_{j=1}^{h} P_n\mathbf{X}_{n+j},$$

and derive the error covariance matrices when $h = 1, 2$, and 3. Compare these results with those obtained in Example 7.6.4.

7.7. Use the program ARVEC to find the minimum AICC AR model of order less than or equal to 20 for the bivariate series $\{(X_{t1}, X_{t2})', t = 1, \ldots, 200\}$ with components filed as APPJK2.DAT. Use the fitted model to predict $(X_{t1}, X_{t2})'$, $t = 201, 202, 203$ and estimate the error covariance matrices of the predictors (assuming the fitted model is appropriate for the data).

7.8. Let $\{X_{t1}, t = 1, \ldots, 63\}$ and $\{X_{t2}, t = 1, \ldots, 63\}$ denote the differenced series $\{\nabla \ln Y_{t1}\}$ and $\{\nabla \ln Y_{t2}\}$, where $\{Y_{t1}\}$ and $\{Y_{t2}\}$ are the annual mink and muskrat trappings filed as APPH.DAT and APPI.DAT, respectively).

a. Compute the sample cross-correlation function of $\{X_{t1}\}$ and $\{X_{t2}\}$ for lags between -30 and 30 using the program TRANS.

b. Test for independence of the two series.

8 State-Space Models

In recent years state-space representations and the associated Kalman recursions have had a profound impact on time series analysis and many related areas. The techniques were originally developed in connection with the control of linear systems (for accounts of this subject see Davis and Vinter, 1985, and Hannan and Deistler, 1988). An extremely rich class of models for time series, including and going well beyond the linear ARIMA and classical decomposition models considered so far in this book, can be formulated as special cases of the general state-space model defined below in Section 8.1. In econometrics the structural time series models developed by Harvey (1990) are formulated (like the classical decomposition model) directly in terms of components of interest such as trend, seasonal component, and noise. However, the rigidity of the classical decomposition model is avoided by allowing the trend and seasonal components to evolve randomly rather than deterministically. An introduction to these structural models is given in Section 8.2 and a state-space representation is developed for a general ARIMA process in Section 8.3. The Kalman recursions, which play a key role in the analysis of state-space models, are derived in Section 8.4. These recursions allow a unified approach to prediction and estimation for all processes that can be given a state-space representation. Following the development of the Kalman recursions we discuss estimation with structural models (Section 8.5) and the formulation of state-space models to deal with missing values (Section 8.6).

In Section 8.7, we introduce the EM algorithm, an iterative procedure for maximizing the likelihood when only a subset of the complete data set is available. The EM algorithm is particularly well suited for estimation problems in the state-space framework. Generalized state-space models are introduced in Section 8.8. These are Bayesian models that can be used to represent time series of many different types, as demonstrated by two applications to time series of count data. Throughout the chapter we shall use the notation

$$\{\mathbf{W}_t\} \sim \mathrm{WN}(\mathbf{0}, \{R_t\})$$

to indicate that the random vectors $\mathbf{W}_t$ have mean $\mathbf{0}$ and that

$$E(\mathbf{W}_s\mathbf{W}_t') = \begin{cases} R_t, & \text{if } s = t, \\ 0, & \text{otherwise.} \end{cases}$$

8.1 State-Space Representations

A state-space model for a (possibly multivariate) time series $\{\mathbf{Y}_t, t = 1, 2, \ldots\}$ consists of two equations. The first, known as the **observation equation**, expresses the w-dimensional observation $\mathbf{Y}_t$ as a linear function of a v-dimensional state variable $\mathbf{X}_t$ plus noise. Thus

$$\mathbf{Y}_t = G_t\mathbf{X}_t + \mathbf{W}_t, \quad t = 1, 2, \ldots, \tag{8.1.1}$$

where $\{\mathbf{W}_t\} \sim \mathrm{WN}(\mathbf{0}, \{R_t\})$ and $\{G_t\}$ is a sequence of $w \times v$ matrices. The second equation, called the **state equation**, determines the state $\mathbf{X}_{t+1}$ at time $t + 1$ in terms of the previous state $\mathbf{X}_t$ and a noise term. The state equation is

$$\mathbf{X}_{t+1} = F_t\mathbf{X}_t + \mathbf{V}_t, \quad t = 1, 2, \ldots, \tag{8.1.2}$$

where $\{F_t\}$ is a sequence of $v \times v$ matrices, $\{\mathbf{V}_t\} \sim \mathrm{WN}(\mathbf{0}, \{Q_t\})$ and $\{\mathbf{V}_t\}$ is uncorrelated with $\{\mathbf{W}_t\}$ (i.e., $E(\mathbf{W}_t\mathbf{V}_s') = 0$ for all s and t). To complete the specification, it is assumed that the initial state $\mathbf{X}_1$ is uncorrelated with all of the noise terms $\{\mathbf{V}_t\}$ and $\{\mathbf{W}_t\}$.

Remark 1. A more general form of the state-space model allows for correlation between $\mathbf{V}_t$ and $\mathbf{W}_t$ (see TSTM, Chapter 12) and for the addition of a control term $H_t\mathbf{u}_t$ in the state-equation. In control theory, $H_t\mathbf{u}_t$ represents the effect of applying a "control" $\mathbf{u}_t$ at time t for the purpose of influencing $\mathbf{X}_{t+1}$. However, the system defined by (8.1.1) and (8.1.2) with $E(\mathbf{W}_t\mathbf{V}_s') = 0$ for all s and t will be adequate for our purposes. □

Remark 2. In many important special cases, the matrices F_t, G_t, Q_t, and R_t will be independent of t, in which case the subscripts will be suppressed. □

Remark 3. It follows from the observation equation (8.1.1) and the state equation (8.1.2) that $\mathbf{X}_t$ and $\mathbf{Y}_t$ have the functional forms, for $t = 2, 3, \ldots,$

$$\begin{aligned}\mathbf{X}_t &= F_{t-1}\mathbf{X}_{t-1} + \mathbf{V}_{t-1}\\ &= F_{t-1}(F_{t-2}\mathbf{X}_{t-2} + \mathbf{V}_{t-2}) + \mathbf{V}_{t-1}\\ &\ \ \vdots\\ &= (F_{t-1}\cdots F_1)\mathbf{X}_1 + (F_{t-1}\cdots F_2)\mathbf{V}_1 + \cdots + F_{t-1}\mathbf{V}_{t-2} + \mathbf{V}_{t-1}\\ &= f_t(\mathbf{X}_1, \mathbf{V}_1, \cdots, \mathbf{V}_{t-1})\end{aligned} \tag{8.1.3}$$

and

$$\mathbf{Y}_t = g_t(\mathbf{X}_1, \mathbf{V}_1, \cdots, \mathbf{V}_{t-1}, \mathbf{W}_t). \tag{8.1.4}$$

□

Remark 4. From Remark 3 and the assumptions on the noise terms, it is clear that

$$E(\mathbf{V}_t\mathbf{X}_s') = 0, \qquad E(\mathbf{V}_t\mathbf{Y}_s') = 0,\ 1 \le s \le t,$$

and

$$E(\mathbf{W}_t\mathbf{X}_s') = 0,\ 1 \le s \le t, \qquad E(\mathbf{W}_t\mathbf{Y}_s') = 0,\ 1 \le s < t.$$

□

Definition 8.1.1 A time series $\{\mathbf{Y}_t\}$ has a **state-space representation** if there exists a state-space model for $\{\mathbf{Y}_t\}$ as specified by equations (8.1.1) and (8.1.2).

As already indicated, it is possible to find a state-space representation for a large number of time-series (and other) models. It is clear also from the definition that neither $\{\mathbf{X}_t\}$ nor $\{\mathbf{Y}_t\}$ is necessarily stationary. The beauty of a state-space representation, when one can be found, lies in the simple structure of the state equation (8.1.2), which permits relatively simple analysis of the process $\{\mathbf{X}_t\}$. The behavior of $\{\mathbf{Y}_t\}$ is then easy to determine from that of $\{\mathbf{X}_t\}$ using the observation equation (8.1.1). If the sequence $\{\mathbf{X}_1, \mathbf{V}_1, \mathbf{V}_2, \ldots\}$ is independent, then $\{\mathbf{X}_t\}$ has the Markov property, i.e., the distribution of $\mathbf{X}_{t+1}$ given $\mathbf{X}_t, \ldots, \mathbf{X}_1$ is the same as the distribution of $\mathbf{X}_{t+1}$ given $\mathbf{X}_t$. This is a property possessed by many physical systems provided we include sufficiently many components in the specification of the state $\mathbf{X}_t$ (for example, we may choose the state-vector in such a way that $\mathbf{X}_t$ includes components of $\mathbf{X}_{t-1}$ for each t).

Example 8.1.1 An AR(1) process

Let $\{Y_t\}$ be the causal AR(1) process given by

$$Y_t = \phi Y_{t-1} + Z_t, \quad \{Z_t\} \sim \mathrm{WN}(0, \sigma^2). \tag{8.1.5}$$

In this case, a state-space representation for $\{Y_t\}$ is easy to construct. We can, for example, define a sequence of state variables X_t by

$$X_{t+1} = \phi X_t + V_t, \quad t = 1, 2, \ldots, \tag{8.1.6}$$

where $X_1 = Y_1 = \sum_{j=0}^{\infty} \phi^j Z_{1-j}$ and $V_t = Z_{t+1}$. The process $\{Y_t\}$ then satisfies the observation-equation

$$Y_t = X_t,$$

which has the form (8.1.1) with $G_t = 1$ and $W_t = 0$. □

Example 8.1.2 An ARMA(1,1) process

Let $\{Y_t\}$ be the causal-invertible ARMA(1,1) process satisfying the equations

$$Y_t = \phi Y_{t-1} + Z_t + \theta Z_{t-1}, \quad \{Z_t\} \sim \text{WN}(0, \sigma^2). \tag{8.1.7}$$

Although the existence of a state-space representation for $\{Y_t\}$ is not obvious, we can find one by observing that

$$Y_t = \theta(B) X_t = [\theta \quad 1] \begin{bmatrix} X_{t-1} \\ X_t \end{bmatrix}, \tag{8.1.8}$$

where $\{X_t\}$ is the causal AR(1) process satisfying

$$\phi(B) X_t = Z_t,$$

or the equivalent equation,

$$\begin{bmatrix} X_t \\ X_{t+1} \end{bmatrix} = \begin{bmatrix} 0 & 1 \\ 0 & \phi \end{bmatrix} \begin{bmatrix} X_{t-1} \\ X_t \end{bmatrix} + \begin{bmatrix} 0 \\ Z_{t+1} \end{bmatrix}. \tag{8.1.9}$$

Noting that $X_t = \sum_{j=0}^{\infty} \phi^j Z_{t-j}$, we see that the equations (8.1.8) and (8.1.9) for $t = 1, 2, \ldots,$ furnish a state-space representation of $\{Y_t\}$ with

$$\mathbf{X}_t = \begin{bmatrix} X_{t-1} \\ X_t \end{bmatrix} \text{ and } \mathbf{X}_1 = \begin{bmatrix} \sum_{j=0}^{\infty} \phi^j Z_{-j} \\ \sum_{j=0}^{\infty} \phi^j Z_{1-j} \end{bmatrix}.$$

The extension of this state-space representation to general ARMA and ARIMA processes is given in Section 8.3. □

In subsequent sections we shall give examples that illustrate the versatility of state-space models. (More examples can be found in Aoki, 1987, Hannan and Deistler, 1988, Harvey, 1990, and West and Harrison, 1989.) Before considering these, we need a slight modification of (8.1.1) and (8.1.2), which allows for series in which the time index runs from $-\infty$ to ∞. This is a more natural formulation for many time series models.

State-Space Models with $t \in \{0, \pm 1, \ldots\}$

Consider the observation and state equations,

$$\mathbf{Y}_t = G\mathbf{X}_t + \mathbf{W}_t, \quad t = 0, \pm 1, \ldots, \tag{8.1.10}$$

$$\mathbf{X}_{t+1} = F\mathbf{X}_t + \mathbf{V}_t, \quad t = 0, \pm 1, \ldots, \tag{8.1.11}$$

where F and G are $v \times v$ and $w \times v$ matrices, respectively, $\{\mathbf{V}_t\} \sim \mathrm{WN}(\mathbf{0}, Q)$, $\{\mathbf{W}_t\} \sim \mathrm{WN}(\mathbf{0}, R)$, and $E(\mathbf{V}_s\mathbf{W}_t') = 0$ for all s, and t.

The state equation (8.1.11) is said to be **stable** if the matrix F has all its eigenvalues in the interior of the unit circle, or equivalently if $\det(I - Fz) \neq 0$ for all z complex such that $|z| \leq 1$. The matrix F is then also said to be stable.

In the stable case the equations (8.1.11) have the unique stationary solution (Problem 8.1) given by

$$\mathbf{X}_t = \sum_{j=0}^{\infty} F^j \mathbf{V}_{t-j-1}.$$

The corresponding sequence of observations,

$$\mathbf{Y}_t = \mathbf{W}_t + \sum_{j=0}^{\infty} GF^j \mathbf{V}_{t-j-1},$$

is also stationary.

8.2 The Basic Structural Model

A structural time series model, like the classical decomposition model defined by (1.5.1), is specified in terms of components such as trend, seasonality, and noise, which are of direct interest in themselves. The deterministic nature of the trend and seasonal components in the classical decomposition model, however, limits its applicability. A natural way in which to overcome this deficiency is to permit random variation in these components. This can be very conveniently done in the framework of a state-space representation and the resulting rather flexible model is called a structural model. Estimation and forecasting with this model can be encompassed in the general procedure for state-space models made possible by the Kalman recursions of Section 8.4.

Example 8.2.1 The random walk plus noise model

One of the simplest structural models is obtained by adding noise to a random walk. It is suggested by the nonseasonal classical decomposition model,

$$Y_t = M_t + W_t, \quad \text{where } \{W_t\} \sim \mathrm{WN}(0, \sigma_w^2), \tag{8.2.1}$$

and $M_t = m_t$, the deterministic "level" or "signal" at time t. We now introduce randomness into the level by supposing that M_t is a random walk satisfying

$$M_{t+1} = M_t + V_t, \quad \text{and} \quad \{V_t\} \sim \text{WN}(0, \sigma_v^2), \tag{8.2.2}$$

with initial value $M_1 = m_1$. The equations (8.2.1) and (8.2.2) constitute the "local level" or "random walk plus noise" model. Figure 8.1 shows a realization of length 100 of this model with $M_1 = 0$, $\sigma_v^2 = 4$ and $\sigma_w^2 = 8$. (The realized values m_t of M_t are plotted as a solid line and the observed data are plotted as square boxes.) The differenced data,

$$D_t := \nabla Y_t = Y_t - Y_{t-1} = V_{t-1} + W_t - W_{t-1}, \; t \geq 2,$$

is a stationary time series with mean 0 and ACF

$$\rho_D(h) = \begin{cases} \dfrac{-\sigma_w^2}{2\sigma_w^2 + \sigma_v^2}, & \text{if } |h| = 1, \\ 0, & \text{if } |h| > 1. \end{cases}$$

Since $\{D_t\}$ is 1-correlated, we conclude from Proposition 2.1.1 that $\{D_t\}$ is an MA(1) process and hence that $\{Y_t\}$ is an ARIMA(0,1,1) process. More specifically,

$$D_t = Z_t + \theta Z_{t-1}, \quad \{Z_t\} \sim \text{WN}(0, \sigma^2), \tag{8.2.3}$$

where θ and σ^2 are found by solving the equations

$$\frac{\theta}{1 + \theta^2} = \frac{-\sigma_w^2}{2\sigma_w^2 + \sigma_v^2} \quad \text{and} \quad \theta\sigma^2 = -\sigma_w^2.$$

For the process $\{Y_t\}$ generating the data in Figure 8.1, the parameters θ and σ^2 of

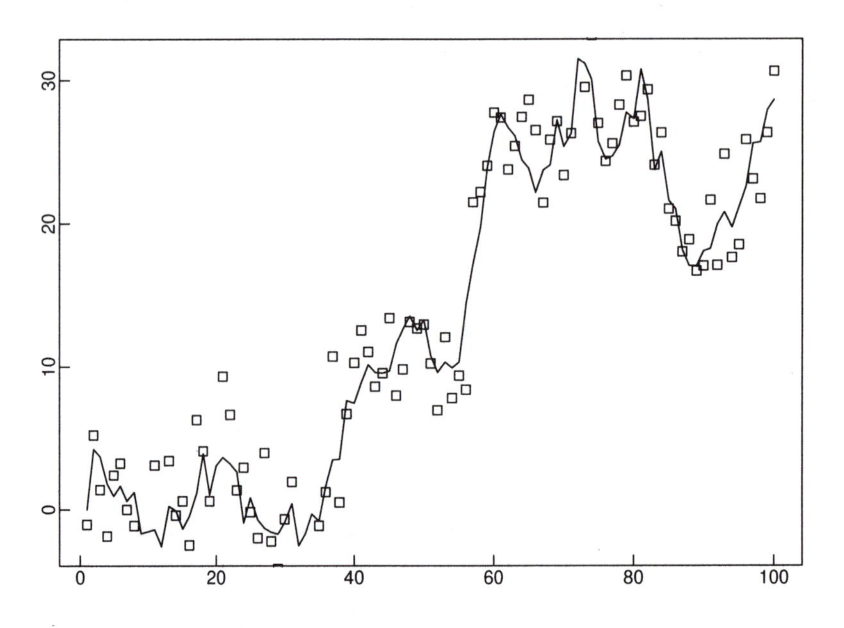

Figure 8-1 Realization from a random walk plus noise model. The random walk is represented by the solid line and the data are represented by boxes.

the differenced series $\{D_t\}$ satisfy $\theta/(1+\theta^2) = -.4$ and $\theta\sigma^2 = -8$. Solving these equations for θ and σ^2, we find that $\theta = -.5$ and $\sigma^2 = 16$ (or $\theta = -2$ and $\sigma^2 = 4$). The sample ACF of the observed differences D_t of the realization of $\{Y_t\}$ in Figure 8.1 is shown in Figure 8.2.

The local level model is often used to represent a measured characteristic of the output of an industrial process for which the unobserved process level $\{M_t\}$ is intended to be within specified limits (to meet the design specifications of the manufactured product). To decide whether or not the process requires corrective attention, it is important to be able to test the hypothesis that the process level $\{M_t\}$ is constant. From the state equation, we see that $\{M_t\}$ is constant (and equal to m_1) when $V_t = 0$ or equivalently when $\sigma_v^2 = 0$. This in turn is equivalent to the moving average model (8.2.3) for $\{D_t\}$ being noninvertible with $\theta = -1$ (see Problem 8.2). Tests of the unit root hypothesis $\theta = -1$ were discussed in Section 6.3.2. □

The local level model can easily be extended to incorporate a locally linear trend with slope β_t at time t. The equation (8.2.2) is replaced by

$$M_t = M_{t-1} + B_{t-1} + V_{t-1}, \tag{8.2.4}$$

where $B_{t-1} = \beta_{t-1}$. Now if we introduce randomness into the slope by replacing it with the random walk,

$$B_t = B_{t-1} + U_{t-1}, \quad \text{where } \{U_t\} \sim \text{WN}(0, \sigma_u^2), \tag{8.2.5}$$

we obtain the "local linear trend" model.

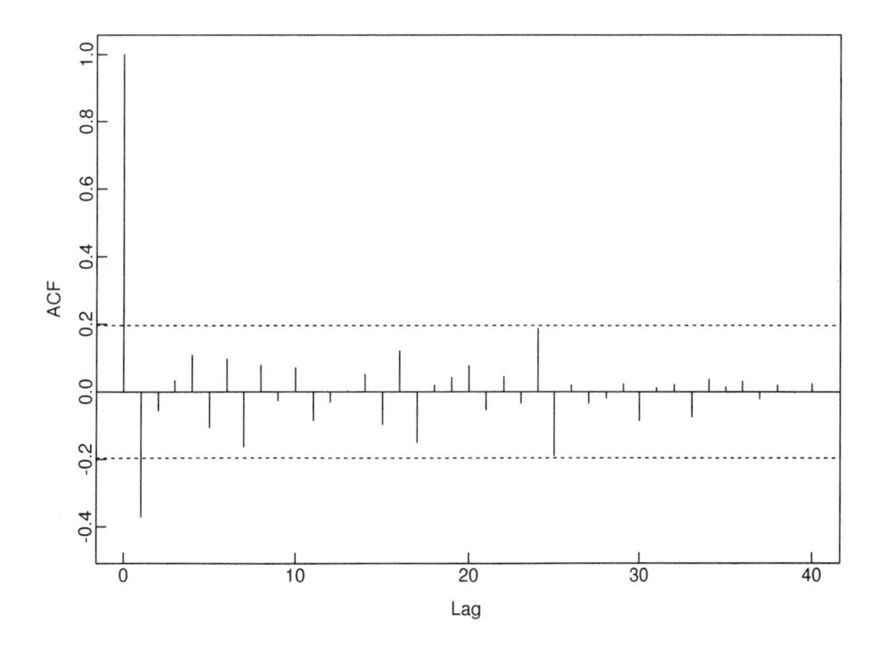

Figure 8-2
Sample ACF of the series obtained by differencing the data in Figure 8.1.

To express the local linear trend model in state-space form we introduce the state vector

$$\mathbf{X}_t = (M_t, B_t)'.$$

Then (8.2.4) and (8.2.5) can be written in the equivalent form,

$$\mathbf{X}_{t+1} = \begin{bmatrix} 1 & 1 \\ 0 & 1 \end{bmatrix} \mathbf{X}_t + \mathbf{V}_t, \quad t = 1, 2, \ldots, \tag{8.2.6}$$

where $\mathbf{V}_t = (V_t, U_t)'$. The process $\{Y_t\}$ is then determined by the observation equation,

$$Y_t = [1 \quad 0]\, \mathbf{X}_t + W_t. \tag{8.2.7}$$

If $\{\mathbf{X}_1, U_1, V_1, W_1, U_2, V_2, W_2, \ldots\}$ is an uncorrelated sequence, the equations (8.2.6) and (8.2.7) constitute a state-space representation of the process $\{Y_t\}$, which is a model for data with randomly varying trend and added noise. For this model we have $v = 2$, $w = 1$,

$$F = \begin{bmatrix} 1 & 1 \\ 0 & 1 \end{bmatrix}, \; G = [1 \quad 0], \; Q = \begin{bmatrix} \sigma_v^2 & 0 \\ 0 & \sigma_u^2 \end{bmatrix} \text{ and } R = \sigma_w^2.$$

Example 8.2.2 A seasonal series with noise

The classical decomposition (1.5.11) expressed the time series $\{X_t\}$ as a sum of trend, seasonal, and noise components. The seasonal component (with period d) was a sequence $\{s_t\}$ with the properties $s_{t+d} = s_t$ and $\sum_{t=1}^{d} s_t = 0$. Such a sequence can be generated, for *any* values of $s_1, s_0, \ldots, s_{-d+3}$, by means of the recursions

$$s_{t+1} = -s_t - \cdots - s_{t-d+2}, \quad t = 1, 2, \ldots. \tag{8.2.8}$$

A somewhat more general seasonal component, $\{Y_t\}$, allowing for random deviations from strict periodicity, is obtained by adding a term V_t to the right side of (8.2.8), where $\{V_t\}$ is white noise with mean zero. This leads to the recursion relations

$$Y_{t+1} = -Y_t - \cdots - Y_{t-d+2} + V_t, \quad t = 1, 2, \ldots. \tag{8.2.9}$$

To find a state-space representation for $\{Y_t\}$ we introduce the $(d-1)$-dimensional state vector,

$$\mathbf{X}_t = (Y_t, Y_{t-1}, \ldots, Y_{t-d+2})'.$$

The series $\{Y_t\}$ is then given by the observation equation

$$Y_t = [1 \quad 0 \quad 0 \cdots 0]\, \mathbf{X}_t, \quad t = 1, 2, \ldots, \tag{8.2.10}$$

where $\{\mathbf{X}_t\}$ satisfies the state equation

$$\mathbf{X}_{t+1} = F\mathbf{X}_t + \mathbf{V}_t, \quad t = 1, 2 \ldots, \tag{8.2.11}$$

$\mathbf{V}_t = (V_t, 0, \ldots, 0)'$ and

$$F = \begin{bmatrix} -1 & -1 & \cdots & -1 & -1 \\ 1 & 0 & \cdots & 0 & 0 \\ 0 & 1 & \cdots & 0 & 0 \\ \vdots & \vdots & \ddots & \vdots & \vdots \\ 0 & 0 & \cdots & 1 & 0 \end{bmatrix}.$$

□

Example 8.2.3 A randomly varying trend with seasonal and noise components

Such a series can be constructed by adding the two series in Examples 8.2.1 and 8.2.2. (Addition of series with state-space representations is in fact always possible by means of the following construction. See Problem 8.9.) We introduce the state-vector

$$\mathbf{X}_t = \begin{bmatrix} \mathbf{X}_t^1 \\ \mathbf{X}_t^2 \end{bmatrix},$$

where $\mathbf{X}_t^1$ and $\mathbf{X}_t^2$ are the state vectors in (8.2.6) and (8.2.11). We then have the following representation for $\{Y_t\}$, the sum of the two series whose state-space representations were given in (8.2.6)–(8.2.7) and (8.2.10)–(8.2.11). The state equation is

$$\mathbf{X}_{t+1} = \begin{bmatrix} F_1 & 0 \\ 0 & F_2 \end{bmatrix} \mathbf{X}_t + \begin{bmatrix} \mathbf{V}_t^1 \\ \mathbf{V}_t^2 \end{bmatrix}, \tag{8.2.12}$$

where F_1, F_2 are the coefficient matrices and $\{\mathbf{V}_t^1\}$, $\{\mathbf{V}_t^2\}$ are the noise vectors in the state equations (8.2.6) and (8.2.11), respectively. The observation equation is

$$Y_t = [1 \quad 0 \quad 1 \quad 0 \cdots 0]\, \mathbf{X}_t + W_t, \tag{8.2.13}$$

where $\{W_t\}$ is the noise sequence in (8.2.7). If the sequence of random vectors, $\{\mathbf{X}_1, \mathbf{V}_1^1, \mathbf{V}_1^2, W_1, \mathbf{V}_2^1, \mathbf{V}_2^2, W_2, \ldots\}$, is uncorrelated, the equations (8.2.12) and (8.2.13) constitute a state-space representation for $\{Y_t\}$. □

8.3 State-Space Representation of ARIMA Models

We begin by establishing a state-space representation for the causal AR(p) process and then build on this example to find representations for the general ARMA and ARIMA processes.

Example 8.3.1 State-space representation of a causal AR(p) process

Consider the AR(p) process defined by

$$Y_{t+1} = \phi_1 Y_t + \phi_2 Y_{t-1} + \cdots + \phi_p Y_{t-p+1} + Z_{t+1}, \quad t = 0, \pm 1, \ldots, \tag{8.3.1}$$

where $\{Z_t\} \sim \mathrm{WN}(0, \sigma^2)$, and $\phi(z) := 1 - \phi_1 z - \cdots - \phi_p z^p$ is nonzero for $|z| \le 1$. To express $\{Y_t\}$ in state-space form we simply introduce the state vectors,

$$\mathbf{X}_t = \begin{bmatrix} Y_{t-p+1} \\ Y_{t-p+2} \\ \vdots \\ Y_t \end{bmatrix}, \quad t = 0, \pm 1, \ldots. \tag{8.3.2}$$

From (8.3.1) and (8.3.2), the observation equation is

$$Y_t = [0 \quad 0 \quad 0 \cdots 1]\mathbf{X}_t, \quad t = 0, \pm 1, \ldots, \tag{8.3.3}$$

while the state equation is given by

$$\mathbf{X}_{t+1} = \begin{bmatrix} 0 & 1 & 0 & \cdots & 0 \\ 0 & 0 & 1 & \cdots & 0 \\ \vdots & \vdots & \vdots & \ddots & \vdots \\ 0 & 0 & 0 & \cdots & 1 \\ \phi_p & \phi_{p-1} & \phi_{p-2} & \cdots & \phi_1 \end{bmatrix} \mathbf{X}_t + \begin{bmatrix} 0 \\ 0 \\ \vdots \\ 0 \\ 1 \end{bmatrix} Z_{t+1}, \quad t = 0, \pm 1, \ldots. \tag{8.3.4}$$

These equations have the required forms (8.1.10) and (8.1.11) with $\mathbf{W}_t = \mathbf{0}$ and $\mathbf{V}_t = (0, 0, \ldots, Z_{t+1})'$, $t = 0, \pm 1, \ldots$. □

Remark 1. In Example 8.3.1 the causality condition, $\phi(z) \ne 0$ for $|z| \le 1$, is equivalent to the condition that the state equation (8.3.4) is stable, since the eigenvalues of the coefficient matrix in (8.3.4) are simply the reciprocals of the zeroes of $\phi(z)$ (Problem 8.3). □

Remark 2. If equations (8.3.3) and (8.3.4) are postulated to hold only for $t = 1, 2, \ldots$, and if $\mathbf{X}_1$ is a random vector such that $\{\mathbf{X}_1, Z_1, Z_2, \ldots\}$ is an uncorrelated sequence, then we have a state-space representation for $\{Y_t\}$ of the type defined earlier by (8.1.1) and (8.1.2). The resulting process $\{Y_t\}$ is well-defined, regardless of whether or not the state equation is stable, but it will not in general be stationary. It will be stationary if the state equation is stable and if $\mathbf{X}_1$ is defined by (8.3.2) with $Y_t = \sum_{j=0}^{\infty} \psi_j Z_{t-j}$, $t = 1, 0, \ldots, 2 - p$, and $\psi(z) = 1/\phi(z)$, $|z| \le 1$. □

Example 8.3.2 State-space form of a causal ARMA(p, q) process

State-space representations are not unique. Here we shall give one of the (infinitely many) possible representations of a causal ARMA(p,q) process which can easily be derived from Example 8.3.1. Consider the ARMA(p,q) process defined by

$$\phi(B)Y_t = \theta(B)Z_t, \quad t = 0, \pm 1, \ldots, \tag{8.3.5}$$

where $\{Z_t\} \sim \mathrm{WN}(0, \sigma^2)$ and $\phi(z) \ne 0$ for $|z| \le 1$. Let

$$r = \max(p, q+1), \quad \phi_j = 0 \quad \text{for } j > p, \quad \theta_j = 0 \quad \text{for } j > q, \quad \text{and} \quad \theta_0 = 1.$$

If $\{U_t\}$ is the causal AR(p) process satisfying

$$\phi(B)U_t = Z_t, \tag{8.3.6}$$

then $Y_t = \theta(B)U_t$ since

$$\phi(B)Y_t = \phi(B)\theta(B)U_t = \theta(B)\phi(B)U_t = \theta(B)Z_t.$$

Consequently,

$$Y_t = [\theta_{r-1} \quad \theta_{r-2} \ \cdots \ \theta_0]\mathbf{X}_t, \tag{8.3.7}$$

where

$$\mathbf{X}_t = \begin{bmatrix} U_{t-r+1} \\ U_{t-r+2} \\ \vdots \\ U_t \end{bmatrix}. \tag{8.3.8}$$

But from Example 8.3.1 we can write

$$\mathbf{X}_{t+1} = \begin{bmatrix} 0 & 1 & 0 & \cdots & 0 \\ 0 & 0 & 1 & \cdots & 0 \\ \vdots & \vdots & \vdots & \ddots & \vdots \\ 0 & 0 & 0 & \cdots & 1 \\ \phi_r & \phi_{r-1} & \phi_{r-2} & \cdots & \phi_1 \end{bmatrix} \mathbf{X}_t + \begin{bmatrix} 0 \\ 0 \\ \vdots \\ 0 \\ 1 \end{bmatrix} Z_{t+1}, \quad t = 0, \pm 1, \ldots. \tag{8.3.9}$$

Equations (8.3.7) and (8.3.9) are the required observation and state equations. As in Example 8.3.1, the observation and state noise vectors are again $\mathbf{W}_t = \mathbf{0}$ and $\mathbf{V}_t = (0, 0, \ldots, Z_{t+1})'$, $t = 0, \pm 1, \ldots$. □

Example 8.3.3 State-space representation of an ARIMA(p, d, q) process

If $\{Y_t\}$ is an ARIMA(p, d, q) process with $\{\nabla^d Y_t\}$ satisfying (8.3.5), then by the preceding example $\{\nabla^d Y_t\}$ has the representation

$$\nabla^d Y_t = G\mathbf{X}_t, \quad t = 0, \pm 1, \ldots, \tag{8.3.10}$$

where $\{\mathbf{X}_t\}$ is the unique stationary solution of the state equation

$$\mathbf{X}_{t+1} = F\mathbf{X}_t + \mathbf{V}_t,$$

F and G are the coefficients of $\mathbf{X}_t$ in (8.3.9) and (8.3.7), respectively, and $\mathbf{V}_t = (0, 0, \ldots, Z_{t+1})'$. Let A and B be the $d \times 1$ and $d \times d$ matrices defined by $A = B = 1$ if $d = 1$ and

$$A = \begin{bmatrix} 0 \\ 0 \\ \vdots \\ 0 \\ 1 \end{bmatrix}, \ B = \begin{bmatrix} 0 & 1 & 0 & \cdots & 0 \\ 0 & 0 & 1 & \cdots & 0 \\ \vdots & \vdots & \vdots & \ddots & \vdots \\ 0 & 0 & 0 & \cdots & 1 \\ (-1)^{d+1}\binom{d}{d} & (-1)^d\binom{d}{d-1} & (-1)^{d-1}\binom{d}{d-2} & \cdots & d \end{bmatrix}$$

if $d > 1$. Then since

$$Y_t = \nabla^d Y_t - \sum_{j=1}^{d} \binom{d}{j} (-1)^j Y_{t-j}, \tag{8.3.11}$$

the vector

$$\mathbf{Y}_{t-1} := (Y_{t-d}, \ldots, Y_{t-1})'$$

satisfies the equation

$$\mathbf{Y}_t = A\nabla^d Y_t + B\mathbf{Y}_{t-1} = AG\mathbf{X}_t + B\mathbf{Y}_{t-1}.$$

Defining a new state vector $\mathbf{T}_t$ by stacking $\mathbf{X}_t$ and $\mathbf{Y}_{t-1}$, we therefore obtain the state equation

$$\mathbf{T}_{t+1} := \begin{bmatrix} \mathbf{X}_{t+1} \\ \mathbf{Y}_t \end{bmatrix} = \begin{bmatrix} F & 0 \\ AG & B \end{bmatrix} \mathbf{T}_t + \begin{bmatrix} \mathbf{V}_t \\ \mathbf{0} \end{bmatrix}, \quad t = 1, 2, \cdots, \tag{8.3.12}$$

and the observation equation, from (8.3.10) and (8.3.11),

$$Y_t = \begin{bmatrix} G & (-1)^{d+1}\binom{d}{d} & (-1)^d \binom{d}{d-1} & (-1)^{d-1}\binom{d}{d-2} & \cdots & d \end{bmatrix} \begin{bmatrix} \mathbf{X}_t \\ \mathbf{Y}_{t-1} \end{bmatrix}$$

$$t = 1, 2, \ldots, \tag{8.3.13}$$

with initial condition

$$\mathbf{T}_1 = \begin{bmatrix} \mathbf{X}_1 \\ \mathbf{Y}_0 \end{bmatrix} = \begin{bmatrix} \sum_{j=0}^{\infty} F^j \mathbf{V}_{-j} \\ \mathbf{Y}_0 \end{bmatrix}, \tag{8.3.14}$$

and the assumption

$$E(\mathbf{Y}_0 Z_t') = 0, \ t = 0, \pm 1, \ldots, \tag{8.3.15}$$

where $\mathbf{Y}_0 = (Y_{1-d}, Y_{2-d}, \ldots, Y_0)'$. The conditions (8.3.15),which are satisfied in particular if $\mathbf{Y}_0$ is considered to be nonrandom and equal to the vector of *observed* values $(y_{1-d}, y_{2-d}, \ldots, y_0)'$, are imposed to ensure that the assumptions of a state-space model given in Section 8.1 are satisfied. They also imply that $E(\mathbf{X}_1 \mathbf{Y}_0') = 0$ and $E(\mathbf{Y}_0 \nabla^d Y_t') = 0, \ t \geq 1$, as required earlier in Section 6.4 for prediction of ARIMA processes.

State-space models for more general ARIMA processes (e.g., $\{Y_t\}$ such that $\{\nabla\nabla_{12} Y_t\}$ is an ARMA(p, q) process) can be constructed in the same way. See Problem 8.4. □

For the ARIMA(1, 1, 1) process defined by

$$(1 - \phi B)(1 - B)Y_t = (1 + \theta B)Z_t, \quad \{Z_t\} \sim \text{WN}(0, \sigma^2),$$

the vectors $\mathbf{X}_t$ and $\mathbf{Y}_{t-1}$ reduce to $\mathbf{X}_t = (X_{t-1}, X_t)'$ and $\mathbf{Y}_{t-1} = Y_{t-1}$. From (8.3.12) and (8.3.13) the state-space representation is therefore (Problem 8.8)

$$Y_t = [\theta \quad 1 \quad 1] \begin{bmatrix} X_{t-1} \\ X_t \\ Y_{t-1} \end{bmatrix}, \tag{8.3.16}$$

where

$$\begin{bmatrix} X_t \\ X_{t+1} \\ Y_t \end{bmatrix} = \begin{bmatrix} 0 & 1 & 0 \\ 0 & \phi & 0 \\ \theta & 1 & 1 \end{bmatrix} \begin{bmatrix} X_{t-1} \\ X_t \\ Y_{t-1} \end{bmatrix} + \begin{bmatrix} 0 \\ Z_{t+1} \\ 0 \end{bmatrix}, \quad t = 1, 2, \ldots, \tag{8.3.17}$$

and

$$\begin{bmatrix} X_0 \\ X_1 \\ Y_0 \end{bmatrix} = \begin{bmatrix} \sum_{j=0}^{\infty} \phi^j Z_{-j} \\ \sum_{j=0}^{\infty} \phi^j Z_{1-j} \\ Y_0 \end{bmatrix}. \tag{8.3.18}$$

8.4 The Kalman Recursions

In this section we shall consider three fundamental problems associated with the state-space model defined by (8.1.1) and (8.1.2) in Section 8.1. These are all concerned with finding best (in the sense of minimum mean-square error) linear estimates of the state-vector $\mathbf{X}_t$ in terms of the observations $\mathbf{Y}_1, \mathbf{Y}_2, \ldots,$ and a random vector $\mathbf{Y}_0$ which is orthogonal to $\mathbf{V}_t$ and $\mathbf{W}_t$ for all $t \geq 1$. In many cases $\mathbf{Y}_0$ will be the constant vector $(1, 1, \ldots, 1)'$. Estimation of $\mathbf{X}_t$ in terms of

a. $\mathbf{Y}_0, \ldots, \mathbf{Y}_{t-1}$ defines the **prediction problem**,
b. $\mathbf{Y}_0, \ldots, \mathbf{Y}_t$ defines the **filtering problem**, and
c. $\mathbf{Y}_0, \ldots, \mathbf{Y}_n$ $(n > t)$ defines the **smoothing problem**.

Each of these problems can be solved recursively using an appropriate set of Kalman recursions, which will be established in this section.

In the following definition of best linear predictor (and throughout this chapter) it should be noted that we do not automatically include the constant 1 among the predictor variables as we did in Sections 2.5 and 7.5. (It can, however, be included by choosing $\mathbf{Y}_0 = (1, 1, \ldots, 1)'$.)

Definition 8.4.1

For the random vector $\mathbf{X} = (X_1, \ldots, X_v)'$,

$$P_t(\mathbf{X}) := (P_t(X_1), \ldots, P_t(X_v))'$$

where $P_t(X_i) := P(X_i|\mathbf{Y}_0, \mathbf{Y}_1, \ldots, \mathbf{Y}_t)$ is the best linear predictor of X_i in terms of all components of $\mathbf{Y}_0, \mathbf{Y}_1, \ldots, \mathbf{Y}_t$.

Remark 1. By the definition of the best predictor of each component X_i of $\mathbf{X}$, $P_t(\mathbf{X})$ is the unique random vector of the form

$$P_t(\mathbf{X}) = A_0\mathbf{Y}_0 + \cdots + A_t\mathbf{Y}_t$$

with $v \times w$ matrices $A_0, \ldots, A_t$ such that

$$[\mathbf{X} - P_t(\mathbf{X})] \perp \mathbf{Y}_s, \quad s = 0, \ldots, t$$

(cf. (7.5.2) and (7.5.3)). Recall that two random vectors $\mathbf{X}$ and $\mathbf{Y}$ are orthogonal (written $\mathbf{X} \perp \mathbf{Y}$) if $E(\mathbf{X}\mathbf{Y}')$ is a matrix of zeroes. □

Remark 2. If all the components of $\mathbf{X}, \mathbf{Y}_1, \ldots, \mathbf{Y}_t$ are jointly normally distributed and $\mathbf{Y}_0 = (1, \ldots, 1)'$, then

$$P_t(\mathbf{X}) = E(\mathbf{X}|\mathbf{Y}_1, \ldots, \mathbf{Y}_t), \quad t \geq 1.$$ □

Remark 3. P_t is linear in the sense that if A is any $k \times v$ matrix and $\mathbf{X}, \mathbf{V}$ are two v-variate random vectors with finite second moments then (Problem 8.10)

$$P_t(A\mathbf{X}) = AP_t(\mathbf{X})$$

and

$$P_t(\mathbf{X} + \mathbf{V}) = P_t(\mathbf{X}) + P_t(\mathbf{V}).$$ □

Remark 4. If $\mathbf{X}$ and $\mathbf{Y}$ are random vectors with v and w components, respectively, each with finite second moments, then

$$P(\mathbf{X}|\mathbf{Y}) = M\mathbf{Y},$$

where M is the $v \times w$ matrix, $M = E(\mathbf{X}\mathbf{Y}')[E(\mathbf{Y}\mathbf{Y}')]^{-1}$ and $[E(\mathbf{Y}\mathbf{Y}')]^{-1}$ is any generalized inverse of $E(\mathbf{Y}\mathbf{Y}')$. (A generalized inverse of a matrix S is a matrix S^{-1} such that $SS^{-1}S = S$. Every matrix has at least one. See Problem 8.11.)

In the notation just developed, the prediction, filtering, and smoothing problems (a), (b), and (c) formulated above reduce to the determination of $P_{t-1}(\mathbf{X}_t)$, $P_t(\mathbf{X}_t)$, and $P_n(\mathbf{X}_t)$ ($n > t$), respectively. We deal first with the prediction problem. □

Kalman Prediction:

For the state-space model (8.1.1)–(8.1.2), the one-step predictors $\hat{\mathbf{X}}_t := P_{t-1}(\mathbf{X}_t)$ and their error covariance matrices $\Omega_t = E[(\mathbf{X}_t - \hat{\mathbf{X}}_t)(\mathbf{X}_t - \hat{\mathbf{X}}_t)']$ are uniquely determined by the initial conditions

$$\hat{\mathbf{X}}_1 = P(\mathbf{X}_1|\mathbf{Y}_0), \qquad \Omega_1 = E[(\mathbf{X}_1 - \hat{\mathbf{X}}_1)(\mathbf{X}_1 - \hat{\mathbf{X}}_1)']$$

and the recursions, for $t = 1, \ldots,$

$$\hat{\mathbf{X}}_{t+1} = F_t\hat{\mathbf{X}}_t + \Theta_t\Delta_t^{-1}(\mathbf{Y}_t - G_t\hat{\mathbf{X}}_t), \tag{8.4.1}$$

$$\Omega_{t+1} = F_t\Omega_tF_t' + Q_t - \Theta_t\Delta_t^{-1}\Theta_t', \tag{8.4.2}$$

where

$$\Delta_t = G_t\Omega_tG_t' + R_t,$$

$$\Theta_t = F_t\Omega_tG_t',$$

and Δ_t^{-1} is any generalized inverse of Δ_t.

Proof We shall make use of the **innovations**, $\mathbf{I}_t$, defined by $\mathbf{I}_0 = \mathbf{Y}_0$ and

$$\mathbf{I}_t = \mathbf{Y}_t - P_{t-1}\mathbf{Y}_t = \mathbf{Y}_t - G_t\hat{\mathbf{X}}_t = G_t(\mathbf{X}_t - \hat{\mathbf{X}}_t) + \mathbf{W}_t, \quad t = 1, 2, \ldots.$$

The sequence $\{\mathbf{I}_t\}$ is orthogonal by Remark 1. Using Remarks 3 and 4 and the relation

$$P_t(\cdot) = P_{t-1}(\cdot) + P(\cdot|\mathbf{I}_t) \tag{8.4.3}$$

(see Problem 8.12), we find that

$$\begin{aligned}\hat{\mathbf{X}}_{t+1} &= P_{t-1}(\mathbf{X}_{t+1}) + P(\mathbf{X}_{t+1}|\mathbf{I}_t)\\ &= P_{t-1}(F_t\mathbf{X}_t + \mathbf{V}_t) + \Theta_t\Delta_t^{-1}\mathbf{I}_t\\ &= F_t\hat{\mathbf{X}}_t + \Theta_t\Delta_t^{-1}\mathbf{I}_t,\end{aligned} \tag{8.4.4}$$

where

$$\begin{aligned}\Delta_t &= E(\mathbf{I}_t\mathbf{I}_t')\\ &= G_t\Omega_tG_t' + R_t,\\ \Theta_t &= E(\mathbf{X}_{t+1}\mathbf{I}_t')\\ &= E[(F_t\mathbf{X}_t + \mathbf{V}_t)([\mathbf{X}_t - \hat{\mathbf{X}}_t]'G_t' + \mathbf{W}_t')]\\ &= F_t\Omega_tG_t'.\end{aligned}$$

To verify (8.4.2), we observe from the definition of Ω_{t+1} that

$$\Omega_{t+1} = E(\mathbf{X}_{t+1}\mathbf{X}_{t+1}') - E(\hat{\mathbf{X}}_{t+1}\hat{\mathbf{X}}_{t+1}').$$

With (8.1.2) and (8.4.4) this gives

$$\Omega_{t+1} = F_t E(\mathbf{X}_t\mathbf{X}_t')F_t' + Q_t - F_t E(\hat{\mathbf{X}}_t\hat{\mathbf{X}}_t')F_t' - \Theta_t\Delta_t^{-1}\Theta_t'$$
$$= F_t\Omega_t F_t' + Q_t - \Theta_t\Delta_t^{-1}\Theta_t'. \quad \blacksquare$$

h-step Prediction of $\{\mathbf{Y}_t\}$ Using the Kalman Recursions

The Kalman prediction equations lead to a very simple algorithm for recursive calculation of the best linear mean-square predictors, $P_t\mathbf{Y}_{t+h}$, $h = 1, 2, \ldots$. From (8.4.4), (8.1.1), (8.1.2), and Remark 3 in Section 8.1, we find that

$$P_t\mathbf{X}_{t+1} = F_t P_{t-1}\mathbf{X}_t + \Theta_t\Delta_t^{-1}(\mathbf{Y}_t - P_{t-1}\mathbf{Y}_t), \tag{8.4.5}$$

$$P_t\mathbf{X}_{t+h} = F_{t+h-1}P_t\mathbf{X}_{t+h-1}$$
$$\vdots$$
$$= (F_{t+h-1}F_{t+h-2}\cdots F_{t+1})\,P_t\mathbf{X}_{t+1}, \quad h = 2, 3, \ldots, \tag{8.4.6}$$

and

$$P_t\mathbf{Y}_{t+h} = G_{t+h}P_t\mathbf{X}_{t+h}, \quad h = 1, 2, \ldots. \tag{8.4.7}$$

From the relation

$$\mathbf{X}_{t+h} - P_t\mathbf{X}_{t+h} = F_{t+h-1}(\mathbf{X}_{t+h-1} - P_t\mathbf{X}_{t+h-1}) + \mathbf{V}_{t+h-1}, \quad h = 2, 3, \ldots,$$

we find that $\Omega_t^{(h)} := E[(\mathbf{X}_{t+h} - P_t\mathbf{X}_{t+h})(\mathbf{X}_{t+h} - P_t\mathbf{X}_{t+h})']$ satisfies the recursions

$$\Omega_t^{(h)} = F_{t+h-1}\Omega_t^{(h-1)}F_{t+h-1}' + Q_{t+h-1}, \quad h = 2, 3, \ldots, \tag{8.4.8}$$

with $\Omega_t^{(1)} = \Omega_{t+1}$. Then from (8.1.1) and (8.4.7), $\Delta_t^{(h)} := E[(\mathbf{Y}_{t+h} - P_t\mathbf{Y}_{t+h})(\mathbf{Y}_{t+h} - P_t\mathbf{Y}_{t+h})']$ is given by

$$\Delta_t^{(h)} = G_{t+h}\Omega_t^{(h)}G_{t+h}' + R_{t+h}, \quad h = 1, 2, \ldots. \tag{8.4.9}$$

Example 8.4.1 Consider the random walk plus noise model of Example 8.2.1 defined by

$$Y_t = X_t + W_t, \quad \{W_t\} \sim \text{WN}(0, \sigma_w^2),$$

where the local level X_t follows the random walk

$$X_{t+1} = X_t + V_t, \quad \{V_t\} \sim \text{WN}(0, \sigma_v^2).$$

Applying the Kalman prediction equations with $Y_0 := 1$, $R = \sigma_w^2$ and $Q = \sigma_v^2$, we obtain

$$\hat{Y}_{t+1} = P_tY_{t+1} = \hat{X}_t + \frac{\Theta_t}{\Delta_t}(Y_t - \hat{Y}_t)$$
$$= (1 - a_t)\hat{Y}_t + a_tY_t$$

where

$$a_t = \frac{\Theta_t}{\Delta_t} = \frac{\Omega_t}{\Omega_t + \sigma_w^2}.$$

For a state-space model (like this one) with time-independent parameters, the solution of the Kalman recursions (8.4.2) is called a **steady-state solution** if Ω_t is independent of t. If $\Omega_t = \Omega$ for all t, then from (8.4.2)

$$\Omega_{t+1} = \Omega = \Omega + \sigma_v^2 - \frac{\Omega^2}{\Omega + \sigma_w^2}$$

$$= \frac{\Omega\sigma_w^2}{\Omega + \sigma_w^2} + \sigma_v^2.$$

Solving this quadratic equation for Ω and noting that $\Omega \geq 0$, we find that

$$\Omega = \left(\sigma_v^2 + \sqrt{\sigma_v^4 + 4\sigma_v^2\sigma_w^2}\right)/2.$$

Since $\Omega_{t+1} - \Omega_t$ is a continuous function of Ω_t on $\Omega_t \geq 0$, positive at $\Omega_t = 0$, negative for large Ω_t and zero only at $\Omega_t = \Omega$, it is clear that $\Omega_{t+1} - \Omega_t$ is negative for $\Omega_t > \Omega$ and positive for $\Omega_t < \Omega$. A similar argument shows (Problem 8.14) that $(\Omega_{t+1} - \Omega)(\Omega_t - \Omega) \geq 0$ for all $\Omega_t \geq 0$. These observations imply that Ω_{t+1} always falls between Ω and Ω_t. Consequently, regardless of the value of Ω_1, Ω_t converges to Ω, the unique solution of $\Omega_{t+1} = \Omega_t$. For *any* initial predictors $\hat{Y}_1 = \hat{X}_1$ and any initial mean squared error, $\Omega_1 = E(X_1 - \hat{X}_1)^2$, the coefficients $a_t := \Omega_t/(\Omega_t + \sigma_w^2)$ converge to

$$a = \frac{\Omega}{\Omega + \sigma_w^2}$$

and the mean squared errors of the predictors defined by

$$\hat{Y}_{t+1} = (1 - a_t)\hat{Y}_t + a_t Y_t$$

converge to $\Omega + \sigma_w^2$.

If, as is often the case, we do not know Ω_1, then we cannot determine the sequence $\{a_t\}$. It is natural, therefore, to consider the behavior of the predictors defined by

$$\hat{Y}_{t+1} = (1 - a)\hat{Y}_t + aY_t$$

with a as above and arbitrary $\hat{Y}_1$. It can be shown (Problem 8.16) that this sequence of predictors is also asymptotically optimal in the sense that the mean squared error converges to $\Omega + \sigma_w^2$ as $t \to \infty$.

As shown in Example 8.2.1, the differenced process $D_t = Y_t - Y_{t-1}$ is the MA(1) process

$$D_t = Z_t + \theta Z_{t-1}, \ \{Z_t\} \sim \mathrm{WN}(0, \sigma^2),$$

where $\theta/(1+\theta^2) = -\sigma_w^2/(2\sigma_w^2 + \sigma_v^2)$. Solving this equation for θ (Problem 8.15), we find that

$$\theta = -\left(2\sigma_w^2 + \sigma_v^2 - \sqrt{\sigma_v^4 + 4\sigma_v^2\sigma_w^2}\right)/(2\sigma_w^2)$$

and that $\theta = a - 1$.

It is instructive to derive the exponential smoothing formula for $\hat{Y}_t$ directly from the ARIMA(0,1,1) structure of $\{Y_t\}$. For $t \ge 2$, we have from Section 6.5 that

$$\begin{aligned}\hat{Y}_{t+1} &= Y_t + \theta_{t1}(Y_t - \hat{Y}_t)\\ &= -\theta_{t1}\hat{Y}_t + (1+\theta_{t1})Y_t\end{aligned}$$

for $t \ge 2$, where θ_{t1} is found by application of the innovations algorithm to an MA(1) process with coefficient θ. It follows that $1 - a_t = -\theta_{t1}$ and since $\theta_{t1} \to \theta$ (see Remark 1 of Section 3.3) and a_t converges to the steady-state solution a, we conclude that

$$1 - a = \lim_{t\to\infty}(1 - a_t) = -\lim_{t\to\infty}\theta_{t1} = -\theta. \qquad \square$$

Kalman Filtering:

The filtered estimates $\mathbf{X}_{t|t} = P_t(\mathbf{X}_t)$ and their error covariance matrices $\Omega_{t|t} = E[(\mathbf{X}_t - \mathbf{X}_{t|t})(\mathbf{X}_t - \mathbf{X}_{t|t})']$ are determined by the relations

$$P_t\mathbf{X}_t = P_{t-1}\mathbf{X}_t + \Omega_t G_t'\Delta_t^{-1}(\mathbf{Y}_t - G_t\hat{\mathbf{X}}_t) \tag{8.4.10}$$

and

$$\Omega_{t|t} = \Omega_t - \Omega_t G_t'\Delta_t^{-1}G_t\Omega_t'. \tag{8.4.11}$$

Proof From (8.4.3) it follows that

$$P_t\mathbf{X}_t = P_{t-1}\mathbf{X}_t + M\mathbf{I}_t,$$

where

$$\begin{aligned}M &= E(\mathbf{X}_t\mathbf{I}_t')[E(\mathbf{I}_t\mathbf{I}_t')]^{-1}\\ &= E\left[\mathbf{X}_t(G_t(\mathbf{X}_t - \hat{\mathbf{X}}_t) + W_t)'\right]\Delta_t^{-1}\\ &= \Omega_t G_t'\Delta_t^{-1}.\end{aligned} \tag{8.4.12}$$

To establish (8.4.11) we write

$$\mathbf{X}_t - P_{t-1}\mathbf{X}_t = \mathbf{X}_t - P_t\mathbf{X}_t + P_t\mathbf{X}_t - P_{t-1}\mathbf{X}_t = \mathbf{X}_t - P_t\mathbf{X}_t + M\mathbf{I}_t.$$

Using (8.4.12) and the orthogonality of $\mathbf{X}_t - P_t\mathbf{X}_t$ and $M\mathbf{I}_t$, we find from the last equation that

$$\Omega_t = \Omega_{t|t} + \Omega_t G_t'\Delta_t^{-1}G_t\Omega_t',$$

as required. ■

Kalman Fixed Point Smoothing:

The smoothed estimates $\mathbf{X}_{t|n} = P_n\mathbf{X}_t$ and the error covariance matrices $\Omega_{t|n} = E[(\mathbf{X}_t - \mathbf{X}_{t|n})(\mathbf{X}_t - \mathbf{X}_{t|n})']$ are determined for fixed t by the following recursions, which can be solved successively for $n = t, t+1, \ldots$:

$$P_n\mathbf{X}_t = P_{n-1}\mathbf{X}_t + \Omega_{t,n}G_n'\Delta_n^{-1}(\mathbf{Y}_n - G_n\hat{\mathbf{X}}_n), \tag{8.4.13}$$

$$\Omega_{t,n+1} = \Omega_{t,n}[F_n - \Theta_n\Delta_n^{-1}G_n]', \tag{8.4.14}$$

$$\Omega_{t|n} = \Omega_{t|n-1} - \Omega_{t,n}G_n'\Delta_n^{-1}G_n\Omega_{t,n}', \tag{8.4.15}$$

with initial conditions, $P_{t-1}\mathbf{X}_t = \hat{\mathbf{X}}_t$ and $\Omega_{t,t} = \Omega_{t|t-1} = \Omega_t$ (found from Kalman prediction).

Proof Using (8.4.3) we can write $P_n\mathbf{X}_t = P_{n-1}\mathbf{X}_t + C\mathbf{I}_n$, where $\mathbf{I}_n = G_n(\mathbf{X}_n - \hat{\mathbf{X}}_n) + \mathbf{W}_n$. By Remark 4 above,

$$C = E[\mathbf{X}_t(G_n(\mathbf{X}_n - \hat{\mathbf{X}}_n) + \mathbf{W}_n)'][E(\mathbf{I}_n\mathbf{I}_n')]^{-1} = \Omega_{t,n}G_n'\Delta_n^{-1}, \tag{8.4.16}$$

where $\Omega_{t,n} := E[(\mathbf{X}_t - \hat{\mathbf{X}}_t)(\mathbf{X}_n - \hat{\mathbf{X}}_n)']$. It follows now from (8.1.2), (8.4.5), the orthogonality of $\mathbf{V}_n$ and $\mathbf{W}_n$ with $\mathbf{X}_t - \hat{\mathbf{X}}_t$, and the definition of $\Omega_{t,n}$ that

$$\Omega_{t,n+1} = E[(\mathbf{X}_t - \hat{\mathbf{X}}_t)(\mathbf{X}_n - \hat{\mathbf{X}}_n)'(F_n - \Theta_n\Delta_n^{-1}G_n)'] = \Omega_{t,n}[F_n - \Theta_n\Delta_n^{-1}G_n]'$$

thus establishing (8.4.14). To establish (8.4.15) we write

$$\mathbf{X}_t - P_n\mathbf{X}_t = \mathbf{X}_t - P_{n-1}\mathbf{X}_t - C\mathbf{I}_n.$$

Using (8.4.16) and the orthogonality of $\mathbf{X}_t - P_n\mathbf{X}_t$ and $\mathbf{I}_n$, the last equation then gives

$$\Omega_{t|n} = \Omega_{t|n-1} - \Omega_{t,n}G_n'\Delta_n^{-1}G_n\Omega_{t,n}', \quad n = t, t+1, \ldots,$$

as required. ■

8.5 Estimation For State-Space Models

Consider the state-space model defined by equations (8.1.1) and (8.1.2) and suppose that the model is completely parameterized by the components of the vector $\boldsymbol{\theta}$. The maximum likelihood estimate of $\boldsymbol{\theta}$ is found by maximizing the likelihood of the observations $\mathbf{Y}_1, \ldots, \mathbf{Y}_n$ with respect to the components of the vector $\boldsymbol{\theta}$. If the conditional probability density of $\mathbf{Y}_t$ given $\mathbf{Y}_{t-1} = \mathbf{y}_{t-1}, \ldots, \mathbf{Y}_0 = \mathbf{y}_0$, is $f_t(\cdot|\mathbf{y}_{t-1}, \ldots, \mathbf{y}_0)$, then the likelihood of $\mathbf{Y}_t, t = 1, \ldots, n$ (conditional on $\mathbf{Y}_0$) can immediately be written as

$$L(\boldsymbol{\theta}; \mathbf{Y}_1, \ldots, \mathbf{Y}_n) = \prod_{t=1}^{n} f_t(\mathbf{Y}_t|\mathbf{Y}_{t-1}, \ldots, \mathbf{Y}_1). \tag{8.5.1}$$

The calculation of the likelihood for any fixed numerical value of $\boldsymbol{\theta}$ is extremely complicated in general, but is greatly simplified if the noise vectors $\mathbf{W}_t, \mathbf{V}_t$, $t = 1, 2, \ldots$ are assumed to be jointly Gaussian. The resulting likelihood is called the Gaussian likelihood and is widely used in time series analysis (cf. Section 5.2) whether the time series is truly Gaussian or not. As before, we shall continue to use the term *likelihood* to mean Gaussian likelihood.

If we assume that $\mathbf{W}_t, \mathbf{V}_t, t = 1, 2, \ldots$ are jointly Gaussian, then the conditional densities in (8.5.1) are given by

$$f_t(\mathbf{Y}_t|\mathbf{Y}_{t-1}, \ldots, \mathbf{Y}_1) = (2\pi)^{-w/2}(\det\Delta_t)^{-1/2}\exp\left[-\frac{1}{2}\mathbf{I}_t'\Delta_t^{-1}\mathbf{I}_t\right],$$

where $\mathbf{I}_t = \mathbf{Y}_t - P_{t-1}\mathbf{Y}_t = \mathbf{Y}_t - G\hat{\mathbf{X}}_t$ and $t \geq 1$, are the one-step predictors and error covariance matrices found from Kalman prediction. The likelihood of the observations $\mathbf{Y}_1, \ldots, \mathbf{Y}_n$ can therefore be expressed as

$$L(\boldsymbol{\theta}; \mathbf{Y}_1, \ldots, \mathbf{Y}_n) = (2\pi)^{-nw/2}\left(\prod_{j=1}^{n}\det\Delta_j\right)^{-1/2}\exp\left[-\frac{1}{2}\sum_{j=1}^{n}\mathbf{I}_j'\Delta_j^{-1}\mathbf{I}_j\right]. \quad (8.5.2)$$

Given the observations $\mathbf{Y}_1, \ldots, \mathbf{Y}_n$, the appropriate initial vector $\mathbf{Y}_0$ (see Section 8.4) and a particular parameter value $\boldsymbol{\theta}$, the numerical value of the likelihood L can be computed from the previous equation with the aid of the Kalman recursions of Section 8.4. To find maximum likelihood estimates of the components of $\boldsymbol{\theta}$, a nonlinear optimization algorithm must be used to search for the value of $\boldsymbol{\theta}$ that maximizes the value of L.

Having estimated the parameter vector $\boldsymbol{\theta}$, we can compute forecasts based on the fitted state-space model and estimated mean squared errors by direct application of equations (8.4.7) and (8.4.9).

Application to Structural Models

The general structural model for a univariate time series $\{Y_t\}$ of which we gave examples in Section 8.2 has the form

$$Y_t = G\mathbf{X}_t + W_t, \quad \{W_t\} \sim \mathrm{WN}(0, \sigma_w^2), \quad (8.5.3)$$

$$\mathbf{X}_{t+1} = F\mathbf{X}_t + \mathbf{V}_t, \quad \{\mathbf{V}_t\} \sim \mathrm{WN}(0, Q), \quad (8.5.4)$$

for $t = 1, 2, \ldots$ where F and G are assumed known. We set $Y_0 = 1$ in order to include constant terms in our predictors and complete the specification of the model by prescribing the mean and covariance matrix of the initial state $\mathbf{X}_1$. A simple and convenient assumption is that $\mathbf{X}_1$ is equal to a deterministic but unknown parameter $\boldsymbol{\mu}$ and that $\hat{\mathbf{X}}_1 = \boldsymbol{\mu}$ so that $\Omega_1 = 0$. The parameters of the model are then $\boldsymbol{\mu}$, Q, and σ_w^2.

Direct maximization of the likelihood (8.5.2) is difficult if the dimension of the state-vector is large. The maximization can, however, be simplified by the following stepwise procedure. For fixed Q we find $\hat{\mu}(Q)$ and $\sigma_w^2(Q)$ that maximize the likelihood $L(\mu, Q, \sigma_w^2)$. We then maximize the "reduced likelihood" $L(\hat{\mu}(Q), Q, \hat{\sigma}_w^2(Q))$ with respect to Q.

To achieve this we define the mean-corrected state vectors, $\mathbf{X}_t^* = \mathbf{X}_t - F^{t-1}\mu$, and apply the Kalman prediction recursions to $\{\mathbf{X}_t^*\}$ with initial condition $\mathbf{X}_1^* = \mathbf{0}$. This gives, from (8.4.1),

$$\hat{\mathbf{X}}_{t+1}^* = F\hat{\mathbf{X}}_t^* + \Theta_t\Delta_t^{-1}(Y_t - G\hat{\mathbf{X}}_t^*), \quad t = 1, 2, \ldots \tag{8.5.5}$$

with $\hat{\mathbf{X}}_1^* = \mathbf{0}$. Since $\hat{\mathbf{X}}_t$ also satisfies (8.5.5), but with initial condition $\hat{\mathbf{X}}_t = \mu$, it follows that

$$\hat{\mathbf{X}}_t = \hat{\mathbf{X}}_t^* + C_t\mu \tag{8.5.6}$$

for some $v \times v$ matrices C_t. (Note that although $\hat{\mathbf{X}}_t = P(\mathbf{X}_t|Y_0, Y_1, \ldots, Y_t)$, the quantity $\hat{\mathbf{X}}_t^*$ is not the corresponding predictor of $\mathbf{X}_t^*$.) The matrices C_t can be determined recursively from (8.5.5), (8.5.6), and (8.4.1). Substituting (8.5.6) into (8.5.5) and using (8.4.1), we have

$$\begin{aligned}\hat{\mathbf{X}}_{t+1}^* &= F(\hat{\mathbf{X}}_t - C_t\mu) + \Theta_t\Delta_t^{-1}(Y_t - G(\hat{\mathbf{X}}_t - C_t\mu)) \\ &= F\hat{\mathbf{X}}_t + \Theta_t\Delta_t^{-1}(Y_t - G\hat{\mathbf{X}}_t) - (F - \Theta_t\Delta_t^{-1}G)C_t\mu \\ &= \hat{\mathbf{X}}_{t+1} - (F - \Theta_t\Delta_t^{-1}G)C_t\mu,\end{aligned}$$

so that

$$C_{t+1} = (F - \Theta_t\Delta_t^{-1}G)C_t \tag{8.5.7}$$

with C_1 equal to the identity matrix. The quadratic form in the likelihood (8.5.2) is therefore

$$S(\mu, Q, \sigma_w^2) = \sum_{t=1}^{n} \frac{(Y_t - G\hat{\mathbf{X}}_t)^2}{\Delta_t} \tag{8.5.8}$$

$$= \sum_{t=1}^{n} \frac{(Y_t - G\hat{\mathbf{X}}_t^* - GC_t\mu)^2}{\Delta_t}. \tag{8.5.9}$$

Now let $Q^* := \sigma_w^{-2}Q$ and define L^* to be the likelihood function with this new parameterization, i.e., $L^*(\mu, Q^*, \sigma_w^2) = L(\mu, \sigma_w^2 Q^*, \sigma_w^2)$. Writing $\Delta_t^* = \sigma_w^{-2}\Delta_t$ and $\Omega_t^* = \sigma_w^{-2}\Omega_t$, we see that the predictors $\hat{X}_t^*$ and the matrices C_t in (8.5.7) depend on the parameters only through Q^*. Thus,

$$S(\mu, Q, \sigma_w^2) = \sigma_w^{-2}S(\mu, Q^*, 1)$$

so that

$$-2\ln L^*(\mu, Q^*, \sigma_w^2) = n\ln(2\pi) + \sum_{t=1}^{n} \ln \Delta_t + \sigma_w^{-2} S(\mu, Q^*, 1)$$

$$= n\ln(2\pi) + \sum_{t=1}^{n} \ln \Delta_t^* + n\ln\sigma_w^2 + \sigma_w^{-2} S(\mu, Q^*, 1).$$

For Q^* fixed, it is easy to show (see Problem 8.18) that this function is minimized when

$$\hat{\mu} = \hat{\mu}(Q^*) = \left[\sum_{t=1}^{n} \frac{C_t' G' G C_t}{\Delta_t^*}\right]^{-1} \sum_{t=1}^{n} \frac{C_t' G'(Y_t - G\hat{\mathbf{X}}_t^*)}{\Delta_t^*} \tag{8.5.10}$$

and

$$\hat{\sigma}_w^2 = \hat{\sigma}_w^2(Q^*) = n^{-1} \sum_{t=1}^{n} \frac{(Y_t - G\hat{\mathbf{X}}_t^* - GC_t\hat{\mu})^2}{\Delta_t^*}. \tag{8.5.11}$$

Replacing μ and σ_w^2 by these values in $-2\ln L^*$ and ignoring constants, the reduced likelihood becomes

$$\ell(Q^*) = \ln\left(n^{-1} \sum_{t=1}^{n} \frac{(Y_t - G\hat{\mathbf{X}}_t^* - GC_t\hat{\mu})^2}{\Delta_t^*}\right) + n^{-1}\sum_{t=1}^{n} \ln(\det \Delta_t^*). \tag{8.5.12}$$

If $\hat{Q}^*$ denotes the minimizer of (8.5.12), then the maximum likelihood estimator of the parameters μ, Q, σ_w^2 are $\hat{\mu}, \hat{\sigma}_w^2\hat{Q}^*, \hat{\sigma}_w^2$, where $\hat{\mu}$ and $\hat{\sigma}_w^2$ are computed from (8.5.10) and (8.5.11) with Q^* replaced by $\hat{Q}^*$.

We can now summarize the steps required for computing the maximum likelihood estimators of μ, Q, and σ_w^2 for the model (8.5.3)–(8.5.4).

1. For a fixed Q^*, apply the Kalman prediction recursions with $\hat{\mathbf{X}}_1^* = \mathbf{0}$, $\Omega_1 = 0$, $Q = Q^*$, and $\sigma_w^2 = 1$ to obtain the *predictors* $\hat{\mathbf{X}}_t^*$. Let Δ_t^* denote the one-step prediction error produced by these recursions.
2. Set $\hat{\mu} = \hat{\mu}(Q^*) = \left[\sum_{t=1}^{n} C_t' G' G C_t/\Delta_t\right]^{-1} \sum_{t=1}^{n} C_t' G'(Y_t - G\hat{\mathbf{X}}_t^*)/\Delta_t^*$.
3. Let $\hat{Q}^*$ be the minimizer of (8.5.12).
4. The maximum likelihood estimators of μ, Q, and σ_w^2 are then given by $\hat{\mu}, \hat{\sigma}_w^2\hat{Q}^*$, and $\hat{\sigma}_w^2$, respectively, where $\hat{\mu}$ and $\hat{\sigma}_w^2$ are found from (8.5.10) and (8.5.11) evaluated at $\hat{Q}^*$.

Example 8.5.1 Random walk plus noise model

In Example 8.2.1, 100 observations were generated from the structural model

$$Y_t = M_t + W_t, \quad \{W_t\} \sim \mathrm{WN}(0, \sigma_w^2)$$

$$M_{t+1} = M_t + V_t, \quad \{V_t\} \sim \mathrm{WN}(0, \sigma_v^2)$$

with initial value $\mu = M_1 = 0$, $\sigma_w^2 = 8$ and $\sigma_v^2 = 4$. The maximum likelihood estimates of the parameters are found by first minimizing (8.5.12) with $\hat{\mu}$ given by (8.5.10). Substituting these values into (8.5.11) gives $\hat{\sigma}_w^2$. The resulting estimates are $\hat{\mu} = .906$, $\hat{\sigma}_v^2 = 5.351$, and $\hat{\sigma}_w^2 = 8.233$ which are in reasonably close agreement with the true values. □

Example 8.5.2 International airline passengers, 1949–1960; AIRPASS.DAT

The monthly totals of international airline passengers from January 1949 to December 1960 (Box and Jenkins, 1976) are displayed in Figure 8.3. The data exhibit both a strong seasonal pattern and a nearly linear trend. Since the variability of the data, $Y_1, \ldots, Y_{144}$, increases for larger values of Y_t, it may be appropriate to consider a logarithmic transformation of the data. For the purpose of this illustration, however, we will fit a structural model incorporating a randomly varying trend and seasonal and noise components (see Example 8.2.3) to the raw data. This model has the form

$$Y_t = G\mathbf{X}_t + W_t, \quad \{W_t\} \sim \mathrm{WN}(0, \sigma_w^2)$$

$$\mathbf{X}_{t+1} = F\mathbf{X}_t + \mathbf{V}_t, \quad \{\mathbf{V}_t\} \sim \mathrm{WN}(0, Q)$$

where $\mathbf{X}_t$ is a 13-dimensional state-vector,

$$F = \begin{bmatrix} 1 & 1 & 0 & 0 & \cdots & 0 & 0 \\ 0 & 1 & 0 & 0 & \cdots & 0 & 0 \\ 0 & 0 & -1 & -1 & \cdots & -1 & -1 \\ 0 & 0 & 1 & 0 & \cdots & 0 & 0 \\ 0 & 0 & 0 & 1 & \cdots & 0 & 0 \\ \vdots & \vdots & \vdots & \vdots & \ddots & \vdots & \vdots \\ 0 & 0 & 0 & 0 & \cdots & 1 & 0 \end{bmatrix},$$

$$G = [1 \quad 0 \quad 1 \quad 0 \quad \cdots \quad 0],$$

and

$$Q = \begin{bmatrix} \sigma_1^2 & 0 & 0 & 0 & \cdots & 0 \\ 0 & \sigma_2^2 & 0 & 0 & \cdots & 0 \\ 0 & 0 & \sigma_3^2 & 0 & \cdots & 0 \\ 0 & 0 & 0 & 0 & \cdots & 0 \\ \vdots & \vdots & \vdots & \vdots & \ddots & \vdots \\ 0 & 0 & 0 & 0 & \cdots & 0 \end{bmatrix}.$$

The parameters of the model are $\boldsymbol{\mu}, \sigma_1^2, \sigma_2^2, \sigma_3^2$, and σ_w^2, where $\boldsymbol{\mu} = \mathbf{X}_1$. Minimizing (8.5.12) with respect to Q^* we find from (8.5.11) and (8.5.12) that

$$(\hat{\sigma}_1^2, \hat{\sigma}_2^2, \hat{\sigma}_3^2, \hat{\sigma}_w^2) = (170.63, .00000, 11.338, .014179)$$

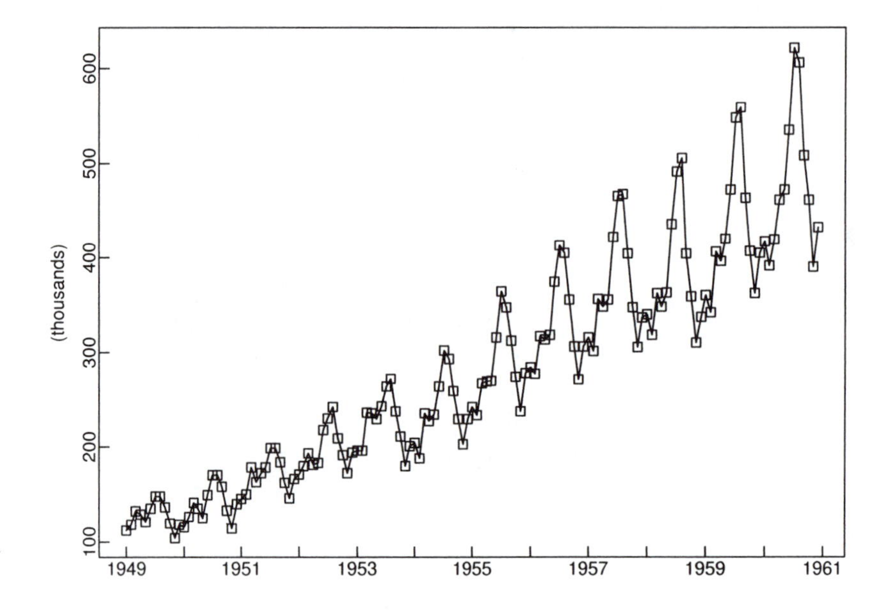

Figure 8-3 International airline passengers; monthly totals from January 1949 to December 1960.

and from (8.5.10) that $\hat{\mu} = (146.9, 2.171, -34.92, -34.12, -47.00, -16.98, 22.99, 53.99, 58.34, 33.65, 2.204, -4.053, -6.894)'$. The first component, X_{t1}, of the state vector corresponds to the local linear trend with slope X_{t2}. Since $\hat{\sigma}_2^2 = 0$, the slope at time t, which satisfies

$$X_{t2} = X_{t-1,2} + V_{t2},$$

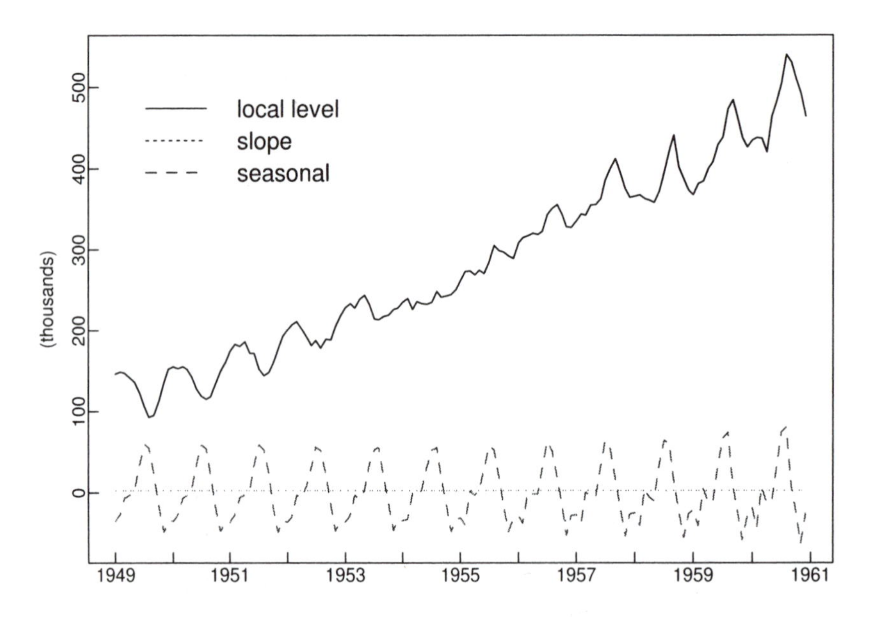

Figure 8-4 The one-step predictors $(\hat{X}_{t1}, \hat{X}_{t2}, \hat{X}_{t3})'$ for the airline passenger data in Example 8.5.2.

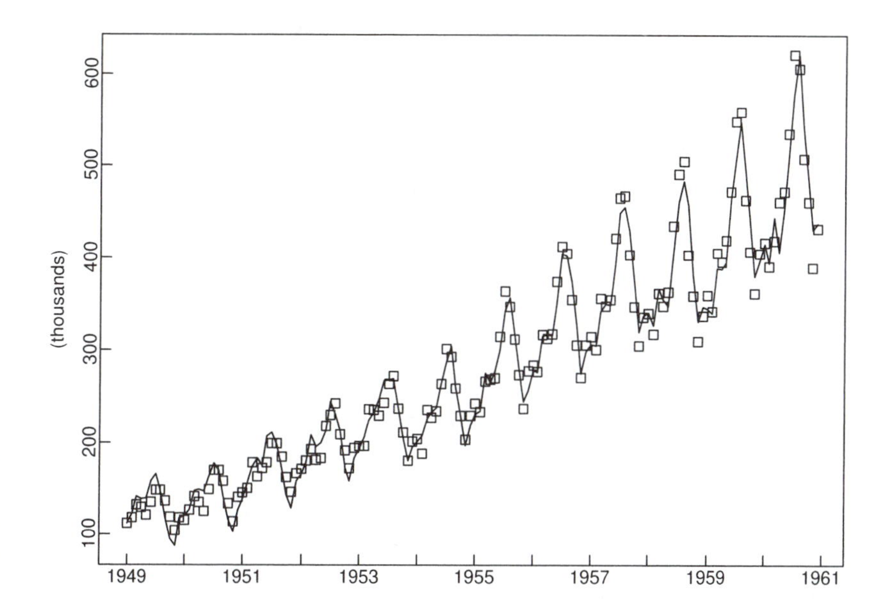

Figure 8-5 The one-step predictors $\hat{Y}_t$ for the airline passenger data (solid line) and the actual data (square boxes).

must be nearly constant and equal to $\hat{X}_{12} = 2.171$. The first three components of the predictors $\hat{\mathbf{X}}_t$ are plotted in Figure 8.4. Notice that the first component varies like a random walk around a straight line while the second component is nearly constant as a result of $\hat{\sigma}_2^2 \approx 0$. The third component, corresponding to the seasonal component, exhibits a clear seasonal cycle that repeats roughly the same pattern throughout the 12 years of data. The one-step predictors, $\hat{X}_{t1} + \hat{X}_{t3}$, of Y_t are plotted in Figure 8.5 (solid line) together with the actual data (square boxes). For this model, the predictors follow the movement of the data quite well. □

8.6 State-Space Models with Missing Observations

State-space representations and the associated Kalman recursions are ideally suited to the analysis of data with missing values, as was pointed out by Jones (1980) in the context of maximum likelihood estimation for ARMA processes. In this section we shall deal with two missing-value problems for state-space models. The first is the evaluation of the (Gaussian) likelihood based on $\{\mathbf{Y}_{i_1}, \ldots, \mathbf{Y}_{i_r}\}$, where $i_1, i_2, \ldots, i_r$ are positive integers such that $1 \le i_1 < i_2 < \cdots < i_r \le n$. (This allows for observation of the process $\{\mathbf{Y}_t\}$ at irregular intervals, or equivalently for the possibility that $(n-r)$ observations are missing from the sequence $\{\mathbf{Y}_1, \ldots, \mathbf{Y}_n\}$.) The solution of this problem will enable us, in particular, to carry out maximum likelihood estimation for ARMA and ARIMA processes with missing values. The second problem to be considered is the minimum mean squared error estimation of the missing values themselves.

The Gaussian Likelihood of $\{\mathbf{Y}_{i_1},\ldots,\mathbf{Y}_{i_r}\}$, $1 \le i_1 < i_2 < \cdots < i_r \le n$

Consider the state-space model defined by equations (8.1.1) and (8.1.2) and suppose that the model is completely parameterized by the components of the vector $\boldsymbol{\theta}$. If there are no missing observations, i.e., if $r = n$ and $i_j = j$, $j = 1,\ldots,n$, then the likelihood of the observations $\{\mathbf{Y}_1,\ldots,\mathbf{Y}_n\}$ is easily found as in Section 8.5 to be

$$L(\boldsymbol{\theta};\mathbf{Y}_1,\ldots,\mathbf{Y}_n) = (2\pi)^{-nw/2}\left(\prod_{j=1}^{n}\det\Delta_j\right)^{-1/2}\exp\left[-\frac{1}{2}\sum_{j=1}^{n}\mathbf{I}_j'\Delta_j^{-1}\mathbf{I}_j\right],$$

where $\mathbf{I}_j = \mathbf{Y}_j - P_{j-1}\mathbf{Y}_j$ and Δ_j, $j \ge 1$, are the one-step predictors and error covariance matrices found from (8.4.7) and (8.4.9) with $\mathbf{Y}_0 = \mathbf{1}$.

To deal with the more general case of possibly irregularly spaced observations $\{\mathbf{Y}_{i_1},\ldots,\mathbf{Y}_{i_r}\}$, we introduce a new series $\{\mathbf{Y}_t^*\}$, related to the process $\{\mathbf{X}_t\}$ by the modified observation equation

$$\mathbf{Y}_t^* = G_t^*\mathbf{X}_t + \mathbf{W}_t^*, \quad t = 1,2,\ldots, \tag{8.6.1}$$

where

$$G_t^* = \begin{cases} G_t & \text{if } t \in \{i_1,\ldots,i_r\}, \\ 0 & \text{otherwise,} \end{cases} \qquad \mathbf{W}_t^* = \begin{cases} \mathbf{W}_t & \text{if } t \in \{i_1,\ldots,i_r\}, \\ \mathbf{N}_t & \text{otherwise,} \end{cases} \tag{8.6.2}$$

and $\{\mathbf{N}_t\}$ is iid with

$$\mathbf{N}_t \sim \mathrm{N}(\mathbf{0}, I_{w\times w}), \quad \mathbf{N}_s \perp \mathbf{X}_1, \quad \mathbf{N}_s \perp \begin{bmatrix}\mathbf{V}_t \\ \mathbf{W}_t\end{bmatrix}, s,t = 0,\pm 1,\ldots. \tag{8.6.3}$$

Equations (8.6.1) and (8.1.2) constitute a state-space representation for the new series $\{\mathbf{Y}_t^*\}$, which coincides with $\{\mathbf{Y}_t\}$ at each $t \in \{i_1, i_2,\ldots,i_r\}$, and at other times takes random values that are independent of $\{\mathbf{Y}_t\}$ with a distribution independent of $\boldsymbol{\theta}$.

Let $L_1(\boldsymbol{\theta};\mathbf{y}_{i_1},\ldots,\mathbf{y}_{i_r})$ be the Gaussian likelihood based on the observed values $\mathbf{y}_{i_1},\ldots,\mathbf{y}_{i_r}$ of $\mathbf{Y}_{i_1},\ldots,\mathbf{Y}_{i_r}$ under the model defined by (8.1.1) and (8.1.2). Corresponding to these observed values, we define a new sequence, $\mathbf{y}_1^*,\ldots,\mathbf{y}_n^*$, by

$$\mathbf{y}_t^* = \begin{cases} \mathbf{y}_t & \text{if } t \in \{i_1,\ldots,i_r\}, \\ \mathbf{0} & \text{otherwise.} \end{cases} \tag{8.6.4}$$

Then it is clear from the preceding paragraph that

$$L_1(\boldsymbol{\theta};\mathbf{y}_{i_1},\ldots,\mathbf{y}_{i_r}) = (2\pi)^{(n-r)w/2}L_2(\boldsymbol{\theta};\mathbf{y}_1^*,\ldots,\mathbf{y}_n^*), \tag{8.6.5}$$

where L_2 denotes the Gaussian likelihood under the model defined by (8.6.1) and (8.1.2).

In view of (8.6.5) we can now compute the required likelihood L_1 of the realized values $\{\mathbf{y}_t, t = i_1,\ldots,i_r\}$ as follows:

i. Define the sequence $\{\mathbf{y}_t^*, t = 1,\ldots,n\}$ as in (8.6.4).

ii. Find the one-step predictors $\hat{\mathbf{Y}}_t^*$ of $\mathbf{Y}_t^*$, and their error covariance matrices Δ_t^*, using Kalman prediction and the equations (8.4.7) and (8.4.9) applied to the state-space representation, (8.6.1) and (8.1.2) of $\{\mathbf{Y}_t^*\}$. Denote the realized values of the predictors, based on the observation sequence $\{\mathbf{y}_t^*\}$, by $\{\hat{\mathbf{y}}_t^*\}$.

iii. The required Gaussian likelihood of the irregularly spaced observations, $\{\mathbf{y}_{i_1}, \ldots, \mathbf{y}_{i_r}\}$, is then, by (8.6.5),

$$L_1(\boldsymbol{\theta}; \mathbf{y}_{i_1}, \ldots, \mathbf{y}_{i_r}) = (2\pi)^{-rw/2} \left(\prod_{j=1}^{n} \det \Delta_j^* \right)^{-1/2} \exp\left\{ -\frac{1}{2} \sum_{j=1}^{n} \mathbf{i}_j^{*\prime} \Delta_j^{*-1} \mathbf{i}_j^* \right\},$$

where $\mathbf{i}_j^*$ denotes the observed innovation $\mathbf{y}_j^* - \hat{\mathbf{y}}_j^*$, $j = 1, \ldots, n$.

Example 8.6.1 An AR(1) series with one missing observation

Let $\{Y_t\}$ be the causal AR(1) process defined by

$$Y_t - \phi Y_{t-1} = Z_t, \qquad \{Z_t\} \sim \mathrm{WN}(0, \sigma^2).$$

To find the Gaussian likelihood of the observations y_1, y_3, y_4, and y_5 of Y_1, Y_3, Y_4, and Y_5 we follow the steps outlined above.

i. Set $y_i^* = y_i$, $i = 1, 3, 4, 5$ and $y_2^* = 0$.

ii. We start with the state-space model for $\{Y_t\}$ from Example 8.1.1, i.e., $Y_t = X_t$, $X_{t+1} = \phi X_t + Z_{t+1}$. The corresponding model for $\{Y_t^*\}$ is then, from (8.6.1),

$$Y_t^* = G_t^* X_t + W_t^*, \; t = 1, 2, \ldots,$$

where

$$X_{t+1} = F_t X_t + V_t, \; t = 1, 2, \ldots,$$

$$F_t = \phi, \quad G_t^* = \begin{cases} 1 & \text{if } t \neq 2, \\ 0 & \text{if } t = 2, \end{cases} \quad V_t = Z_{t+1}, \quad W_t^* = \begin{cases} 0 & \text{if } t \neq 2, \\ N_t & \text{if } t = 2, \end{cases}$$

$$Q_t = \sigma^2, \qquad R_t^* = \begin{cases} 0 & \text{if } t \neq 2, \\ 1 & \text{if } t = 2, \end{cases} \qquad S_t^* = 0,$$

and $X_1 = \sum_{j=0}^{\infty} \phi^j Z_{1-j}$. Starting from the initial conditions,

$$\hat{X}_1 = 0, \qquad \Omega_1 = \sigma^2/(1 - \phi^2),$$

and applying the recursions (8.4.1) and (8.4.2), we find (Problem 8.19) that

$$\Theta_t \Delta_t^{-1} = \begin{cases} \phi & \text{if } t = 1, 3, 4, 5, \\ 0 & \text{if } t = 2, \end{cases} \qquad \Omega_t = \begin{cases} \sigma^2/(1-\phi^2) & \text{if } t = 1, \\ \sigma^2(1+\phi^2) & \text{if } t = 3, \\ \sigma^2 & \text{if } t = 2, 4, 5, \end{cases}$$

and

$$\hat{X}_1 = 0, \quad \hat{X}_2 = \phi Y_1, \quad \hat{X}_3 = \phi^2 Y_1, \quad \hat{X}_4 = \phi Y_3, \quad \hat{X}_5 = \phi Y_4.$$

From (8.4.7) and (8.4.9) with $h = 1$, we find that

$$\hat{Y}_1^* = 0, \quad \hat{Y}_2^* = 0, \quad \hat{Y}_3^* = \phi^2 Y_1, \quad \hat{Y}_4^* = \phi Y_3, \quad \hat{Y}_5^* = \phi Y_4,$$

with corresponding mean squared errors,

$$\Delta_1^* = \sigma^2/(1-\phi^2), \quad \Delta_2^* = 1, \quad \Delta_3^* = \sigma^2(1+\phi^2), \quad \Delta_4^* = \sigma^2, \quad \Delta_5^* = \sigma^2.$$

iii. From the preceding calculations we can now write the likelihood of the original data as

$$L_1(\phi, \sigma^2; y_1, y_3, y_4, y_5) = \sigma^{-4}(2\pi)^{-2}[(1-\phi^2)/(1+\phi^2)]^{1/2}$$
$$\times \exp\left\{-\frac{1}{2\sigma^2}\left[y_1^2(1-\phi^2) + \frac{(y_3-\phi^2 y_1)^2}{1+\phi^2} + (y_4-\phi y_3)^2 + (y_5 - \phi y_4)^2\right]\right\}. \quad \square$$

Remark 1. If we are given observations $y_{1-d}, y_{2-d}, \ldots, y_0, y_{i_1}, y_{i_2}, \ldots, y_{i_r}$ of an ARIMA(p, d, q) process at times $1-d, 2-d, \ldots, 0, i_1, \ldots, i_r$, where $1 \le i_1 < i_2 < \cdots < i_r \le n$, a similar argument can be used to find the Gaussian likelihood of $y_{i_1}, \ldots, y_{i_r}$ *conditional on* $Y_{1-d} = y_{1-d}, Y_{2-d} = y_{2-d}, \ldots, Y_0 = y_0$. Missing values among the first d observations $y_{1-d}, y_{2-d}, \ldots, y_0$ can be handled by treating them as unknown parameters for likelihood maximization. For more on ARIMA series with missing values see TSTM and Ansley and Kohn (1985). $\square$

Estimation of Missing Values for State-Space Models

Given that we observe only $\mathbf{Y}_{i_1}, \mathbf{Y}_{i_2}, \ldots, \mathbf{Y}_{i_r}$, $1 \le i_1 < i_2 < \cdots < i_r \le n$, where $\{\mathbf{Y}_t\}$ has the state-space representation (8.1.1) and (8.1.2), we now consider the problem of finding the minimum mean squared error estimators $P(\mathbf{Y}_t | \mathbf{Y}_0, \mathbf{Y}_{i_1}, \ldots, \mathbf{Y}_{i_r})$ of $\mathbf{Y}_t$, $1 \le t \le n$ where $\mathbf{Y}_0 = \mathbf{1}$. To handle this problem we again use the modified process $\{\mathbf{Y}_t^*\}$ defined by (8.6.1) and (8.1.2) with $\mathbf{Y}_0^* = \mathbf{1}$. Since $\mathbf{Y}_s^* = \mathbf{Y}_s$ for $s \in \{i_1, \ldots, i_r\}$ and $\mathbf{Y}_s^* \perp \mathbf{X}_t, \mathbf{Y}_0$ for $1 \le t \le n$ and $s \notin \{0, i_1, \ldots, i_r\}$, we immediately obtain the minimum mean squared error state estimators

$$P(\mathbf{X}_t | \mathbf{Y}_0, \mathbf{Y}_{i_1}, \ldots, \mathbf{Y}_{i_r}) = P(\mathbf{X}_t | \mathbf{Y}_0^*, \mathbf{Y}_1^*, \ldots, \mathbf{Y}_n^*), \quad 1 \le t \le n. \tag{8.6.6}$$

The right-hand side can be evaluated by application of the Kalman fixed point smoothing algorithm to the state-space model (8.6.1) and (8.1.2). For computational purposes the observed values of $\mathbf{Y}_t^*$, $t \notin \{0, i_1, \ldots, i_r\}$ are quite immaterial. They may for example all be set equal to zero, giving the sequence of *observations* of $\mathbf{Y}_t^*$ defined in (8.6.4).

To evaluate $P(\mathbf{Y}_t|\mathbf{Y}_0, \mathbf{Y}_{i_1}, \ldots, \mathbf{Y}_{i_r})$, $1 \le t \le n$, we use (8.6.6) and the relation

$$\mathbf{Y}_t = G_t\mathbf{X}_t + \mathbf{W}_t. \tag{8.6.7}$$

Since $E(\mathbf{V}_t\mathbf{W}_t') = S_t = 0, \quad t = 1, \ldots, n$, we find from (8.6.7) that

$$P(\mathbf{Y}_t|\mathbf{Y}_0, \mathbf{Y}_{i_1}, \ldots, \mathbf{Y}_{i_r}) = G_t P(\mathbf{X}_t|\mathbf{Y}_0^*, \mathbf{Y}_1^*, \ldots, \mathbf{Y}_n^*). \tag{8.6.8}$$

Example 8.6.2 An AR(1) series with one missing observation

Consider the problem of estimating the missing value Y_2 in Example 8.6.1 in terms of $Y_0 = 1, Y_1, Y_3, Y_4$, and Y_5. We start from the state-space model, $X_{t+1} = \phi X_t + Z_{t+1}$, $Y_t = X_t$, for $\{Y_t\}$. The corresponding model for $\{Y_t^*\}$ is the one used in Example 8.6.1. Applying the Kalman smoothing equations to the latter model, we find that

$$\begin{aligned} &P_1X_2 = \phi Y_1, \quad &&P_2X_2 = \phi Y_1, \quad &&P_3X_2 = \tfrac{\phi(Y_1+Y_3)}{(1+\phi^2)}, \\ &P_4X_2 = P_3X_2, \quad &&P_5X_2 = P_3X_2, && \\ &\Omega_{2,2} = \sigma^2, \quad &&\Omega_{2,3} = \phi\sigma^2, \quad &&\Omega_{2,t} = 0, \quad t \ge 4, \end{aligned}$$

and

$$\Omega_{2|1} = \sigma^2, \quad \Omega_{2|2} = \sigma^2, \quad \Omega_{2|t} = \tfrac{\sigma^2}{(1+\phi^2)}, \quad t \ge 3,$$

where $P_t(\cdot)$ here denotes $P(\cdot|Y_0^*, \ldots, Y_t^*)$ and $\Omega_{t,n}$, $\Omega_{t|n}$ are defined correspondingly. We deduce from (8.6.8) that the minimum mean squared error estimator of the missing value Y_2 is

$$P_5Y_2 = P_5X_2 = \frac{\phi(Y_1 + Y_3)}{(1 + \phi^2)},$$

with mean squared error

$$\Omega_{2|5} = \frac{\sigma^2}{(1 + \phi^2)}.$$

□

Remark 2. Suppose we have observations $Y_{1-d}, Y_{2-d}, \ldots, Y_0, Y_{i_1}, \ldots, Y_{i_r}$ $(1 \le i_1 < i_2 \cdots < i_r \le n)$ of an ARIMA(p, d, q) process. Determination of the best linear estimates of the missing values Y_t, $t \notin \{i_1, \ldots, i_r\}$, in terms of Y_t, $t \in \{i_1, \ldots, i_r\}$ and the components of $\mathbf{Y}_0 := (Y_{1-d}, Y_{2-d}, \ldots, Y_0)'$ can be carried out as in Example 8.6.2 using the state-space representation of the ARIMA series $\{Y_t\}$ from Example 8.3.3 and the Kalman recursions for the corresponding state-space model for $\{Y_t^*\}$ defined by (8.6.1) and (8.1.2). See TSTM for further details. □

We close this section with a brief discussion of a direct approach to estimating missing observations. This approach is often more efficient than the methods just described, especially if the number of missing observations is small and we have a simple (e.g., autoregressive) model. Consider the general problem of computing $E(\mathbf{X}|\mathbf{Y})$ when the random vector $(\mathbf{X}', \mathbf{Y}')'$ has a multivariate normal distribution with mean $\mathbf{0}$ and covariance matrix Σ. (In the missing observation problem, think of $\mathbf{X}$ as the vector of the missing observations and $\mathbf{Y}$ as the vector of observed values.) Then the joint probability density function of $\mathbf{X}$ and $\mathbf{Y}$ can be written as

$$f_{\mathbf{X},\mathbf{Y}}(\mathbf{x}, \mathbf{y}) = f_{\mathbf{X}|\mathbf{Y}}(\mathbf{x}|\mathbf{y}) f_{\mathbf{Y}}(\mathbf{y}), \tag{8.6.9}$$

where $f_{\mathbf{X}|\mathbf{Y}}(\mathbf{x}|\mathbf{y})$ is a multivariate normal density with mean $E(\mathbf{X}|\mathbf{Y})$ and covariance matrix $\Sigma_{\mathbf{X}|\mathbf{Y}}$ (see Proposition A.3.1). In particular,

$$\begin{aligned} & f_{\mathbf{X}|\mathbf{Y}}(\mathbf{x}|\mathbf{y}) \\ & = \frac{1}{\sqrt{(2\pi)^q \det \Sigma_{\mathbf{X}|\mathbf{Y}}}} \exp\{-\frac{1}{2}(\mathbf{x} - E(\mathbf{X}|\mathbf{y}))' \Sigma_{\mathbf{X}|\mathbf{Y}}^{-1} (\mathbf{x} - E(\mathbf{X}|\mathbf{y}))\}, \end{aligned} \tag{8.6.10}$$

where $q = \dim(\mathbf{X})$. It is clear from (8.6.10) that $f_{\mathbf{X}|\mathbf{Y}}(\mathbf{x}|\mathbf{y})$ (and also $f_{\mathbf{X},\mathbf{Y}}(\mathbf{x}, \mathbf{y})$) is maximum when $\mathbf{x} = E(\mathbf{X}|\mathbf{y})$. Thus, the best estimator of $\mathbf{X}$ in terms of $\mathbf{Y}$ can be found by maximizing the joint density of $\mathbf{X}$ and $\mathbf{Y}$ with respect to $\mathbf{x}$. For autoregressive processes it is relatively straightforward to carry out this optimization, as shown in the following example.

Example 8.6.3 Estimating missing observations in an AR process

Suppose $\{Y_t\}$ is the AR(p) process defined by

$$Y_t = \phi_1 Y_{t-1} + \cdots + \phi_p Y_{t-p} + Z_t, \quad \{Z_t\} \sim \mathrm{WN}(0, \sigma^2),$$

and $\mathbf{Y} = (Y_{i_1}, \ldots, Y_{i_r})'$, with $1 \le i_1 < \cdots < i_r \le n$, are the observed values. If there are no missing observations in the first p observations, then the best estimates of the missing values are found by minimizing

$$\sum_{t=p+1}^{n} (Y_t - \phi_1 Y_{t-1} - \cdots - \phi_p Y_{t-p})^2 \tag{8.6.11}$$

with respect to the missing values (see Problem 8.20). For the AR(1) model in Example 8.6.2, minimization of (8.6.11) is equivalent to minimizing

$$(Y_2 - \phi Y_1)^2 + (Y_3 - \phi Y_2)^2$$

with respect to Y_2. Setting the derivative of this expression with respect to Y_2 equal to 0 and solving for Y_2 we obtain $E(Y_2|Y_1, Y_3, Y_4, Y_5) = \phi(Y_1 + Y_3)/(1 + \phi^2)$. □

8.7 The EM Algorithm

The expectation-maximization (EM) algorithm is an iterative procedure for computing the maximum likelihood estimator when only a subset of the complete data set is available. Dempster, Laird, and Rubin (1977) demonstrated the wide applicability of the EM algorithm and are largely responsible for popularizing this method in statistics. Details regarding the convergence and performance of the EM algorithm can be found in Wu (1983).

In the usual formulation of the EM algorithm, the "complete" data vector $\mathbf{W}$ is made up of "observed" data $\mathbf{Y}$ (sometimes called incomplete data) and "unobserved" data $\mathbf{X}$. In many applications, $\mathbf{X}$ consists of values of a "latent" or unobserved process occurring in the specification of the model. For example, in the state-space model of Section 8.1, $\mathbf{Y}$ could consist of the observed vectors $\mathbf{Y}_1, \ldots, \mathbf{Y}_n$ and $\mathbf{X}$ of the unobserved state vectors $\mathbf{X}_1, \ldots, \mathbf{X}_n$. The EM algorithm provides an iterative procedure for computing the maximum likelihood estimator based only on the observed data $\mathbf{Y}$. Each iteration of the EM algorithm consists of two steps. If $\theta^{(i)}$ denotes the estimated value of the parameter θ after i iterations, then the two steps in the $(i+1)^{\text{th}}$ iteration are

$$\textbf{E}-\textbf{step}. \qquad \text{Calculate } Q(\theta|\theta^{(i)}) = E_{\theta^{(i)}}\left[\ell(\theta; \mathbf{X}, \mathbf{Y})|\mathbf{Y}\right]$$

and

$$\textbf{M}-\textbf{step}. \qquad \text{Maximize } Q(\theta|\theta^{(i)}) \text{ with respect to } \theta.$$

$\theta^{(i+1)}$ is then set equal to the maximizer of Q in the M-step. In the E-step, $\ell(\theta; \mathbf{x}, \mathbf{y}) = \ln f(\mathbf{x}, \mathbf{y}; \theta)$, and $E_{\theta^{(i)}}(\cdot|\mathbf{Y})$ denotes the conditional expectation relative to the conditional density $f(\mathbf{x}|\mathbf{y}; \theta^{(i)}) = f(\mathbf{x}, \mathbf{y}; \theta^{(i)})/f(\mathbf{y}; \theta^{(i)})$.

It can be shown that $l(\theta^{(i)}; \mathbf{Y})$ is non-decreasing in i and a simple heuristic argument shows that if $\theta^{(i)}$ has a limit $\hat{\theta}$, then $\hat{\theta}$ must be a solution of the likelihood equations, $\ell'(\hat{\theta}; \mathbf{Y}) = 0$. To see this, observe that $\ln f(\mathbf{x}, \mathbf{y}; \theta) = \ln f(\mathbf{x}|\mathbf{y}; \theta) + \ell(\theta; \mathbf{y})$, from which we obtain

$$Q(\theta|\theta^{(i)}) = \int (\ln f(\mathbf{x}|\mathbf{Y}; \theta))\, f(\mathbf{x}|\mathbf{Y}; \theta^{(i)})\, d\mathbf{x} + \ell(\theta; \mathbf{Y})$$

and

$$Q'(\theta|\theta^{(i)}) = \int \left[\frac{\partial}{\partial\theta} f(\mathbf{x}|\mathbf{Y}; \theta)\right] / f(\mathbf{x}|\mathbf{Y}; \theta) f(\mathbf{x}|\mathbf{Y}; \theta^{(i)})\, d\mathbf{x} + \ell'(\theta; \mathbf{Y}).$$

Now replacing θ with $\theta^{(i+1)}$, noticing that $Q'(\theta^{(i+1)}|\theta^{(i)}) = 0$, and letting $i \to \infty$, we find that

$$0 = \int \frac{\partial}{\partial\theta}\left[f(\mathbf{x}|\mathbf{Y}; \theta)\right]_{\theta=\hat{\theta}}\, d\mathbf{x} + \ell'(\hat{\theta}; \mathbf{Y}) = \ell'(\hat{\theta}; \mathbf{Y}).$$

The last equality follows from the fact that

$$0 = \frac{\partial}{\partial\theta}(1) = \frac{\partial}{\partial\theta}\left[\int (f(\mathbf{x}|\mathbf{Y};\theta)\,d\mathbf{x}\right]_{\theta=\hat{\theta}} = \int \left[\frac{\partial}{\partial\theta} f(\mathbf{x}|\mathbf{Y};\theta)\right]_{\theta=\hat{\theta}} d\mathbf{x}.$$

The computational advantage of the EM algorithm over direct maximization of the likelihood is most pronounced when the calculation and maximization of the exact likelihood is difficult as compared with the maximization of Q in the M-step. (There are some applications in which the maximization of Q can easily be carried out explicitly.)

Missing Data

The EM algorithm is particularly useful for estimation problems in which there are missing observations. Suppose the complete data set consists of $Y_1, \ldots, Y_n$ of which r are observed and $n - r$ are missing. Denote the observed and missing data by $\mathbf{Y} = (Y_{i_1}, \ldots, Y_{i_r})'$, and $\mathbf{X} = (Y_{j_1}, \ldots, Y_{j_{n-r}})'$, respectively. Assuming that $\mathbf{W} = (\mathbf{X}', \mathbf{Y}')'$ has a multivariate normal distribution with mean $\mathbf{0}$ and covariance matrix Σ, which depends on the parameter $\boldsymbol{\theta}$, the log-likelihood of the complete data is given by

$$\ell(\boldsymbol{\theta}; \mathbf{W}) = -\frac{n}{2}\ln(2\pi) - \frac{1}{2}\ln\det(\Sigma) - \frac{1}{2}\mathbf{W}'\Sigma\mathbf{W}.$$

The E-step requires that we compute the expectation of $\ell(\boldsymbol{\theta}; \mathbf{W})$ with respect to the conditional distribution of $\mathbf{W}$ given $\mathbf{Y}$ with $\boldsymbol{\theta} = \boldsymbol{\theta}^{(i)}$. Writing $\Sigma(\boldsymbol{\theta})$ as the block matrix

$$\Sigma = \begin{bmatrix} \Sigma_{11} & \Sigma_{12} \\ \Sigma_{21} & \Sigma_{22} \end{bmatrix},$$

which is conformable with $\mathbf{X}$ and $\mathbf{Y}$, the conditional distribution of $\mathbf{W}$ given $\mathbf{Y}$ is multivariate normal with mean $\left[\begin{smallmatrix} \hat{\mathbf{X}} \\ \mathbf{0} \end{smallmatrix}\right]$ and covariance matrix $\left[\begin{smallmatrix} \Sigma_{11|2}(\boldsymbol{\theta}) & 0 \\ 0 & 0 \end{smallmatrix}\right]$, where $\hat{\mathbf{X}} = E_{\boldsymbol{\theta}}(\mathbf{X}|\mathbf{Y}) = \Sigma_{12}\Sigma_{22}^{-1}\mathbf{Y}$ and $\Sigma_{11|2}(\boldsymbol{\theta}) = \Sigma_{11} - \Sigma_{12}\Sigma_{22}^{-1}\Sigma_{21}$ (see Proposition A.3.1). Using Problem A.8, we have

$$E_{\boldsymbol{\theta}^{(i)}}\left[(\mathbf{X}', \mathbf{Y}')\Sigma^{-1}(\boldsymbol{\theta})(\mathbf{X}', \mathbf{Y}')'|\mathbf{Y}\right] = \operatorname{trace}\left(\Sigma_{11|2}(\boldsymbol{\theta}^{(i)})\Sigma_{11|2}^{-1}(\boldsymbol{\theta})\right) + \hat{\mathbf{W}}'\Sigma^{-1}(\boldsymbol{\theta})\hat{\mathbf{W}},$$

where $\hat{\mathbf{W}} = (\hat{\mathbf{X}}', \mathbf{Y}')'$. It follows that

$$Q(\boldsymbol{\theta}|\boldsymbol{\theta}^{(i)}) = \ell(\boldsymbol{\theta}, \hat{\mathbf{W}}) - \frac{1}{2}\operatorname{trace}\left(\Sigma_{11|2}(\boldsymbol{\theta}^{(i)})\Sigma_{11|2}^{-1}(\boldsymbol{\theta})\right).$$

The first term on the right is the log-likelihood based on the complete data, but with $\mathbf{X}$ replaced by its "best estimate," $\hat{\mathbf{X}}$, calculated from the previous iteration. If the increments $\boldsymbol{\theta}^{(i+1)} - \boldsymbol{\theta}^{(i)}$ are small, then the second term on the right is nearly constant ($\approx n - r$) and can be ignored. For ease of computation in this application, we shall use the modified version,

$$\tilde{Q}(\boldsymbol{\theta}|\boldsymbol{\theta}^{(i)}) = \ell(\boldsymbol{\theta}; \hat{\mathbf{W}}).$$

With this adjustment, the steps in the EM algorithm are as follows:

E-step. Calculate $E_{\theta^{(i)}}(\mathbf{X}|\mathbf{Y})$ (e.g., with the Kalman fixed-point smoother) and form $\ell(\boldsymbol{\theta}; \hat{\mathbf{W}})$.

M-step. Find the maximum likelihood estimator for the "complete" data problem, i.e., maximize $\ell(\boldsymbol{\theta} : \hat{\mathbf{W}})$. For ARMA processes, PEST can be used directly, with the missing values replaced with their best estimates computed in the E-step.

Example 8.7.1 The lake data

It was found in Example 5.2.5 that the AR(2) model

$$W_t - 1.0415W_{t-1} + 0.2494W_{t-2} = Z_t, \quad \{Z_t\} \sim \text{WN}(0, .4790)$$

was a good fit to the mean-corrected lake data $\{W_t\}$. To illustrate the use of the EM algorithm for missing data, consider fitting an AR(2) model to the mean-corrected data assuming there are 10 missing values at times $t = 17, 24, 31, 38, 45, 52, 59, 66, 73$, and 80. We start the algorithm at iteration 0 with $\hat{\phi}_1^{(0)} = \hat{\phi}_2^{(0)} = 0$. Since this initial model represents white noise, the first E-step gives, in the notation used above, $\hat{W}_{17} = \cdots = \hat{W}_{80} = 0$. Replacing the "missing" values of the mean-corrected lake data with 0 and fitting a mean-zero AR(2) model to the resulting complete data set using the maximum likelihood option in PEST, we find that $\hat{\phi}_1^{(1)} = .7252$, $\hat{\phi}_2^{(1)} = .0236$. (Examination of the plots of the ACF and PACF of this new data set suggests an AR(1) as a better model. This is also borne out by the small estimated value of ϕ_2.) The updated missing values at times $t = 17, 24, \ldots, 80$ are found (see Section 8.6 and Problem 8.21) by minimizing

$$\sum_{j=0}^{2}(W_{t+j} - \hat{\phi}_1^{(1)}W_{t+j-1} - \hat{\phi}_2^{(1)}W_{t+j-2})^2$$

with respect to W_t. The solution is given by

$$\hat{W}_t = \left(\hat{\phi}_2^{(1)}(W_{t-2} + W_{t+2}) + (\hat{\phi}_1^{(1)} - \hat{\phi}_1^{(1)}\hat{\phi}_2^{(1)})(W_{t-1} + W_{t+1})\right) \Big/ (1 + (\hat{\phi}_1^{(1)})^2 + (\hat{\phi}_2^{(1)})^2).$$

The M-step of iteration 1 is then carried out by fitting an AR(2) model using PEST applied to the updated data set. As seen in the summary of the results reported in Table 8.1, the EM algorithm converges in four iterations with the final parameter estimates reasonably close to the fitted model based on the complete data set. (In Table 8.1, estimates of the missing values are only recorded for the first three.) Also notice how $-2\ell(\theta^{(i)}, \mathbf{W})$ decreases at every iteration. The standard errors of the parameter estimates produced from the last iteration of PEST are based on a "complete" ' data set and, as such, underestimate the true sampling errors. Formulae for adjusting the

Table 8.1 Estimates of the missing observations at times $t = 17$, 24, 31 and the AR estimates using the EM algorithm in Example 8.7.1.

iteration i	$\hat{W}_{17}$	$\hat{W}_{24}$	$\hat{W}_{31}$	$\hat{\phi}_1^{(i)}$	$\hat{\phi}_2^{(i)}$	$-2\ell(\theta^{(i)}, \mathbf{W})$
0				0	0	322.60
1	0	0	0	.7252	.0236	244.76
2	.534	.205	.746	1.0729	−.2838	203.57
3	.458	.393	.821	1.0999	−.3128	202.25
4	.454	.405	.826	1.0999	−.3128	202.25

standard errors to reflect the true sampling error based on the observed data can be found in Dempster, Laird, and Rubin (1977). □

8.8 Generalized State-Space Models

As in Section 8.1, we consider a sequence of state-variables $\{X_t, t \geq 1\}$ and a sequence of observations $\{Y_t, t \geq 1\}$. For simplicity, we only consider one-dimensional state and observation variables since extensions to higher dimensions can be carried out with little change. Throughout this section, it will be convenient to write $\mathbf{Y}^{(t)}$ and $\mathbf{X}^{(t)}$ for the t–dimensional column vectors $\mathbf{Y}^{(t)} = (Y_1, Y_2, \ldots, Y_t)'$ and $\mathbf{X}^{(t)} = (X_1, X_2, \ldots, X_t)'$.

There are two important types of state-space models, "parameter driven" and "observation driven," both of which are frequently used in time series analysis. The observation equation is the same for both, but the state vectors of a parameter driven model evolve independently of the past history of the observation process, while the state vectors of an observation driven model depend on past observations.

8.8.1 Parameter-Driven Models

In place of the observation and state equations (8.1.1) and (8.1.2), we now make the assumptions that Y_t given $(X_t, \mathbf{X}^{(t-1)}, \mathbf{Y}^{(t-1)})$ is independent of $(\mathbf{X}^{(t-1)}, \mathbf{Y}^{(t-1)})$ with conditional probability density,

$$p(y_t|x_t) := p(y_t|x_t, \mathbf{x}^{(t-1)}, \mathbf{y}^{(t-1)}), \quad t = 1, 2, \ldots, \tag{8.8.1}$$

and that X_{t+1} given $(X_t, \mathbf{X}^{(t-1)}, \mathbf{Y}^{(t)})$ is independent of $(\mathbf{X}^{(t-1)}, \mathbf{Y}^{(t)})$ with conditional density function,

$$p(x_{t+1}|x_t) := p(x_{t+1}|x_t, \mathbf{x}^{(t-1)}, \mathbf{y}^{(t)}) \quad t = 1, 2, \ldots. \tag{8.8.2}$$

We shall also assume that the initial state X_1 has probability density p_1. The joint density of the observation and state variables can be computed directly from (8.8.1)–(8.8.2) as

$$\begin{aligned} p(y_1, \ldots, y_n, x_1, \ldots, x_n) &= p(y_n|x_n, \mathbf{x}^{(n-1)}, \mathbf{y}^{(n-1)})p(x_n, \mathbf{x}^{(n-1)}, \mathbf{y}^{(n-1)}) \\ &= p(y_n|x_n)p(x_n|\mathbf{x}^{(n-1)}, \mathbf{y}^{(n-1)})p(\mathbf{y}^{(n-1)}, \mathbf{x}^{(n-1)}) \\ &= p(y_n|x_n)p(x_n|x_{n-1})p(\mathbf{y}^{(n-1)}, \mathbf{x}^{(n-1)}) \\ &= \cdots \\ &= \left(\prod_{j=1}^{n} p(y_j|x_j)\right)\left(\prod_{j=2}^{n} p(x_j|x_{j-1})\right) p_1(x_1), \end{aligned}$$

and since (8.8.2) implies that $\{X_t\}$ is Markov (see Problem 8.22),

$$p(y_1, \ldots, y_n|x_1, \ldots, x_n) = \left(\prod_{j=1}^{n} p(y_j|x_j)\right). \tag{8.8.3}$$

We conclude that $Y_1, \ldots, Y_n$ are conditionally independent given the state variables $X_1, \ldots, X_n$, so that the dependence structure of $\{Y_t\}$ is inherited from that of the state-process $\{X_t\}$. The sequence of state-variables $\{X_t\}$ is often referred to as the **hidden** or **latent** generating process associated with the observed process.

In order to solve the **filtering** and **prediction** problems in this setting, we shall determine the conditional densities $p(x_t|\mathbf{y}^{(t)})$ of X_t given $\mathbf{Y}^{(t)}$, and $p(x_t|\mathbf{y}^{(t-1)})$ of X_t given $\mathbf{Y}^{(t-1)}$, respectively. The minimum mean squared error estimates of X_t based on $\mathbf{Y}^{(t)}$ and $\mathbf{Y}^{(t-1)}$ can then be computed as the conditional expectations, $E(X_t|\mathbf{Y}^{(t)})$ and $E(X_t|\mathbf{Y}^{(t-1)})$.

An application of Bayes's theorem, using the assumption that the distribution of Y_t given $(X_t, \mathbf{X}^{(t-1)}, \mathbf{Y}^{(t-1)})$ does not depend on $(\mathbf{X}^{(t-1)}, \mathbf{Y}^{(t-1)})$, yields

$$p(x_t|\mathbf{y}^{(t)}) = p(y_t|x_t)p(x_t|\mathbf{y}^{(t-1)})/p(y_t|\mathbf{y}^{(t-1)}) \tag{8.8.4}$$

and

$$p(x_{t+1}|\mathbf{y}^{(t)}) = \int p(x_t|\mathbf{y}^{(t)})p(x_{t+1}|x_t)\, d\mu(x_t). \tag{8.8.5}$$

(The integral relative to $d\mu(x_t)$ in (8.8.4) is interpreted as the integral relative to dx_t in the continuous case and as the sum over all values of x_t in the discrete case.) The initial condition needed to solve these recursions is

$$p(x_1|\mathbf{y}^{(0)}) := p_1(x_1). \tag{8.8.6}$$

The factor $p(y_t|\mathbf{y}^{(t-1)})$ appearing in the denominator of (8.8.4) is just a scale factor, determined by the condition $\int p(x_t|\mathbf{y}^{(t)})\,d\mu(x_t) = 1$. In the generalized state-space setup, prediction of a future state variable is less important than forecasting a future value of the observations. The relevant forecast density can be computed from (8.8.5) as

$$p(y_{t+1}|\mathbf{y}^{(t)}) = \int p(y_{t+1}|x_{t+1})p(x_{t+1}|\mathbf{y}^{(t)})\,d\mu(x_{t+1}). \tag{8.8.7}$$

Equations (8.8.1)–(8.8.2) can be regarded as a Bayesian model specification. A classical Bayesian model has two key assumptions. The first is that the data $Y_1, \ldots, Y_t$, given an unobservable parameter ($\mathbf{X}^{(t)}$ in our case), are independent with specified conditional distribution. This corresponds to (8.8.3). The second specifies a **prior distribution** for the parameter-value. This corresponds to (8.8.2). The **posterior distribution** is then the conditional distribution of the parameter given the data. In the present setting the posterior distribution of the component X_t of $\mathbf{X}^{(t)}$ is determined by the solution (8.8.4) of the filtering problem.

Example 8.8.1 Consider the simplified version of the linear state-space model of Section 8.1,

$$Y_t = GX_t + W_t, \quad \{W_t\} \sim \text{IID N}(0, R), \tag{8.8.8}$$

$$X_{t+1} = FX_t + V_t, \quad \{V_t\} \sim \text{IID N}(0, Q), \tag{8.8.9}$$

where the noise sequences $\{W_t\}$ and $\{V_t\}$ are independent of one another. For this model, the probability densities in (8.8.1)–(8.8.2) become

$$p_1(x_1) = n(x_1; EX_1, \text{Var}(X_1)), \tag{8.8.10}$$

$$p(y_t|x_t) = n(y_t; Gx_t, R), \tag{8.8.11}$$

$$p(x_{t+1}|x_t) = n(x_{t+1}; Fx_t, Q), \tag{8.8.12}$$

where $n(x; \mu, \sigma^2)$ is the normal density with mean μ and variance σ^2 defined in Example (a) of Section A.1.

To solve the filtering and prediction problems in this new framework, we first observe that the filtering and prediction densities in (8.8.4) and (8.8.5) are both normal. We shall write them, using the notation of Section 8.4, as

$$p(x_t|\mathbf{Y}^{(t)}) = n(x_t; X_{t|t}, \Omega_{t|t}) \tag{8.8.13}$$

and

$$p(x_{t+1}|\mathbf{Y}^{(t)}) = n(x_{t+1}; \hat{X}_{t+1}, \Omega_{t+1}). \tag{8.8.14}$$

From (8.8.5), (8.8.12), (8.8.13), and (8.8.14), we find that

$$\begin{aligned}
\hat{X}_{t+1} &= \int_{-\infty}^{\infty} x_{t+1} p(x_{t+1}|\mathbf{Y}^{(t)}) dx_{t+1} \\
&= \int_{-\infty}^{\infty} x_{t+1} \int_{-\infty}^{\infty} p(x_t|\mathbf{Y}^{(t)}) p(x_{t+1}|x_t)\, dx_t\, dx_{t+1} \\
&= \int_{-\infty}^{\infty} p(x_t|\mathbf{Y}^{(t)}) \left[\int_{-\infty}^{\infty} x_{t+1} p(x_{t+1}|x_t)\, dx_{t+1} \right] dx_t \\
&= \int_{-\infty}^{\infty} F x_t p(x_t|\mathbf{Y}^{(t)})\, dx_t \\
&= F X_{t|t}
\end{aligned}$$

and (see Problem 8.23)

$$\Omega_{t+1} = F^2 \Omega_{t|t} + Q.$$

Substituting the corresponding densities (8.8.11) and (8.8.14) into (8.8.4), we find by equating the coefficient of x_t^2 on both sides of (8.8.4) that

$$\begin{aligned}
\Omega_{t|t}^{-1} &= G^2 R^{-1} + \Omega_t^{-1} \\
&= G^2 R^{-1} + (F^2 \Omega_{t-1|t-1} + Q)^{-1}
\end{aligned}$$

and

$$X_{t|t} = \hat{X}_t + \Omega_{t|t} G R^{-1} (Y_t - G \hat{X}_t).$$

Also, from (8.8.4) with $p(x_1|\mathbf{y}^{(0)}) = n(x_1; EX_1, \Omega_1)$ we obtain the initial conditions

$$X_{1|1} = EX_1 + \Omega_{1|1} G R^{-1} (Y_1 - EX_1)$$

and

$$\Omega_{1|1}^{-1} = G^2 R^{-1} + \Omega_1^{-1}.$$

The Kalman prediction and filtering recursions of Section 8.4 give the same results for $\hat{X}_t$ and $X_{t|t}$, since for Gaussian systems best linear mean-square estimation is equivalent to best mean-square estimation. □

Example 8.8.2 A non-Gaussian example

In general the solution of the recursions (8.8.4) and (8.8.5) presents substantial computational problems. Numerical methods for dealing with non-Gaussian models are discussed by Sorenson and Alspach (1971) and Kitagawa (1987). Here we shall illustrate the recursions (8.8.4) and (8.8.5) in a very simple special case. Consider the state equation

$$X_t = a X_{t-1}, \tag{8.8.15}$$

with observation density

$$p(y_t|x_t) = \frac{(\pi x_t)^{y_t} e^{-\pi x_t}}{y_t!}, \quad y_t = 0, 1, \ldots, \tag{8.8.16}$$

where π is a constant between 0 and 1. The relationship in (8.8.15) implies that the transition density (in the discrete sense—see the comment after (8.8.5)) for the state variables is

$$p(x_{t+1}|x_t) = \begin{cases} 1, & \text{if } x_{t+1} = ax_t, \\ 0, & \text{otherwise.} \end{cases}$$

We shall assume that X_1 has the gamma density function,

$$p_1(x_1) = g(x_1; \alpha, \lambda) = \frac{\lambda^\alpha x_1^{\alpha-1} e^{-\lambda x_1}}{\Gamma(\alpha)}, \quad x_1 > 0.$$

(This is a simplified model for the evolution of the number X_t of individuals at time t infected with a rare disease in which X_t is treated as a continuous rather than an integer-valued random variable. The observation Y_t represents the number of infected individuals observed in a random sample consisting of a small fraction π of the population at time t.) Because the transition distribution of $\{X_t\}$ is not continuous, we use the integrated version of (8.8.5) to compute the prediction density. Thus,

$$\begin{aligned} P(X_t \le x|\mathbf{y}^{(t-1)}) &= \int_0^\infty P(X_t \le x|x_{t-1}) p(x_{t-1}|\mathbf{y}^{(t-1)})\, dx_{t-1} \\ &= \int_0^{x/a} p(x_{t-1}|\mathbf{y}^{(t-1)})\, dx_{t-1}. \end{aligned}$$

Differentiation with respect to x gives

$$p(x_t|\mathbf{y}^{(t-1)}) = a^{-1} p_{X_{t-1}|\mathbf{Y}^{(t-1)}}(a^{-1}x_t|\mathbf{y}^{(t-1)}). \tag{8.8.17}$$

Now applying (8.8.4), we find that

$$\begin{aligned} p(x_1|y_1) &= p(y_1|x_1)p_1(x_1)/p(y_1) \\ &= \left(\frac{(\pi x_1)^{y_1} e^{-\pi x_1}}{y_1!}\right)\left(\frac{\lambda^\alpha x_1^{\alpha-1} e^{-\lambda x_1}}{\Gamma(\alpha)}\right)\left(\frac{1}{p(y_1)}\right) \\ &= c(y_1) x_1^{\alpha+y_1-1} e^{-(\pi+\lambda)x_1}, \quad x_1 > 0, \end{aligned}$$

where $c(y_1)$ is an integration factor ensuring that $p(\cdot|y_1)$ integrates to 1. Since $p(\cdot|y_1)$ has the form of a gamma density, we deduce (see Example (d) of Section A.1) that

$$p(x_1|y_1) = g(x_1; \alpha_1, \lambda_1) \tag{8.8.18}$$

where $\alpha_1 = \alpha + y_1$ and $\lambda_1 = \lambda + \pi$. The prediction density, calculated from (8.8.5) and (8.8.18), is

$$\begin{aligned} p(x_2|\mathbf{y}^{(1)}) &= a^{-1} p_{X_1|\mathbf{Y}^{(1)}}(a^{-1}x_2|\mathbf{y}^{(1)}) \\ &= a^{-1} g(a^{-1}x_2; \alpha_1, \lambda_1) \\ &= g(x_2; \alpha_1, \lambda_1/a). \end{aligned}$$

Iterating the recursions (8.8.4) and (8.8.5) and using (8.8.17), we find that for $t \geq 1$,

$$p(x_t|\mathbf{y}^{(t)}) = g(x_t; \alpha_t, \lambda_t) \tag{8.8.19}$$

and

$$\begin{aligned} p(x_{t+1}|\mathbf{y}^{(t)}) &= a^{-1} g(a^{-1}x_{t+1}; \alpha_t, \lambda_t) \\ &= g(x_{t+1}; \alpha_t, \lambda_t/a), \end{aligned} \tag{8.8.20}$$

where $\alpha_t = \alpha_{t-1} + y_t = \alpha + y_1 + \cdots + y_t$ and $\lambda_t = \lambda_{t-1}/a + \pi = \lambda a^{1-t} + \pi(1 - a^{-t})/(1 - a^{-1})$. In particular the minimum mean squared error estimate of x_t based on $\mathbf{y}^{(t)}$ is the conditional expectation α_t/λ_t with conditional variance α_t/λ_t^2. From (8.8.7) the probability density of Y_{t+1} given $\mathbf{Y}^{(t)}$ is

$$\begin{aligned} p(y_{t+1}|\mathbf{y}^{(t)}) &= \int_0^\infty \left(\frac{(\pi x_{t+1})^{y_{t+1}} e^{-\pi x_{t+1}}}{y_{t+1}!} \right) g(x_{t+1}; \alpha_t, \lambda_t/a)\, dx_{t+1} \\ &= \frac{\Gamma(\alpha_t + y_{t+1})}{\Gamma(\alpha_t)\Gamma(y_{t+1} + 1)} (1 - \frac{\pi}{\lambda_{t+1}})^{\alpha_t} (\frac{\pi}{\lambda_{t+1}})^{y_{t+1}}, \\ &= nb(y_{t+1}; \alpha_t, 1 - \pi/\lambda_{t+1}), \quad y_{t+1} = 0, 1, \ldots, \end{aligned}$$

where $nb(y; \alpha, p)$ is the negative-binomial density defined in example (i) of Section A.1. Conditional on $\mathbf{Y}^{(t)}$, the best one-step predictor of Y_{t+1} is therefore the mean, $\alpha_t \pi/(\lambda_{t+1} - \pi)$, of this negative binomial distribution. The conditional mean squared error of the predictor is $\mathrm{Var}(Y_{t+1}|\mathbf{Y}^{(t)}) = \alpha_t \pi \lambda_{t+1}/(\lambda_{t+1} - \pi)^2$ (see Problem 8.25). □

Example 8.8.3 A model for time series of counts

We often encounter time series in which the observations represent count data. One such example is the monthly number of newly recorded cases of poliomyelitis in the U.S. for the years 1970–1983 plotted in Figure 8.6. Unless the actual counts are large and can be approximated by continuous variables, Gaussian and linear time series models are generally inappropriate for analyzing such data. The parameter-driven specification provides a flexible class of models for modelling count data. We now discuss a specific model based on a Poisson observation density. This model is similar to the one presented by Zeger (1988) for analyzing the polio data. The observation density is assumed to be Poisson with mean $\exp\{x_t\}$, i.e.,

$$p(y_t|x_t) = \frac{e^{x_t y_t} e^{-e^{x_t}}}{y_t!}, \qquad y_t = 0, 1, \ldots, \tag{8.8.21}$$

while the state-variables are assumed to follow a regression model with Gaussian AR(1) noise. If $\mathbf{u}_t = (u_{t1}, \ldots, u_{tk})'$ are the regression variables, then

$$X_t = \boldsymbol{\beta}'\mathbf{u}_t + W_t \tag{8.8.22}$$

where $\boldsymbol{\beta}$ is a k-dimensional regression parameter and

$$W_t = \phi W_{t-1} + Z_t, \quad \{Z_t\} \sim \text{IID N}(0, \sigma^2).$$

The transition density function for the state-variables is then

$$p(x_{t+1}|x_t) = n(x_{t+1}; \boldsymbol{\beta}'\mathbf{u}_{t+1} + \phi(x_t - \boldsymbol{\beta}'\mathbf{u}_t), \sigma^2). \tag{8.8.23}$$

The case $\sigma^2 = 0$ corresponds to a log-linear model with Poisson noise.

Estimation of the parameters $\boldsymbol{\theta} = (\boldsymbol{\beta}', \phi, \sigma^2)'$ in the model by direct numerical maximization of the likelihood function is difficult since the likelihood cannot be written down in closed form. (From (8.8.3) the likelihood is the n-fold integral,

$$\int_{-\infty}^{\infty} \cdots \int_{-\infty}^{\infty} \exp\left\{\sum_{t=1}^{n}(x_t y_t - e^{x_t})\right\} L(\boldsymbol{\theta}; \mathbf{x}^{(n)})\,(dx_1 \cdots dx_n) \Big/ \prod_{i=1}^{n}(y_i!),$$

where $L(\boldsymbol{\theta}; \mathbf{x})$ is the likelihood based on $X_1, \ldots, X_n$.) To overcome this difficulty, Chan and Ledolter (1995) proposed an algorithm, called Monte Carlo EM (MCEM), whose iterates $\theta^{(i)}$ converge to the maximum likelihood estimate. To apply this algorithm, first note that the conditional distribution of $\mathbf{Y}^{(n)}$ given $\mathbf{X}^{(n)}$ does not depend

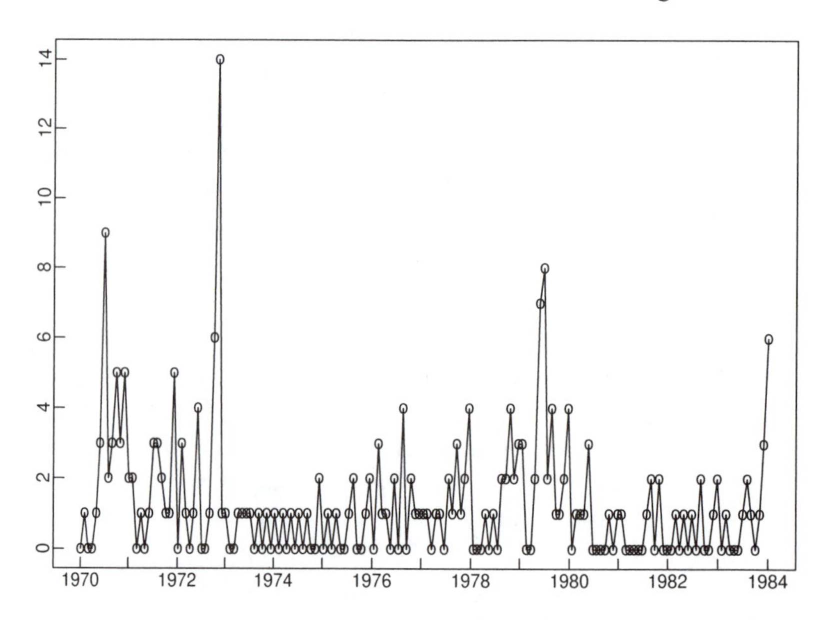

Figure 8-6 Monthly number of U.S. cases of polio, Jan. '70–Dec. '83.

on $\boldsymbol{\theta}$ so that the likelihood based on the complete data $(\mathbf{X}^{(n)\prime}, \mathbf{Y}^{(n)\prime})'$ is given by

$$L(\boldsymbol{\theta}; \mathbf{X}^{(n)}, \mathbf{Y}^{(n)}) = f(\mathbf{Y}^{(n)}|\mathbf{X}^{(n)})L(\boldsymbol{\theta}; \mathbf{X}^{(n)}).$$

The E-step of the algorithm (see Section 8.7) requires calculation of

$$\begin{aligned} Q(\boldsymbol{\theta}|\boldsymbol{\theta}^{(i)}) &= E_{\boldsymbol{\theta}^{(i)}}\left(\ln L(\boldsymbol{\theta}; \mathbf{X}^{(n)}, \mathbf{Y}^{(n)})|\mathbf{Y}^{(n)}\right) \\ &= E_{\boldsymbol{\theta}^{(i)}}\left(\ln f(\mathbf{Y}^{(n)}|\mathbf{X}^{(n)})|\mathbf{Y}^{(n)}\right) + E_{\boldsymbol{\theta}^{(i)}}\left(\ln L(\boldsymbol{\theta}; \mathbf{X}^{(n)})|\mathbf{Y}^{(n)}\right). \end{aligned}$$

We delete the first term from the definition of Q, since it is independent of $\boldsymbol{\theta}$ and hence plays no role in the M-step of the EM algorithm. The new Q is redefined as

$$Q(\boldsymbol{\theta}|\boldsymbol{\theta}^{(i)}) = E_{\boldsymbol{\theta}^{(i)}}\left(\ln L(\boldsymbol{\theta}; \mathbf{X}^{(n)})|\mathbf{Y}^{(n)}\right). \tag{8.8.24}$$

Even with this simplification, direct calculation of Q is still intractable. Suppose for the moment that it is possible to generate replicates of $\mathbf{X}^{(n)}$ from the conditional distribution of $\mathbf{X}^{(n)}$ given $\mathbf{Y}^{(n)}$ when $\boldsymbol{\theta} = \boldsymbol{\theta}^{(i)}$. If we denote m independent replicates of $\mathbf{X}^{(n)}$ by $\mathbf{X}_1^{(n)}, \ldots, \mathbf{X}_m^{(n)}$, then a Monte Carlo approximation to Q in (8.8.24) is given by

$$Q_m(\boldsymbol{\theta}|\boldsymbol{\theta}^{(i)}) = \frac{1}{m}\sum_{j=1}^{m} \ln L(\boldsymbol{\theta}; \mathbf{X}_j^{(n)}).$$

The M-step is easy to carry out using Q_m in place of Q (especially if we condition on $X_1 = 0$ in all the simulated replicates) since L is just the Gaussian likelihood of the regression model with AR(1) noise treated in Section 6.6. The difficult steps in the algorithm are the generation of replicates of $\mathbf{X}^{(n)}$ given $\mathbf{Y}^{(n)}$ and the choice of m. Chan and Ledolter (1995) discuss the use of the Gibb's sampler for generating the desired replicates and give some guidelines on the choice of m.

In their analyses of the polio data, Zeger (1988) and Chan and Ledolter (1995) included as regression components an intercept, a slope, and harmonics at periods of 6 and 12 months. Specifically, they took

$$\mathbf{u}_t = (1, t/1000, \cos(2\pi t/12), \sin(2\pi t/12), \cos(2\pi t/6), \sin(2\pi t/6))'.$$

The implementation of Chan and Ledolter's MCEM method by Kuk and Cheng (1994) gave estimates $\hat{\beta} = (.247, -3.871, .162, -.482, .414, -.011)'$, $\hat{\phi} = .648$ and $\hat{\sigma}^2 = .281$. The estimated trend function $\hat{\beta}'\mathbf{u}_t$ is displayed in Figure 8.7. The negative coefficient of $t/1000$ indicates a slight downward trend in the monthly number of polio cases. □

8.8.2 Observation-Driven Models

In an observation-driven model it is again assumed that Y_t, conditional on $(X_t, \mathbf{X}^{(t-1)}, \mathbf{Y}^{(t-1)})$, is independent of $(\mathbf{X}^{(t-1)}, \mathbf{Y}^{(t-1)})$. The model is specified by the conditional densities

$$p(y_t|x_t) = p(y_t|\mathbf{x}^{(t)}, \mathbf{y}^{(t-1)}), \quad t = 1, 2, \ldots, \tag{8.8.25}$$

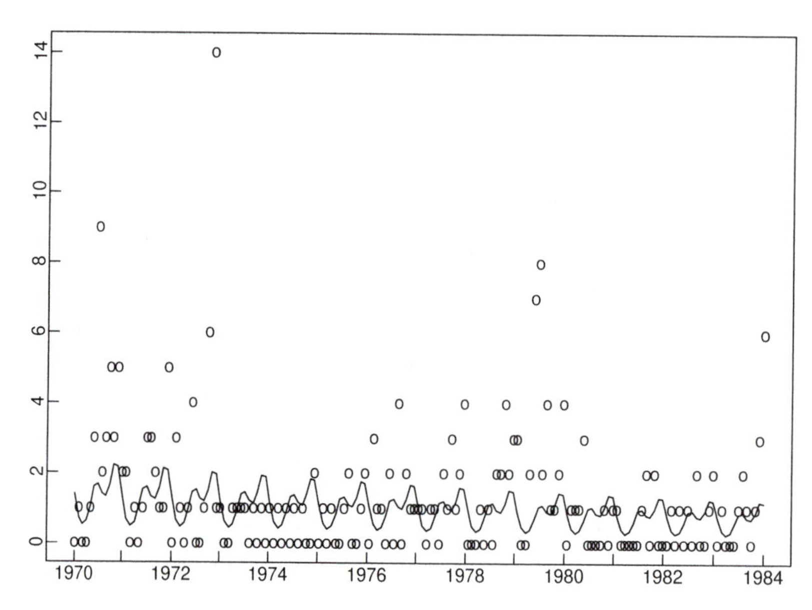

Figure 8-7
Trend estimate for the monthly number of U.S. cases of polio, Jan. '70–Dec. '83.

$$p(x_{t+1}|\mathbf{y}^{(t)}) = p_{X_{t+1}|\mathbf{Y}^{(t)}}(x_{t+1}|\mathbf{y}^{(t)}), \quad t = 0, 1, \ldots, \tag{8.8.26}$$

where $p(x_1|\mathbf{y}^{(0)}) := p_1(x_1)$ for some prespecified initial density $p_1(x_1)$. The advantage of the observation driven state equation (8.8.26) is that the posterior distribution of X_t given $\mathbf{Y}^{(t)}$ can be computed directly from (8.8.4) without the use of the updating formula (8.8.5). This then allows for easy computation of the forecast function in (8.8.7) and hence of the joint density function of $(Y_1, \ldots, Y_n)'$,

$$p(y_1, \ldots, y_n) = \prod_{t=1}^{n} p(y_t|\mathbf{y}^{(t-1)}). \tag{8.8.27}$$

On the other hand, the mechanism by which the state X_{t-1} makes the transition to X_t is not explicitly defined. In fact, without further assumptions there may be state-sequences $\{X_t\}$ and $\{X_t^*\}$ with different distributions, for which both (8.8.25) and (8.8.26) hold (see Example 8.8.6). Both sequences, however, lead to the same joint distribution, given by (8.8.27), for $Y_1, \ldots, Y_n$. The ambiguity in the specification of the distribution of the state-variables can be removed by assuming that X_{t+1} given $(\mathbf{X}^{(t)}, \mathbf{Y}^{(t)})$ is independent of $\mathbf{X}^{(t)}$, with conditional distribution (8.8.26), i.e.,

$$p(x_{t+1}|\mathbf{x}^{(t)}, \mathbf{y}^{(t)}) = p_{X_{t+1}|\mathbf{Y}^{(t)}}(x_{t+1}|\mathbf{y}^{(t)}). \tag{8.8.28}$$

With this modification, the joint density of $\mathbf{Y}^{(n)}$ and $\mathbf{X}^{(n)}$ is given by (cf. (8.8.3))

$$\begin{aligned} p(\mathbf{y}^{(n)}, \mathbf{x}^{(n)}) &= p(y_n|x_n)p(x_n|\mathbf{y}^{(n-1)})p(\mathbf{y}^{(n-1)}, \mathbf{x}^{(n-1)}) \\ &= \cdots \\ &= \prod_{t=1}^{n} \left(p(y_t|x_t)p(x_t|\mathbf{y}^{(t-1)})\right). \end{aligned}$$

Example 8.8.4 An AR(1) process

An AR(1) process with IID noise can be expressed as an observation driven model. Suppose $\{Y_t\}$ is the AR(1) process

$$Y_t = \phi Y_{t-1} + Z_t,$$

where $\{Z_t\}$ is an IID sequence of random variables with mean 0 and some probability density function $f(x)$. Then with $X_t := Y_{t-1}$ we have

$$p(y_t|x_t) = f(y_t - \phi x_t)$$

and

$$p(x_{t+1}|\mathbf{y}^{(t)}) = \begin{cases} 1, & \text{if } x_{t+1} = y_t, \\ 0, & \text{otherwise.} \end{cases}$$

□

Example 8.8.5 Suppose the observation-equation density is given by

$$p(y_t|x_t) = \frac{x_t^{y_t} e^{-x_t}}{y_t!}, \quad y_t = 0, 1, \ldots. \tag{8.8.29}$$

and the state-equation (8.8.26) is

$$p(x_{t+1}|\mathbf{y}^{(t)}) = g(x_t; \alpha_t, \lambda_t) \tag{8.8.30}$$

where $\alpha_t = \alpha + y_1 + \cdots + y_t$ and $\lambda_t = \lambda + t$. It is possible to give a parameter-driven specification that gives rise to the same state-equation (8.8.30). Let $\{X_t^*\}$ be the parameter-driven state-variables, where $X_t^* = X_{t-1}^*$ and X_1^* has a gamma distribution with parameters α and λ. (This corresponds to the model in Example 8.8.2 with $\pi = a = 1$.) Then from (8.8.19), we see that $p(x_t^*|\mathbf{y}^{(t)}) = g(x_t^*; \alpha_t, \lambda_t)$, which coincides with the state-equation (8.8.30). If $\{X_t\}$ are the state variables whose joint distribution is specified through (8.8.28), then $\{X_t\}$ and $\{X_t^*\}$ cannot have the same joint distributions. To see this, note that

$$p(x_{t+1}^*|x_t^*) = \begin{cases} 1, & \text{if } x_{t+1}^* = x_t^*, \\ 0, & \text{otherwise,} \end{cases}$$

while

$$p(x_{t+1}|\mathbf{x}^{(t)}, \mathbf{y}^{(t)}) = p(x_{t+1}|\mathbf{y}^{(t)}) = g(x_t; \alpha_t, \lambda_t).$$

If the two sequences had the same joint distribution, then the latter density could only take the values 0 or 1, which contradicts the continuity (as a function of x_t) of this density. □

Exponential Family Models

The exponential family of distributions provides a large and flexible class of distributions for use in the observation equation. The density in the observation equation is said to belong to an **exponential family** (in natural parameterization) if

$$p(y_t|x_t) = \exp\{y_t x_t - b(x_t) + c(y_t)\} \tag{8.8.31}$$

where $b(\cdot)$ is a twice continuously differentiable function and $c(y_t)$ does not depend on x_t. This family includes the normal, exponential, gamma, Poisson, binomial, and many other distributions frequently encountered in statistics. Detailed properties of the exponential family can be found in Barndorff-Nielsen (1978) and an excellent treatment of its use in the analysis of linear models is given by McCullagh and Nelder (1989). We shall need only the following important facts:

$$e^{b(x_t)} = \int \exp\{y_t x_t + c(y_t)\}\, \nu(dy_t), \tag{8.8.32}$$

$$b'(x_t) = E(Y_t|x_t), \tag{8.8.33}$$

$$\begin{aligned} b''(x_t) &= \operatorname{Var}(Y_t|x_t) \\ &:= \int y_t^2 p(y_x|x_t)\, \nu(dy_t) - [b'(x_t)]^2, \end{aligned} \tag{8.8.34}$$

where integration with respect to $\nu(dy_t)$ means integration with respect to dy_t in the continuous case and summation over all values of y_t in the discrete case.

Proof of (8.8.32)–(8.8.34)

The first relation is simply the statement that $p(y_t|x_t)$ integrates to 1. The second relation is established by differentiating both sides of (8.8.32) with respect to x_t and then multiplying through by $e^{-b(x_t)}$ (for justification of the differentiation under the integral sign see Barndorff-Nielson (1978)). The last relation is obtained by differentiating (8.8.32) twice with respect to x_t and simplifying. ■

Example 8.8.6 The Poisson case

If the observation Y_t, given $X_t = x_t$, has a Poisson distribution of the form (8.8.21), then

$$p(y_t|x_t) = \exp\{y_t x_t - e^{x_t} - \ln y_t!\}, \quad y_t = 0, 1, \ldots, \tag{8.8.35}$$

which has the form (8.8.31) with $b(x_t) = e^{x_t}$ and $c(y_t) = -\ln y_t!$. From (8.8.33) we easily find that $E(Y_t|x_t) = b'(x_t) = e^{x_t}$. This parameterization is slightly different from the one used in Examples 8.8.2 and 8.8.5 where the conditional mean of Y_t given x_t was πx_t and not e^{x_t}. For this observation equation, define the family of densities

$$f(x;\alpha,\lambda) = \exp\{\alpha x - \lambda b(x) + A(\alpha,\lambda)\}, \quad -\infty < x < \infty, \tag{8.8.36}$$

where $\alpha > 0$ and $\lambda > 0$ are parameters and $A(\gamma,\lambda) = -\ln\Gamma(\alpha) + \alpha\ln\lambda$. Now consider state densities of the form

$$p(x_{t+1}|\mathbf{y}^{(t)}) = f(x_{t+1};\alpha_{t+1|t},\lambda_{t+1|t}) \tag{8.8.37}$$

where $\alpha_{t+1|t}$ and $\lambda_{t+1|t}$ are, for the moment, unspecified functions of $\mathbf{y}^{(t)}$. (The subscript $t+1|t$ on the parameters is a shorthand way to indicate dependence on the conditional distribution of X_{t+1} given $\mathbf{Y}^{(t)}$.) With this specification of the state-densities, the parameters $\alpha_{t+1|t}$ are related to the best one-step predictor of Y_t through the formula

$$\alpha_{t+1|t}/\lambda_{t+1|t} = \hat{Y}_{t+1} := E(Y_{t+1}|\mathbf{y}^{(t)}). \tag{8.8.38}$$

Proof We have from (8.8.7) and (8.8.33) that

$$\begin{aligned} E(Y_{t+1}|\mathbf{y}^{(t)}) &= \sum_{y_{t+1}=0}^{\infty} \int_{-\infty}^{\infty} y_{t+1} p(y_{t+1}|x_{t+1}) p(x_{t+1}|\mathbf{y}^{(t)})\, dx_{t+1} \\ &= \int_{-\infty}^{\infty} b'(x_{t+1}) p(x_{t+1}|\mathbf{y}^{(t)})\, dx_{t+1}. \end{aligned}$$

Addition and subtraction of $\frac{\alpha_{t+1|t}}{\lambda_{t+1|t}}$ then gives

$$\begin{aligned} E(Y_{t+1}|\mathbf{y}^{(t)}) &= \int_{-\infty}^{\infty} \left(b'(x_{t+1}) - \frac{\alpha_{t+1|t}}{\lambda_{t+1|t}}\right) p(x_{t+1}|\mathbf{y}^{(t)})\, dx_{t+1} + \frac{\alpha_{t+1|t}}{\lambda_{t+1|t}} \\ &= \int_{-\infty}^{\infty} -\lambda_{t+1|t}^{-1} p'(x_{t+1}|\mathbf{y}^{(t)})\, dx_{t+1} + \frac{\alpha_{t+1|t}}{\lambda_{t+1|t}} \\ &= \left[-\lambda_{t+1|t}^{-1} p(x_{t+1}|\mathbf{y}^{(t)})\right]_{x_{t+1}=-\infty}^{x_{t+1}=\infty} + \frac{\alpha_{t+1|t}}{\lambda_{t+1|t}} \\ &= \frac{\alpha_{t+1|t}}{\lambda_{t+1|t}}. \end{aligned}$$

■

Letting $A_{t|t-1} = A(\alpha_{t|t-1}, \lambda_{t|t-1})$, we can write the posterior density of X_t given $\mathbf{Y}^{(t)}$ as

$$\begin{aligned} p(x_t|\mathbf{y}^{(t)}) &= \exp\{y_t x_t - b(x_t) - c(y_t)\} \exp\{\alpha_{t|t-1} x_t - \lambda_{t|t-1} b(x_t) \\ &\quad + A_{t|t-1}\}/p(y_t|\mathbf{y}^{(t-1)}) \\ &= \exp\{\lambda_{t|t}\left(\alpha_{t|t} x_t - b(x_t)\right) - A_{t|t}\}, \\ &= f(x_t; \alpha_t, \lambda_t), \end{aligned}$$

where we find, by equating coefficients of x_t and $b(x_t)$, that the coefficients λ_t and α_t are determined by

$$\lambda_t = 1 + \lambda_{t|t-1} \tag{8.8.39}$$

$$\alpha_t = y_t + \alpha_{t|t-1} \tag{8.8.40}$$

The family of prior densities in (8.8.37) is called a **conjugate family of priors** for the observation equation (8.8.35), since the resulting posterior densities are again members of the same family.

As mentioned earlier, the parameters $\alpha_{t|t-1}$ and $\lambda_{t|t-1}$ can be quite arbitrary—any nonnegative functions of $\mathbf{y}^{(t-1)}$ will lead to a consistent specification of the state densities. One convenient choice is to link these parameters with the corresponding parameters of the posterior distribution at time $t-1$ through the relations

$$\lambda_{t+1|t} = \delta\lambda_t \left(= \delta(1 + \lambda_{t|t-1})\right) \tag{8.8.41}$$

$$\alpha_{t+1|t} = \delta\alpha_t \left(= \delta(y_t + \alpha_{t|t-1})\right) \tag{8.8.42}$$

where $0 < \delta < 1$ (see Remark 4 below). Iterating the relation (8.8.41), we see that

$$\begin{aligned}\lambda_{t+1|t} = \delta(1 + \lambda_{t|t-1}) &= \delta + \delta\lambda_{t|t-1} \\ &= \delta + \delta(\delta + \delta\lambda_{t-2|t-2}) \\ &= \cdots \\ &= \delta + \delta^2 + \cdots + \delta^t + \delta^t\lambda_{1|0} \\ &\to \delta/(1-\delta)\end{aligned} \tag{8.8.43}$$

as $t \to \infty$. Similarly,

$$\begin{aligned}\alpha_{t+1|t} &= \delta y_t + \delta\alpha_{t|t-1} \\ &= \cdots \\ &= \delta y_t + \delta^2 y_{t-1} + \cdots + \delta^t y_1 + \delta^t \alpha_{1|0}.\end{aligned} \tag{8.8.44}$$

For large t, we have the approximations

$$\lambda_{t+1|t} = \delta/(1-\delta) \tag{8.8.45}$$

and

$$\alpha_{t+1|t} = \delta \sum_{j=0}^{t-1} \delta^j y_{t-j}, \tag{8.8.46}$$

which are exact if $\lambda_{1|0} = \delta/(1-\delta)$ and $\alpha_{1|0} = 0$. From (8.8.38) the one-step predictors are linear and given by

$$\hat{Y}_{t+1} = \frac{\alpha_{t+1|t}}{\lambda_{t+1|t}} = \frac{\sum_{j=0}^{t-1} \delta^j y_{t-j} + \delta^{t-1}\alpha_{1|0}}{\sum_{j=0}^{t-1} \delta^j + \delta^{t-1}\lambda_{1|0}}. \tag{8.8.47}$$

Replacing the denominator with its limiting value, or starting with $\lambda_{1|0} = \delta/(1-\delta)$, we find that $\hat{Y}_{t+1}$ is the solution of the recursions

$$\hat{Y}_{t+1} = (1-\delta)y_t + \delta\hat{Y}_t, \quad t = 1, 2, \ldots \tag{8.8.48}$$

with initial condition $\hat{Y}_1 = (1-\delta)\delta\alpha_0$. In other words, under the restrictions of (8.8.41) and (8.8.42), the best one-step predictors can be found by exponential smoothing. □

Remark 1. The preceding analysis for the Poisson-distributed observation equation holds, almost verbatim, for the general family of exponential densities (8.8.31). (One only needs to take care in specifying the correct range for x and the allowable parameter space for α and λ in (8.8.37).) The relations (8.8.43)–(8.8.44), as well as the exponential smoothing formula (8.8.48), continue to hold even in the more general setting provided the parameters $\alpha_{t|t-1}$ and $\lambda_{t|t-1}$ satisfy the relations (8.8.41)–(8.8.42). □

Remark 2. Equations (8.8.41)–(8.8.42) are equivalent to the assumption that the prior density of X_t given $\mathbf{y}^{(t-1)}$ is proportional to the δ-power of the posterior distribution of X_{t-1} given $\mathbf{Y}^{(t-1)}$ or more succinctly that

$$\begin{aligned} f(x_t; \alpha_{t|t-1}, \lambda_{t|t-1}) &= f(x_t; \delta\alpha_{t-1|t-1}, \delta\lambda_{t-1|t-1}) \\ &\propto f^{\delta}(x_t; \alpha_{t-1|t-1}, \lambda_{t-1|t-1}). \end{aligned}$$

This power relationship is sometimes referred to as the **power steady model** (Grunwald, Raftery, and Guttorp, 1993, and Smith, 1979). □

Remark 3. The transformed state-variables $W_t = e^{X_t}$ have a gamma state-density given by

$$p(w_{t+1}|\mathbf{y}^{(t)}) = g(w_{t+1}; \alpha_{t+1|t}, \lambda_{t+1|t})$$

(see Problem 8.26). The mean and variance of this conditional density are

$$E(W_{t+1}|\mathbf{y}^{(t)}) = \alpha_{t+1|t} \quad \text{and} \quad \operatorname{Var}(W_{t+1}|\mathbf{y}^{(t)}) = \alpha_{t+1|t}/\lambda^2_{t+1|t}.$$ □

Remark 4. If we regard the random walk plus noise model of Example 8.2.1 as the prototypical state-space model, then from the calculations in Example 8.8.1 with $G = F = 1$, we have

$$E(X_{t+1}|\mathbf{Y}^{(t)}) = E(X_t|\mathbf{Y}^{(t)})$$

and

$$\operatorname{Var}(X_{t+1}|\mathbf{Y}^{(t)}) = \operatorname{Var}(X_t|\mathbf{Y}^{(t)}) + Q > \operatorname{Var}(X_t|\mathbf{Y}^{(t)}).$$

The first of these equations implies that the best estimate of the next state is the same as the best estimate of the current state while the second implies that the variance increases. Under the conditions (8.8.41), and (8.8.42), the same is also true for the state-variables in the above model (see Problem 8.26). This was, in part, the rationale behind these conditions given in Harvey and Fernandes (1989). □

Remark 5. While the calculations work out neatly for the power steady model, Grunwald, Hyndman, and Hamza (1994) have shown that such processes have degenerate sample paths for large t. In the Poisson example above, they argue that the observations Y_t converge to 0 as $t \to \infty$ (see Figure 8.12). Although, such models may still be useful in practice for modelling series of moderate length, the efficacy of using such models for describing long term behavior is doubtful. □

Example 8.8.7 Goals scored by England against Scotland

The time series of the number of goals scored by England against Scotland in soccer matches played at Hampden Park in Glasgow is graphed in Figure 8.8. The matches are played nearly every second year with interruptions during the war years. We will treat the data, $y_1, \ldots, y_{52}$, as coming from an equally spaced time series model $\{Y_t\}$. Since the number of goals scored is small (see the frequency histogram in Figure 8.9), a model based on the Poisson distribution might be deemed appropriate. The observed relative frequencies and those based on a Poisson distribution with mean equal to $\bar{y}_{52} = 1.269$ is contained in Table 8.2. The standard chi-squared goodness of fit test, comparing the observed frequencies with expected frequencies based on a Poisson model, has a p-value of .02. The lack of fit with a Poisson distribution is hardly unexpected since the sample variance (1.652) is much larger than the sample mean while the mean and variance of the Poisson distribution are equal. In this case the data are said to be *overdispersed* in the sense that there is more variability in the data than one would expect from a sample of independent Poisson-distributed variables. Overdispersion can sometimes be explained by serial dependence in the data.

Dependence in count data can often be revealed by estimating the probabilities of transition from one state to another. Table 8.3 contains estimates of these probabilities, computed as the average number of one-step transitions from state y_t to state y_{t+1}. If the data were independent, then in each column, the entries should be nearly the same. This is certainly not the case in Table 8.3. For example, England is very unlikely to be shut out or score 3 or more goals in the next match after scoring at least 3 goals in the previous encounter.

Harvey and Fernandes (1989) model the dependence in this data using an observation-driven model of the type described in Example 8.8.6. Their model assumes a

Table 8.2 Relative frequency and fitted Poisson distribution of goals scored by England against Scotland

	Number of goals					
	0	1	2	3	4	5
Relative frequency	.288	.423	.154	.019	.096	.019
Poisson distribution	.281	.356	.226	.096	.030	.008

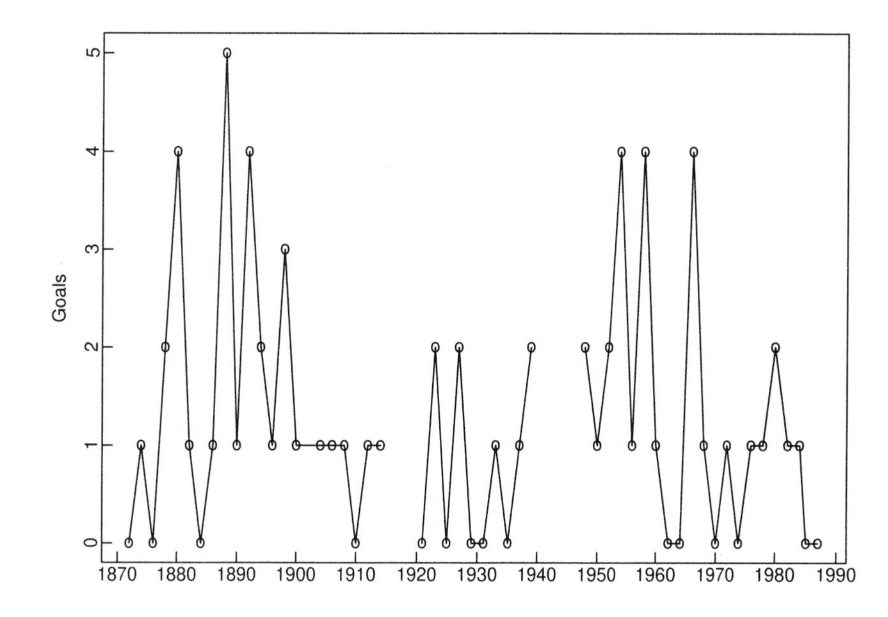

Figure 8-8
Goals scored by England against Scotland at Hampden Park, Glasgow, 1872–1987.

Poisson observation equation and a log-gamma state equation:

$$p(y_t|x_t) = \frac{\exp\{y_t x_t - e^{x_t}\}}{y_t!}, \qquad y_t = 0, 1, \ldots,$$

$$p(x_t|\mathbf{y}^{(t-1)}) = f(x_t; \alpha_{t|t-1}, \lambda_{t|t-1}), \qquad -\infty < x < \infty,$$

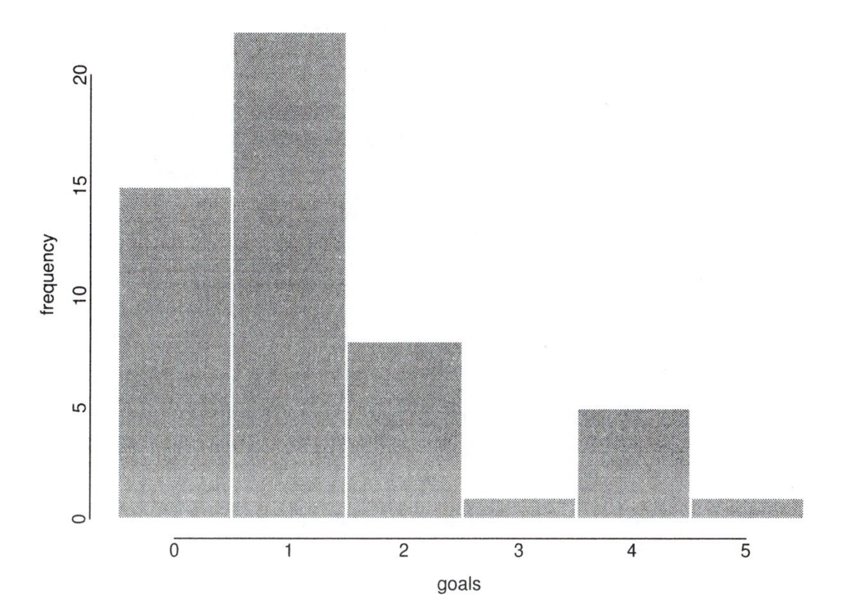

Figure 8-9
Histogram of the data in Figure 8.7.

Table 8.3 Transition probabilities for the number of goals scored by England against Scotland.

		y_{t+1}			
	$p(y_{t+1}\|y_t)$	0	1	2	≥ 3
	0	.214	.500	.214	.072
y_t	1	.409	.272	.136	.182
	2	.250	.375	.125	.250
	≥ 3	0	.857	.143	0

for $t = 1, 2, \ldots$ where f is given by (8.8.36) and $\alpha_{1|0} = 0, \lambda_{1|0} = 0$. The power steady conditions (8.8.41)–(8.8.42) are assumed to hold for $\alpha_{t|t-1}$ and $\lambda_{t|t-1}$. The only unknown parameter in the model is δ. The log-likelihood function for δ based on the conditional distribution of $y_1, \ldots, y_{52}$ given y_1 is given by (see (8.8.27))

$$\ell(\delta, \mathbf{y}^{(n)}) = \sum_{t=1}^{n-1} \ln p(y_{t+1}|\mathbf{y}^{(t)}), \tag{8.8.49}$$

where $p(y_{t+1}|\mathbf{y}^{(t)}))$ is the negative binomial density (see Problem 8.25(c))

$$p(y_{t+1}|\mathbf{y}^{(t)}) = nb(y_{t+1}; \alpha_{t+1|t}, (1 + \lambda_{t+1|t})^{-1}),$$

with $\alpha_{t+1|t}$ and $\lambda_{t+1|t}$ as defined in (8.8.44) and (8.8.43). (For the goal data, $y_1 = 0$, which implies $\alpha_{2|1} = 0$ and hence that $p(y_2|y^{(1)})$ is a degenerate density with unit mass at $y_2 = 0$. Harvey and Fernandes avoid this complication by conditioning the likelihood on $y^{(\tau)}$ where τ is the time of the first nonzero data value.)

Maximizing this likelihood with respect to δ, we obtain $\hat{\delta} = .844$. (Starting the equations (8.8.43)–(8.8.44) with $\alpha_{1|0} = 0$ and $\lambda_{1|0} = \delta/(1 - \delta)$, we obtain $\hat{\delta} = .732$.) Using .844 as our estimate of δ, the prediction density of the next observation Y_{53} given $\mathbf{y}^{(52)}$ is $nb(y_{53}; \alpha_{53|52}, (1 + \lambda_{53|52})^{-1}$. The first five values of this distribution are given in Table 8.4. Under this model, the probability that England will be held scoreless in the next match is .471. The one-step predictors, $\hat{Y}_1 = 0, \hat{Y}_2, \ldots, Y_{52}$ are graphed in Figure 8.10. (This graph can be obtained by using the exponential smoothing option in SMOOTH with parameter $a = .156$.)

Figures 8.11 and 8.12 contain two realizations from the fitted model for the goal data. The general appearance of the first realization is somewhat compatible with the goal data, while the second realization illustrates the convergence of the sample path to 0 in accordance with the result of Grunwald et al. (1994). □

Table 8.4 Prediction density of Y_{53} given $\mathbf{Y}^{(52)}$ for data in Figure 8.7.

	Number of goals					
	0	1	2	3	4	5
$p(y_{53}\|\mathbf{y}^{(52)})$	.472	.326	.138	.046	.013	.004

Example 8.8.8 The exponential case

Suppose Y_t given X_t has an exponential density with mean $-1/X_t$ ($X_t < 0$). The observation density is given by

$$p(y_t|x_t) = \exp\{y_t x_t + \ln(-x_t)\}, \quad y_t > 0,$$

which has the form (8.8.31) with $b(x) = -\ln(-x)$ and $c(y) = 0$. The state-densities corresponding to the family of conjugate priors (see (8.8.37)) are given by

$$p(x_{t+1}|\mathbf{y}^{(t)}) = \exp\{\alpha_{t+1|t}x_{t+1} - \lambda_{t+1|t}b(x_{t+1}) + A_{t+1|t}\}, \quad -\infty < x < 0.$$

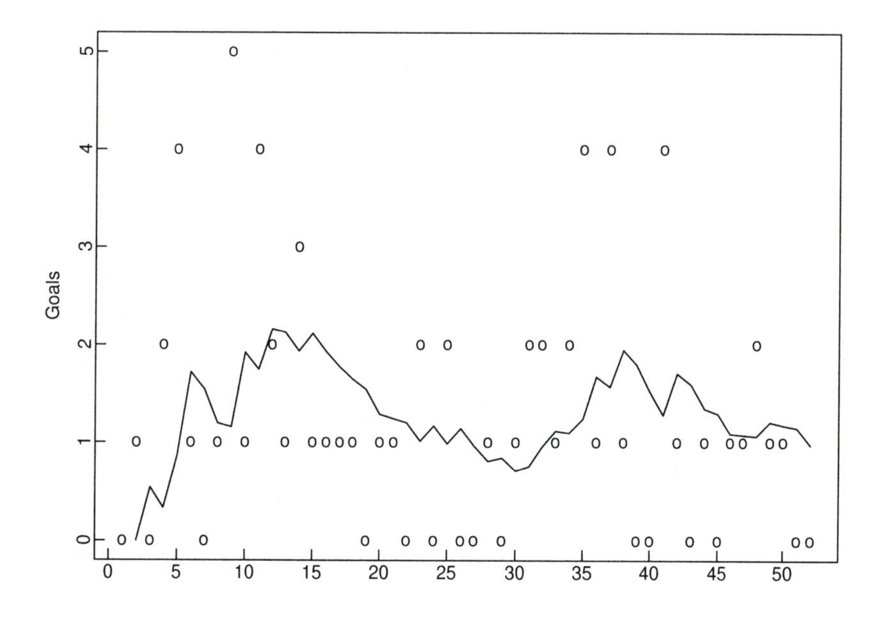

Figure 8-10 One-step predictors of the goal data.

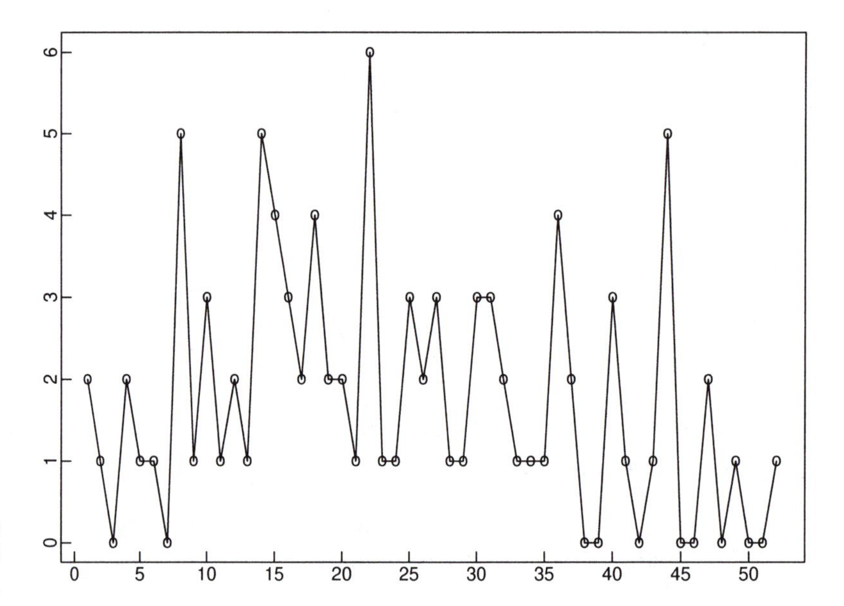

Figure 8-11
A simulated time series from the fitted model to the goal data.

(Here $p(x_{t+1}|\mathbf{y}^{(t)})$ is a probability density when $\alpha_{t+1|t} > 0$ and $\lambda_{t+1|t} > -1$.) The one-step prediction density is

$$p(y_{t+1}|\mathbf{y}^{(t)}) = \int_{-\infty}^{0} e^{x_{t+1}y_{t+1}+\ln(-x_{t+1})+\alpha_{t+1|t}x-\lambda_{t+1|t}b(x)+A_{t+1|t}}\, dx_{t+1}$$

$$= (\lambda_{t+1|t}+1)\alpha_{t+1|t}^{\lambda_{t+1|t}+1}(y_{t+1}+\alpha_{t+1|t})^{-\lambda_{t+1|t}-2},\ y_{t+1} > 0$$

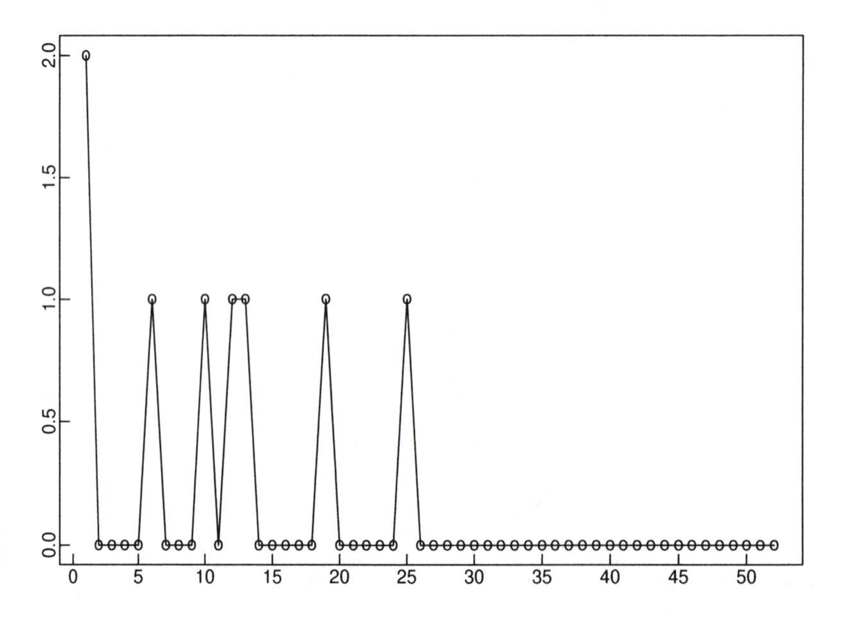

Figure 8-12
A second simulated time series from the fitted model to the goal data.

(see Problem 8.28). While $E(Y_{t+1}|\mathbf{y}^{(t)}) = \alpha_{t+1|t}/\lambda_{t+1|t}$, the conditional variance is finite if and only if $\lambda_{t+1|t} > 1$. Under assumptions (8.8.41)–(8.8.42), and starting with $\lambda_{1|0} = \delta/(1-\delta)$, the exponential smoothing formula (8.8.48) remains valid. □

Problems

8.1. Show that if all the eigenvalues of F are less than 1 in absolute value (or equivalently that $F^k \to 0$ as $k \to \infty$), the unique stationary solution of equation (8.1.11) is given by the infinite series

$$\mathbf{X}_t = \sum_{j=0}^{\infty} F^j V_{t-j-1}$$

and that the corresponding observation vectors are

$$\mathbf{Y}_t = \mathbf{W}_t + \sum_{j=0}^{\infty} G F^j \mathbf{V}_{t-j-1}.$$

Deduce that $\{(\mathbf{X}_t', \mathbf{Y}_t')'\}$ is a multivariate stationary process. (Hint: Use a vector analogue of the argument in Example 2.2.1.)

8.2. In Example 8.2.1, show that $\theta = -1$ if and only if $\sigma_v^2 = 0$, which in turn is equivalent to the signal M_t being constant.

8.3. Let F be the coefficient of $\mathbf{X}_t$ in the state equation (8.3.4) for the causal AR(p) process

$$X_t - \phi_1 X_{t-1} - \cdots - \phi_p X_{t-p} = Z_t, \quad \{Z_t\} \sim \mathrm{WN}(0, \sigma^2).$$

Establish the stability of (8.3.4) by showing that

$$\det(zI - F) = z^p \phi(z^{-1}),$$

and hence that the eigenvalues of F are the reciprocals of the zeros of the autoregressive polynomial $\phi(z) = 1 - \phi_1 z - \cdots - \phi_p z^p$.

8.4. By following the argument in Example 8.3.3, find a state-space model for $\{Y_t\}$ when $\{\nabla\nabla_{12} Y_t\}$ is an ARMA(p, q) process.

8.5. For the local linear trend model defined by equations (8.2.6)–(8.2.7), show that $\nabla^2 Y_t = (1-B)^2 Y_t$ is a 2-correlated sequence and hence, by Proposition 2.1.1, is an MA(2) process. Show that this MA(2) process is noninvertible if $\sigma_u^2 = 0$.

8.6. a. For the seasonal model of Example 8.2.2, show that $\nabla_d Y_t = Y_t - Y_{t-d}$ is an MA(1) process.

b. Show that $\nabla\nabla_{12}Y_t$ is an MA(13) process where $\{Y_t\}$ follows the seasonal model with a local linear trend as described in Example 8.2.3.

8.7. Let $\{Y_t\}$ be the MA(1) process,

$$Y_t = Z_t + \theta Z_{t-1}, \quad \{Z_t\} \sim \mathrm{WN}(0, \sigma^2).$$

Show that $\{Y_t\}$ has the state-space representation

$$Y_t = [1 \quad 0]\mathbf{X}_t,$$

where $\{\mathbf{X}_t\}$ is the unique stationary solution of

$$\mathbf{X}_{t+1} = \begin{bmatrix} 0 & 1 \\ 0 & 0 \end{bmatrix}\mathbf{X}_t + \begin{bmatrix} 1 \\ \theta \end{bmatrix} Z_{t+1}.$$

In particular, show that the state vector $\mathbf{X}_t$ can written as

$$\mathbf{X}_t = \begin{bmatrix} 1 & \theta \\ \theta & 0 \end{bmatrix}\begin{bmatrix} Z_t \\ Z_{t-1} \end{bmatrix}.$$

8.8. Verify equations (8.3.16) – (8.3.18) for an ARIMA(1,1,1) process.

8.9. Consider the two state-space models,

$$\begin{cases} \mathbf{X}_{t+1,1} = F_1\mathbf{X}_{t1} + \mathbf{V}_{t1}, \\ \mathbf{Y}_{t1} \quad = G_1\mathbf{X}_{t1} + \mathbf{W}_{t1}, \end{cases}$$

and

$$\begin{cases} \mathbf{X}_{t+1,2} = F_2\mathbf{X}_{t2} + \mathbf{V}_{t2}, \\ \mathbf{Y}_{t2} \quad = G_2\mathbf{X}_{t2} + \mathbf{W}_{t2}. \end{cases}$$

where $\{(\mathbf{V}'_{t1}, \mathbf{W}'_{t1}, \mathbf{V}'_{t2}, \mathbf{W}'_{t2})'\}$ is white noise. Derive a state-space representation for $\{(\mathbf{Y}'_{t1}, \mathbf{Y}'_{t2})'\}$.

8.10. Use Remark 1 of Section 8.4 to establish the linearity properties of the operator P_t stated in Remark 3.

8.11. a. Show that if the matrix equation $XS = B$ can be solved for X then $X = BS^{-1}$ is a solution for *any* generalized inverse S^{-1} of S.

b. Use the result of (a) to derive the expression for $P(\mathbf{X}|\mathbf{Y})$ in Remark 4 of Section 8.4.

8.12. In the notation of the Kalman prediction equations, show that every vector of the form

$$\mathbf{Y} = A_1\mathbf{X}_1 + \cdots + A_t\mathbf{X}_t,$$

can be expressed as

$$\mathbf{Y} = B_1\mathbf{X}_1 + \cdots + B_{t-1}\mathbf{X}_{t-1} + C_t\mathbf{I}_t,$$

where $B_1, \ldots B_{t-1}$ and C_t are matrices that depend on the matrices $A_1, \ldots, A_t$. Show also that the converse is true. Use these results and the fact that $E(\mathbf{X}_s \mathbf{I}_t) = 0$ for all $s < t$, to establish (8.4.3).

8.13. In Example 8.4.1, verify that the steady-state solution of the Kalman recursions (8.1.2) is given by $\Omega_t = \left(\sigma_v^2 + \sqrt{\sigma_v^4 + 4\sigma_v^2\sigma_w^2}\right)/2$.

8.14. Show from the difference equations for Ω_t in Example 8.4.1 that $(\Omega_{t+1} - \Omega)(\Omega_t - \Omega) \geq 0$ for all $\Omega_t \geq 0$, where Ω is the steady-state solution for Ω_t given in Problem 8.13.

8.15. Show directly that for the MA(1) model (8.2.3), the parameter θ is equal to $-\left(2\sigma_w^2 + \sigma_v^2 - \sqrt{\sigma_v^4 + 4\sigma_v^2\sigma_w^2}\right)/(2\sigma_w^2)$ which in turn is equal to $-\sigma_w^2/(\Omega + \sigma_w^2)$, where Ω is the steady-state solution for Ω_t given in Problem 8.13.

8.16. Use the ARMA(0,1,1) representation of the series $\{Y_t\}$ in Example 8.4.1 to show that the predictors defined by

$$\hat{Y}_{n+1} = aY_n + (1-a)\hat{Y}_n, \quad n = 1, 2, \ldots,$$

where $a = \Omega/(\Omega + \sigma_w^2)$, satisfy

$$Y_{n+1} - \hat{Y}_{n+1} = Z_{n+1} + (1-a)^n(Y_0 - Z_0 - \hat{Y}_1).$$

Deduce that if $0 < a < 1$, the mean squared error of $\hat{Y}_{n+1}$ converges to $\Omega + \sigma_w^2$ for any initial predictor $\hat{Y}_1$ with finite mean squared error.

8.17. a. Using equations (8.4.1) and (8.4.10), show that $\hat{\mathbf{X}}_{t+1} = F_t \mathbf{X}_{t|t}$.

b. From (a) and (8.4.10) show that $\mathbf{X}_{t|t}$ satisfies the recursions

$$\mathbf{X}_{t|t} = F_{t-1}\mathbf{X}_{t-1|t-1} + \Omega_t G_t' \Delta_t^{-1}(\mathbf{Y}_t - G_t F_{t-1}\mathbf{X}_{t-1|t-1})$$

for $t = 2, 3, \ldots$, with $\mathbf{X}_{1|1} = \hat{\mathbf{X}}_1 + \Omega_1 G_1' \Delta_1^{-1}(\mathbf{Y}_1 - G_1\hat{\mathbf{X}}_1)$.

8.18. In Section 8.5, show that for fixed Q^*, $-2\ln L(\mu, Q^*, \sigma_w^2)$ is minimized when μ and σ_w^2 are given by (8.5.10) and (8.5.11), respectively.

8.19. Verify the calculation of $\Theta_t \Delta_t^{-1}$ and Ω_t in Example 8.6.1.

8.20. Verify that the best estimates of missing values in an AR(p) process are found by minimizing (8.6.11) with respect to the missing values.

8.21. Suppose that $\{Y_t\}$ is the AR(2) process

$$Y_t = \phi_1 Y_{t-1} + \phi_2 Y_{t-2} + Z_t, \quad \{Z_t\} \sim \text{WN}(0, \sigma^2)$$

and that we observe $Y_1, Y_2, Y_4, Y_5, Y_6, Y_7$. Show that the best estimator of Y_3 is

$$(\phi_2(Y_1 + Y_5) + (\phi_1 - \phi_1\phi_2)(Y_2 + Y_4))/(1 + \phi_1^2 + \phi_2^2).$$

8.22. Let X_t be the state at time t of a parameter driven model (see (8.8.2)). Show that $\{X_t\}$ is a Markov chain and that (8.8.3) holds.

8.23. For the generalized state space model of Example 8.8.1, show that $\Omega_{t+1} = F^2\Omega_{t|t} + Q$.

8.24. If Y and X are random variables, show that

$$\text{Var}(Y) = E(\text{Var}(Y|X)) + \text{Var}(E(Y|X)).$$

8.25. Suppose that Y and X are two random variables such that the distribution of Y given X is Poisson with mean πX, $0 < \pi \le 1$, and X has the gamma density, $g(x; \alpha, \lambda)$.

a. Show that the posterior distribution of X given Y also has a gamma density and determine its parameters.
b. Compute $E(X|Y)$ and $\text{Var}(X|Y)$.
c. Show that Y has a negative-binomial density and determine its parameters.
d. Use (c) to compute $E(Y)$ and $\text{Var}(Y)$.
e. Verify in Example 8.8.2 that $E(Y_{t+1}|\mathbf{Y}^{(t)}) = \alpha_t\pi/(\lambda_{t+1} - \pi)$ and $\text{Var}(Y_{t+1}|\mathbf{Y}^{(t)}) = \alpha_t\pi\lambda_{t+1}/(\lambda_{t+1} - \pi)^2$.

8.26. For the model of Example 8.8.6, show that

a. $E(X_{t+1}|\mathbf{Y}^{(t)}) = E(X_t|\mathbf{Y}^{(t)})$, $\text{Var}(X_{t+1}|\mathbf{Y}^{(t)}) > \text{Var}(X_t|\mathbf{Y}^{(t)})$, and
b. the transformed sequence $W_t = e^{X_t}$ has a gamma state-density.

8.27. Let $\{V_t\}$ be a sequence of independent exponential random variables with $EV_t = t^{-1}$ and suppose that $\{X_t, t \ge 1\}$ and $\{Y_t, t \ge 1\}$ are the state and observation random variables, respectively, of the parameter driven state-space system

$$X_1 = V_1$$

$$X_t = X_{t-1} + V_t, \quad t = 2, 3, \ldots,$$

where the distribution of the observation Y_t, conditional on the random variables $Y_1, Y_2, \ldots, Y_{t-1}, X_t$, is Poisson with mean X_t.

a. Determine the observation and state transition density functions, $p(y_t|x_t)$ and $p(x_{t+1}|x_t)$, in the parameter driven model for $\{Y_t\}$.
b. Show, using (8.8.4)–(8.8.6), that

$$p(x_1|y_1) = g(x_1; y_1 + 1, 2)$$

and

$$p(x_2|y_1) = g(x_2; y_1 + 2, 2),$$

where $g(x; \alpha, \lambda)$ is the gamma density function (see Example (d) of Section A.1).

c. Show that

$$p(x_t|\mathbf{y}^{(t)}) = g(x_t; \alpha_t + t, t + 1)$$

and

$$p(x_{t+1}|\mathbf{y}^{(t)}) = g(x_{t+1}; \alpha_t + t + 1, t + 1),$$

where $\alpha_t = y_1 + \cdots + y_t$.

d. Conclude from (c) that the minimum mean squared error estimates of X_t and X_{t+1} based on $Y_1, \ldots, Y_t$, are

$$X_{t|t} = \frac{t + Y_1 + \cdots + Y_t}{t + 1}$$

and

$$\hat{X}_{t+1} = \frac{t + 1 + Y_1 + \cdots + Y_t}{t + 1},$$

respectively.

8.28. Let Y and X be two random variables such that Y given X is exponential with mean $1/X$, and X has the gamma density function with

$$g(x; \lambda + 1, \alpha) = \frac{\alpha^{\lambda+1} x^{\lambda} \exp\{-\alpha x\}}{\Gamma(\lambda + 1)} \quad , x > 0,$$

where $\lambda > -1$ and $\alpha > 0$.

a. Determine the posterior distribution of X given Y.

b. Show that Y has a Pareto distribution,

$$p(y) = (\lambda + 1)\alpha^{\lambda+1}(y + \alpha)^{-\lambda-2}, \quad y > 0.$$

c. Find the mean of variance of Y. Under what conditions on α and λ does the latter exist?

d. Verify the calculation of $p(y_{t+1}|\mathbf{y}^{(t)})$ and $E(Y_{t+1}|\mathbf{y}^{(t)})$ for the model in Example 8.8.8.

8.29. Consider an observation driven model in which Y_t given X_t is binomial with parameters n and X_t, i.e.,

$$p(y_t|x_t) = \binom{n}{y_t} x_t^{y_t}(1 - x_t)^{n-y_t}, \quad y_t = 0, 1, \ldots, n.$$

a. Show that the observation equation with state-variable transformed by the logit transformation $W_t = \ln(X_t/(1 - X_t))$ follows an exponential family,

$$p(y_t|w_t) = \exp\{y_t w_t - b(w_t) + c(y_t)\}.$$

Determine the functions $b(\cdot)$ and $c(\cdot)$.

b. Suppose that the state X_t has the beta density

$$p(x_{t+1}|\mathbf{y}^{(t)}) = f(x_{t+1}; \alpha_{t+1|t}, \lambda_{t+1|t}),$$

where

$$f(x; \alpha, \lambda) = [B(\alpha, \lambda)]^{-1} x^{\alpha-1}(1-x)^{\lambda-1}, \quad 0 < x < 1,$$

$B(\alpha, \lambda) := \Gamma(\alpha)\Gamma(\lambda)/\Gamma(\alpha+\lambda)$ is the beta function, and $\alpha, \lambda > 0$. Show that the posterior distribution of X_t given Y_t is also beta and express its parameters in terms of y_t and $\alpha_{t|t-1}, \lambda_{t|t-1}$.

c. Under the assumptions made in (b), show that $E(X_t|\mathbf{Y}^{(t)}) = E(X_{t+1}|\mathbf{Y}^{(t)})$ and $\text{Var}(X_t|\mathbf{Y}^{(t)}) < \text{Var}(X_{t+1}|\mathbf{Y}^{(t)})$.

d. Assuming that the parameters in (b) satisfy (8.8.41)–(8.8.42), show that the one-step prediction density $p(y_{t+1}|\mathbf{y}^{(t)})$ is beta-binomial,

$$p(y_{t+1}|\mathbf{y}^{(t)}) = \frac{B(\alpha_{t+1|t} + y_{t+1}, \lambda_{t+1|t} + n - y_{t+1})}{(n+1)B(y_{t+1}+1, n - y_{t+1} + 1)B(\alpha_{t+1|t}, \lambda_{t+1|t})},$$

and verify that $\hat{Y}_{t+1}$ is given by (8.8.47).

9 Forecasting Techniques

We have focused until now on the construction of time series models for stationary and nonstationary series and the determination, assuming the appropriateness of these models, of minimum mean-squared error predictors. If the observed series had in fact been generated by the fitted model, this procedure would give minimum mean-squared error forecasts. In this chapter we discuss three forecasting techniques that have less emphasis on the explicit construction of a model for the data. Each of the three selects, from a limited class of algorithms, the one that is optimal according to specified criteria.

The three techniques have been found in practice to be effective on wide ranges of real data sets (for example, the economic time series used in the forecasting competition described by Makridakis et al., 1984).

The ARAR algorithm described in Section 9.1 is an adaptation of the ARARMA algorithm (Newton and Parzen, 1984, Parzen, 1982) in which the idea is to apply automatically selected "memory-shortening" transformations (if necessary) to the data and then to fit an ARMA model to the transformed series. The ARAR algorithm we describe is a version of this in which the ARMA fitting step is replaced by the fitting of a subset AR model to the transformed data.

The Holt-Winters (HW) algorithm described in Section 9.2 uses a set of simple recursions that generalize the exponential smoothing recursions of Section 1.5.1 to generate forecasts of series containing a locally linear trend.

The Holt-Winters seasonal (HWS) algorithm extends the HW algorithm to handle data in which there are both trend and seasonal variation of known period. It is described in Section 9.3.

The algorithms can be applied to specific data sets with the aid of the ITSM programs ARAR and HW.

9.1 The ARAR Algorithm

9.1.1 Memory Shortening

Given a data set $\{Y_t, t = 1, 2, \ldots, n\}$, the first step is to decide whether the underlying process is "long-memory," and if so to apply a memory-shortening transformation before attempting to fit an autoregressive model. The differencing operations permitted by PEST are examples of memory-shortening transformations; however, the ones allowed by ARAR are more general. There are two types allowed:

$$\tilde{Y}_t = Y_t - \hat{\phi}(\hat{\tau})Y_{t-\hat{\tau}} \tag{9.1.1}$$

and

$$\tilde{Y}_t = Y_t - \hat{\phi}_1 Y_{t-1} - \hat{\phi}_2 Y_{t-2}. \tag{9.1.2}$$

With the aid of the five-step algorithm described below, we classify $\{Y_t\}$ and take one of the following three courses of action:

- **L.** Declare $\{Y_t\}$ to be long-memory and form $\{\tilde{Y}_t\}$ using (9.1.1).
- **M.** Declare $\{Y_t\}$ to be moderately long-memory and form $\{\tilde{Y}_t\}$ using (9.1.2).
- **S.** Declare $\{Y_t\}$ to be short-memory.

If the alternatives L or M are chosen then the transformed series $\{\tilde{Y}_t\}$ is again checked. If it is found to be long-memory or moderately long-memory, then a further transformation is performed. The process continues until the transformed series is classified as short-memory. The program ARAR allows at most three memory-shortening transformations. It is very rare to require more than two. The algorithm for deciding among L, M, and S can be described as follows:

1. For each $\tau = 1, 2, \ldots, 15$, we find the value $\hat{\phi}(\tau)$ of ϕ that minimizes

$$\text{ERR}(\phi, \tau) = \frac{\sum_{t=\tau+1}^{n}[Y_t - \phi Y_{t-\tau}]^2}{\sum_{t=\tau+1}^{n} Y_t^2}.$$

We then define

$$\text{Err}(\tau) = \text{ERR}(\hat{\phi}(\tau), \tau)$$

and choose the lag $\hat{\tau}$ to be the value of τ that minimizes $\text{Err}(\tau)$.

2. If $\text{Err}(\hat{\tau}) \le 8/n$, go to L.
3. If $\hat{\phi}(\hat{\tau}) \ge .93$ and $\hat{\tau} > 2$, go to L.
4. If $\hat{\phi}(\hat{\tau}) \ge .93$ and $\hat{\tau} = 1$ or 2, determine the values $\hat{\phi}_1$ and $\hat{\phi}_2$ of ϕ_1 and ϕ_2 that minimize $\sum_{t=3}^{n}[Y_t - \phi_1 Y_{t-1} - \phi_2 Y_{t-2}]^2$, then go to M.
5. If $\hat{\phi}(\hat{\tau}) < .93$, go to S.

9.1.2 Fitting a Subset Autoregression

Let $\{S_t, t = k+1, \ldots, n\}$ denote the memory-shortened series derived from $\{Y_t\}$ by the algorithm of the previous section and let $\overline{S}$ denote the sample mean of $S_{k+1}, \ldots, S_n$.

The next step in the modelling procedure is to fit an autoregressive process to the mean-corrected series,

$$X_t = S_t - \overline{S}, t = k+1, \ldots, n.$$

The fitted model has the form

$$X_t = \phi_1 X_{t-1} + \phi_{l_1} X_{t-l_1} + \phi_{l_2} X_{t-l_2} + \phi_{l_3} X_{t-l_3} + Z_t,$$

where $\{Z_t\} \sim \text{WN}(0, \sigma^2)$, and, for given lags, l_1, l_2, and l_3, the coefficients ϕ_j and the white noise variance σ^2 are found from the Yule-Walker equations,

$$\begin{bmatrix} 1 & \hat{\rho}(l_1 - 1) & \hat{\rho}(l_2 - 1) & \hat{\rho}(l_3 - 1) \\ \hat{\rho}(l_1 - 1) & 1 & \hat{\rho}(l_2 - l_1) & \hat{\rho}(l_3 - l_1) \\ \hat{\rho}(l_2 - 1) & \hat{\rho}(l_2 - l_1) & 1 & \hat{\rho}(l_3 - l_2) \\ \hat{\rho}(l_3 - 1) & \hat{\rho}(l_3 - l_1) & \hat{\rho}(l_3 - l_2) & 1 \end{bmatrix} \begin{bmatrix} \phi_1 \\ \phi_{l_1} \\ \phi_{l_2} \\ \phi_{l_3} \end{bmatrix} = \begin{bmatrix} \hat{\rho}(1) \\ \hat{\rho}(l_1) \\ \hat{\rho}(l_2) \\ \hat{\rho}(l_3) \end{bmatrix}$$

and

$$\sigma^2 = \hat{\gamma}(0)[1 - \phi_1 \hat{\rho}(1) - \phi_{l_1} \hat{\rho}(l_1) - \phi_{l_2} \hat{\rho}(l_2) - \phi_{l_3} \hat{\rho}(l_3)],$$

where $\hat{\gamma}(j)$ and $\hat{\rho}(j)$, $j = 0, 1, 2, \ldots$, are the sample autocovariances and autocorrelations of the series $\{X_t\}$.

The program computes the coefficients ϕ_j for each set of lags such that

$$1 < l_1 < l_2 < l_3 \le m$$

where m can be chosen to be either 13 or 26. It then selects the model for which the Yule-Walker estimate σ^2 is minimum and prints out the lags, coefficients, and white noise variance for the fitted model.

A slower procedure chooses the lags and coefficients (computed from the Yule-Walker equations as above) that maximize the Gaussian likelihood of the observations. For this option the maximum lag m is 13.

The options are displayed in the Subset AR Menu which appears on the screen when memory-shortening has been completed, or when you opt to bypass memory shortening and fit a subset AR to the original (mean-corrected) data.

9.1.3 Forecasting

If the memory-shortening filter found in the first step has coefficients $\psi_0(= 1)$, $\psi_1, \ldots, \psi_k$, $(k \geq 0)$, then the memory-shortened series can be expressed as

$$S_t = \psi(B)Y_t = Y_t + \psi_1 Y_{t-1} + \cdots + \psi_k Y_{t-k}, \tag{9.1.3}$$

where $\psi(B)$ is the polynomial in the backward shift operator,

$$\psi(B) = 1 + \psi_1 B + \cdots + \psi_k B^k.$$

Similarly, if the coefficients of the subset autoregression found in the second step are $\phi_1, \phi_{l_1}, \phi_{l_2}$, and ϕ_{l_3}, then the subset AR model for the mean-corrected series $\{X_t = S_t - \overline{S}\}$ is

$$\phi(B)X_t = Z_t, \tag{9.1.4}$$

where $\{Z_t\} \sim \text{WN}(0, \sigma^2)$ and

$$\phi(B) = 1 - \phi_1 B - \phi_{l_1} B^{l_1} - \phi_{l_2} B^{l_2} - \phi_{l_3} B^{l_3}.$$

From (9.1.3) and (9.1.4) we obtain the equations

$$\xi(B)Y_t = \phi(1)\overline{S} + Z_t, \tag{9.1.5}$$

where

$$\xi(B) = \psi(B)\phi(B) = 1 + \xi_1 B + \cdots + \xi_{k+l_3} B^{k+l_3}.$$

Assuming that the fitted model (9.1.5) is appropriate and that the white noise term Z_t is uncorrelated with $\{Y_j, j < t\}$ for each t, we can determine the minimum mean squared error linear predictors $P_n Y_{n+h}$ of Y_{n+h} in terms of $\{1, Y_1, \ldots, Y_n\}$, for $n > k + l_3$, from the recursions

$$P_n Y_{n+h} = -\sum_{j=1}^{k+l_3} \xi_j P_n Y_{n+h-j} + \phi(1)\overline{S}, \; h \geq 1, \tag{9.1.6}$$

with the initial conditions

$$P_n Y_{n+h} = Y_{n+h}, \text{ for } h \leq 0. \tag{9.1.7}$$

The mean squared error of the predictor $P_n Y_{n+h}$ is found to be (Problem 9.1)

$$E[(Y_{n+h} - P_n Y_{n+h})^2] = \sum_{j=1}^{h-1} \tau_j^2 \sigma^2, \tag{9.1.8}$$

where $\sum_{j=0}^{\infty} \tau_j z^j$ is the Taylor expansion of $1/\xi(z)$ in a neighborhood of $z = 0$. Equivalently the sequence $\{\tau_j\}$ can be found from the recursion

$$\tau_0 = 1, \sum_{j=0}^{n} \tau_j \xi_{n-j} = 0, \; n = 1, 2, \ldots. \tag{9.1.9}$$

9.1.4 Running the Program ARAR

To determine an ARAR model for the given data set $\{Y_t\}$ and to use it to forecast future values of the series, we first read in the data set. Following the appearance on the screen of the Main Menu, we type **D** $\hookleftarrow$ to select the option [`Determine the memory-shortening polynomial` ...], which then finds the best memory-shortening filter. After a short time delay the coefficients $1, \psi_1, \ldots, \psi_k$ of the chosen filter will be displayed on the screen. The memory shortened series is

$$S_t = Y_t + \psi_1 Y_{t-1} + \cdots + \psi_k Y_{t-k}.$$

Type $\hookleftarrow$ and the Subset AR Menu will appear. The first option (selected by typing **F**) fits an autoregression with four nonzero coefficients to the mean-corrected series $X_t = S_t - \overline{S}$, choosing the lags and coefficients that minimize the Yule-Walker estimate of white noise variance. Type **F** and the optimal lags and corresponding coefficients in the model

$$X_t = \phi_1 X_{t-1} + \phi_{l_1} X_{t-l_1} + \phi_{l_2} X_{t-l_2} + \phi_{l_3} X_{t-l_3} + Z_t,$$

will be printed on the screen. The coefficients ξ_j of B^j in the overall whitening filter (B is the backward shift operator),

$$\xi(B) = (1 + \psi_1 B + \cdots + \psi_k B^k)(1 - \phi_1 B - \phi_{l_1} B^{l_1} - \phi_{l_2} B^{l_2} - \phi_{l_3} B^{l_3}),$$

are also printed.

Type $\hookleftarrow$ again and you will be asked for the number of future values of $\{Y_t\}$ to be predicted. Enter the required number and type $\hookleftarrow\hookleftarrow$ to see the graph of the original data. Type $\hookleftarrow$ again and the predicted values will be added to the graph. Type $\hookleftarrow$ **C N** and you will be asked if you wish to file the predictors. Following this you will be returned to the Subset AR Menu, from which you may either select one of the other fitting options or return to the Main Menu from which you may leave the program.

Example 9.1.1 To use the program ARAR to predict 24 values of the data file DEATHS.DAT, proceed as follows (starting from the Main Menu, immediately after having read in the data file DEATHS.DAT).

Type **D** and the coefficients of the selected memory-shortening filter will appear. The chosen filter is $(1 - .9779B^{12})$.

To continue, type $\hookleftarrow$ **F**. The program then fits a four-coefficient AR model to the mean-corrected memory-shortened data, with maximum lag 13, selecting the model with minimum estimated white-noise variance. The results of the fit are displayed on the screen in the following format:

```
Optimal lags           1         3         12        13
Optimal coeffs      .5915    -.3822    -.3022     .2970
WN Variance:   .12314E+06
COEFFICIENTS OF OVERALL WHITENING FILTER:
1.0000    -.5915     .0000    -.2093     .0000
 .0000     .0000     .0000     .0000     .0000
 .0000     .0000    -.6757     .2814     .0000
 .2047     .0000     .0000     .0000     .0000
 .0000     .0000     .0000     .0000    -.2955
 .2904
```

To predict 24 future values of the series, type ↩ 24↩. At this stage the screen will show the square root of the *observed* mean squared error of the one-step predictors of the data itself. To plot the predictors of future values, type ↩↩ and you will see the original data with the 24 predictors plotted on the same graph as in Figure 9.1.

To exit from the program, type ↩ and follow the program prompts. □

In Table 9.1 we compare the predictors of the next six values of the accidental deaths series with the actual observed values. The predicted values obtained from ARAR as described in the example are shown together with the predictors obtained by fitting ARIMA models as described in Chapter 6 (see Table 9.1). The observed root mean squared errors (i.e., $\sqrt{\sum_{h=1}^{6}(Y_{72+h} - P_{72}Y_{72+h})^2/6}$) for the three prediction methods are easily calculated to be 253 for ARAR, 583 for the ARIMA model (6.5.8), and 501 for the ARIMA model (6.5.9). The ARAR algorithm thus performs very

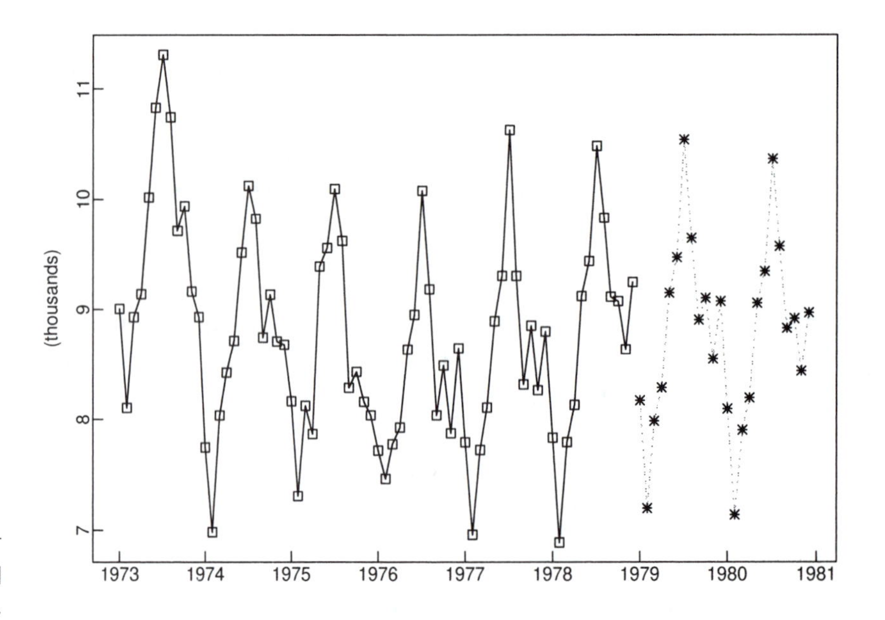

Figure 9-1
The data set DEATHS.DAT with 24 values predicted by the ARAR algorithm.

Table 9.1 Predicted and observed values of the accidental deaths series for $t = 73, \ldots, 78$.

t	73	74	75	76	77	78
Observed Y_t	7798	7406	8363	8460	9217	9316
Predicted by ARAR	8168	7196	7982	8284	9144	9465
Predicted by (6.5.8)	8441	7704	8549	8885	9843	10279
Predicted by (6.5.9)	8345	7619	8356	8742	9795	10179

well here. Notice that in this particular example the ARAR algorithm effectively fits a causal AR model to the data, but this is not always the case.

9.2 The Holt-Winters Algorithm

Given observations $Y_1, Y_2, \ldots, Y_n$ from the "trend plus noise" model (1.5.2), the exponential smoothing recursions (1.5.7) allowed us to compute estimates $\hat{m}_t$, of the trend at times $t = 1, 2, \ldots, n$. If the series is stationary, then m_t is constant and the exponential smoothing forecast of Y_{n+h} based on the observations $Y_1, \ldots, Y_n$ is

$$P_n Y_{n+h} = \hat{m}_n, h = 1, 2, \ldots. \tag{9.2.1}$$

If the data have a (nonconstant) trend, then a natural generalization of the forecast function (9.2.1) that takes this into account is

$$P_n Y_{n+h} = \hat{a}_n + \hat{b}_n h, h = 1, 2, \ldots, \tag{9.2.2}$$

where $\hat{a}_n$ and $\hat{b}_n$ can be thought of as estimates of the "level," a_n, and "slope," b_n, of the trend function at time n. Holt (1957) suggested a recursive scheme for computing the quantities $\hat{a}_n$ and $\hat{b}_n$ in (9.2.2). Denoting by $\hat{Y}_{n+1}$ the one-step forecast $P_n Y_{n+1}$, we have from (9.2.2)

$$\hat{Y}_{n+1} = \hat{a}_n + \hat{b}_n.$$

Now, as in exponential smoothing, we suppose that the estimated level at time $n + 1$ is a linear combination of the observed value at time $n + 1$ and the forecast value at time $n + 1$. Thus,

$$\hat{a}_{n+1} = \alpha Y_{n+1} + (1 - \alpha)(\hat{a}_n + \hat{b}_n). \tag{9.2.3}$$

We can then estimate the slope at time $n + 1$ as a linear combination of $\hat{a}_{n+1} - \hat{a}_n$ and the estimated slope $\hat{b}_n$ at time n. Thus,

$$\hat{b}_{n+1} = \beta(\hat{a}_{n+1} - \hat{a}_n) + (1 - \beta)\hat{b}_n. \tag{9.2.4}$$

In order to solve the recursions (9.2.3) and (9.2.4) we need initial conditions. A natural choice is to set

$$\hat{a}_2 = Y_2 \tag{9.2.5}$$

and

$$\hat{b}_2 = Y_2 - Y_1. \tag{9.2.6}$$

Then (9.2.3) and (9.2.4) can be solved successively for $\hat{a}_i$ and $\hat{b}_i$, $i = 3, \ldots, n$ and the predictors $P_n Y_{n+h}$ found from (9.2.2).

The forecasts depend on the "smoothing parameters" α and β. These can either be prescribed arbitrarily (with values between 0 and 1) or chosen in a more systematic way to minimize the sum of squares of the one-step errors, $\sum_{i=3}^{n}(Y_i - P_{i-1}Y_i)^2$, obtained when the algorithm is applied to the already observed data. Both options are available in the program HW.

Before illustrating the use of the Holt-Winters forecasting procedure, we discuss the connection between the recursions (9.2.3)–(9.2.4) and the steady-state solution of the Kalman filtering equations for a local linear trend model. Suppose $\{Y_t\}$ follows the local linear structural model with observation equation

$$Y_t = M_t + W_t$$

and state equation

$$\begin{bmatrix} M_{t+1} \\ B_{t+1} \end{bmatrix} = \begin{bmatrix} 1 & 1 \\ 0 & 1 \end{bmatrix} \begin{bmatrix} M_t \\ B_t \end{bmatrix} + \begin{bmatrix} V_t \\ U_t \end{bmatrix}$$

(see (8.2.4)–(8.2.7)). Now define $\hat{a}_n$ and $\hat{b}_n$ to be the filtered estimates of M_n and B_n respectively, i.e.,

$$\hat{a}_n = M_{n|n} := P_n M_n$$

$$\hat{b}_n = B_{n|n} := P_n B_n.$$

Using Problem 8.17 and the Kalman recursion (8.4.10), we find that

$$\begin{bmatrix} \hat{a}_{n+1} \\ \hat{b}_{n+1} \end{bmatrix} = \begin{bmatrix} \hat{a}_n + \hat{b}_n \\ \hat{b}_n \end{bmatrix} + \Delta_n^{-1}\Omega_n G'(Y_n - \hat{a}_n - \hat{b}_n), \tag{9.2.7}$$

where $G = [\,1 \quad 0\,]$. Assuming that $\Omega_n = \Omega_1 = [\Omega_{ij}]_{i,j=1}^{2}$ is the steady-state solution of (8.4.2) for this model, then $\Delta_n = \Omega_{11} + \sigma_w^2$ for all n so that (9.2.7) simplifies to the equations

$$\hat{a}_{n+1} = \hat{a}_n + \hat{b}_n + \frac{\Omega_{11}}{\Omega_{11} + \sigma_w^2}(Y_n - \hat{a}_n - \hat{b}_n) \tag{9.2.8}$$

and

$$\hat{b}_{n+1} = \hat{b}_n + \frac{\Omega_{12}}{\Omega_{11} + \sigma_w^2}(Y_n - \hat{a}_n - \hat{b}_n). \tag{9.2.9}$$

Solving (9.2.8) for $(Y_n - \hat{a}_n - \hat{b}_n)$ and substituting into (9.2.9), we find that

$$\hat{a}_{n+1} = \alpha Y_{n+1} + (1 - \alpha)(\hat{a}_n + \hat{b}_n) \tag{9.2.10}$$

$$\hat{b}_{n+1} = \beta(\hat{a}_{n+1} - \hat{a}_n) + (1 - \beta)\hat{b}_n \tag{9.2.11}$$

with $\alpha = \Omega_{11}/(\Omega_{11} + \sigma_w^2)$ and $\beta = \Omega_{21}/\Omega_{11}$. These equations coincide with the Holt-Winters recursions (9.2.3)–(9.2.4). An explicit solution for α and β in terms of the variances σ_u^2, σ_v^2, and σ_w^2 can be found in Harvey (1990).

Example 9.2.1 To apply the algorithm specified by (9.2.2)–(9.2.6) to predict 24 values of the data file DEATHS.DAT, we use the program HW. On running the program you will first be asked to enter the data. After selecting the data file DEATHS.DAT, the data will automatically be plotted. Type ↩ **C** and the sample autocorrelation function will also be plotted. It is evident from these two graphs that the data contain a substantial periodic component with period 12. For the moment, however, we shall ignore this (Section 9.3 shows how to modify the recursions (9.2.2)–(9.2.6) to allow for seasonality). After inspecting the sample autocorrelation function, type ↩ and you will be asked the question

```
Period of seasonality (0 if no seasonality)?
```

Since we are not taking account of seasonality, type 0 ↩ and you will be asked if you wish to optimize the coefficients α and β. Type **Y** and the coefficients will be chosen to minimize the observed one-step mean squared error, $\sum_{i=3}^{72}(Y_i - P_{i-1}Y_i)^2/70$. The square root of this mean squared error will then appear on the screen with the corresponding estimates of α and β. In response to the request for the number of future values to be predicted, type 24 ↩. Then type ↩ to see a graph of the original data and ↩ to append the required 24 forecasts. Type ↩ **C N** to return to the initial menu options and then **X** to exit from the program. Figure 9.2 shows the data and 24

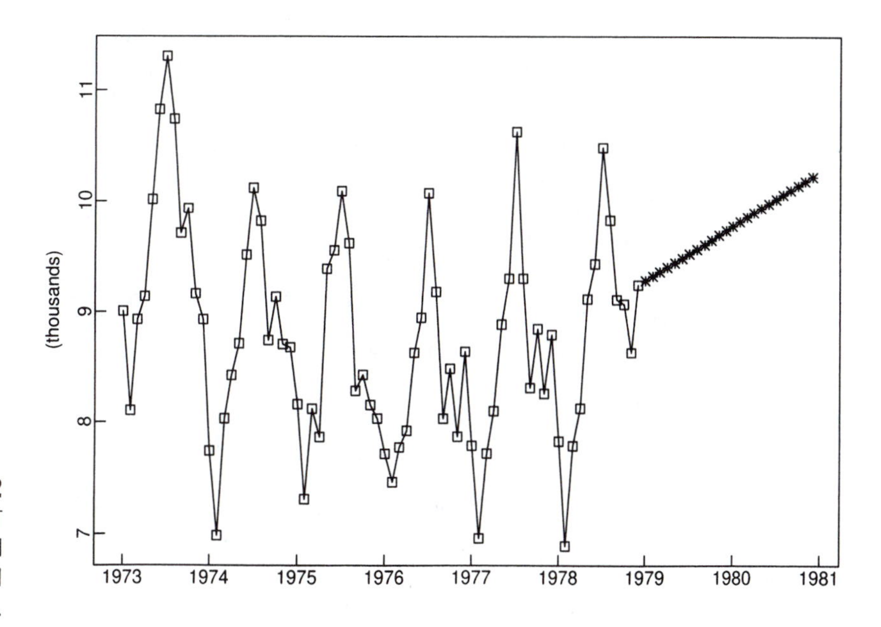

Figure 9-2 The data set DEATHS.DAT with 24 values predicted by the nonseasonal Holt-Winters algorithm.

Table 9.2 Predicted and observed values of the accidental deaths series for $t = 73, \ldots, 78$ from the (nonseasonal) Holt-Winters algorithm.

t	73	74	75	76	77	78
Observed Y_t	7798	7406	8363	8460	9217	9316
Predicted by HW	9281	9322	9363	9404	9445	9486

predictors found in this way and Table 9.2 compares the predictors of $Y_{73}, \ldots, Y_{78}$ with the corresponding observed values. □

The root mean squared error $(\sqrt{\sum_{h=1}^{6}(Y_{72+h} - P_{72}Y_{72+h})^2/6}\,)$ for the nonseasonal Holt-Winters forecasts is found to be 1143. Not surprisingly, since we have not taken seasonality into account, this is a much larger value than for the three sets of forecasts shown in Table 9.1. In the next section we show how to modify the Holt-Winters algorithm to allow for seasonality.

9.3 The Holt-Winters Seasonal Algorithm

If the series $Y_1, Y_2, \ldots, Y_n$, contains not only trend, but also seasonality with period d (as in the model (1.5.11)), then a further generalization of the forecast function (9.2.2) that takes this into account is

$$P_n Y_{n+h} = \hat{a}_n + \hat{b}_n h + \hat{c}_{n+h}, \quad h = 1, 2, \ldots, \tag{9.3.1}$$

where $\hat{a}_n$, $\hat{b}_n$, and $\hat{c}_n$ can be thought of as estimates of the "trend level," a_n, "trend slope," b_n, and "seasonal component," c_n, at time n. The values of $\hat{c}_{n+h}$, $h = 1, 2, \ldots$ are found from the recursion

$$\hat{c}_{n+h} = \hat{c}_{n+h-d}, \quad h \geq 1, \tag{9.3.2}$$

while the values of $\hat{a}_i$, $\hat{b}_i$, and $\hat{c}_i$, $i = 1, \ldots, n$, are found from recursions analogous to (9.2.3) and (9.2.4), namely,

$$\hat{a}_{n+1} = \alpha(Y_{n+1} - \hat{c}_{n+1-d}) + (1 - \alpha)(\hat{a}_n + \hat{b}_n), \tag{9.3.3}$$

$$\hat{b}_{n+1} = \beta(\hat{a}_{n+1} - \hat{a}_n) + (1 - \beta)\hat{b}_n \tag{9.3.4}$$

and

$$\hat{c}_{n+1} = \gamma(Y_{n+1} - \hat{a}_{n+1}) + (1 - \gamma)\hat{c}_{n+1-d}. \tag{9.3.5}$$

A natural set of initial conditions for these recursions is as follows:

$$\hat{a}_{d+1} = Y_{d+1}, \tag{9.3.6}$$

$$\hat{b}_{d+1} = (Y_{d+1} - Y_1)/d, \tag{9.3.7}$$

and

$$\hat{c}_i = Y_i - (Y_1 + \hat{b}_{d+1}(i-1)), i = 1, \ldots, d. \tag{9.3.8}$$

Then (9.3.3)–(9.3.5) can be solved successively for $\hat{a}_i$, $\hat{b}_i$, and $\hat{c}_i$, $i = d+1, \ldots, n$, and the predictors $P_n Y_{n+h}$ found from (9.3.1).

As in the nonseasonal case of Section 9.2, the forecasts depend on the parameters α, β, and γ. These can either be prescribed arbitrarily (with values between 0 and 1) or chosen in a more systematic way to minimize the sum of squares of the one-step errors, $\sum_{i=d+2}^{n}(Y_i - P_{i-1}Y_i)^2$, obtained when the algorithm is applied to the already observed data. To apply the algorithm we again use the program HW.

Example 9.3.1 As in the example in Section 9.2, run the program HW and select the data file DEATHS.DAT. The time series graph of the data and the sample autocorrelation function both clearly indicate a substantial periodic component with period 12. In response to the question,

```
Period of seasonality (0 if no seasonality)?
```

type `12` ←. You will then be asked if you wish to optimize the coefficients α, β, and γ. Type **Y** and the coefficients will be chosen to minimize the observed one-step mean squared error, $\sum_{i=14}^{72}(Y_i - P_{i-1}Y_i)^2/59$. The square root of this mean squared error will then appear on the screen with the corresponding estimates of α, β, and γ. In response to the request for the number of future values to be predicted, type 24 ←. Then type ← to see a graph of the original data and ← to append the required 24 forecasts. Type ← **C N N** to return to the initial menu options and then **X** to exit

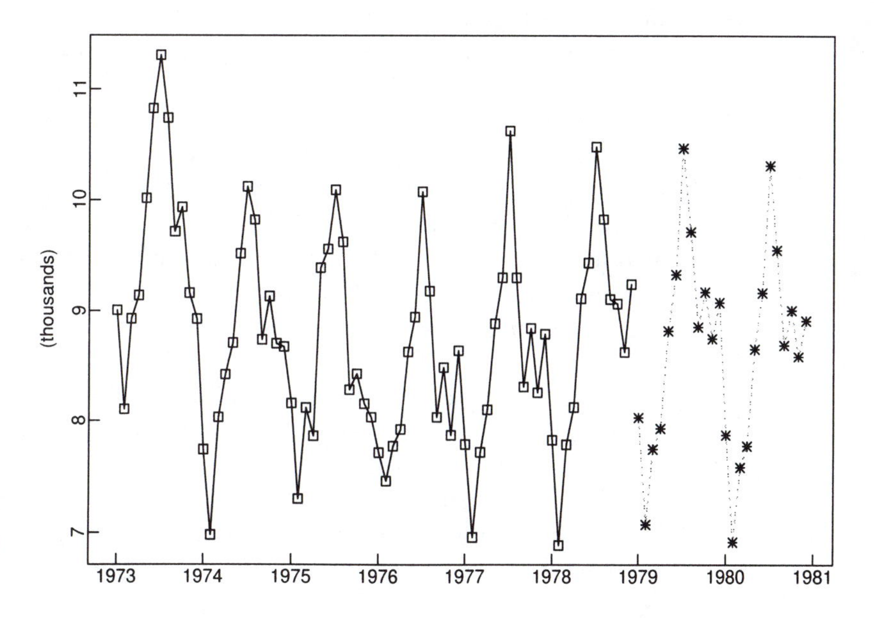

Figure 9-3 The data set DEATHS.DAT with 24 values predicted by the seasonal Holt-Winters algorithm.

Table 9.3 Predicted and observed values of the accidental deaths series for $t = 73, \ldots, 78$ from the seasonal Holt-Winters algorithm.

t	73	74	75	76	77	78
Observed Y_t	7798	7406	8363	8460	9217	9316
Predicted by HWS	8039	7077	7750	7941	8824	9329

from the program. Figure 9.3 shows the data and 24 predictors found in this way and Table 9.3 compares the predictors of $Y_{73}, \ldots, Y_{78}$ with the corresponding observed values. □

The root mean squared error ($\sqrt{\sum_{h=1}^{6}(Y_{72+h} - P_{72}Y_{72+h})^2/6}$) for the seasonal Holt-Winters forecasts is found to be 401. This is not as good as the value 253 achieved by ARAR in this example but is substantially better than the values achieved by the nonseasonal Holt-Winters algorithm (1143) and the ARIMA models (6.5.8) and (6.5.9) (583 and 501, respectively).

9.4 Choosing a Forecasting Algorithm

Real data is rarely if ever generated by a simple mathematical model such as an ARIMA process. Forecasting methods that are predicated on the assumption of such a model are therefore not necessarily the best, even in the mean squared error sense. Nor is the measurement of error in terms of mean squared error necessarily always the most appropriate one in spite of its mathematical convenience. Even within the framework of minimum mean-squared-error forecasting, we may ask (for example) if we wish to minimize the one-step, two-step, or twelve-step mean squared error.

The use of more heuristic algorithms such as those discussed in this chapter is therefore well worth serious consideration in practical forecasting problems. But how do we decide which method to use? A relatively simple solution to this problem, given the availability of a substantial historical record, is to choose between competing algorithms by comparing the relevant errors when the algorithms are applied to the data already observed (e.g., by comparing the mean absolute percentage errors of the twelve-step predictors of the historical data if twelve-step prediction is of primary concern).

It is extremely difficult to make general theoretical statements about the relative merits of the various techniques we have discussed (ARIMA modelling, exponential smoothing, ARAR, and HW methods). For the series DEATHS.DAT we found on the basis of average mean squared error for predicting the series at times 73–78 that the ARAR method was best, followed by the seasonal Holt-Winters algorithm, and then the ARIMA models fitted in Chapter 6. This ordering is by no means universal. For example, if we consider the natural logarithms $\{Y_t\}$ of the first 130 observations in the series WINE.DAT (Figure 1.1) and compare the average mean-squared errors of the forecasts of $Y_{131}, \ldots, Y_{142}$, we find (Problem 9.2) that an MA(12) model fitted to the differenced series $\{Y_t - Y_{t-12}\}$ does better than seasonal Holt-Winters (with period 12), which in turn does better than ARAR and (not surprisingly) dramatically better than nonseasonal Holt-Winters. An interesting empirical comparison of these and other methods applied to a variety of economic time series is contained in Makridakis et al. (1984).

The versions of the Holt-Winters algorithms we have discussed in Sections 9.2 and 9.3 are referred to as "additive" since the seasonal and trend components enter the forecasting function in an additive manner. "Multiplicative" versions of the algorithms can also be constructed to deal directly with processes of the form

$$Y_t = m_t s_t Z_t, \tag{9.4.1}$$

where m_t, s_t, and Z_t are trend, seasonal, and noise factors, respectively (see e.g., Makridakis et al., 1983). An alternative approach (provided $Y_t > 0$ for all t) is to apply the linear Holt-Winters algorithms to $\{\ln Y_t\}$ (as in the WINE.DAT example above). Because of the rather general memory-shortening permitted by the ARAR algorithm, it gives reasonable results when applied directly to series of the form (9.4.1), even without preliminary transformations. In particular, if we consider the first 132 observations in the series AIRPASS.DAT and apply ARAR to predict the last 12, values in the series, we obtain (Problem 9.4) an observed root mean squared error of 18.21. On the other hand if we use the same data, take logarithms, difference at lag 12, and then fit an AR(13) model by maximum likelihood using PEST and

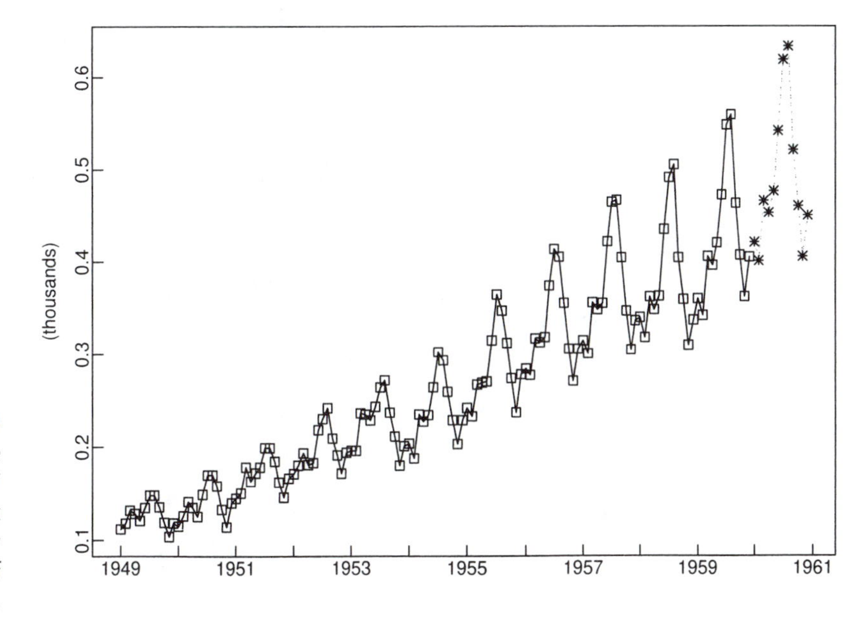

Figure 9-4
The first 132 values of the data set AIRPASS.DAT and predictors of the last 12 values obtained by direct application of the ARAR algorithm.

use it to predict the last 12 values, we obtain an observed root mean squared error of 21.67. The data and predicted values from ARAR are shown in Figure 9.4.

Problems

9.1. Establish the formula (9.1.8) for the mean-squared error of the h-step forecast based on the ARAR algorithm.

9.2. Use a text editor or WORD6 to create a new data set WSHORT.DAT which consists of the data in WINE.DAT with the last 12 values deleted. Denote these observations by $\{X_1, \ldots, X_{130}\}$ and their natural logarithms by $\{Y_1, \ldots, Y_{130}\}$.

a. Use the program PEST to find the maximum likelihood MA(12) model for the differenced series $\{Y_t - Y_{t-12}, t = 13, \ldots, 130\}$.

b. Use the model in (a) to file the forecasts of $\{Y_{131}, \ldots, Y_{142}\}$.

c. Tabulate the forecast errors, $\{Y_t - P_{130}Y_t, t = 131, \ldots, 142\}$.

d. Compute the average squared error for the 12 forecasts.

e. Repeat steps (b), (c), and (d) for the corresponding forecasts obtained by applying the seasonal Holt-Winters algorithm (with period 12) to the logged data $\{Y_t, t = 1, \ldots, 130\}$.

f. Repeat steps (b), (c), and (d) for the corresponding forecasts obtained by applying the ARAR algorithm to the logged data $\{Y_t, t = 1, \ldots, 130\}$.

g. Repeat steps (b), (c), and (d) for the corresponding forecasts obtained by applying the nonseasonal Holt-Winters algorithm to the logged data $\{Y_t, t = 1, \ldots, 130\}$.

h. Compare the average squared errors obtained by the four methods.

9.3. In equations (9.2.10)–(9.2.11), show that $\alpha = \Omega_{11}/(\Omega_{11}+\sigma_w^2)$ and $\beta = \Omega_{21}/\Omega_{11}$.

9.4. Verify the assertions made in the last paragraph of Section 9.4 comparing the forecasts of the last 12 values of the series AIRPASS.DAT obtained from ARAR (with no log transformation) and the corresponding forecasts obtained from PEST by taking logarithms, fitting an AR(13) model to the logarithms differenced at lag 12, forecasting the last 12 values, then transforming these forecasts back to forecasts of the original data.

10 Further Topics

In this final chapter we touch on a variety of topics of special interest. In Section 10.1 we consider transfer function models, designed to exploit for predictive purposes the relationship between two time series when one acts as a leading indicator for the other. Section 10.2 deals with intervention analysis, which allows for possible changes in the mechanism generating a time series, causing it to have different properties over different time intervals. In Section 10.3 we introduce the very fast growing area of nonlinear time series analysis, and in Section 10.4 we briefly discuss continuous-time ARMA processes, which, besides being of interest in their own right, are very useful also for modelling irregularly spaced data. In Section 10.5 we discuss fractionally integrated ARMA processes, sometimes called "long-memory" processes on account of the slow rate of convergence of their autocorrelation functions to zero as the lag increases.

10.1 Transfer Function Models

In this section we consider the problem of estimating the transfer function of a linear filter when the output includes added uncorrelated noise. Suppose that $\{X_{t1}\}$ and $\{X_{t2}\}$

are, respectively, the input and output of the transfer function model

$$X_{t2} = \sum_{j=0}^{\infty} \tau_j X_{t-j,1} + N_t, \tag{10.1.1}$$

where $T = \{\tau_j, j = 0, 1, \ldots\}$ is a causal time-invariant linear filter and $\{N_t\}$ is a zero-mean stationary process, uncorrelated with the input process $\{X_{t1}\}$. We further assume that $\{X_{t1}\}$ is a zero-mean stationary time series. Then the bivariate process $\{(X_{t1}, X_{t2})'\}$ is also stationary. Multiplying each side of (10.1.1) by $X_{t-k,1}$ and then taking expectations gives the equation

$$\gamma_{21}(k) = \sum_{j=0}^{\infty} \tau_j \gamma_{11}(k-j). \tag{10.1.2}$$

Equation (10.1.2) simplifies a great deal if the input process happens to be white noise. For example, if $\{X_{t1}\} \sim \text{WN}(0, \sigma_1^2)$, then we can immediately identify t_k from (10.1.2) as

$$\tau_k = \gamma_{21}(k)/\sigma_1^2. \tag{10.1.3}$$

This observation suggests that "pre-whitening" of the input process might simplify the identification of an appropriate transfer-function model and at the same time provide simple preliminary estimates of the coefficients t_k.

If $\{X_{t1}\}$ can be represented as an invertible ARMA(p, q) process,

$$\phi(B)X_{t1} = \theta(B)Z_t, \qquad \{Z_t\} \sim \text{WN}(0, \sigma_Z^2), \tag{10.1.4}$$

then application of the filter $\pi(B) = \phi(B)\theta^{-1}(B)$ to $\{X_{t1}\}$ will produce the whitened series $\{Z_t\}$. Now applying the operator $\pi(B)$ to each side of (10.1.1) and letting $Y_t = \pi(B)X_{t2}$, we obtain the relation

$$Y_t = \sum_{j=0}^{\infty} \tau_j Z_{t-j} + N_t',$$

where

$$N_t' = \pi(B)N_t,$$

and $\{N_t'\}$ is a zero-mean stationary process, uncorrelated with $\{Z_t\}$. The same arguments that led to (10.1.3) therefore yield the equation

$$\tau_j = \rho_{YZ}(j)\sigma_Y/\sigma_Z, \tag{10.1.5}$$

where ρ_{YZ} is the cross-correlation function of $\{Y_t\}$ and $\{Z_t\}$, $\sigma_Z^2 = \text{Var}(Z_t)$ and $\sigma_Y^2 = \text{Var}(Y_t)$.

Given the observations $\{(X_{t1}, X_{t2})', t = 1, \ldots, n\}$, the results of the previous paragraph suggest the following procedure for estimating $\{\tau_j\}$ and analyzing the noise $\{N_t\}$ in the model (10.1.1):

1. Fit an ARMA model to $\{X_{t1}\}$ and file the residuals $(\hat{Z}_1, \ldots, \hat{Z}_n)$. Let $\hat{\phi}$ and $\hat{\theta}$ denote the maximum likelihood estimates of the autoregressive and moving average parameters and let $\hat{\sigma}_Z^2$ be the maximum likelihood estimate of the variance of $\{Z_t\}$.
2. Apply the operator $\hat{\pi}(B) = \hat{\phi}(B)\hat{\theta}^{-1}(B)$ to $\{X_{t2}\}$, giving the series $(\hat{Y}_1, \ldots, \hat{Y}_n)$. The values $\hat{Y}_t$ can be computed as the residuals obtained by running the computer program PEST with initial coefficients $\hat{\phi}$, $\hat{\theta}$ and choosing the option *ARMA parameter estimation* then the suboption *Likelihood of Model* (*no optimization*). Let $\hat{\sigma}_Y^2$ denote the sample variance of $\hat{Y}_t$ (the "white noise" variance computed by PEST when run as just described).
3. Compute the sample cross-correlation function $\rho_{YZ}(h)$ between $\{\hat{Y}_t\}$ and $\{\hat{Z}_t\}$. Comparison of $\hat{\rho}_{YZ}(h)$ with the bounds $\pm 1.96 n^{-1/2}$ gives a preliminary indication of the lags h at which $\rho_{YZ}(h)$ is significantly different from zero. A more refined check can be carried out by using Bartlett's formula in Section 7.3.4 for the asymptotic variance of $\hat{\rho}_{YZ}(h)$. Under the assumptions that $\{\hat{Z}_t\} \sim \mathrm{WN}(0, \hat{\sigma}_Z^2)$ and $\{(\hat{Y}_t, \hat{Z}_t)'\}$ is a stationary Gaussian process,

$$n\mathrm{Var}(\hat{\rho}_{YZ}(h)) \sim 1 - \rho_{YZ}^2(h)\left[1.5 - \sum_{k=-\infty}^{\infty} (\rho_{YZ}^2(k) + \rho_{YY}^2(k)/2)\right]$$
$$+ \sum_{k=-\infty}^{\infty} \left[\rho_{YZ}(h+k)\rho_{YZ}(h-k) - 2\rho_{YZ}(h)\rho_{YZ}(k+h)\rho_{YY}^2(k)\right].$$

In order to check the hypothesis H_0 that $\rho_{YZ}(h) = 0$, $h \notin [a, b]$, where a and b are integers, we note from Corollary 7.3.1 that under H_0,

$$\mathrm{Var}(\hat{\rho}_{YZ}(h)) \sim n^{-1} \qquad \text{for } h \notin [a, b].$$

We can therefore check the hypothesis H_0 by comparing $\hat{\rho}_{YZ}$, $h \notin [a, b]$ with the bounds $\pm\ 1.96 n^{-1/2}$. Observe that $\rho_{YZ}(h)$ should be zero for $h < 0$ if the model (10.1.1) is valid.
4. Preliminary estimates of τ_h for lags h at which $\rho_{YZ}(h)$ is found to be significantly different from zero are

$$\hat{\tau}_h = \hat{\rho}_{YZ}(h)\hat{\sigma}_Y/\hat{\sigma}_Z.$$

For other values of h the preliminary estimates are $\hat{\tau}_h = 0$. Let $m \geq 0$ be the largest value of j such that $\hat{\tau}_j$ is nonzero and let $b \geq 0$ be the smallest such value. Then b is known as the *delay parameter* of the filter $\{\hat{\tau}_j\}$. If m is very large and if the coefficients $\{\hat{\tau}_j\}$ are approximately related by difference equations of the form

$$\hat{\tau}_j - v_1\hat{\tau}_{j-1} - \cdots - v_p\hat{\tau}_{j-p} = 0, \qquad j \geq b + p,$$

then $\hat{T}(B) = \sum_{j=b}^{m} \hat{\tau}_j B^j$ can be represented approximately, using fewer parameters, as

$$\hat{T}(B) = w_0(1 - v_1 B - \cdots - v_p B_p)^{-1} B^b.$$

In particular if $\hat{\tau}_j = 0$, $j < b$, and $\hat{\tau}_j = w_0 v_1^{j-b}$, $j \geq b$, then

$$\hat{T}(B) = w_0(1 - v_1 B)^{-1} B^b. \tag{10.1.6}$$

Box and Jenkins (1976) recommend choosing $\hat{T}(B)$ to be a ratio of two polynomials, however, the degrees of the polynomials are often difficult to estimate from $\{\hat{\tau}_j\}$. The primary objective at this stage is to find a parametric function that provides an adequate approximation to $\hat{T}(B)$ without introducing too large a number of parameters. If $\hat{T}(B)$ is represented as $\hat{T}(B) = B^b w(B) v^{-1}(B) = B^b(w_0 + w_1 B + \cdots + w_q B^q)(1 - v_1 B - \cdots - v_p B^p)^{-1}$ with $v(z) \neq 0$ for $|z| \leq 1$, then we define $m = \max(q + b, p)$.

5. The noise sequence $\{N_t, t = m + 1, \ldots, n\}$ is estimated by

$$\hat{N}_t = X_{t2} - \hat{T}(B) X_{t1}.$$

(We set $\hat{N}_t = 0, t \leq m$, in order to compute $\hat{N}_t, t > m = \max(b + q, p)$.)

6. Preliminary identification of a suitable model for the noise sequence is carried out by fitting a causal invertible ARMA model

$$\phi^{(N)}(B) N_t = \theta^{(N)}(B) W_t, \qquad \{W_t\} \sim \mathrm{WN}(0, \sigma_W^2), \tag{10.1.7}$$

to the estimated noise $\hat{N}_{m+1}, \ldots, \hat{N}_n$.

7. Selection of the parameters b, p, and q and the orders p_2 and q_2 of $\phi^{(N)}(\cdot)$ and $\theta^{(N)}(\cdot)$ gives the preliminary model

$$\phi^{(N)}(B) v(B) X_{t2} = B^b \phi^{(N)}(B) w(B) X_{t1} + \theta^{(N)}(B) v(B) W_t,$$

where $\hat{T}(B) = B^b w(B) v^{-1}(B)$ as in step (4). For this model we can compute $\hat{W}_t(\mathbf{w}, \mathbf{v}, \boldsymbol{\phi}^{(N)}, \boldsymbol{\theta}^{(N)})$, $t > m^* = \max(p_2 + p, b + p_2 + q)$, by setting $\hat{W}_t = 0$ for $t \leq m^*$. The parameters $\mathbf{w}$, $\mathbf{v}$, $\boldsymbol{\phi}^{(N)}$, and $\boldsymbol{\theta}^{(N)}$ can then be estimated by minimizing

$$\sum_{t=m^*+1}^{n} \hat{W}_t^2(\mathbf{w}, \mathbf{v}, \boldsymbol{\phi}^{(N)}, \boldsymbol{\theta}^{(N)}).$$

The preliminary estimates from steps (4) and (6) can be used as initial values in the minimization and the minimization carried out using program TRANS. Alternatively, maximum likelihood estimation may be carried out using a state-space representation for the transfer function model (see TSTM, Section 13.1).

8. From the least squares estimators of the parameters of $T(B)$, a new estimated noise sequence can be computed as in step (5) and checked for compatibility with the ARMA model for $\{N_t\}$ fitted by the least squares procedure. If the new estimated noise sequence suggests different orders for $\phi^{(N)}(\cdot)$ and $\theta^{(N)}(\cdot)$, the least squares procedure in step (7) can be repeated using the new orders.

9. To test for goodness of fit, the residuals from the ARMA fitting in steps (1) and (6) should both be checked as described in Section 5.3. The sample cross-correlations of the two residual series $\{\hat{Z}_t, t > m^*\}$ and $\{\hat{W}_t, t > m^*\}$ should also be compared with the bounds $\pm 1.96/\sqrt{n}$ in order to check the hypothesis that the sequences $\{N_t\}$ and $\{Z_t\}$ are uncorrelated.

Example 10.1.1 Sales with a leading indicator

In this example we fit a transfer function model to the bivariate time series of Example 7.1.2. Let

$$X_{t1} = (1 - B)Y_{t1} - .0228, \qquad t = 1, \ldots, 149,$$

$$X_{t2} = (1 - B)Y_{t2} - .420, \qquad t = 1, \ldots, 149,$$

where $\{Y_{t1}\}$ and $\{Y_{t2}\}$, $t = 0, \ldots, 149$, are the leading indicator and sales data, respectively. It was found in Example 7.1.2 that $\{X_{t1}\}$ and $\{X_{t2}\}$ can be modelled as low-order zero mean ARMA processes. In particular we fitted the model

$$X_{t1} = (1 - .474B)Z_t, \qquad \{Z_t\} \sim \text{WN}(0, .0779),$$

to the series $\{X_{t1}\}$. We can therefore whiten the series by application of the filter $\hat{\pi}(B) = (1 - .474B)^{-1}$. Applying $\hat{\pi}(B)$ to both $\{X_{t1}\}$ and $\{X_{t2}\}$ we obtain

$$\hat{Z}_t = (1 - .474B)^{-1}X_{t1}, \quad \hat{\sigma}_Z^2 = .0779,$$

$$\hat{Y}_t = (1 - .474B)^{-1}X_{t2}, \quad \hat{\sigma}_Y^2 = 4.0217.$$

These calculations and the filing of the series $\{\hat{Z}_t\}$ and $\{\hat{Y}_t\}$ were carried out using the program PEST as described in step (2). The sample correlation function $\hat{\rho}_{YZ}(h)$ of $\{\hat{Y}_t\}$ and $\{\hat{Z}_t\}$, computed using the program TRANS, is shown in Figure 10.1. Comparison of $\hat{\rho}_{YZ}(h)$ with the bounds $\pm 1.96(149)^{-1/2} = \pm .161$ suggests that $\rho_{YZ}(h) = 0$ for $h < 3$. Since $\hat{\tau}_j = \hat{\rho}_{YZ}(j)\hat{\sigma}_Y/\hat{\sigma}_Z$ is decreasing approximately geometrically for $j \geq 3$, we take $T(B)$ to have the form (10.1.6), i.e.,

$$T(B) = w_0(1 - v_1 B)^{-1}B^3.$$

Preliminary estimates of w_0 and v_1 are given by $\hat{w}_0 = \hat{\tau}_3 = 4.86$ and $\hat{v}_1 = \hat{\tau}_4/\hat{\tau}_3 = .698$. The estimated noise sequence is obtained from the equation

$$\hat{N}_t = X_{t2} - 4.86B^3(1 - .698B)^{-1}X_{t1}, \qquad t = 4, 5, \ldots, 149.$$

Examination of this sequence using the program PEST leads to the MA(1) model

$$N_t = (1 - .364B)W_t, \qquad \{W_t\} \sim \text{WN}(0, .0590).$$

Substituting these preliminary noise and transfer function models into equation (10.1.1) then gives

$$X_{t2} = 4.86B^3(1 - .698B)^{-1}X_{t1} + (1 - .364B)W_t, \quad \{W_t\} \sim \text{WN}(0, .0590).$$

Now minimizing the sum of squares in step (7) with respect to the parameters $(w_0, v_1, \theta_1^{(N)})$ using the program TRANS, we obtain the least squares model

$$X_{t2} = 4.717B^3(1 - .724B)^{-1}X_{t1} + (1 - .582B)W_t,$$
$$\{W_t\} \sim \text{WN}(0, .0486), \qquad (10.1.8)$$

where

$$X_{t1} = (1 - .474B)Z_t, \quad \{Z_t\} \sim \text{WN}(0, .0779).$$

Notice the reduced white noise variance of $\{W_t\}$ in the least squares model as compared with the preliminary model.

The sample autocorrelation function of the series

$$\hat{N}_t = X_{t2} - 4.717B^3(1 - .724B)^{-1}X_{t1}$$

is shown in Figure 10.2. It suggests that the MA(1) model is appropriate for the noise process. Moreover the residuals $\hat{W}_t$ obtained from the least squares model (10.1.8) pass the diagnostic tests for white noise as described in Section 5.3, and the sample cross-correlations between the residuals $\hat{W}_t$ and $\hat{Z}_t$, $t = 4, \cdots, 149$, are found to lie between the bounds $\pm 1.96/\sqrt{144}$ for all lags between ± 20. □

10.1.1 Prediction Based on a Transfer-Function Model

When predicting $X_{n+h,2}$ on the basis of the transfer-function model defined by (10.1.1), (10.1.4), and (10.1.7), with observations of X_{t1} and X_{t2}, $t = 1, \ldots, n$, our aim is to find the linear combination of $1, X_{11}, \ldots, X_{n1}, X_{12}, \ldots, X_{n2}$ that predicts $X_{n+h,2}$ with

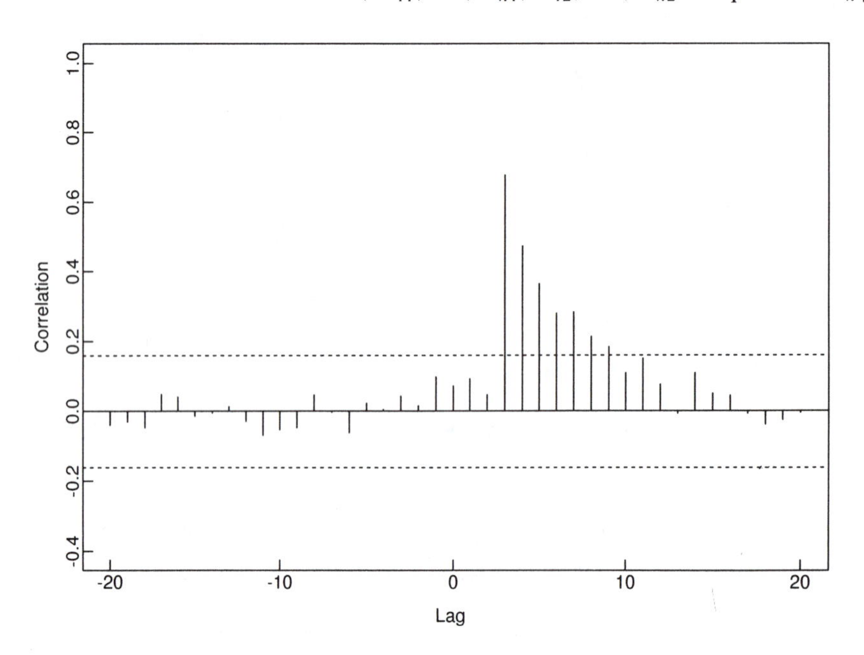

Figure 10-1 The sample cross-correlation function $\hat{\rho}_{12}(h)$, $-20 \le h \le 20$, of Example 10.1.1.

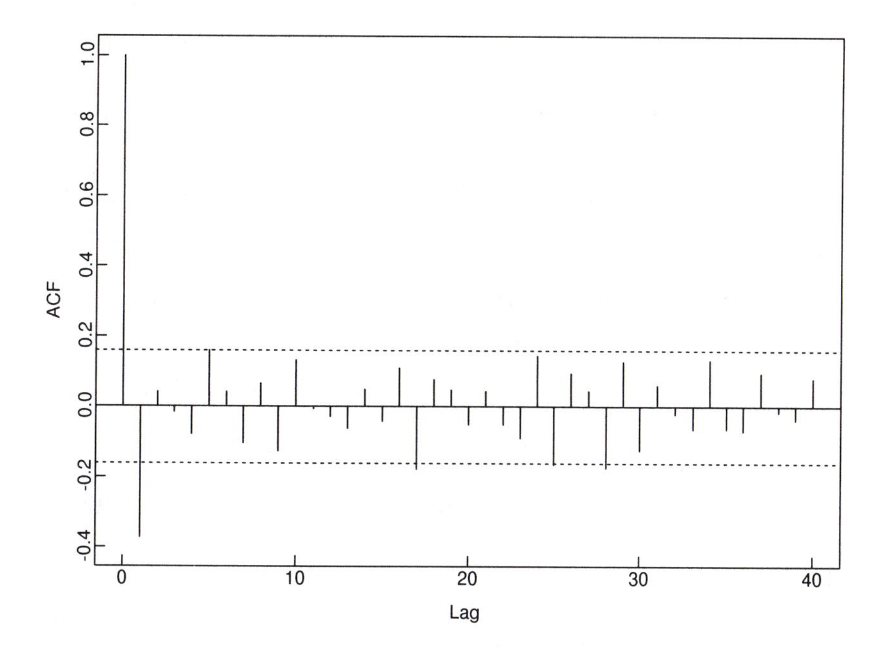

Figure 10-2 The sample ACF of the estimated noise sequence, $\hat{N}_t$, of Example 10.1.1.

minimum mean squared error. The exact solution of this problem can be found with the help of the Kalman recursions (see TSTM, Section 13.1 for details). The program TRANS uses these recursions to compute the predictors and their mean squared errors.

In order to provide a little more insight, we give here the predictors $\tilde{P}_n X_{n+h}$ and mean squared errors based on infinitely many past observations, X_{t1} and X_{t2}, $-\infty < t \le n$. These predictors and their mean squared errors will be close to those based on X_{t1} and X_{t2}, $1 \le t \le n$ if n is sufficiently large.

The transfer function model defined by (10.1.1), (10.1.4), and (10.1.7), can be rewritten as

$$X_{t2} = T(B)X_{t1} + \beta(B)W_t \tag{10.1.9}$$

$$X_{t1} = \theta(B)\phi^{-1}(B)Z_t, \tag{10.1.10}$$

where $\beta(B) = \theta^{(N)}(B)/\phi^{(N)}(B)$. Eliminating X_{t1} gives

$$X_{t2} = \sum_{j=0}^{\infty} \alpha_j Z_{t-j} + \sum_{j=0}^{\infty} \beta_j W_{t-j}, \tag{10.1.11}$$

where $\alpha(B) = T(B)\theta(B)/\phi(B)$.

Noting that each limit of linear combinations of $\{X_{t1}, X_{t2}, -\infty < t \le n\}$ is a limit of linear combinations of $\{Z_t, W_t, -\infty < t \le n\}$ and conversely and that $\{Z_t\}$ and $\{W_t\}$ are uncorrelated, we see at once from (10.1.11) that

$$\tilde{P}_n X_{n+h,2} = \sum_{j=h}^{\infty} \alpha_j Z_{n+h-j} + \sum_{j=h}^{\infty} \beta_j W_{n+h-j}. \tag{10.1.12}$$

Setting $t = n+h$ in (10.1.11) and subtracting (10.1.12) gives the mean squared error,

$$E(X_{n+h,2} - \tilde{P}nX_{n+h,2})^2 = \sigma_Z^2 \sum_{j=0}^{h-1} \alpha_j^2 + \sigma_W^2 \sum_{j=0}^{h-1} \beta_j^2. \quad (10.1.13)$$

To compute the predictors $\tilde{P}_n X_{n+h,2}$ we proceed as follows. Rewrite (10.1.9) as

$$A(B)X_{t2} = B^b U(B)X_{t1} + V(B)W_t, \quad (10.1.14)$$

where A, U, and V are polynomials of the form

$$A(B) = 1 - A_1 B - \cdots - A_a B^a,$$

$$U(B) = U_0 + U_1 B + \cdots + U_u B^u,$$

$$V(B) = 1 + V_1 B + \cdots V_v B^v.$$

Applying the operator $\tilde{P}_n$ to equation (10.1.14) with $t = n + h$, we obtain

$$\tilde{P}_n X_{n+h,2} = \sum_{j=1}^{a} A_j \tilde{P}_n X_{n+h-j,2} + \sum_{j=0}^{u} U_j \tilde{P}_n X_{n+h-b-j,1} + \sum_{j=h}^{v} V_j W_{n+h-j}, \quad (10.1.15)$$

where the last sum is zero if $h > v$.

Since $\{X_{t1}\}$ is uncorrelated with $\{W_t\}$, the predictors appearing in the second sum in (10.1.15) are therefore obtained by predicting the univariate series $\{X_{t1}\}$ as described in Section 3.3 using the model (10.1.10). In keeping with our assumption that n is large, we can replace $\tilde{P}_n X_{j1}$ for each j by the finite-past predictor obtained from the program PEST. The values W_j, $j \le n$, are replaced by their estimated values $\hat{W}_j$ from the least squares estimation in step (7) of the modelling procedure.

Equations (10.1.15) can now be solved recursively for the predictors $\tilde{P}_n X_{n+1,2}$, $\tilde{P}_n X_{n+2,2}$, $\tilde{P}_n X_{n+3,2}$, ...

Example 10.1.2 Sales with a leading indicator

Applying the preceding results to the series $\{X_{t1}, X_{t2}, 1 \le t \le 149\}$ of Example 10.1.1, and using the values $X_{147,1} = -.093$, $X_{149,2} = .08$, $\hat{W}_{149} = -.0706$, $\hat{W}_{148} = .1449$, we find from (10.1.8) and (10.1.15) that

$$\tilde{P}_{149} X_{150,2} = .724 X_{149,2} + 4.717 X_{147,1} - 1.306 W_{149} + .421 W_{148}$$

$$= -.228$$

and, using the value $X_{148,1} = .237$, that

$$\tilde{P}_{149} X_{151,2} = .724 \tilde{P}_{149} X_{150,2} + 4.717 X_{148,1} + .421 W_{149}$$

$$= .923.$$

In terms of the original sales data $\{Y_{t2}\}$ we have $Y_{149,2} = 262.7$ and

$$Y_{t2} = Y_{t-1,2} + X_{t2} + .420.$$

Hence the predictors of actual sales are

$$P_{149}^{*}Y_{150,2} = 262.70 - .228 + .420 = 262.89,$$

$$P_{149}^{*}Y_{151,2} = 262.89 + .923 + .420 = 264.23,$$

where P_{149}^{*} is based on $\{1, Y_{01}, Y_{02}, X_{s1}, X_{s2}, -\infty < s \le 149\}$, and it is assumed that Y_{01} and Y_{02} are uncorrelated with $\{X_{s1}\}$ and with $\{X_{s2}\}$. The predicted values are in close agreement with those based on the finite number of available observations that are computed by TRANS. Since our model for the sales data is

$$(1 - B)Y_{t2} = .420 + 4.717B^3(1 - .474B)(1 - .724B)^{-1}Z_t + (1 - .582B)W_t,$$

it can be shown, using an argument analogous to that which gave (10.1.13), that the mean squared errors are given by

$$E(Y_{149+h,2} - P_{149}Y_{149+h,2})^2 = \sigma_Z^2 \sum_{j=0}^{h-1} \alpha_j^{*2} + \sigma_W^2 \sum_{j=0}^{h-1} \beta_j^{*2},$$

where

$$\sum_{j=0}^{\infty} \alpha_j^{*} z^j = 4.717z^3(1 - .474z)(1 - .724z)^{-1}(1 - z)^{-1}$$

and

$$\sum_{j=0}^{\infty} \beta_j^{*} z^j = (1 - .582z)(1 - z)^{-1}.$$

For $h = 1$ and 2 we obtain

$$E(Y_{150,2} - P_{149}^{*}Y_{150,2})^2 = .0486,$$

$$E(Y_{151,2} - P_{149}^{*}Y_{151,2})^2 = .0570,$$

in close agreement with the finite-past mean squared errors obtained by TRANS.

It is interesting to examine the improvement obtained by using the transfer function model rather than fitting a univariate model to the sales data alone. If we adopt the latter course we find the model

$$X_{t2} - .249X_{t-1,2} - .199X_{t-2,2} = U_t,$$

where $\{U_t\} \sim \text{WN}(0, 1.794)$ and $X_{t2} = Y_{t2} - Y_{t-1,2} - .420$. The corresponding predictors of $Y_{150,2}$ and $Y_{151,2}$ are easily found from the program PEST to be 263.14 and 263.58 with mean squared errors 1.794 and 4.593 respectively. These mean squared errors are dramatically worse than those obtained using the transfer function model. □

10.2 Intervention Analysis

During the period for which a time series is observed, it is sometimes the case that a change occurs which affects the level of the series. A change in the tax laws may, for example, have a continuing effect on the daily closing prices of shares on the stock market. In the same way construction of a dam on a river may have a dramatic effect on the time series of streamflows below the dam. In the following we shall assume that the time T at which the change (or "intervention") occurs is known.

To account for such changes, Box and Tiao (1975) introduced a model for intervention analysis that has the same form as the transfer function model,

$$Y_t = \sum_{j=0}^{\infty} \tau_j X_{t-j} + N_t, \tag{10.2.1}$$

except that the input series $\{X_t\}$ is not a random series but a deterministic function of t. It is clear from (10.2.1) that $\sum_{j=0}^{\infty} \tau_j X_{t-j}$ is then the mean of Y_t. The function $\{X_t\}$ and the coefficients $\{\tau_j\}$ are therefore chosen in such a way that the changing level of the observations of $\{Y_t\}$ is well represented by the sequence $\sum_{j=0}^{\infty} \tau_j X_{t-j}$. For a series $\{Y_t\}$ with $EY_t = 0$ for $t \le T$ and $EY_t \to 0$ as $t \to \infty$, a suitable input series is

$$X_t = I_t(T) = \begin{cases} 1 & \text{if } t = T, \\ 0 & \text{if } t \ne T. \end{cases} \tag{10.2.2}$$

For a series $\{Y_t\}$ with $EY_t = 0$ for $t \le T$ and $EY_t \to a \ne 0$ as $t \to \infty$, a suitable input series is

$$X_t = H_t(T) = \sum_{k=T}^{\infty} I_t(k) = \begin{cases} 1 & \text{if } t \ge T, \\ 0 & \text{if } t < T. \end{cases} \tag{10.2.3}$$

(Other deterministic input functions $\{X_t\}$ can also be used, for example when interventions occur at more than one time.) Having selected the function $\{X_t\}$ by inspection of the data, the determination of the coefficients $\{\tau_j\}$ in (10.2.1) then reduces to a regression problem in which the errors $\{N_t\}$ constitute an ARMA process. This problem can be solved using the program TRANS as described below.

The goal of intervention analysis is to estimate the effect of the intervention as indicated by the term $\sum_{j=0}^{\infty} \tau_j X_{t-j}$ and to use the resulting model (10.2.1) for forecasting. For example, Wichern and Jones (1978) used intervention analysis to investigate the effect of the American Dental Association's endorsement of Crest toothpaste on Crest's market share. Other applications of intervention analysis can be found in Box and Tiao (1975), Atkins (1979), and Bhattacharyya and Layton (1979). A more general approach can also be found in West and Harrison (1989), Harvey (1990), and Pole, West, and Harrison (1994).

As in the case of transfer function modelling, once $\{X_t\}$ has been chosen (usually as either (10.2.2) or (10.2.3)), estimation of the linear filter $\{\tau_j\}$ in (10.2.1) is simplified by approximating the operator $T(B) = \sum_{j=0}^{\infty} \tau_j B^j$ with a rational operator of the form

$$T(B) = \frac{B^b W(B)}{V(B)}, \tag{10.2.4}$$

where b is the delay parameter and $W(B)$ and $V(B)$ are polynomials of the form

$$W(B) = w_0 + w_1 B + \cdots + w_q B^q$$

and

$$V(B) = 1 - v_1 B - \cdots - v_p B^p.$$

By suitable choice of the parameters b, q, p, and the coefficients w_i and v_j, the intervention term $T(B)X_t$ can made to take a great variety of functional forms.

For example, if $T(B) = wB^2/(1 - vB)$ and $X_t = I_t(T)$ as in (10.2.2), the resulting intervention term is

$$\frac{wB^2}{(1 - vB)} I_t(T) = \sum_{j=0}^{\infty} v^j w I_{t-j-2}(T) = \sum_{j=0}^{\infty} v^j w I_t(T + 2 + j),$$

a series of pulses of sizes $v^j w$ at times $T + 2 + j$, $j = 0, 1, 2, \ldots$. If $|v| < 1$ the effect of the intervention is to add a series of pulses with size w at time $T + 2$, decreasing to zero at a geometric rate depending on v as $t \to \infty$. Similarly, with $X_t = H_t(T)$ as in (10.2.3)

$$\frac{wB^2}{(1 - vB)} H_t(T) = \sum_{j=0}^{\infty} v^j w H_{t-j-2}(T) = \sum_{j=0}^{\infty} (1 + v + \cdots + v^j) w I_t(T + 2 + j),$$

a series of pulses of sizes $(1 + v + \cdots + v^j)w$ at times $T + 2 + j$, $j = 0, 1, 2, \ldots$. If $|v| < 1$ the effect of the intervention is to bring about a shift in level of the series X_t, the size of the shift converging to $w/(1 - v)$ as $t \to \infty$.

Having chosen an appropriate form for X_t and possible values of b, q, and p by inspection of the data, the estimation of the parameters in (10.2.4) and the fitting of the model for $\{N_t\}$ can be carried out using steps (6) – (9) of the transfer function modelling procedure described in Section 10.1. Start with step (7) and assume $\{N_t\}$ is white noise to get preliminary estimates of the coefficients w_i and v_j by least squares. The residuals are filed and used as estimates of $\{N_t\}$. Then go to step (6) and continue exactly as for transfer function modelling with input series $\{X_t\}$ and output series $\{Y_t\}$.

Example 10.2.1 Car deaths and injuries; CDID.DAT

The series CDID.DAT shown in Figure 10.3 is $\{Y_t = (1 - B^4)D_t - 305.2\}$, where $\{D_t\}$ is the number of car drivers killed or seriously injured (quarterly) in Great Britain in the years 1969–1984. The data $\{D_t\}$ are from Pole, West, and Harrison (1994). There is an intervention here at the sixteenth value of Y_t, corresponding to the imposition

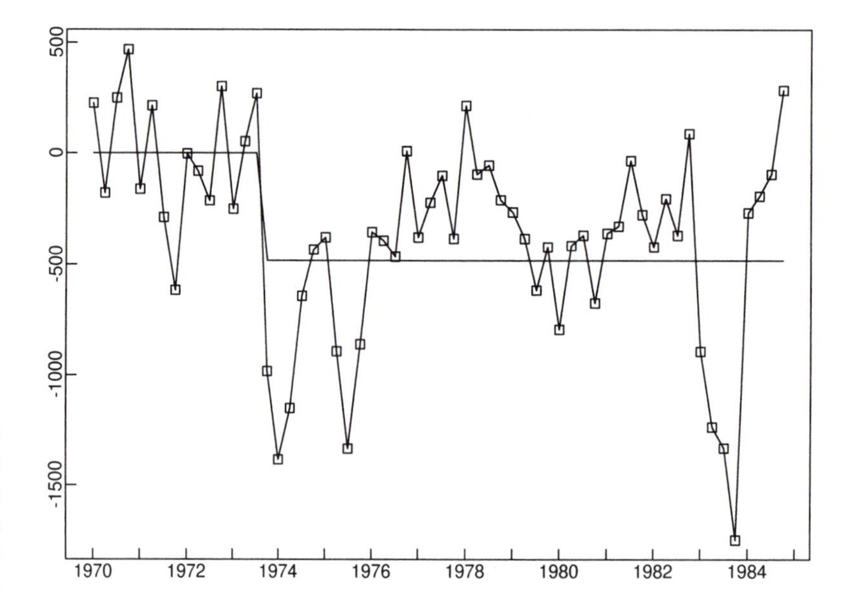

Figure 10-3
The time series of Example 10.2.1 (showing also the fitted intervention term to account for the Arab oil embargo of 1973).

of the Arab oil embargo in November 1973. The data suggest a simple level change that can be modelled by an intervention term of the form $wX(t)$ with $\{X_t\}$ the step function $\{H_t(16)\}$ defined by (10.2.3). To fit an intervention model we first use a text editor or WORD6 to create a data file (CDIN.DAT) containing the input series

$$X_t = \begin{cases} 0, & \text{if } 0 \le t \le 15, \\ 1, & \text{if } 16 \le t \le 60. \end{cases} \tag{10.2.5}$$

(You will find that this series has already been created and stored on the ITSM diskette.)

The parameter w and a model for the noise term N_t in (10.2.1) are obtained as follows.

1. Run the program TRANS and select the option *Transfer function modelling*. Select CDIN.DAT as the first series and CDID.DAT as the second series and do not mean-correct the data. For the input model enter the values, $p_1 = q_1 = 0$ and WNV=1 (the input model is irrelevant for intervention analysis). Then enter the delay parameter $b = 0$ and $r = 0$, $w(0) = -500$ and $s = 0$. Enter $p = q = 0$ to indicate that the noise term is initially assumed to be white noise. Then choose the option *Least squares estimation* with step sizes 0.1, then 0.01, and finally 0.001 to get the estimated value -486 for $w(0)$. File the residuals as NOISE.DAT. This series estimates $\{N_t\}$ in the model (10.2.1).
2. Return to the main menu of TRANS and exit from the program.

3. Run PEST to fit an ARMA model to the series NOISE.DAT (with no mean correction). Maximum likelihood estimation gives the MA(4) model

$$N_t = Z_t + 0.576Z_{t-1} + 0.520Z_{t-2} + 0.420Z_{t-3} - 0.523Z_{t-4},$$

where $\{Z_t\} \sim \text{WN}(0, 73673)$.

4. Run TRANS again as in step (1) using the estimate of $w(0)$ found in (1) and the model for $\{N_t\}$ found in (3). Least squares estimation of the parameters with step sizes .1, .01, and then .001 leads to the model

$$Y_t = -483X_t + N_t,$$

where $\{X_t\}$ is defined by (10.2.3) with $T = 16$ and

$$N_t = Z_t + 0.609Z_{t-1} + 0.477Z_{t-2} + 0.358Z_{t-3} - 0.369Z_{t-4},$$

with $\{Z_t\} \sim \text{WN}(0, 103560)$. File the residuals (which are estimates of $\{Z_t\}$) as RES.DAT and exit from TRANS.

5. Using PEST we test the residual series RES.DAT for whiteness (the sample ACF of RES.DAT is displayed in Figure 10.4) and, since it passes all the tests, we conclude that the model found in Step 4 is satisfactory. □

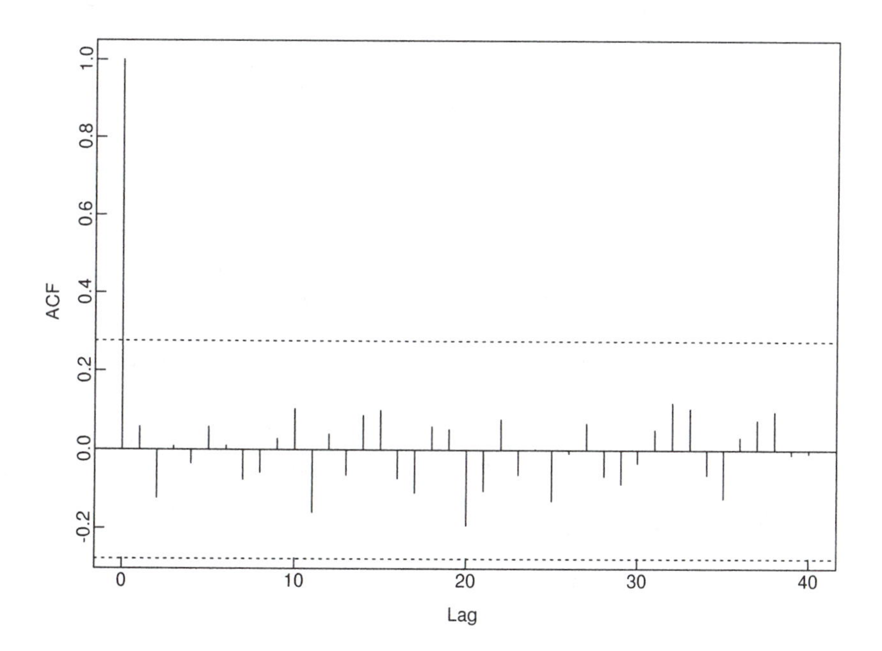

Figure 10-4 The sample ACF of the series RES.DAT of residuals from the model found in Step 4 of Example 10.2.1.

10.3 Nonlinear Models

A time series of the form

$$X_t = \sum_{j=0}^{\infty} \psi_j Z_{t-j}, \quad \{Z_t\} \sim \text{IID}(0, \sigma^2), \tag{10.3.1}$$

where Z_t is expressible as a mean square limit of linear combinations of $\{X_s, \infty < s \leq t\}$, has the property that the best mean square predictor, $E(X_{t+h}|X_s, -\infty < s \leq t)$ and the best linear predictor, $\tilde{P}_t X_{t+h}$, in terms of $\{X_s, -\infty < s \leq t\}$ are identical. It can be shown that if IID is replaced by WN in (10.3.1), then the two predictors are identical if and only if $\{Z_t\}$ is a **martingale difference sequence** relative to $\{X_t\}$, i.e., if and only if $E(Z_t|X_s, -\infty < s \leq t) = 0$ for all t.

The Wold decomposition (Section 2.6) ensures that every purely nondeterministic stationary process can be expressed in the form (10.3.1) with $\{Z_t\} \sim \text{WN}(0, \sigma^2)$. The process $\{Z_t\}$ in the Wold decomposition, however, is generally not an IID sequence and the best mean square predictor of X_{t+h} may be quite different from the best linear predictor.

In the case when $\{X_t\}$ is a purely nondeterministic Gaussian stationary process, the sequence $\{Z_t\}$ in the Wold decomposition is Gaussian and therefore IID. Every stationary purely nondeterministic Gaussian process can therefore be generated by applying a causal linear filter to an IID Gaussian sequence. We shall therefore refer to such a process as a **Gaussian linear process**.

In this section we shall use the term linear process to mean a process $\{X_t\}$ of the form (10.3.1). This is a more restrictive use of the term than in Definition 2.2.1.

10.3.1 Deviations from Linearity

Many of the time series encountered in practice exhibit characteristics not shown by linear processes, and so to obtain good models and predictors it is necessary to look to models more general than those satisfying (10.3.1) with IID noise. As indicated above, this will mean that the minimum mean squared error predictors are not in general linear functions of the past observations.

Gaussian linear processes have a number of properties that are often found to be violated by observed time series. The former are reversible in the sense that $(X_{t_1}, \cdots, X_{t_n})'$ has the same distribution as $(X_{t_n}, \cdots, X_{t_1})'$. (Except in a few special cases, ARMA processes are reversible if and only if they are Gaussian (Breidt and Davis, 1992).) Deviations from this property by observed time series are suggested by sample paths that rise to their maxima and fall away at different rates (see, for example, the sunspot numbers filed as SUNSPOTS.DAT). Bursts of outlying values are frequently observed in practical time series and are seen also in the sample paths of nonlinear (and infinite-variance) models. They are rarely seen, however, in the sample paths of Gaussian linear processes. Other characteristics suggesting deviation from a Gaussian linear model are discussed by Tong (1990).

Many observed time series, particularly financial time series, exhibit periods during which they are "less predictable" (or "more volatile"), depending on the past history of the series. This dependence of the predictability (i.e., the size of the prediction mean squared error) on the past of the series cannot be modelled with a linear time series, since for a linear process the minimum h-step mean squared error is independent of the past history. Linear models thus fail to take account of the possibility that certain past histories may permit more accurate forecasting than others, and cannot identify the circumstances under which more accurate forecasts can be expected. Nonlinear models on the other hand do allow for this. The ARCH models considered below are in fact constructed around the dependence of the conditional variance of the process on its past history.

10.3.2 Chaotic Deterministic Sequences

To distinguish between linear and nonlinear processes, we need to be able to decide in particular when a white noise sequence is also IID. Sequences generated by nonlinear deterministic difference equations can exhibit sample correlation functions that are very close to those of samples from a white noise sequence. However, the deterministic nature of the recursions implies the strongest possible dependence between successive observations. For example, the celebrated logistic equation (see May, 1976, and Tong, 1990) defines a sequence $\{x_n\}$, for any given x_0, via the equations

$$x_n = 4x_{n-1}(1 - x_{n-1}), \quad 0 < x_0 < 1.$$

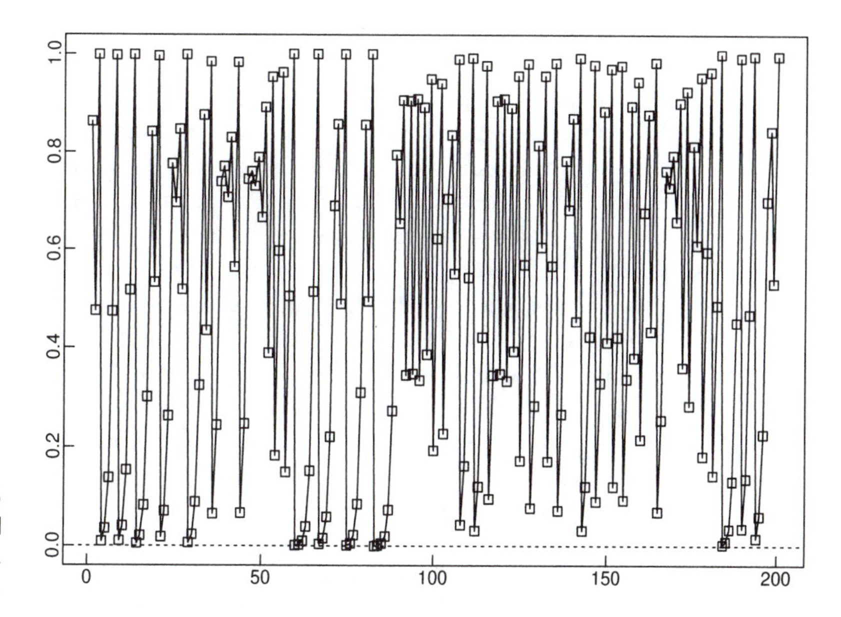

Figure 10-5
A sequence generated by the recursions, $x_n = 4x_{n-1}(1 - x_{n-1})$.

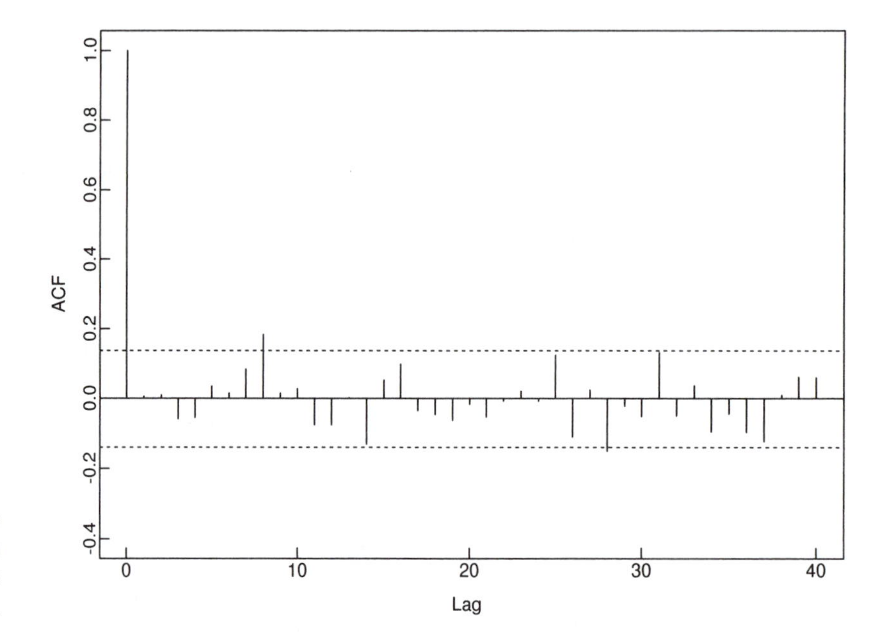

Figure 10-6
The sample autocorrelation function of the sequence in Figure 10.5.

The values of x_n are, for even moderately large values of n, extremely sensitive to small changes in x_0. This is clear from the fact that the sequence can be expressed explicitly as

$$x_n = \sin^2\left(2^n \arcsin(\sqrt{x_0})\right), \quad n = 0, 1, 2, \ldots.$$

A very small change δ in $\arcsin(\sqrt{x_0})$ leads to a change $2^n\delta$ in the argument of the sin function defining x_n. If we generate a sequence numerically, the generated sequence will, for most values of x_0 in the interval (0,1), be random in appearance, with a sample autocorrelation function similar to that of a sample from white noise. The data file CHAOS.DAT contains the sequence $x_1, \ldots, x_{200}$ (correct to nine decimal places) generated by the logistic equation with $x_0 = \pi/10$. The calculation requires specification of x_0 to at least 70 decimal places and the use of correspondingly high precision arithmetic. The series and its sample autocorrelation function are shown in Figures 10.5 and 10.6. The sample ACF and the AICC criterion both suggest white noise with mean .4954 as a model for the series. Under this model the best linear predictor of X_{201} would be .4954. However the best predictor of X_{201} to nine decimal places is in fact $4x_{200}(1 - x_{200}) = 0.016286669$, with zero mean-squared error.

Distinguishing between IID and non-IID white noise is clearly not possible on the basis of second-order properties, so we must consider moments of higher order. In the following paragraphs we consider several approaches to this problem.

10.3.3 Distinguishing Between White Noise and IID Sequences

If $\{X_t\} \sim \text{WN}(0, \sigma^2)$ and $E|X_t|^4 < \infty$, a useful tool for deciding whether or not $\{X_t\}$ is IID is the ACF $\rho_{X^2}(h)$ of the process $\{X_t^2\}$. If $\{X_t\}$ is IID then $\rho_{X^2}(h) = 0$ for all $h \neq 0$, whereas this is not necessarily the case otherwise. This is the basis for the test of McLeod and Li described in Section 1.6.

Now suppose that $\{X_t\}$ is a strictly stationary time series such that $E|X_t|^k \le K < \infty$ for some integer $k \ge 3$. The kth order cumulant $C_k(r_1, \ldots, r_{k-1})$ of $\{X_t\}$ is then defined as the joint cumulant of the random variables, $X_t, X_{t+r_1}, \ldots, X_{t+r_{k-1}}$, i.e., as the coefficient of $i^k z_1 z_2 \cdots z_k$ in the Taylor expansion about $(0, \ldots, 0)$ of

$$\chi(z_1, \ldots, z_k) := \ln E[\exp(iz_1 X_t + iz_2 X_{t+r_1} + \cdots + iz_k X_{t+r_{k-1}})]. \quad (10.3.2)$$

(Since $\{X_t\}$ is strictly stationary, this quantity does not depend on t.) In particular, the third-order cumulant function C_3 of $\{X_t\}$ coincides with the third-order central moment function, i.e.,

$$C_3(r, s) = E[(X_t - \mu)(X_{t+r} - \mu)(X_{t+s} - \mu)], \ \ r, s \in \{0, \pm 1, \cdots\},$$

where $\mu = EX_t$. If $\sum_r \sum_s |C_3(r, s)| < \infty$, we define the third-order polyspectral density (or bispectral density) of $\{X_t\}$ to be the Fourier transform

$$f_3(\omega_1, \omega_2) = \frac{1}{(2\pi)^2} \sum_{r=-\infty}^{\infty} \sum_{s=-\infty}^{\infty} C_3(r, s) e^{-ir\omega_1 - is\omega_2}, \quad -\pi \le \omega_1, \omega_2 \le \pi,$$

in which case

$$C_3(r, s) = \int_{-\pi}^{\pi} \int_{-\pi}^{\pi} e^{ir\omega_1 + is\omega_2} f_3(\omega_1, \omega_2) d\omega_1 \, d\omega_2.$$

[More generally, if the kth order cumulants $C_k(r_1, \cdots, r_{k-1})$, of $\{X_t\}$ are absolutely summable, we define the kth order polyspectral density as the Fourier transform of C_k. For details see Rosenblatt (1985) and Priestley (1988).]

If $\{X_t\}$ is a Gaussian linear process, it follows from Problem 10.3 that the cumulant function C_3 of $\{X_t\}$ is identically zero. (The same is also true of all the cumulant functions C_k with $k > 3$.) Consequently $f_3(\omega_1, \omega_2) = 0$ for all $\omega_1, \omega_2 \in [-\pi, \pi]$. Appropriateness of a Gaussian linear model for a given data set can therefore be checked by using the data to test the null hypothesis, $f_3 = 0$. For details of such a test, see Subba-Rao and Gabr (1984).

If $\{X_t\}$ is a linear process of the form (10.3.1) with $E|Z_t|^3 < \infty$, $EZ_t^3 = \eta$ and $\sum_{j=0}^{\infty} |\psi_j| < \infty$, it can be shown from (10.3.2) (see Problem 10.3) that the third-order cumulant function of $\{X_t\}$ is given by

$$C_3(r, s) = \eta \sum_{i=-\infty}^{\infty} \psi_i \psi_{i+r} \psi_{i+s} \quad (10.3.3)$$

(with $\psi_j = 0$ for $j < 0$), and hence that $\{X_t\}$ has bispectral density,

$$f_3(\omega_1, \omega_2) = \frac{\eta}{4\pi^2}\psi(e^{i(\omega_1+\omega_2)})\psi(e^{-i\omega_1})\psi(e^{-i\omega_2}), \tag{10.3.4}$$

where $\psi(z) := \sum_{j=0}^{\infty}\psi_j z^j$. By Proposition 4.3.1, the spectral density of $\{X_t\}$ is

$$f(\omega) = \frac{\sigma^2}{2\pi}|\psi(e^{-i\omega})|^2.$$

Hence,

$$\phi(\omega_1, \omega_2) := \frac{|f_3(\omega_1, \omega_2)|^2}{f(\omega_1)f(\omega_2)f(\omega_1+\omega_2)} = \frac{\eta^2}{2\pi\sigma^6}.$$

Appropriateness of the linear process (10.3.1) for modelling a given data set can therefore be checked by using the data to test for constancy of $\phi(\omega_1, \omega_2)$ (see Subba-Rao and Gabr, 1984).

10.3.4 Three Useful Classes of Nonlinear Models

If it is decided that a linear Gaussian model is not appropriate, there is a choice of several families of nonlinear processes that have been found useful for modelling purposes. These include bilinear models, autoregressive models with random coefficients, and threshold models. Excellent accounts of these are available in Subba-Rao and Gabr (1984), Nicholls and Quinn (1982), and Tong (1990), respectively.

The bilinear model of order (p, q, r, s) is defined by the equations

$$X_t = Z_t + \sum_{i=1}^{p} a_i X_{t-i} + \sum_{j=1}^{q} b_j Z_{t-j} + \sum_{i=1}^{r}\sum_{j=1}^{s} c_{ij} X_{t-i} Z_{t-j},$$

where $\{Z_t\} \sim \text{IID}(0, \sigma^2)$. A sufficient condition for the existence of a strictly stationary solution of these equations is given by Liu and Brockwell (1988).

A random coefficient autoregressive process $\{X_t\}$ of order p satisfies an equation of the form

$$X_t = \sum_{i=1}^{p}(\phi_i + U_t^{(i)})X_{t-i} + Z_t,$$

where $\{Z_t\} \sim \text{IID}(0, \sigma^2)$, $\{U_t^{(i)}\} \sim \text{IID}(0, \nu^2)$, $\{Z_t\}$ is independent of $\{U_t\}$ and $\phi_1, \ldots, \phi_p \in \mathbb{R}$.

Threshold models can be regarded as piecewise linear models in which the linear relationship varies with the values of the process. For example, if $R^{(i)}$, $i = 1, \ldots, k$ is a partition of $\mathbb{R}^p$, and $\{Z_t\} \sim \text{IID}(0, 1)$, then the k difference equations

$$X_t = \sigma^{(i)} Z_t + \sum_{j=1}^{p}\phi_j^{(i)} X_{t-j}, \quad (X_{t-1}, \cdots, X_{t-p}) \in R^{(i)}, \quad i = 1, \cdots, k, \tag{10.3.5}$$

define a threshold AR(p) model. Model identification and parameter estimation for threshold models can be carried out in a manner similar to that for linear models using maximum likelihood and the AIC criterion.

10.3.5 Modelling Volatility

For modelling changing volatility as discussed above under deviations from linearity, Engle (1982) introduced the **ARCH(p) process** $\{X_t\}$ as a solution of the equations

$$X_t = \sigma_t Z_t, \quad \{Z_t\} \sim \text{IID N}(0, 1), \tag{10.3.6}$$

where σ_t is the (positive) function of $\{X_s, s < t\}$, defined by

$$\sigma_t^2 = \alpha_0 + \sum_{i=1}^{p} \alpha_i X_{t-i}^2, \tag{10.3.7}$$

with $\alpha_0 > 0$ and $\alpha_j \geq 0$, $j = 1, \ldots, p$. The name ARCH signifies autoregressive conditional heteroscedasticity.

The simplest such process is the ARCH(1) process. In this case the recursions (10.3.6) and (10.3.7) give

$$\begin{aligned} X_t^2 &= \alpha_0 Z_t^2 + \alpha_1 X_{t-1}^2 Z_t^2 \\ &= \alpha_0 Z_t^2 + \alpha_1 \alpha_0 Z_t^2 Z_{t-1}^2 + \alpha_1^2 X_{t-2}^2 Z_t^2 Z_{t-1}^2 \\ &= \cdots \\ &= \alpha_0 \sum_{j=0}^{n} \alpha_1^j Z_t^2 Z_{t-1}^2 \cdots Z_{t-j}^2 + \alpha_1^{n+1} X_{t-n-1}^2 Z_t^2 Z_{t-1}^2 \cdots Z_{t-n}^2. \end{aligned}$$

If $|\alpha_1| < 1$ and $\{X_t\}$ is stationary and causal (i.e., X_t is a function of $\{Z_s, s \leq t\}$), then the expectation of the last term converges to 0 and the expectation of the sum converges to $\alpha_0 \sum_{j=0} \alpha_1^j = \alpha_0/(1 - \alpha_1) < \infty$ as $n \to \infty$. It follows that the series, $\alpha_0 \sum_{j=0}^{n} \alpha_1^j Z_t^2 Z_{t-1}^2 \cdots Z_{t-j}^2$, converges (see Proposition 3.1.1 of TSTM) and hence

$$X_t^2 = \alpha_0 \sum_{j=0}^{\infty} \alpha_1^j Z_t^2 Z_{t-1}^2 \cdots Z_{t-j}^2. \tag{10.3.8}$$

From (10.3.8) we immediately find that

$$EX_t^2 = \alpha_0/(1 - \alpha_1). \tag{10.3.9}$$

Since

$$X_t = Z_t \sqrt{\alpha_0 \left(1 + \sum_{j=1}^{\infty} \alpha_1^j Z_{t-1}^2 \cdots Z_{t-j}^2\right)}, \tag{10.3.10}$$

it is clear that $\{X_t\}$ is strictly stationary and hence, since $EX_t^2 < \infty$, also stationary in the weak sense. We have now established the following result.

Solution of the ARCH(1) Equations:

If $|\alpha_1| < 1$, the unique causal stationary solution of the ARCH(1) equations is given by (10.3.10). It has the properties

$$E(X_t) = E(E(X_t|Z_s, s < t)) = 0,$$

$$\text{Var}(X_t) = \alpha_0/(1 - \alpha_1),$$

and

$$E(X_{t+h}X_t) = E(E(X_{t+h}X_t|Z_s, s < t + h)) = 0 \text{ for } h > 0.$$

Thus the ARCH(1) process with $|\alpha_1| < 1$ is strictly stationary white noise. However, it is not an IID sequence since, from (10.3.6) and (10.3.7),

$$E(X_t^2|X_{t-1}) = (\alpha_0 + \alpha_1 X_{t-1}^2)E(Z_t^2|X_{t-1}) = \alpha_0 + \alpha_1 X_{t-1}^2.$$

This also shows that $\{X_t\}$ is not Gaussian since strictly stationary Gaussian white noise is necessarily IID. From (10.3.10) it is clear that the distribution of X_t is symmetric, i.e., that X_t and $-X_t$ have the same distribution. From (10.3.8) it is easy to calculate $E(X_t^4)$ (Problem 10.4) and hence to show that $E(X_t^4)$ is finite if and only if $3\alpha_1^2 < 1$. More generally (see Engle, 1982), it can be shown that for every α_1 in the interval $(0, 1)$, $E(X^{2k}) = \infty$ for some positive integer k. This indicates the "heavy-tailed" nature of the marginal distribution of X_t. If $EX_t^4 < \infty$, the squared process, $Y_t = X_t^2$ has the same ACF as the AR(1) process $W_t = \alpha_1 W_{t-1} + Z_t$, a result that extends also to ARCH(p) processes (see Problem 10.5).

The ARCH(p) process is conditionally Gaussian, in the sense that for given values of $\{X_s, s = t-1, t-2, \ldots, t-p\}$, X_t is Gaussian with known distribution. This makes it easy to write down the likelihood of $X_{p+1}, \ldots, X_n$ conditional on $\{X_1, \ldots, X_p\}$ and hence, by numerical maximization, to compute conditional maximum likelihood estimates of the parameters. For example, the conditional likelihood of observations $\{x_2, \ldots, x_n\}$ of observations of the ARCH(1) process given $X_1 = x_1$ is

$$L = \prod_{t=2}^{n} \frac{1}{\sqrt{2\pi(\alpha_0 + \alpha_1 x_{t-1}^2)}} \exp\left\{-\frac{x_t^2}{2(\alpha_0 + \alpha_1 x_{t-1}^2)}\right\}.$$

Example 10.3.1 An ARCH(1) series

Figure 10.7 shows a realization of the ARCH(1) process with $\alpha_0 = 1$ and $\alpha_1 = 0.5$. The graph of the realization and the sample autocorrelation function shown in Figure 10.8 suggest that the process is white noise. This conclusion is correct from a second-order point of view. However, the fact that the series is not a realization of IID noise is very strongly indicated by Figure 10.9, which shows the sample autocorrelation function of the series $\{X_t^2\}$. It is instructive to apply the Ljung-Box and McLeod-Li

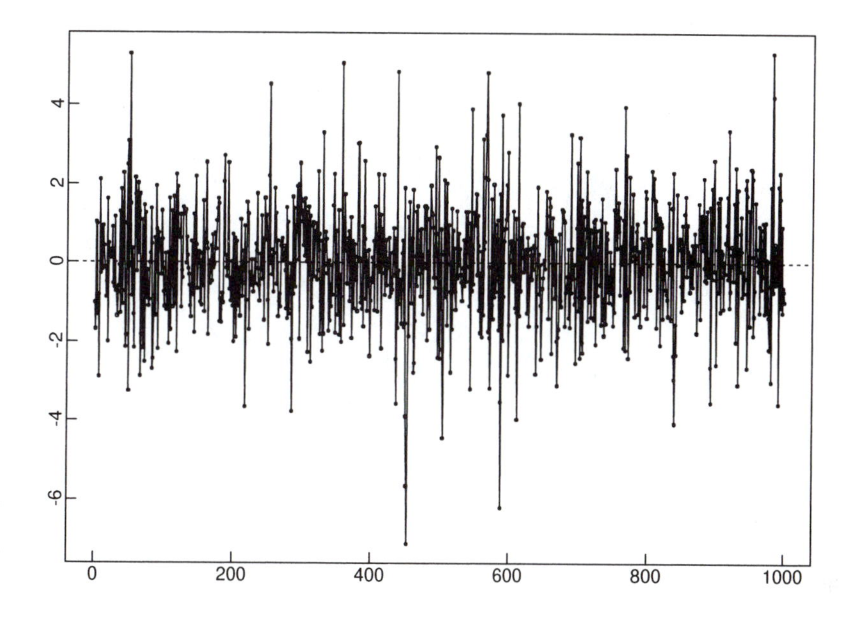

Figure 10-7
A realization of the process $X_t = Z_t\sqrt{1 + 0.5X_{t-1}^2}$.

portmanteau tests for white noise to this series (see Section 1.6). To do this using PEST, read the series ARCH.DAT into the program, do not subtract the mean, then go to the option *ARMA parameter estimation* and compute the likelihood of the data with no optimization. The residuals from this calculation will be the same as the original data. Then choose the options *File and Analyze Residuals* followed by *Tests of randomness*. We find (with $h = 20$) the Ljung-Box portmanteau statistic has the value

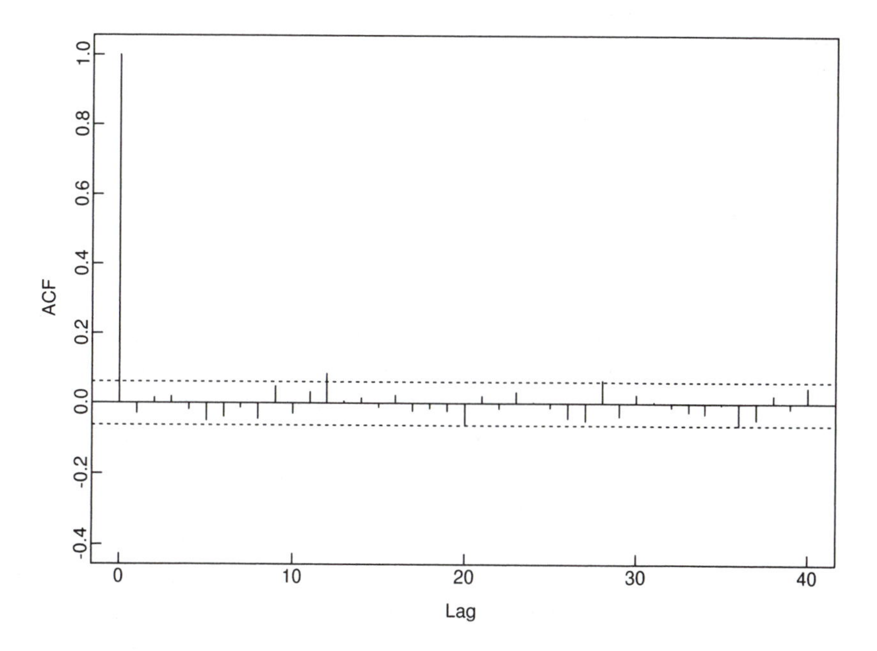

Figure 10-8
The sample autocorrelation function of the series in Figure 10.7.

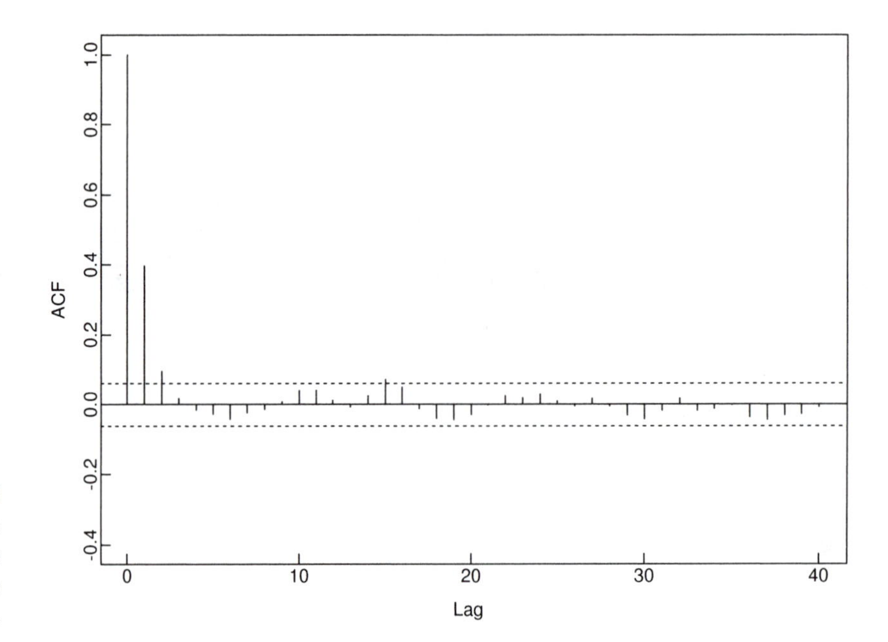

Figure 10-9
The sample autocorrelation function of the squares of the data shown in Figure 10.7.

24.95 while the McLeod-Li statistic is 187.71. Under the null hypothesis of Gaussian IID white noise, both of these statistics are approximately distributed as chi-squared with 20 degrees of freedom. Unlike the Ljung-Box test, the McLeod-Li statistic very clearly rejects this hypothesis (with p-value much smaller than 10^{-6}). □

The GARCH(p, q) process (see Bollerslev, 1986) is a generalization of the ARCH(p) process in which the variance equation (10.3.7) is replaced by

$$\sigma_t^2 = \alpha_0 + \sum_{i=1}^{p} \alpha_i X_{t-i}^2 + \sum_{j=1}^{q} \beta_j \sigma_{t-j}^2, \tag{10.3.11}$$

with $\alpha_0 > 0$ and $\alpha_j, \beta_j \geq 0,\ j = 1, 2, \ldots$.

10.4 Continuous-Time Models

Discrete time series are often obtained by observing a continuous-time process at a discrete sequence of observation times. It is then natural, even though the observations are made at discrete times, to model the underlying process as a continuous-time series. Even if there is no underlying continuous-time process, it may still be advantageous to model the data as observations of a continuous-time process at discrete times. The analysis of time series data observed at irregularly spaced times can be handled very conveniently via continuous-time models as pointed out by Jones (1980).

Continuous-time ARMA processes are defined in terms of stochastic differential equations analogous to the difference equations that are used to define discrete-time

ARMA processes. Here we shall confine attention to the continuous-time AR(1) process, which is defined as a stationary solution of the first-order stochastic differential equation

$$DX(t) + aX(t) = \sigma DB(t) + b, \tag{10.4.1}$$

where the operator D denotes differentiation with respect to t, $\{B(t)\}$ is standard Brownian motion, and a, b, and σ are parameters. The derivative $DB(t)$ does not exist in the usual sense, so equation (10.4.1) is interpreted as an Itô differential equation,

$$dX(t) + aX(t)dt = \sigma dB(t) + b\,dt, \quad t > 0, \tag{10.4.2}$$

with $dX(t)$ and $dB(t)$ denoting the increments of X and B in the time interval $(t, t+dt)$ and $X(0)$ a random variable with finite variance, independent of $\{B(t)\}$ (see e.g., Chung and Williams, 1990, Karatzas and Shreve, 1991, and Oksendal, 1992). The solution of (10.4.2) can be written as

$$X(t) = e^{-at}X(0) + \sigma \int_0^t e^{-a(t-u)}\,dB(u) + b\int_0^t e^{-a(t-u)}\,du,$$

or equivalently,

$$X(t) = e^{-at}X(0) + e^{-at}I(t) + be^{-at}\int_0^t e^{au}\,du, \tag{10.4.3}$$

where $I(t) = \sigma\int_0^t e^{au}dB(u)$ is an Itô integral (see Chung and Williams, 1990) satisfying $E(I(t)) = 0$ and $\text{Cov}(I(t+h), I(t)) = \sigma^2\int_0^t e^{2au}\,du$ for all $t \geq 0$ and $h \geq 0$.

If $a > 0$ and $X(0)$ has mean $\frac{b}{a}$ and variance $\frac{\sigma^2}{2a}$, it is easy to check (Problem 10.6) that $\{X(t)\}$ as defined by (10.4.3) is stationary with

$$E(X(t)) = \frac{b}{a} \quad \text{and} \quad \text{Cov}(X(t+h), X(t)) = \frac{\sigma^2}{2a}e^{-ah}, \ t, h \geq 0. \tag{10.4.4}$$

Conversely if $\{X(t)\}$ is stationary, then by equating the variances of both sides of (10.4.3), we find that $(1 - e^{-2at})\text{Var}(X(0)) = \sigma^2\int_0^t e^{-2au}\,du$ for all $t \geq 0$, and hence that $a > 0$ and $\text{Var}(X(0)) = \frac{\sigma^2}{2a}$. Equating the means of both sides of (10.4.3) then gives $E(X(0)) = \frac{b}{a}$. Necessary and sufficient conditions for $\{X(t)\}$ to be stationary are therefore $a > 0$, $E(X(0)) = \frac{b}{a}$, and $\text{Var}(X(0)) = \frac{\sigma^2}{2a}$. If $a > 0$ and $X(0)$ is $N(\frac{b}{a}, \frac{\sigma^2}{2a})$, then the CAR(1) process will also be Gaussian and strictly stationary.

If $a > 0$ and $0 \leq s \leq t$, it follows from (10.4.3) that X_t can be expressed as

$$X(t) = e^{-a(t-s)}X(s) + \frac{b}{a}(1 - e^{-a(t-s)}) + e^{-at}(I(t) - I(s)). \tag{10.4.5}$$

This shows that the process is Markovian, i.e., that the distribution of $X(t)$ given $X(u), u \leq s$ is the same as the distribution of $X(t)$ given $X(s)$. It also shows that the conditional mean and variance of $X(t)$ given $X(s)$ are

$$E(X(t)|X(s)) = e^{-a(t-s)}X(s) + \frac{b}{a}(1 - e^{-a(t-s)})$$

and

$$\text{Var}(X(t)|X(s)) = \frac{\sigma^2}{2a}\left[1 - e^{-2a(t-s)}\right].$$

We can now use the Markov property and the moments of the stationary distribution to write down the Gaussian likelihood of observations $x(t_1), \ldots, x(t_n)$ at times $t_1, \ldots, t_n$ of a CAR(1) process satisfying (10.4.1). This is just the joint density of $(X(t_1), \ldots, X(t_n))'$ at $(x(t_1), \ldots, x(t_n))'$, which can be expressed as the product of the stationary density at $x(t_1)$ and the transition densities of $X(t_i)$ given $X(t_{i-1}) = x(t_{i-1})$, $i = 2, \ldots, n$. The joint density g is therefore given by

$$g(x(t_1), \ldots, x(t_n); a, b, \sigma^2) = \prod_{i=1}^{n} \frac{1}{\sqrt{v_i}} f\left(\frac{x(t_i) - m_i}{\sqrt{v_i}}\right), \tag{10.4.6}$$

where $f(x) = n(x; 0, 1)$ is the standard normal density, $m_1 = \frac{b}{a}$, $v_1 = \frac{\sigma^2}{2a}$ and for $i > 1$,

$$m_i = e^{-a(t_i - t_{i-1})}x(t_{i-1}) + \frac{b}{a}\left(1 - e^{-a(t_i - t_{i-1})}\right)$$

and

$$v_i = \frac{\sigma^2}{2a}\left[1 - e^{-2a(t_i - t_{i-1})}\right].$$

The maximum likelihood estimators of a, b, and σ^2 are the values that maximize $g(x(t_1), \ldots, x(t_n); a, b, \sigma^2)$. These can be found with the aid of a nonlinear maximization algorithm. Notice that the times t_i appearing in (10.4.6) are quite arbitrarily spaced. It is this feature that makes the CAR(1) process so useful for modelling irregularly spaced data.

If the observations are regularly spaced, say $t_i = i$, $i = 1, \ldots, n$ then the joint density g is exactly the same as the joint density of observations of the discrete-time Gaussian AR(1) process,

$$Y_n - \frac{b}{a} = e^{-a}\left(Y_{n-1} - \frac{b}{a}\right) + Z_n, \quad \{Z_t\} \sim \text{WN}\left(0, \frac{\sigma^2(1 - e^{-2a})}{2a}\right).$$

This shows that the "embedded" discrete-time process $\{X(i), i = 1, 2, \ldots\}$ of the CAR(1) process is a discrete-time AR(1) process with coefficient e^{-a}. This coefficient is clearly positive, immediately raising the question of whether there is a continuous-time ARMA process for which the embedded process is a discrete-time AR(1) process with negative coefficient. It can be shown (Chan and Tong, 1987) that the answer is yes and that given a discrete-time AR(1) process with negative coefficient, it can always be embedded in a suitably chosen continuous-time ARMA(2,1) process.

We define a zero-mean CARMA(p, q) process $\{Y(t)\}$ (with $0 \le q < p$) to be a stationary solution of the pth order linear differential equation,

$$D^p Y(t) + a_1 D^{p-1} Y(t) + \cdots + a_p Y(t) =$$
$$b_0 DB(t) + b_1 D^2 B(t) + \cdots + b_q D^{q+1} B(t), \tag{10.4.7}$$

where $D^{(j)}$ denotes j-fold differentiation with respect to t, $\{B(t)\}$ is standard Brownian motion and $a_1, \ldots, a_p$, $b_0, \ldots, b_q$, and c are constants. We assume that $b_q \neq 0$ and define $b_j := 0$ for $j > q$. Since the derivatives $D^j B(t)$, $j > 0$, do not exist in the usual sense, we interpret (10.4.7) as being equivalent to the *observation* and *state* equations,

$$Y(t) = \mathbf{b}'\mathbf{X}(t), \quad t \ge 0, \tag{10.4.8}$$

and

$$d\mathbf{X}(t) = A\mathbf{X}(t)\,dt + \mathbf{e}\,dB(t), \tag{10.4.9}$$

where

$$A = \begin{bmatrix} 0 & 1 & 0 & \cdots & 0 \\ 0 & 0 & 1 & \cdots & 0 \\ \vdots & \vdots & \vdots & \ddots & \vdots \\ 0 & 0 & 0 & \cdots & 1 \\ -a_p & -a_{p-1} & -a_{p-2} & \cdots & -a_1 \end{bmatrix},$$

$\mathbf{e} = [0 \;\; 0 \;\; \cdots \;\; 0 \;\; 1]'$, $\mathbf{b} = [b_0 \;\; b_1 \;\; \cdots \;\; b_{p-2} \;\; b_{p-1}]'$, and (10.4.8) is an Ito differential equation for the state vector $\mathbf{X}(t)$. (We assume also that $\mathbf{X}(0)$ is independent of $\{B(t)\}$.)

The solution of (10.4.9) can be written as

$$\mathbf{X}(t) = e^{At}\mathbf{X}(0) + \int_0^t e^{A(t-u)}\mathbf{e}\,dB(u),$$

which is stationary if and only if

$$E(\mathbf{X}(0)) = [0 \;\; 0 \;\; \cdots \;\; 0]',$$

$$\text{Cov}(\mathbf{X}(0)) = \int_0^\infty e^{Ay}\mathbf{e}\mathbf{e}'e^{A'y}\,dy,$$

and all the eigenvalues of A (i.e., the roots of $z^p + a_1 z^{p-1} + \cdots + a_p = 0$) have negative real parts.

Then $\{Y(t), t \ge 0\}$ is said to be a zero-mean **CARMA(p,q) process with parameters** $(a_1, \ldots, a_p, b_0, \ldots, b_q)$ if $q < p$ and

$$Y(t) = [b_0 \;\; b_1 \;\; \cdots \;\; b_{p-2} \;\; b_{p-1}]\mathbf{X}(t),$$

where $\{\mathbf{X}(t)\}$ is a stationary solution of (10.4.9) and $b_j := 0$ for $j > q$.

The autocovariance function of the process $\mathbf{X}(t)$ at lag h is easily found to be

$$\text{Cov}(\mathbf{X}(t+h), \mathbf{X}(t)) = e^{Ah}\Sigma, \quad h \geq 0,$$

where

$$\Sigma := \int_0^\infty e^{Ay}\mathbf{e}\,\mathbf{e}'e^{A'y}dy.$$

The mean and autocovariance function of the CARMA(p, q) process $\{Y(t)\}$ are therefore given by

$$EY(t) = 0$$

and

$$\text{Cov}(Y(t+h), Y(t)) = \mathbf{b}'e^{Ah}\Sigma\mathbf{b}.$$

Inference for continuous-time ARMA processes is more complicated than for continuous-time AR(1) processes because higher-order processes are not Markovian so the simple calculation that led to (10.4.6) must be modified. However, the likelihood of observations at times $t_1, \ldots, t_n$ can still easily be computed using the discrete-time Kalman recursions (see Jones, 1980).

Continuous-time ARMA processes with thresholds constitute a useful class of nonlinear time-series models. For example the continuous-time threshold AR(1) process with threshold at r is defined as a solution of the stochastic differential equations

$$dX(t) + a_1X(t)dt = b_1\,dt + \sigma_1 dB(t), \quad X(t) \leq r$$

and

$$dX(t) + a_2X(t)dt = b_2\,dt + \sigma_2 dB(t), \quad X(t) > r.$$

For a detailed discussion of such processes, see Stramer, Brockwell, and Tweedie (1996). Continuous-time threshold ARMA processes are discussed in Brockwell (1994). For more on continuous-time models see Bergstrom (1990) and Harvey (1990).

10.5 Long-Memory Models

The autocorrelation function $\rho(\cdot)$ of an ARMA process at lag h converges rapidly to zero as $h \to \infty$ in the sense that there exists $r > 1$ such that

$$r^h\rho(h) \to 0, \qquad \text{as} \quad h \to \infty. \tag{10.5.1}$$

Stationary processes with much more slowly decreasing autocorrelation function, known as **fractionally integrated ARMA processes**, or more precisely as ARIMA (p, d, q) processes with $0 < |d| < 0.5$, satisfy difference equations of the form

$$(1 - B)^d\phi(B)X_t = \theta(B)Z_t, \tag{10.5.2}$$

where $\phi(z)$ and $\theta(z)$ are polynomials of degrees p and q, respectively, satisfying

$$\phi(z) \neq 0 \quad \text{and} \quad \theta(z) \neq 0 \qquad \text{for all } z \text{ such that } |z| \leq 1,$$

B is the backward shift operator, and $\{Z_t\}$ is a white noise sequence with mean 0 and variance σ^2. The operator $(1 - B)^d$ is defined by the binomial expansion

$$(1 - B)^d = \sum_{j=0}^{\infty} \pi_j B^j,$$

where $n_0 = 1$ and

$$\pi_j = \prod_{0<k\leq j} \frac{k - 1 - d}{k}, \qquad j = 1, 2, \ldots .$$

The autocorrelation $\rho(h)$ at lag h of an ARIMA(p, d, q) process with $0 < |d| < 0.5$ has the property

$$\rho(h)h^{1-2d} \to c \qquad \text{as} \quad h \to \infty. \tag{10.5.3}$$

This implies (see (10.5.1)) that $\rho(h)$ converges to zero as $h \to \infty$ at a much slower rate than $\rho(h)$ for an ARMA. Consequently fractionally integrated ARMA processes are said to have "long memory." In contrast, stationary processes whose ACF converges to 0 rapidly, such as ARMA processes, are said to have "short memory."

A fractionally integrated ARIMA(p, d, q) process can be regarded as an ARMA (p, q) process driven by fractionally integrated noise, i.e., we can replace equation (10.5.2) by the two equations

$$\phi(B)X_t = \theta(B)W_t, \tag{10.5.4}$$

and

$$(1 - B)^d W_t = Z_t. \tag{10.5.5}$$

The process $\{W_t\}$ is called **fractionally integrated white noise** and can be shown (see e.g., TSTM, Section 13.2) to have variance and autocorrelations given by

$$\gamma_W(0) = \sigma^2 \frac{\Gamma(1 - 2d)}{\Gamma^2(1 - d)} \tag{10.5.6}$$

and

$$\rho_W(h) = \frac{\Gamma(h + d)\Gamma(1 - d)}{\Gamma(h - d + 1)\Gamma(d)} = \prod_{0<k\leq h} \frac{k - 1 + d}{k - d}, \qquad h = 1, 2, \ldots, \tag{10.5.7}$$

where $\Gamma(\cdot)$ is the gamma function (see Example (d) of Section A.1). The exact autocovariance function of the ARIMA(p, d, q) process $\{X_t\}$ defined by (10.5.2) can therefore be expressed, by Proposition 2.2.1, as

$$\gamma_X(h) = \sum_{j=0}^{\infty} \sum_{k=0}^{\infty} \psi_j \psi_k \gamma_W(h + j - k), \tag{10.5.8}$$

where $\sum_{i=0}^{\infty} \psi_i z^i = \theta(z)/\phi(z)$, $|z| \le 1$, and $\gamma_W(\cdot)$ is the autocovariance function of fractionally integrated white noise with parameters d and σ^2, i.e.,

$$\gamma_W(h) = \gamma_W(0)\rho_W(h),$$

with $\gamma_W(0)$ and $\rho_W(h)$ as in (10.5.6) and (10.5.7). The series (10.5.8) converges rapidly as long as $\phi(z)$ does not have zeroes with absolute value close to 1.

The spectral density of $\{X_t\}$ is given by

$$f(\lambda) = \frac{\sigma^2}{2\pi} \frac{|\theta(e^{-i\lambda})|^2}{|\phi(e^{-i\lambda})|^2} |1 - e^{-i\lambda}|^{-2d}. \tag{10.5.9}$$

Calculation of the exact Gaussian likelihood of observations $\{x_1, \ldots, x_n\}$ of a fractionally integrated ARMA process is very slow and demanding in terms of computer memory. Instead of estimating the parameters $d, \phi_1, \ldots, \phi_p, \theta_1, \ldots, \theta_q$ and σ^2 by maximizing the exact Gaussian likelihood, it is much simpler to maximize the Whittle approximation L_W, defined by

$$-2\ln(L_W) = n\ln(2\pi) + 2n\ln\sigma + \sigma^{-2}\sum_j \frac{I_n(\omega_j)}{g(\omega_j)} + \sum_j \ln g(\omega_j), \tag{10.5.10}$$

where I_n is the periodogram, $\sigma^2 g(= f)$ is the model spectral density, and $\sum_j$ denotes the sum over all nonzero Fourier frequencies, $\omega_j = 2\pi j/n \in (-\pi, \pi]$. The program LONGMEM estimates parameters for ARIMA(p, d, q) models in this way. It can also be used to predict and simulate fractionally integrated ARMA series and to compute the autocovariance function of any specified fractionally integrated ARMA model.

Example 10.5.1 Estimated Australian residents, 1971–1994; AUSTRES.DAT

The data file AUSTRES.DAT contains quarterly estimates of the number of Australian residents (in thousands) from March 1971, through March 1994. The file RESPC.DAT contains the corresponding quarterly percentage changes $\{x_1, \ldots, x_{88}\}$. These values are plotted in Figure 10.10 with the corresponding sample autocorrelations shown in Figure 10.11. The rather slow decay of the sample autocorrelation function suggests the possibility of a fractionally integrated model for the mean-corrected series $Y_t = X_t - 0.343$.

The program LONGMEM estimates $d, \phi_1, \ldots, \phi_p, \theta_1, \ldots, \theta_q$ and σ^2 by minimizing $-2\ln L_W$ as defined in (10.5.10). Minimization can be carried out either with respect to all the parameters or with d constrained to be zero (or some other value). For the purpose of comparing different candidate models an AIC value is computed as $-2\ln L_W + 2(p+q+2)$ if d is estimated and $-2\ln L_W + 2(p+q+1)$ otherwise.

The minimum AIC model for $\{Y_t\}$ obtained from LONGMEM with d constrained to be zero is the AR(4) defined by

$$X_t - 0.431X_{t-1} - 0.033X_{t-2} - 0.114X_{t-3} - 0.162X_{t-4} = Z_t, \tag{10.5.11}$$

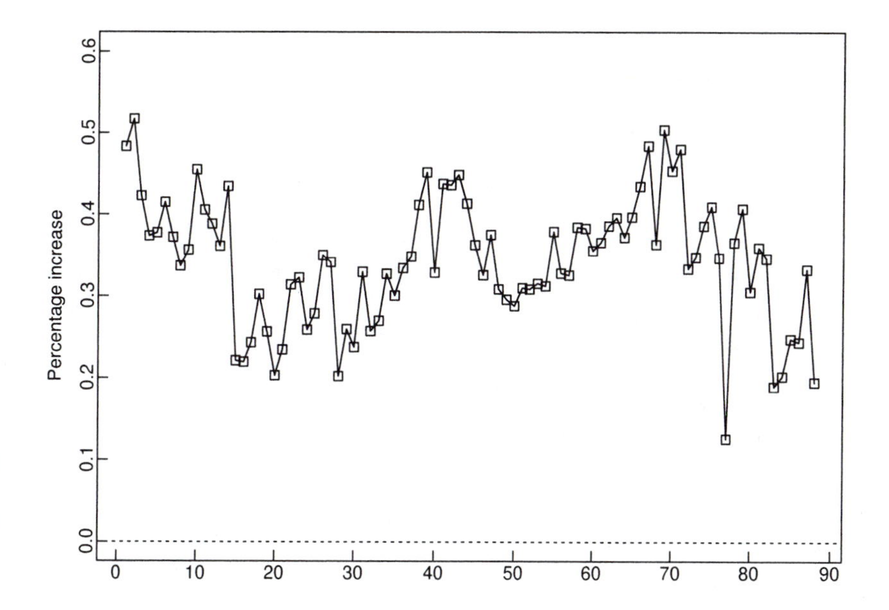

Figure 10-10 The quarterly percentage increase $\{x_t\}$ in estimated residents of Australia, March 1971–March 1994.

where $\{Z_t\} \sim \text{WN}(0, 0.00434)$, and AIC$= -221.6$. (A very similar AR(4) model is also obtained from PEST by minimizing the AIC based on the *exact* likelihood rather than the Whittle approximation used by LONGMEM.)

If on the other hand we minimize with respect to d also we find that the minimum AIC model for $\{Y_t\}$ is the fractionally integrated noise process defined by

$$(1 - B)^{0.468} Y_t = Z_t, \tag{10.5.12}$$

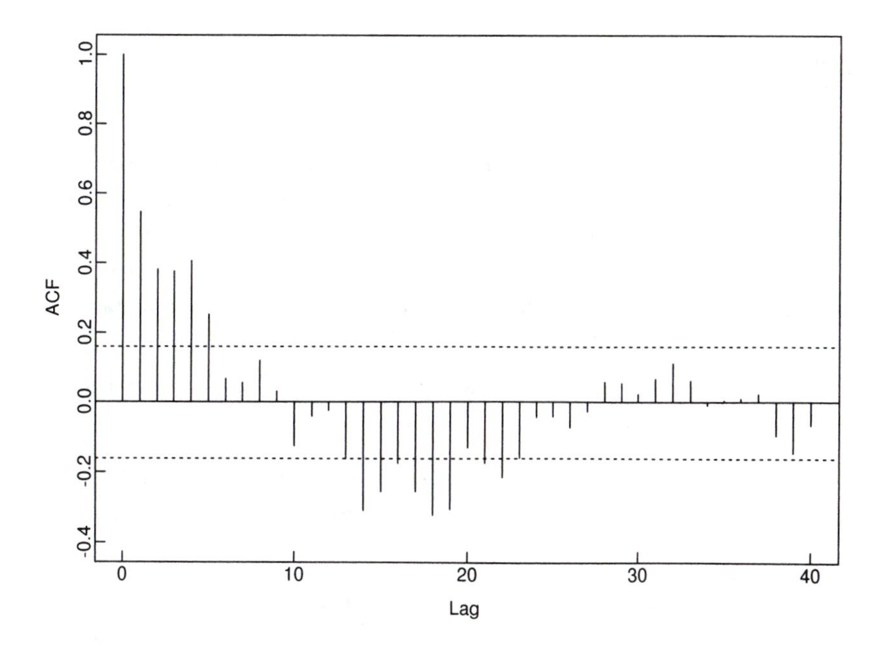

Figure 10-11 The sample correlation function of the data in Figure 10.10.

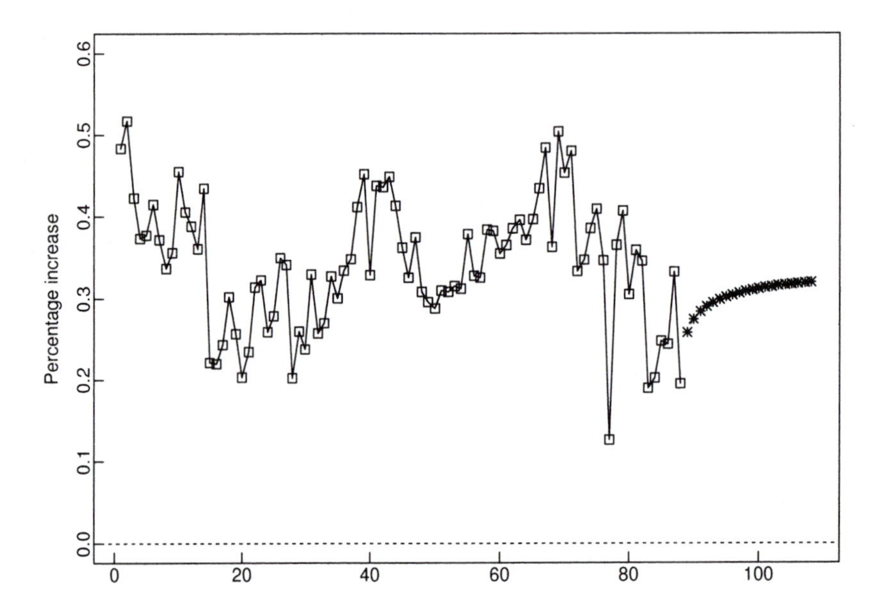

Figure 10-12 The quarterly percentage increase $\{x_t\}$ in estimated residents of Australia, March 1971–March 1994, with 20 forecasts based on the model (10.5.12).

where $\{Z_t\} \sim \text{WN}(0, 0.00464)$, and $\text{AIC} = -223.4$.

Figure 10.12 shows the graph of $\{x_1, \ldots, x_{88}\}$ with predictors of the next 20 values obtained from the model (10.5.12) for the mean-corrected series. Notice the slowness of the approach of the predictors toward the mean value 0.343. This is characteristic of predictors based on fractionally integrated models. □

Problems

10.1. Find a transfer function model relating the input and output series X_{t1} and X_{t2}, $t = 1, \ldots, 200$, contained in the ITSM data files APPJ.DAT and APPK.DAT, respectively. Use the fitted model to predict $X_{201,2}$, $X_{202,2}$, and $X_{203,2}$. Compare the predictors and their mean squared errors with the corresponding predictors and mean squared errors obtained by modelling $\{X_{t2}\}$ as a univariate ARMA process and with the results of Problem 7.7.

10.2. Verify the calculations of Example 10.2.1 to fit an intervention model to the series CDID.DAT.

10.3. If $\{X_t\}$ is the linear process (10.3.1) with $\{Z_t\} \sim \text{IID}(0, \sigma^2)$ and $\eta = EZ_t^3$, show that the third-order cumulant function of $\{X_t\}$ is given by

$$C_3(r, s) = \eta \sum_{i=-\infty}^{\infty} \psi_i \psi_{i+r} \psi_{i+s}.$$

Use this result to establish equation (10.3.4). Conclude that if $\{X_t\}$ is a Gaussian linear process, then $C_3(r, s) \equiv 0$ and $f_3(\omega_1, \omega_2) \equiv 0$.

10.4. Evaluate EX_t^4 for the ARCH(1) process (10.3.10) with $0 < \alpha_1 < 1$. Deduce that $EX_t^4 < \infty$ if and only if $3\alpha_1^2 < 1$.

10.5. Let $\{X_t\}$ be a causal stationary solution of the ARCH(p) equations (10.3.6) and (10.3.7) with $EX_t^4 < \infty$. Assuming such a process exists, show that $Y_t = X_t^2/\alpha_0$ satisfies the equations

$$Y_t = Z_t^2(1 + \sum_{i=1}^{p} \alpha_i Y_{t-i})$$

and deduce that $\{Y_t\}$ has the same autocorrelation function as the AR(p) process

$$W_t = \sum_{i=1}^{p} \alpha_i W_{t-i} + Z_t, \quad \{Z_t\} \sim \text{WN}(0, 1).$$

(In the case $p = 1$, a necessary and sufficient condition for existence of a causal stationary solution of (10.3.6) and (10.3.7) with $EX_t^4 < \infty$ is $3\alpha_1^2 < 1$, as shown by the results of Section 10.3 and Problem 10.4.)

10.6. If $a > 0$ and $X(0)$ has mean $\frac{b}{a}$ and variance $\frac{\sigma^2}{2a}$, show that the process defined by (10.4.3) is stationary and evaluate its mean and autocovariance function.

Random Variables and Probability Distributions

A.1 Distribution Functions and Expectation

The distribution function F of a random variable X is defined by

$$F(x) = P[X \le x] \tag{A.1.1}$$

for all real x. The following properties are direct consequences of (A.1.1).

1. F is nondecreasing, i.e., $F(x) \le F(y)$ if $x \le y$.
2. F is right continuous, i.e., $F(y) \downarrow F(x)$ as $y \downarrow x$.
3. $F(x) \to 1$ and $F(y) \to 0$ as $x \to \infty$ and $y \to -\infty$, respectively.

Conversely, any function that satisfies properties 1–3 is the distribution function of some random variable.

Most of the commonly encountered distribution functions F can be expressed either as

$$F(x) = \int_{-\infty}^{x} f(y)dy \tag{A.1.2}$$

or

$$F(x) = \sum_{j:x_j \le x} p(x_j),$$

where $\{x_0, x_1, x_2, \ldots\}$ is a finite or countably infinite set. In the former case we shall say that the random variable X is **continuous**. The function f is called the **probability density function** (pdf) of X and can be found from the relation

$$f(x) = F'(x).$$

In the latter case, the possible values of X are restricted to the set $\{x_0, x_1, \ldots\}$ and we shall say that the random variable X is **discrete**. The function p is called the **probability mass function** (pmf) of X and F is constant except for upward jumps of sizes $p(x_j)$ at the points x_j. Thus $p(x_j)$ is the size of the jump in F at x_j, i.e.,

$$p(x_j) = F(x_j) - F(x_j^-) = P[X = x_j],$$

where $F(x_j^-) = \lim_{y \uparrow x_j} F(y)$.

Examples of Continuous Distributions

(a) *The normal distribution with mean μ and variance σ^2.* We say that a random variable X has the normal distribution with mean μ and variance σ^2 (written more concisely as $X \sim N(\mu, \sigma^2)$) if X has the pdf given by

$$n(x; \mu, \sigma^2) = (2\pi)^{-1/2}\sigma^{-1}e^{-\frac{1}{2\sigma^2}(x-\mu)^2}, \qquad -\infty < x < \infty.$$

It follows then that $Z = (X - \mu)/\sigma \sim N(0, 1)$ and that

$$P[X \le x] = P[Z \le \frac{x-\mu}{\sigma}] = \Phi\left(\frac{x-\mu}{\sigma}\right)$$

where $\Phi(x) = \int_{-\infty}^{x}(2\pi)^{-1/2}e^{-\frac{1}{2}z^2}\,dz$ is known as the **standard normal distribution function**. The significance of the terms *mean* and *variance* for the parameters μ and σ^2 is explained below (see Example A.1.1).

(b) *The uniform distribution on $[a, b]$.* The pdf of a random variable uniformly distributed on the interval $[a, b]$ is given by

$$u(x; a, b) = \begin{cases} \dfrac{1}{b-a}, & \text{if } a \le x \le b, \\ 0, & \text{otherwise.} \end{cases}$$

(c) *The exponential distribution with parameter λ.* The pdf of an exponentially distributed random variable with parameter $\lambda > 0$ is

$$e(x; \lambda) = \begin{cases} 0, & \text{if } x < 0, \\ \lambda e^{-\lambda x}, & \text{if } x \ge 0. \end{cases}$$

The corresponding distribution function is

$$F(x) = \begin{cases} 0, & \text{if } x < 0, \\ 1 - e^{-\lambda x}, & \text{if } x \ge 0. \end{cases}$$

(d) *The gamma distribution with parameters α and λ.* The pdf of a gamma distributed random variable is

$$g(x;\alpha,\lambda) = \begin{cases} 0, & \text{if } x < 0, \\ x^{\alpha-1}\lambda^{\alpha}e^{-\lambda x}/\Gamma(\alpha), & \text{if } x \geq 0, \end{cases}$$

where the parameters α and λ are both positive and Γ is the gamma function defined as

$$\Gamma(\alpha) = \int_0^{\infty} x^{\alpha-1}e^{-x}\,dx.$$

Note that f is the exponential pdf when $\alpha = 1$ and that

$$\Gamma(\alpha) = (\alpha-1)(\alpha-2)\cdots 1 \text{ when } \alpha \text{ is a positive integer.}$$

(e) *The chi-squared distribution with ν degrees of freedom.* For each positive integer ν, the chi-squared distribution with ν degrees of freedom is defined to be the distribution of the sum

$$X = Z_1^2 + \cdots + Z_{\nu}^2,$$

where $Z_1, \ldots, Z_{\nu}$ are independent normally distributed random variables with mean 0 and variance 1. This distribution is the same as the gamma distribution with parameters $\alpha = \nu/2$ and $\lambda = 1/2$.

Examples of Discrete Distributions

(f) *The binomial distribution with parameters n and p.* The pmf of a binomially distributed random variable X with parameters n and p is

$$b(j;n,p) = P[X=j] = \binom{n}{j}p^j(1-p)^{n-j}, \; j = 0, 1, \ldots, n,$$

where n is a positive integer and $0 \leq p \leq 1$.

(g) *The uniform distribution on $\{1, 2, \ldots, k\}$.* The pmf of a random variable X uniformly distributed on $\{1, 2, \ldots, k\}$ is

$$p(j) = P[X=j] = \frac{1}{k}, \; j = 1, 2 \ldots, k,$$

where k is a positive integer.

(h) *The Poisson distribution with parameter λ.* A random variable X is said to have a Poisson distribution with parameter $\lambda > 0$ if

$$p(j;\lambda) = P[X=j] = \frac{\lambda^j}{j!}e^{-\lambda}, \; j = 0, 1, \ldots.$$

We shall see in Example A.1.2 below that λ is the mean of X.

(i) *The negative binomial distribution with parameters α and p.* The random variable X is said to have a negative binomial distribution with parameters $\alpha > 0$ and $p \in [0, 1]$ if it has pmf

$$nb(j; \alpha, p) = \left(\prod_{k=1}^{j} \frac{k - 1 + \alpha}{k} \right) (1 - p)^j p^\alpha, \quad j = 0, 1, \ldots,$$

where the product is defined to be 1 if $j = 0$.

Not all random variables can be neatly categorized as either continuous or discrete. For example consider the time you spend waiting to be served at a checkout counter and suppose that the probability of finding no customers ahead of you is $1/2$. Then the time you spend waiting for service can be expressed as

$$W = \begin{cases} 0, & \text{with probability } 1/2, \\ W_1, & \text{with probability } 1/2, \end{cases}$$

where W_1 is a continuous random variable. If the distribution of W_1 is exponential with parameter 1, then the distribution function of W is

$$F(x) = \begin{cases} 0, & \text{if } x < 0, \\ \frac{1}{2} + \frac{1}{2}(1 - e^{-x}) = 1 - \frac{1}{2}e^{-x}, & \text{if } x \geq 0. \end{cases}$$

This distribution function is neither continuous (since it has a discontinuity at $x = 0$) nor discrete (since it increases continuously for $x > 0$). It is expressible as a *mixture*,

$$F = pF_d + (1 - p)F_c,$$

with $p = 1/2$, of a discrete distribution function,

$$F_d = \begin{cases} 0, & x < 0, \\ 1, & x \geq 0, \end{cases}$$

and a continuous distribution function,

$$F_c = \begin{cases} 0, & x < 0, \\ 1 - e^{-x}, & x \geq 0. \end{cases}$$

Every distribution function can in fact be expressed in the form

$$F = p_1 F_d + p_2 F_c + p_3 F_{sc},$$

where $0 \leq p_1, p_2, p_3 \leq 1$, $p_1 + p_2 + p_3 = 1$, F_d is discrete, F_c is continuous, and F_{sc} is *singular continuous* (continuous but not of the form A.1.2). Distribution functions with a singular continuous component are rarely encountered.

Expectation, Mean, and Variance

The **expectation** of a function g of a random variable X is defined by

$$E(g(X)) = \int g(x)\, dF(x),$$

where

$$\int g(x)\, dF(x) := \begin{cases} \int_{-\infty}^{\infty} g(x) f(x)\, dx & \text{in the continuous case,} \\ \sum_{j=0}^{\infty} g(x_j) p(x_j) & \text{in the discrete case,} \end{cases}$$

and g is any function such that $E|g(x)| < \infty$. (If F is the mixture, $F = pF_c + (1-p)F_d$, then $E(g(X)) = p\int g(x)\, dF_c(x) + (1-p)\int g(x)\, dF_d(x)$.) The **mean** and **variance** of X are defined as $\mu = EX$ and $\sigma^2 = E(X-\mu)^2$, respectively. They are evaluated by setting $g(x) = x$ and $g(x) = (x-\mu)^2$ in the definition of $E(g(X))$.

It is clear from the definition that expectation has the **linearity property**,

$$E(aX + b) = aE(X) + b,$$

for any real constants a and b (provided $E|X| < \infty$).

Example A.1.1 The normal distribution

If X has the normal distribution with pdf $n(x; \mu, \sigma^2)$ as defined in Example (a) above, then

$$E(X-\mu) = \int_{-\infty}^{\infty} (x-\mu) n(x; \mu, \sigma^2)\, dx = -\sigma^2 \int_{-\infty}^{\infty} n'(x : \mu, \sigma^2)\, dx = 0.$$

This shows, with the help of the linearity property of E, that

$$E(X) = \mu,$$

i.e., that the parameter μ *is* in fact the mean of the normal distribution defined in Example (a). Similarly,

$$E(X-\mu)^2 = \int_{-\infty}^{\infty} (x-\mu)^2 n(x; \mu, \sigma^2)\, dx = -\sigma^2 \int_{-\infty}^{\infty} (x-\mu) n'(x; \mu, \sigma^2)\, dx.$$

Integrating by parts and using the fact that f is a pdf, we find that the variance of X is

$$E(X-\mu)^2 = \sigma^2 \int_{-\infty}^{\infty} n(x; \mu, \sigma^2)\, dx = \sigma^2. \qquad \square$$

Example A.1.2 The Poisson distribution

The mean of the Poisson distribution with parameter λ (see Example (h) above) is given by

$$\mu = \sum_{j=0}^{\infty} \frac{j\lambda^j}{j!} e^{-\lambda} = \sum_{j=1}^{\infty} \frac{\lambda\lambda^{j-1}}{(j-1)!} e^{-\lambda} = \lambda e^{\lambda} e^{-\lambda} = \lambda.$$

A similar calculation shows that the variance is also equal to λ (see Problem A.2). □

Remark. Functions and parameters associated with a random variable X will be labelled with the subscript X whenever it is necessary to identify the particular random variable to which they refer. For example, the distribution function, pdf, mean, and variance of X will be written as F_X, f_X, μ_X, and σ_X^2, respectively, whenever it is necessary to distinguish them from the corresponding quantities F_Y, f_Y, μ_Y, and σ_Y^2 associated with a different random variable Y. □

A.2 Random Vectors

An n-dimensional random vector is a column vector $\mathbf{X} = (X_1, \ldots, X_n)'$, each of whose components is a random variable. The distribution function F of $\mathbf{X}$, also called the **joint distribution** of $X_1, \ldots, X_n$, is defined by

$$F(x_1, \ldots, x_n) = P[X_1, \le x_1, \ldots, X_n \le x_n] \tag{A.2.1}$$

for all real numbers $x_1, \ldots, x_n$. This can be reexpressed in a more compact form as

$$F(\mathbf{x}) = P[\mathbf{X} \le \mathbf{x}], \quad \mathbf{x} = (x_1, \ldots, x_n)',$$

for all real vectors $\mathbf{x} = (x_1, \ldots, x_n)'$. The joint distribution of any subcollection $X_{i_1}, \ldots, X_{i_k}$ of these random variables can be obtained from F by setting $x_j = \infty$ in (A.2.1) for all $j \notin \{i_1, \ldots, i_k\}$. In particular the distributions of X_1 and $(X_1, X_n)'$ are given by

$$F_{X_1}(x_1) = P[X_1 \le x_1] = F(x_1, \infty, \ldots, \infty)$$

and

$$F_{X_1, X_n}(x_1, x_n) = P[X_1 \le x_1, X_n \le x_n] = F(x_1, \infty, \ldots, \infty, x_n).$$

As in the univariate case, a random vector with distribution function F is said to be continuous if F has a density function, i.e., if

$$F(x_1, \ldots, x_n) = \int_{-\infty}^{x_n} \cdots \int_{-\infty}^{x_2} \int_{-\infty}^{x_1} f(y_1, \ldots, y_n)\, dy_1 dy_2 \cdots dy_n.$$

The probability density of $\mathbf{X}$ is then found from

$$f(x_1, \ldots, x_n) = \frac{\partial^n F(x_1, \ldots, x_n)}{\partial x_1 \cdots \partial x_n}.$$

The random vector $\mathbf{X}$ is said to be discrete if there exist real-valued vectors $\mathbf{x}_0, \mathbf{x}_1, \ldots$ and a probability mass function $p(\mathbf{x}_j) = P[\mathbf{X} = \mathbf{x}_j]$ such that

$$\sum_{j=0}^{\infty} p(\mathbf{x}_j) = 1.$$

The expectation of a function g of a random vector $\mathbf{X}$ is defined by

$$E(g(\mathbf{X})) = \int g(\mathbf{x})\, dF(\mathbf{x}) = \int g(x_1, \ldots, x_n)\, dF(x_1, \ldots, x_n)$$

where

$$\int g(x_1, \ldots, x_n)\, dF(x_1, \ldots, x_n) =$$

$$\begin{cases} \displaystyle\int \cdots \int g(x_1, \ldots, x_n) f(x_1, \ldots, x_n)\, dx_1 \cdots dx_n, & \text{in the continuous case,} \\ \displaystyle\sum_{j_1} \cdots \sum_{j_n} g(x_{j_1}, \ldots, x_{j_n}) p(x_{j_1}, \ldots, x_{j_n}), & \text{in the discrete case,} \end{cases}$$

and g is any function such that $E|g(\mathbf{X})| < \infty$.

The random variables $X_1, \ldots, X_n$ are said to be **independent** if and only if

$$P[X_1 \le x_1, \ldots, X_n \le x_n] = P[X_1 \le x_1] \cdots P[X_n \le x_n],$$

i.e.,

$$F(x_1, \ldots, x_n) = F_{X_1}(x_1) \cdots F_{X_n}(x_n)$$

for all real numbers $x_1, \ldots, x_n$. In the continuous and discrete cases, independence is equivalent to the factorization of the joint density function or probability mass function into the product of the respective marginal densities or mass functions, i.e.,

$$f(x_1, \ldots, x_n) = f_{X_1}(x_1) \cdots f_{X_n}(x_n) \tag{A.2.2}$$

or

$$p(x_1, \ldots, x_n) = p_{X_1}(x_1) \cdots p_{X_n}(x_n). \tag{A.2.3}$$

For two random vectors $\mathbf{X} = (X_1, \ldots, X_n)'$ and $\mathbf{Y} = (Y_1, \ldots, Y_m)'$, with joint density function $f_{\mathbf{X},\mathbf{Y}}$, the conditional density of $\mathbf{Y}$ given $\mathbf{X} = \mathbf{x}$ is

$$f_{\mathbf{Y}|\mathbf{X}}(\mathbf{y}|\mathbf{x}) = \begin{cases} \dfrac{f_{\mathbf{X},\mathbf{Y}}(\mathbf{x}, \mathbf{y})}{f_{\mathbf{X}}(\mathbf{x})}, & \text{if } f_{\mathbf{X}}(\mathbf{x}) > 0, \\ f_{\mathbf{Y}}(\mathbf{y}), & \text{if } f_{\mathbf{X}}(\mathbf{x}) = 0. \end{cases}$$

The conditional expectation of $g(\mathbf{Y})$ given $\mathbf{X} = \mathbf{x}$ is then

$$E(g(\mathbf{Y})|\mathbf{X} = \mathbf{x}) = \int_{-\infty}^{\infty} g(\mathbf{y}) f_{\mathbf{Y}|\mathbf{X}}(\mathbf{y}|\mathbf{x})\, d\mathbf{y}$$

If $\mathbf{X}$ and $\mathbf{Y}$ are independent, then $f_{\mathbf{Y}|\mathbf{X}}(\mathbf{y}|\mathbf{x}) = f_{\mathbf{Y}}(\mathbf{y})$ by (A.2.2) and so the conditional expectation of $g(\mathbf{Y})$ given $\mathbf{X} = \mathbf{x}$ is

$$E(g(\mathbf{Y})|\mathbf{X} = \mathbf{x}) = E(g(\mathbf{Y})),$$

which, as expected, does not depend on $\mathbf{x}$. The same ideas hold in the discrete case with the probability mass function assuming the role of the density function.

Means and Covariances

If $E|X_i| < \infty$ for each i, then we define the mean or expected value of $\mathbf{X} = (X_1, \ldots, X_n)'$ to be the column vector

$$\mu_X = E\mathbf{X} = (EX_1, \ldots, EX_n)'.$$

In the same way we define the expected value of any array whose elements are random variables (e.g. a matrix of random variables) to be the same array with each random variable replaced by its expected value (if the expectation exists).

If $\mathbf{X} = (X_1, \ldots, X_n)'$ and $\mathbf{Y} = (Y_1, \ldots, Y_m)'$ are random vectors such that each X_i and Y_j has a finite variance, then the **covariance matrix** of $\mathbf{X}$ and $\mathbf{Y}$ is defined to be the matrix

$$\begin{aligned}\Sigma_{\mathbf{XY}} = \mathrm{Cov}(\mathbf{X}, \mathbf{Y}) &= E[(\mathbf{X} - E\mathbf{X})(\mathbf{Y} - E\mathbf{Y})'] \\ &= E(\mathbf{XY}') - (E\mathbf{X})(E\mathbf{Y})'.\end{aligned}$$

The (i, j)-element of $\Sigma_{\mathbf{XY}}$ is the covariance $\mathrm{Cov}(X_i, Y_j) = E(X_iY_j) - E(X_i)E(Y_j)$. In the special case when $\mathbf{Y} = \mathbf{X}$, $\mathrm{Cov}(\mathbf{X}, \mathbf{Y})$ reduces to the covariance matrix of the random vector $\mathbf{X}$.

Now suppose that $\mathbf{Y}$ and $\mathbf{X}$ are linearly related through the equation

$$\mathbf{Y} = \mathbf{a} + B\mathbf{X}$$

where $\mathbf{a}$ is an m-dimensional column vector and B is an $m \times n$ matrix. Then $\mathbf{Y}$ has mean

$$E\mathbf{Y} = \mathbf{a} + BE\mathbf{X}, \tag{A.2.4}$$

and covariance matrix

$$\Sigma_{\mathbf{YY}} = B\Sigma_{\mathbf{XX}}B', \tag{A.2.5}$$

(see Problem A.3).

Proposition A.2.1 *The covariance matrix $\Sigma_{\mathbf{XX}}$ of a random vector $\mathbf{X}$ is symmetric and nonnegative definite, i.e., $\mathbf{b}'\Sigma_{\mathbf{XX}}\mathbf{b} \geq 0$ for all vectors $\mathbf{b} = (b_1, \ldots, b_n)'$ with real components.*

Proof Since the (i,j)-element of $\Sigma_{\mathbf{XX}}$ is $\text{Cov}(X_i, X_j) = \text{Cov}(X_j, X_i)$, it is clear that $\Sigma_{\mathbf{XX}}$ is symmetric. To prove nonnegative definiteness, let $\mathbf{b} = (b_1, \ldots, b_n)'$ be an arbitrary vector. Then applying (A.2.5) with $\mathbf{a} = \mathbf{0}$ and $B = \mathbf{b}$, we have

$$\mathbf{b}'\Sigma_{\mathbf{XX}}\mathbf{b} = \text{Var}(\mathbf{b}'\mathbf{X}) = \text{Var}(b_1X_1 + \cdots + b_nX_n) \geq 0.$$ ■

Proposition A.2.2 *Every $n \times n$ covariance matrix Σ can be factorized as*

$$\Sigma = P\Lambda P',$$

where P is an orthogonal matrix (i.e., $P' = P^{-1}$) whose columns are an orthonormal set of right eigenvectors corresponding to the (nonnegative) eigenvalues $\lambda_1, \ldots, \lambda_n$ of Σ, and Λ is the diagonal matrix,

$$\Lambda = \begin{bmatrix} \lambda_1 & 0 & \cdots & 0 \\ 0 & \lambda_2 & \cdots & 0 \\ \vdots & \vdots & \ddots & \vdots \\ 0 & 0 & \cdots & \lambda_n \end{bmatrix}.$$

In particular Σ is nonsingular if and only if all the eigenvalues are strictly positive.

Proof Every covariance matrix is symmetric and nonnegative definite by Proposition A.2.1, and for such matrices the specified factorization is a standard result (see Graybill, 1983 for a proof). The determinant of an orthogonal matrix is 1, so that $\det(\Sigma) = \det(P)\det(\Lambda)\det(P) = \lambda_1 \cdots \lambda_n$. It follows that Σ is nonsingular if and only if $\lambda_i > 0$ for all i. ■

Remark. Given a covariance matrix Σ, it is sometimes useful to be able to find a square root, $A = \Sigma^{1/2}$, with the property that $AA' = \Sigma$. It is clear from Proposition A.2.2 and the orthogonality of P that one such matrix is given by

$$A = \Sigma^{1/2} = P\Lambda^{1/2}P'.$$

If Σ is nonsingular then we can define

$$\Sigma^s = P\Lambda^s P', \quad -\infty < s < \infty.$$

The matrix $\Sigma^{-1/2}$ defined in this way is then a square root of Σ^{-1} and also the inverse of $\Sigma^{1/2}$. □

A.3 The Multivariate Normal Distribution

The multivariate normal distribution is one of the most commonly encountered and important distributions in statistics. It plays a key role in the modelling of time series data. Let $\mathbf{X} = (X_1, \ldots, X_n)'$ be a random vector.

Definition A.3.1

> $\mathbf{X}$ has a **multivariate normal distribution** with mean $\boldsymbol{\mu}$ and nonsingular covariance matrix $\Sigma = \Sigma_{\mathbf{XX}}$, written as $\mathbf{X} \sim \mathrm{N}(\boldsymbol{\mu}, \Sigma)$, if
>
> $$f_{\mathbf{X}}(\mathbf{x}) = (2\pi)^{-n/2}(\det \Sigma)^{-1/2} \exp\{-\frac{1}{2}(\mathbf{x} - \boldsymbol{\mu})'\Sigma^{-1}(\mathbf{x} - \boldsymbol{\mu})\}.$$

If $\mathbf{X} \sim \mathrm{N}(\boldsymbol{\mu}, \Sigma)$ we can define a *standardized* random vector $\mathbf{Z}$ by applying the linear transformation,

$$\mathbf{Z} = \Sigma^{-1/2}(\mathbf{X} - \boldsymbol{\mu}), \tag{A.3.1}$$

where $\Sigma^{-1/2}$ is defined as in the remark of Section A.2. Then by (A.2.4) and (A.2.5), $\mathbf{Z}$ has mean $\mathbf{0}$ and covariance matrix $\Sigma_{\mathbf{ZZ}} = \Sigma^{-1/2}\Sigma\Sigma^{-1/2} = I_n$ where I_n is the $n \times n$ identity matrix. Using the change of variables formula for probability densities (see Mood, Graybill, and Boes, 1974), we find that the probability density of $\mathbf{Z}$ is

$$\begin{aligned}
f_{\mathbf{Z}}(\mathbf{z}) &= (\det \Sigma)^{1/2} f_{\mathbf{X}}(\Sigma^{1/2}\mathbf{z} + \boldsymbol{\mu}) \\
&= (\det \Sigma)^{1/2}(2\pi)^{-n/2}(\det \Sigma)^{-1/2} \exp\{-\frac{1}{2}(\Sigma^{-1/2}\mathbf{z})'\Sigma^{-1}\Sigma^{-1/2}\mathbf{z}\} \\
&= (2\pi)^{-n/2} \exp\{-\frac{1}{2}\mathbf{z}'\mathbf{z}\} \\
&= \left((2\pi)^{-1/2} \exp\{-\frac{1}{2}z_1^2\}\right) \cdots \left((2\pi)^{-1/2} \exp\{-\frac{1}{2}z_n^2\}\right)
\end{aligned}$$

showing, by (A.2.2), that $Z_1, \ldots, Z_n$ are independent N(0, 1) random variables. Thus the standardized random vector $\mathbf{Z}$ defined by (A.3.1) has independent standard normal random components. Conversely, given any $n \times 1$ mean vector $\boldsymbol{\mu}$, a nonsingular $n \times n$ covariance matrix Σ, and an $n \times 1$ vector of standard normal random variables, we can construct a normally distributed random vector with mean $\boldsymbol{\mu}$ and covariance matrix Σ by defining

$$\mathbf{X} = \Sigma^{1/2}\mathbf{Z} + \boldsymbol{\mu}. \tag{A.3.2}$$

(See Problem A.4.)

Remark 1. The multivariate normal distribution with mean $\boldsymbol{\mu}$ and covariance matrix Σ can be defined, even when Σ is singular, as the distribution of the vector $\mathbf{X}$ in (A.3.2). The **singular multivariate normal distribution** does not have a joint density since the possible values of $\mathbf{X} - \boldsymbol{\mu}$ are constrained to lie in a subspace of $\mathbb{R}^n$ with dimension equal to rank(Σ). □

Remark 2. If $\mathbf{X} \sim \mathrm{N}(\boldsymbol{\mu}, \Sigma)$, B is an $m \times n$ matrix, and $\mathbf{a}$ is a real $m \times 1$ vector, then the random vector

$$\mathbf{Y} = \mathbf{a} + B\mathbf{X}$$

is also multivariate normal (see Problem A.5). Note that from (A.2.4) and (A.2.5), $\mathbf{Y}$ has mean $\mathbf{a} + B\boldsymbol{\mu}$ and covariance matrix $B\Sigma B'$. In particular, by taking B to be the row vector $\mathbf{b}' = (b_1, \ldots, b_n)$, we see that any linear combination of the components of a multivariate normal random vector is normal. Thus $\mathbf{b}'\mathbf{X} = b_1 X_1 + \cdots + b_n X_n \sim \mathrm{N}(\mathbf{b}'\boldsymbol{\mu}_{\mathbf{X}}, \mathbf{b}'\Sigma_{\mathbf{XX}}\mathbf{b})$. □

Example A.3.1 The Bivariate Normal Distribution

Suppose that $\mathbf{X} = (X_1, X_2)'$ is a bivariate normal random vector with mean $\boldsymbol{\mu} = (\mu_1, \mu_2)'$ and covariance matrix

$$\Sigma = \begin{bmatrix} \sigma_1^2 & \rho\sigma_1\sigma_2 \\ \rho\sigma_1\sigma_2 & \sigma_2^2 \end{bmatrix}, \quad \sigma > 0, \sigma_2 > 0, \ -1 < \rho < 1. \tag{A.3.3}$$

The parameters σ_1, σ_2, and ρ are the standard deviations and correlation of the components X_1 and X_2. Every nonsingular 2-dimensional covariance matrix can be expressed in the form (A.3.3). The inverse of Σ is

$$\Sigma^{-1} = (1 - \rho^2)^{-1} \begin{bmatrix} \sigma_1^{-2} & \rho\sigma_1^{-1}\sigma_2^{-1} \\ \rho\sigma_1^{-1}\sigma_2^{-1} & \sigma_2^{-2} \end{bmatrix},$$

and so the pdf of $\mathbf{X}$ is given by

$$f_{\mathbf{X}}(\mathbf{x}) = (2\pi\sigma_1\sigma_2(1 - \rho^2)^{1/2})^{-1} \exp\left\{ \frac{-1}{2(1 - \rho^2)} \left[\left(\frac{x_1 - \mu_1}{\sigma_1} \right)^2 - 2\rho \left(\frac{x_1 - \mu_1}{\sigma_1} \right) \left(\frac{x_2 - \mu_2}{\sigma_2} \right) + \left(\frac{x_2 - \mu_2}{\sigma_2} \right)^2 \right] \right\}. \quad \square$$

Multivariate normal random vectors have the important property that the conditional distribution of any set of components, given any other set, is again multivariate normal. In the following proposition we shall suppose that the nonsingular normal random vector $\mathbf{X}$ is partitioned into two subvectors,

$$\mathbf{X} = \begin{bmatrix} \mathbf{X}^{(1)} \\ \mathbf{X}^{(2)} \end{bmatrix}.$$

Correspondingly, we shall write the mean and covariance matrix of $\mathbf{X}$ as

$$\boldsymbol{\mu} = \begin{bmatrix} \boldsymbol{\mu}^{(1)} \\ \boldsymbol{\mu}^{(2)} \end{bmatrix} \quad \text{and} \quad \Sigma = \begin{bmatrix} \Sigma_{11} & \Sigma_{12} \\ \Sigma_{21} & \Sigma_{22} \end{bmatrix}$$

where $\boldsymbol{\mu}^{(i)} = E\mathbf{X}^{(i)}$ and $\Sigma_{ij} = E(\mathbf{X}^{(i)} - \boldsymbol{\mu}^{(i)})(\mathbf{X}^{(j)} - \boldsymbol{\mu}^{(i)})'$.

Proposition A.3.1

i. $\mathbf{X}^{(1)}$ *and* $\mathbf{X}^{(2)}$ *are independent if and only if* $\Sigma_{12} = 0$.

ii. *The conditional distribution of* $\mathbf{X}^{(1)}$ *given* $\mathbf{X}^{(2)} = \mathbf{x}^{(2)}$ *is* $\mathrm{N}(\boldsymbol{\mu}^{(1)} + \Sigma_{12}\Sigma_{22}^{-1}(\mathbf{x}^{(2)} - \boldsymbol{\mu}^{(2)}), \Sigma_{11} - \Sigma_{12}\Sigma_{22}^{-1}\Sigma_{21})$. *In particular,*

$$E(\mathbf{X}^{(1)}|\mathbf{X}^{(2)} = \mathbf{x}^{(2)}) = \boldsymbol{\mu}^{(1)} + \Sigma_{12}\Sigma_{22}^{-1}(\mathbf{x}^{(2)} - \boldsymbol{\mu}^{(2)}).$$

The proof of this proposition involves routine algebraic manipulations of the multivariate normal density function and is left as an exercise (see Problem A.6).

Example A.3.2 For the bivariate normal random vector $\mathbf{X}$ in Example A.3.1, we immediately deduce from Proposition A.3.1 that X_1 and X_2 are independent if and only if $\rho\sigma_1\sigma_2 = 0$ (or $\rho = 0$ since σ_1 and σ_2 are both positive). The conditional distribution of X_1 given $X_2 = x_2$ is normal with mean

$$E(X_1|X_2 = x_2) = \mu_1 + \rho\sigma_1\sigma_2^{-1}(x_2 - \mu_2)$$

and variance

$$\mathrm{Var}(X_1|X_2 = x_2) = \sigma_1^2(1 - \rho^2).$$

□

Definition A.3.2 $\{X_t\}$ is a **Gaussian time series** if all of its joint distributions are multivariate normal, i.e., if for any collection of integers $i_1, \ldots, i_n$, the random vector $(X_{i_1}, \ldots, X_{i_n})'$ has a multivariate normal distribution.

Remark 3. If $\{X_t\}$ is a Gaussian time series, then all of its joint distributions are completely determined by the mean function $\mu(t) = EX_t$ and the autocovariance function $\kappa(s, t) = \mathrm{Cov}(X_s, X_t)$. If the process also happens to be stationary, then the mean function is constant ($\mu_t = \mu$ for all t) and $\kappa(t + h, t) = \gamma(h)$ for all t. In this case, the joint distribution of $X_1, \ldots, X_n$ is the same as that of $X_{1+h}, \ldots, X_{n+h}$ for all integers h and $n > 0$. Hence for a Gaussian time series strict stationarity is equivalent to weak stationarity (see Section 2.1). □

Problems

A.1. Let X have a negative binomial distribution with parameters α and p, where $\alpha > 0$ and $0 \le p < 1$.

a. Show that the probability generating function of X (defined as $M(s) = E(s^X)$) is

$$M(s) = p^\alpha(1 - s + sp)^{-\alpha}, \quad 0 \le s \le 1.$$

b. Using the property that $M'(1) = E(X)$ and $M''(1) = E(X^2) - E(X)$, show that

$$E(X) = \alpha(1-p)/p \quad \text{and} \quad \text{Var}(X) = \alpha(1-p)/p^2.$$

A.2. If X has the Poisson distribution with mean λ, show that the variance of X is also λ.

A.3. Use the linearity of the expectation operator for real-valued random variables to establish (A.2.4) and (A.2.5).

A.4. If Σ is an $n \times n$ covariance matrix, $\Sigma^{1/2}$ is the square root of Σ defined in the remark of Section A.2, $\mathbf{Z}$ is an n-vector whose components are independent normal random variables with mean 0 and variance 1, show that

$$X = \Sigma^{1/2}\mathbf{Z} + \boldsymbol{\mu}$$

is a normally distributed random vector with mean $\boldsymbol{\mu}$ and covariance matrix Σ.

A.5. Show that if $\mathbf{X}$ is an n-dimensional random vector such that $\mathbf{X} \sim \text{N}(\boldsymbol{\mu}, \Sigma)$, B is a real $m \times n$ matrix and $\mathbf{a}$ is a real-valued m-vector, then

$$\mathbf{Y} = \mathbf{a} + B\mathbf{X}$$

is a multivariate normal random vector. Specify the mean and covariance matrix of $\mathbf{Y}$.

A.6. Prove Proposition A.3.1.

A.7. Suppose that $\mathbf{X} = (X_1, \ldots, X_n)' \sim \text{N}(\mathbf{0}, \Sigma)$ with Σ nonsingular. Using the fact that $\mathbf{Z}$, as defined in (A.3.1) has independent standard normal components, show that $(\mathbf{X} - \boldsymbol{\mu})'\Sigma^{-1}(\mathbf{X} - \boldsymbol{\mu})$ has the chi-squared distribution with n degrees of freedom (Section A.1, Example (e)).

A.8. Suppose that $\mathbf{X} = (X_1, \ldots, X_n)' \sim \text{N}(\boldsymbol{\mu}, \Sigma)$ with Σ nonsingular. If A is a symmetric $n \times n$ matrix, show that $E(\mathbf{X}'A\mathbf{X}) = \text{trace}(A\Sigma) + \boldsymbol{\mu}'\Sigma\boldsymbol{\mu}$.

A.9. Suppose that $\{X_t\}$ is a stationary Gaussian time series with mean 0 and autocovariance function $\gamma(h)$. Find $E(X_t|X_s)$ and $\text{Var}(X_t|X_s)$, $s \neq t$.

B Statistical Complements

B.1 Least Squares Estimation

Consider the problem of finding the "best" straight line,

$$y = \theta_0 + \theta_1 x,$$

to approximate observations $y_1, \ldots, y_n$ of a dependent variable y taken at fixed values $x_1, \ldots, x_n$ of the independent variable x. The **(ordinary) least squares estimates** $\hat{\theta}_0$, $\hat{\theta}_1$ are defined to be values of θ_0, θ_1 that minimize the sum

$$S(\theta_0, \theta_1) = \sum_{i=1}^{n} (y_i - \theta_0 - \theta_1 x_i)^2$$

of squared deviations of the observations y_i from the fitted values $\theta_0 + \theta_1 x_i$. (The "sum of squares," $S(\theta_0, \theta_1)$, is identical to the Euclidean squared distance between $\mathbf{y}$ and $\theta_0 \mathbf{1} + \theta_1 \mathbf{x}$, i.e.,

$$S(\theta_0, \theta_1) = \|\mathbf{y} - \theta_0 \mathbf{1} - \theta_1 \mathbf{x}\|^2,$$

where $\mathbf{x} = (x_1, \ldots, x_n)'$, $\mathbf{1} = (1, \ldots, 1)'$, and $\mathbf{y} = (y_1, \ldots, y_n)'$.) Setting the partial derivatives of S with respect to θ_0 and θ_1 both equal to zero shows that the vector $\hat{\boldsymbol{\theta}} = (\hat{\theta}_0, \hat{\theta}_1)'$ satisfies the "normal equations,"

$$X'X\hat{\boldsymbol{\theta}} = X'\mathbf{y},$$

where X is the $n \times 2$ matrix $X = [\mathbf{1}, \mathbf{x}]$. Since $0 \le S(\boldsymbol{\theta})$ and $S(\boldsymbol{\theta}) \to \infty$ as $\|\boldsymbol{\theta}\| \to \infty$, the normal equations have at least one solution. If $\hat{\boldsymbol{\theta}}^{(1)}$ and $\hat{\boldsymbol{\theta}}^{(2)}$ are two solutions of the normal equations, then a simple calculation shows that

$$(\hat{\boldsymbol{\theta}}^{(1)} - \hat{\boldsymbol{\theta}}^{(2)})' X' X (\hat{\boldsymbol{\theta}}^{(1)} - \hat{\boldsymbol{\theta}}^{(2)}) = 0,$$

i.e., that $X\hat{\boldsymbol{\theta}}^{(1)} = X\hat{\boldsymbol{\theta}}^{(2)}$. The solution of the normal equations is unique if and only if the matrix $X'X$ is nonsingular. But the preceding calculations show that even if $X'X$ is singular, the vector $\hat{\mathbf{y}} = X\hat{\boldsymbol{\theta}}$ of fitted values is the same for *any* solution $\hat{\boldsymbol{\theta}}$ of the normal equations.

The argument just given applies equally well to least squares estimation for the general linear model. Given a set of data points

$$(x_{i1}, x_{i2}, \ldots, x_{im}, y_i), \qquad i = 1, \ldots, n \text{ with } m \le n,$$

the least squares estimate, $\hat{\boldsymbol{\theta}} = (\hat{\theta}_1, \ldots, \hat{\theta}_m)'$ of $\boldsymbol{\theta} = (\theta_1, \ldots, \theta_m)'$ minimizes

$$\begin{aligned} S(\boldsymbol{\theta}) &= \sum_{i=1}^{n} (y_i - \theta_1 x_{i1} - \cdots - \theta_m x_{im})^2 \\ &= \|\mathbf{y} - \theta_1 \mathbf{x}^{(1)} - \cdots - \theta_m \mathbf{x}^{(m)}\|^2, \end{aligned}$$

where $\mathbf{y} = (y_1, \ldots, y_n)'$ and $\mathbf{x}^{(j)} = (x_{1j}, \ldots, x_{nj})'$, $j = 1, \ldots, m$. As in the previous special case, $\hat{\boldsymbol{\theta}}$ satisfies the equations

$$X'X\hat{\boldsymbol{\theta}} = X'\mathbf{y},$$

where X is the $n \times m$ matrix $X = [\mathbf{x}^{(1)}, \ldots, \mathbf{x}^{(m)}]$. The solution of this equation is unique if and only if $X'X$ nonsingular, in which case

$$\hat{\boldsymbol{\theta}} = (X'X)^{-1} X'\mathbf{y}.$$

If $X'X$ is singular there are infinitely many solutions $\hat{\boldsymbol{\theta}}$ but the vector of fitted values $X\hat{\boldsymbol{\theta}}$ is the same for all of them.

Example B.1.1 To illustrate the general case, let us fit a quadratic function

$$y = \theta_0 + \theta_1 x + \theta_2 x^2$$

to the data

x	0	1	2	3	4
y	1	0	3	5	8

The matrix X for this problem is

$$X = \begin{bmatrix} 1 & 0 & 0 \\ 1 & 1 & 1 \\ 1 & 2 & 4 \\ 1 & 3 & 9 \\ 1 & 4 & 16 \end{bmatrix}, \text{ giving } (X'X)^{-1} = \frac{1}{140} \begin{bmatrix} 124 & -108 & 20 \\ -108 & 174 & -40 \\ 20 & -40 & 10 \end{bmatrix}.$$

The least squares estimate $\hat{\boldsymbol{\theta}} = (\hat{\theta}_0, \hat{\theta}_1, \hat{\theta}_2)'$ is therefore unique and given by

$$\hat{\boldsymbol{\theta}} = (X'X)^{-1}X'\mathbf{y} = \begin{bmatrix} 0.6 \\ -0.1 \\ 0.5 \end{bmatrix}.$$

The vector of fitted values is given by

$$\hat{\mathbf{y}} = X\hat{\boldsymbol{\theta}} = (0.6, 1, 2.4, 4.8, 8.2)'$$

as compared with the observed values

$$\mathbf{y} = (1, 0, 3, 5, 8)'.$$

□

B.1.1 The Gauss-Markov Theorem

Suppose now that the observations $y_1, \ldots, y_n$ are realized values of random variables $Y_1, \ldots, Y_n$ satisfying

$$Y_i = \theta_1 x_{i1} + \cdots + \theta_m x_{im} + Z_i,$$

where $Z_i \sim \text{WN}(0, \sigma^2)$. Letting $\mathbf{Y} = (Y_1, \ldots, Y_n)'$ and $\mathbf{Z} = (Z_1, \ldots, Z_n)'$, we can write these equations as

$$\mathbf{Y} = X\boldsymbol{\theta} + \mathbf{Z}.$$

Assume for simplicity that the matrix $X'X$ is nonsingular (for the general case see e.g., Silvey, 1975). Then the least squares estimator of $\boldsymbol{\theta}$ is, as above,

$$\hat{\boldsymbol{\theta}} = (X'X)^{-1}X'\mathbf{Y}$$

and the least squares estimator of the parameter σ^2 is the unbiased estimator

$$\hat{\sigma}^2 = \frac{1}{n-m}\|\mathbf{Y} - X\hat{\boldsymbol{\theta}}\|^2.$$

It is easy to see that $\hat{\boldsymbol{\theta}}$ is also unbiased, i.e., that

$$E(\hat{\boldsymbol{\theta}}) = \boldsymbol{\theta}.$$

It follows at once that if $\mathbf{c}'\boldsymbol{\theta}$ is any linear combination of the parameters θ_i, $i = 1, \ldots, m$, then $\mathbf{c}'\hat{\boldsymbol{\theta}}$ is an unbiased estimator of $\mathbf{c}'\boldsymbol{\theta}$. The Gauss-Markov theorem says that, of all unbiased estimators of $\mathbf{c}'\boldsymbol{\theta}$ of the form $\sum_{i=1}^{n} a_i Y_i$, the estimator $\mathbf{c}'\hat{\boldsymbol{\theta}}$ has the smallest variance.

In the special case when $Z_1, \ldots, Z_n$ are IID $\text{N}(0, \sigma^2)$, the least squares estimator $\hat{\boldsymbol{\theta}}$ has the distribution $\text{N}(\boldsymbol{\theta}, \sigma^2(X'X)^{-1})$ and $(n-m)\hat{\sigma}^2/\sigma^2$ has the χ^2 distribution with $n-m$ degrees of freedom.

B.1.2 Generalized Least Squares

The Gauss-Markov theorem depends on the assumption that the errors $Z_1, \ldots, Z_n$ are uncorrelated with constant variance. If on the other hand $\mathbf{Z} = (Z_1, \ldots, Z_n)'$ has mean

$\mathbf{0}$ and nonsingular covariance matrix $\sigma^2\Sigma$ where $\Sigma \neq I$, we consider the transformed observation vector $U = R^{-1}\mathbf{Y}$ where R is a nonsingular matrix such that $RR' = \Sigma$. Then

$$\mathbf{U} = R^{-1}X\boldsymbol{\theta} + \mathbf{W} = M\boldsymbol{\theta} + \mathbf{W},$$

where $M = R^{-1}X$ and $\mathbf{W}$ has mean $\mathbf{0}$ and covariance matrix $\sigma^2 I$. The Gauss-Markov theorem now implies that the best linear estimate of any linear combination $\mathbf{c}'\boldsymbol{\theta}$ is $\mathbf{c}'\hat{\boldsymbol{\theta}}$ where $\hat{\boldsymbol{\theta}}$ is the **generalized least squares estimator**, which minimizes

$$\|\mathbf{U} - M\boldsymbol{\theta}\|^2.$$

In the special case when $Z_1, \ldots, Z_n$ are uncorrelated and Z_i has mean 0 and variance $\sigma^2 r_i^2$, the generalized least squares estimator minimizes the weighted sum of squares

$$\sum_{i=1}^{n} \frac{1}{r_i^2}(Y_i - \theta_1 x_{i1} - \cdots - \theta_m x_{im})^2.$$

In the general case, if $X'X$ and Σ are both nonsingular, the generalized least squares estimator is given by

$$\hat{\boldsymbol{\theta}} = (M'M)^{-1}M'\mathbf{U}.$$

Although the least squares estimator $(X'X)^{-1}X'\mathbf{Y}$ is unbiased if $E(\mathbf{Z}) = \mathbf{0}$, even when the covariance matrix of $\mathbf{Z}$ is not equal to $\sigma^2 I$, the variance of the corresponding estimate of any linear combination of $\theta_1, \ldots, \theta_m$ is greater than or equal to the estimator based on the generalized least squares estimator.

B.2 Maximum Likelihood Estimation

The method of least squares has an appealing intuitive interpretation. Its application depends on knowledge only of the means and covariances of the observations. Maximum likelihood estimation depends on the assumption of a particular distributional form for the observations, known apart from the values of parameters $\theta_1, \ldots, \theta_m$. We can regard the estimation problem as that of selecting the most appropriate value of a parameter vector $\boldsymbol{\theta}$, taking values in a subset Θ of $\mathbb{R}^m$. We suppose that these distributions have probability densities $p(\mathbf{x}; \boldsymbol{\theta})$, $\boldsymbol{\theta} \in \Theta$. For a fixed vector of observations $\mathbf{x}$, the function $L(\boldsymbol{\theta}) = p(\mathbf{x}; \boldsymbol{\theta})$ on Θ is called the **likelihood function**. A maximum likelihood estimate $\hat{\boldsymbol{\theta}}(\mathbf{x})$ of $\boldsymbol{\theta}$ is a value of $\boldsymbol{\theta} \in \Theta$ that maximizes the value of $L(\boldsymbol{\theta})$ for the given observed value $\mathbf{x}$, i.e.,

$$L(\hat{\boldsymbol{\theta}}) = p(x; \hat{\boldsymbol{\theta}}(\mathbf{x})) = \max_{\boldsymbol{\theta} \in \Theta} p(\mathbf{x}; \boldsymbol{\theta}).$$

Example B.2.1 If $\mathbf{x} = (x_1, \ldots, x_n)'$ is a vector of observations of independent $\mathrm{N}(\mu, \sigma^2)$ random variables, the likelihood function is

$$L(\mu, \sigma^2) = \frac{1}{(2\pi\sigma^2)^{n/2}} \exp\left[-\frac{1}{2\sigma^2}\sum_{i=1}^{n}(x_i - \mu)^2\right], -\infty < \mu < \infty, \sigma > 0.$$

Maximization of L with respect to μ and σ is equivalent to minimization of

$$-2\ln L(\mu, \sigma^2) = n\ln(2\pi) + 2n\ln(\sigma) + \frac{1}{\sigma^2}\sum_{i=1}^{n}(x_i - \mu)^2.$$

Setting the partial derivatives of $-2\ln L$ with respect to μ and σ both equal to zero gives the maximum likelihood estimates,

$$\hat{\mu} = \bar{x} = \frac{1}{n}\sum_{i=1}^{n}x_i \quad \text{and} \quad \hat{\sigma}^2 = \frac{1}{n}\sum_{i=1}^{n}(x_i - \bar{x})^2.$$ □

B.2.1 Properties of Maximum Likelihood Estimators

The Gauss-Markov theorem lent support to the use of least squares estimation by showing its property of minimum variance among unbiased linear estimators. Maximum likelihood estimators are not generally unbiased, but in particular cases they can be shown to have small mean squared error relative to other competing estimators. Their main justification, however, lies in their good large-sample behavior.

For independent and identically distributed observations with true probability density $p(\cdot; \boldsymbol{\theta}_0)$ satisfying certain regularity conditions, it can be shown that the maximum likelihood estimator $\hat{\boldsymbol{\theta}}$ of $\boldsymbol{\theta}_0$ converges in probability to $\boldsymbol{\theta}_0$ and that the distribution of $\sqrt{n}(\hat{\boldsymbol{\theta}} - \boldsymbol{\theta}_0)$ is approximately normal with mean 0 and covariance matrix $I(\boldsymbol{\theta}_0)^{-1}$ where $I(\boldsymbol{\theta})$ is Fisher's information matrix with (i, j)-component

$$E_{\boldsymbol{\theta}}\left[\frac{\partial \ln p(X; \boldsymbol{\theta})}{\partial \theta_i}\frac{\partial \ln p(X; \boldsymbol{\theta})}{\partial \theta_j}\right].$$

In time series analysis the situation is rather more complicated than in the case of iid observations. "Likelihood" in the time series context is almost always used in the sense of Gaussian likelihood, i.e., the likelihood computed under the (possibly false) assumption that the series is Gaussian. Nevertheless estimators of ARMA coefficients computed by maximization of the Gaussian likelihood have good large-sample properties analogous to those described in the preceding paragraph. For details see TSTM, Section 10.8.

B.3 Confidence Intervals

Estimation of a parameter or parameter vector by least squares or maximum likelihood leads to a particular value, often referred to as a **point estimate**. It is clear that this will rarely be exactly equal to the true value and so it is important to convey some idea of the probable accuracy of the estimator. This can be done using the notion of confidence interval, which specifies a random set covering the true parameter value with some specified (high) probability.

Example B.3.1 If $\mathbf{X} = (X_1, \ldots, X_n)'$ is a vector of independent $\text{N}(\mu, \sigma^2)$ random variables, we saw in Section B.2 that the random variable $\overline{X}_n = \frac{1}{n}\sum_{i=1}^n X_i$ is the maximum likelihood estimator of μ. This is a point estimator of μ. To construct a confidence interval for μ from $\overline{X}_n$, we observe that the random variable $\frac{\overline{X}_n - \mu}{S/\sqrt{n}}$ has Student's t-distribution with $n-1$ degrees of freedom, where S is the sample standard deviation, i.e., $S^2 = \frac{1}{n-1}\sum_{i=1}^n (X_i - \overline{X}_n)^2$. Hence,

$$P\left[-t_{1-\alpha/2} < \frac{\overline{X}_n - \mu}{S/\sqrt{n}} < t_{1-\alpha/2}\right] = 1 - \alpha,$$

where $t_{1-\alpha/2}$ denotes the $1-\alpha/2$-quantile of the t-distribution with $n-1$ degrees of freedom. This probability statement can be reexpressed in the form

$$P\left[\overline{X}_n - t_{1-\alpha/2}S/\sqrt{n} < \mu < \overline{X}_n + t_{1-\alpha/2}S/\sqrt{n}\right] = 1 - \alpha,$$

which shows that the random interval bounded by $\overline{X}_n \pm t_{1-\alpha/2}S/\sqrt{n}$ includes the true value μ with probability $1-\alpha$. This interval is called a $(1-\alpha)$ confidence interval for the mean μ. □

B.3.1 Large-Sample Confidence Regions

Many estimators of a vector-valued parameter $\boldsymbol{\theta}$ are approximately normally distributed when the sample size n is large. For example, under mild regularity conditions, the maximum likelihood estimator $\hat{\boldsymbol{\theta}}(\mathbf{X})$ of $\boldsymbol{\theta} = (\theta_1, \ldots, \theta_m)'$ is approximately $\text{N}(\mathbf{0}, \frac{1}{n}I(\hat{\boldsymbol{\theta}})^{-1})$, where $I(\boldsymbol{\theta})$ is the Fisher information defined in Section B.2. Consequently,

$$n(\hat{\boldsymbol{\theta}} - \boldsymbol{\theta})'I(\hat{\boldsymbol{\theta}})(\hat{\boldsymbol{\theta}} - \boldsymbol{\theta})$$

is approximately distributed as χ^2 with m degrees of freedom and the random set of $\boldsymbol{\theta}$-values defined by

$$n(\boldsymbol{\theta} - \hat{\boldsymbol{\theta}})'I(\hat{\boldsymbol{\theta}})(\boldsymbol{\theta} - \hat{\boldsymbol{\theta}}) \le \chi^2_{1-\alpha}(m)$$

covers the true value of $\boldsymbol{\theta}$ with probability approximately equal to $1-\alpha$.

Example B.3.2 For iid observations $X_1, \ldots, X_n$ from $\text{N}(\mu, \sigma^2)$, a straightforward calculation gives, for $\boldsymbol{\theta} = (\mu, \sigma^2)'$,

$$I(\theta) = \begin{bmatrix} \sigma^{-2} & 0 \\ 0 & \sigma^{-4}/2 \end{bmatrix}.$$

Thus we obtain the large-sample confidence region for $(\mu, \sigma^2)'$,

$$n(\mu - \overline{X}_n)^2/\hat{\sigma}^2 + n(\sigma^2 - \hat{\sigma}^2)^2/(2\hat{\sigma}^4) \le \chi^2_{1-\alpha}(2),$$

which covers the true value of $\boldsymbol{\theta}$ with probability approximately equal to $1 - \alpha$. This region is an ellipse centered at $(\overline{X}_n, \hat{\sigma}^2)$. □

B.4 Hypothesis Testing

Parameter estimation can be regarded as choosing one from infinitely many possible decisions regarding the value of a parameter vector, $\boldsymbol{\theta}$. Hypothesis testing on the other hand involves a choice between two alternative hypotheses, a "null" hypothesis, H_0, and an "alternative" hypothesis, H_1, regarding the parameter vector $\boldsymbol{\theta}$. The hypotheses H_0 and H_1 correspond to subsets Θ_0 and Θ_1 of the parameter set Θ. The problem is to decide, on the basis of an observed data vector $\mathbf{X}$, whether or not we should reject the null hypothesis, H_0. A statistical test of H_0 can therefore be regarded as a partition of the *sample* space into one set of values of $\mathbf{X}$ for which we reject H_0 and another for which we do not. The problem is to specify a test (i.e., a subset of the sample space called the "rejection region") for which the corresponding decision rule performs well in practice.

Example B.4.1 If $\mathbf{X} = (X_1, \ldots, X_n)'$ is a vector of independent $N(\mu, 1)$ random variables, we may wish to test the null hypothesis, H_0: $\mu = 0$, against the alternative, H_1: $\mu \ne 0$. A plausible choice of rejection region in this case is the set of all samples $\mathbf{X}$ for which $|\overline{X}_n| > c$ for some suitably chosen constant c. We shall return to this example after considering those factors that should be taken into account in the systematic selection of a "good" rejection region. □

B.4.1 Error Probabilities

There are two types of error that may be incurred in the application of a statistical test.

- type I error is the rejection of H_0 when it is true.
- type II error is the acceptance of H_0 when it is false.

For a given test (i.e., for a given rejection region R), the probabilities of error can both be found from the **power function** of the test, defined as

$$P_{\boldsymbol{\theta}}(R), \quad \boldsymbol{\theta} \in \Theta,$$

where $P_{\boldsymbol{\theta}}$ is the distribution of $\mathbf{X}$ when the true parameter value is $\boldsymbol{\theta}$. The probabilities of a type I error are

$$\alpha(\boldsymbol{\theta}) = P_{\boldsymbol{\theta}}(R), \quad \boldsymbol{\theta} \in \Theta_0,$$

and the probabilities of a type II error are

$$\beta(\boldsymbol{\theta}) = 1 - P_{\boldsymbol{\theta}}(R), \quad \boldsymbol{\theta} \in \Theta_1.$$

It is not generally possible to find a test that simultaneously minimizes $\alpha(\boldsymbol{\theta})$ and $\beta(\boldsymbol{\theta})$ for all values of their arguments. Instead, therefore, we seek to limit the probability of type I error and then, subject to this constraint, to minimize the probability of type II error uniformly on Θ_1. Given a **significance level** α, an optimum level-α test is a test satisfying

$$\alpha(\boldsymbol{\theta}) \leq \alpha, \quad \text{for all } \boldsymbol{\theta} \in \Theta_0,$$

which minimizes $\beta(\boldsymbol{\theta})$ for every $\boldsymbol{\theta} \in \Theta_1$. Such a test is called a **uniformly most powerful (U.M.P.) test of level** $\boldsymbol{\alpha}$. The quantity $\sup_{\boldsymbol{\theta} \in \Theta_0} \alpha(\boldsymbol{\theta})$ is called the **size** of the test.

In the special case of a simple hypothesis vs a simple hypothesis, e.g., H_0: $\boldsymbol{\theta} = \boldsymbol{\theta}_0$ vs H_1: $\boldsymbol{\theta} = \boldsymbol{\theta}_1$, an optimal test based on the likelihood ratio statistic can be constructed (see Silvey, 1975). Unfortunately, it is usually not possible to find a uniformly most powerful test of a simple hypothesis against a composite (more than one value of $\boldsymbol{\theta}$) alternative. This problem can sometimes be solved by searching for uniformly most powerful tests within the smaller classes of unbiased or invariant tests. For further information see Lehmann (1986).

B.4.2 Large-Sample Tests Based on Confidence Regions

There is a natural link between between the testing of a simple hypothesis H_0: $\boldsymbol{\theta} = \boldsymbol{\theta}_0$ vs H_1: $\boldsymbol{\theta} \neq \boldsymbol{\theta}_0$ and the construction of confidence regions. To illustrate this connection, suppose that $\hat{\boldsymbol{\theta}}$ is an estimator of $\boldsymbol{\theta}$ whose distribution is approximately $\mathrm{N}(\boldsymbol{\theta}, n^{-1}I^{-1}(\boldsymbol{\theta}))$, where $I(\boldsymbol{\theta})$ is a positive definite matrix. This is usually the case, for example, when $\hat{\boldsymbol{\theta}}$ is a maximum likelihood estimator and $I(\boldsymbol{\theta})$ is the Fisher information. As in Section B.3.1, we have

$$P_{\boldsymbol{\theta}}\left(n(\boldsymbol{\theta} - \hat{\boldsymbol{\theta}})' I(\hat{\boldsymbol{\theta}})(\boldsymbol{\theta} - \hat{\boldsymbol{\theta}}) \leq \chi^2_{1-\alpha}(m)\right) \approx 1 - \alpha.$$

Consequently, an approximate α-level test is to reject H_0 if

$$n(\boldsymbol{\theta}_0 - \hat{\boldsymbol{\theta}})' I(\hat{\boldsymbol{\theta}})(\boldsymbol{\theta}_0 - \hat{\boldsymbol{\theta}}) > \chi^2_{1-\alpha}(m),$$

or equivalently, if the confidence region determined by those $\boldsymbol{\theta}$'s satisfying

$$n(\boldsymbol{\theta} - \hat{\boldsymbol{\theta}})' I(\hat{\boldsymbol{\theta}})(\boldsymbol{\theta} - \hat{\boldsymbol{\theta}}) \leq \chi^2_{1-\alpha}(m),$$

does not include $\boldsymbol{\theta}_0$.

Example B.4.2 Consider again the problem described in Example B.4.1. Since $\overline{X}_n \sim N(\mu, n^{-1})$, the hypothesis H_0: $\mu = 0$ is rejected at level α if

$$n(\overline{X}_n)^2 > \chi^2_{1-\alpha,1},$$

or equivalently, if

$$|\overline{X}_n| > \frac{\Phi_{1-\alpha/2}}{n^{1/2}}.$$

□

C Mean Square Convergence

The sequence S_n of random variables is said to converge in mean square to the random variable S if and only if

$$E(S_n - S)^2 \to 0 \text{ as } n \to \infty.$$

In particular we say that the sum $\sum_{k=1}^{n} X_k$ converges (in mean square) if and only if there exists a random variable S such that $E(\sum_{k=1}^{n} X_k - S)^2 \to 0$ as $n \to \infty$. If this is the case, then we use the notation $S = \sum_{k=1}^{\infty} X_k$.

C.1 The Cauchy Criterion

For a given sequence S_n of random variables to converge in mean square to *some* random variable, it is necessary and sufficient that

$$E(S_m - S_n)^2 \to 0 \text{ as } m, n \to \infty$$

(for a proof of this see TSTM, Chapter 2). The point of the criterion is that it permits checking for mean square convergence without having to identify the limit of the sequence.

Example C.1.1 Consider the sequence of partial sums $S_n = \sum_{t=-n}^{n} a_t Z_t$, $n = 1, 2, \ldots$, where $\{Z_t\} \sim \text{WN}(0, \sigma^2)$. Under what conditions on the coefficients a_i does this sequence converge in mean square? To answer this question we apply the Cauchy criterion as follows. For $n > m > 0$,

$$E(S_n - S_m)^2 = E\left(\sum_{m<|i|\le n} a_i Z_i\right)^2 = \sigma^2 \sum_{m<|i|\le n} a_i^2.$$

Consequently $E(S_n - S_m)^2 \to 0$ if and only if $\sum_{m<|i|\le n} a_i^2 \to 0$. Since the Cauchy criterion applies also to real-valued sequences, this last condition is equivalent to convergence of the sequence $\sum_{i=-n}^{n} a_i^2$, or equivalently to the condition

$$\sum_{i=-\infty}^{\infty} a_i^2 < \infty.$$

□

Properties of Mean Square Convergence:

If $X_n \to X$ and $Y_n \to Y$, in mean square as $n \to \infty$, then

(a) $E(X_n^2) \to E(X^2)$,

(b) $E(X_n) \to E(X)$,

and

(c) $E(X_n Y_n) \to E(XY)$.

Proof See TSTM, Proposition 2.1.2. ■

D An ITSM Tutorial

Examples to familiarize the reader with the use of the programs in the package ITSM96 are contained in the file `README.DOC` on the diskette. Instructions for installing the package and printing hard copies of the output are also given in `README.DOC`. In this final appendix we give a more detailed example of the analysis of a univariate data set using PEST. Further information on the use of the programs can be found in the book *ITSM for Windows*, Springer-Verlag (1994).

D.1 Getting Started

D.1.1 Running PEST

Double click on the icon labelled *pest* in the *itsm96* window (or in DOS type `PEST`↵) and you will see the PEST title screen. Then press any key and you will see the Main Menu of PEST, which at this stage has seven available options. Further options will appear in the Main Menu after data are entered.

PEST is menu-driven so that you are required only to make choices between options specified by the program. For example, you can choose the first option of the Main Menu [**D**`ata entry; statistics; transformations`] by typing the highlighted letter **D**. (In the text, the letter corresponding to the "hot" key for immediate selection of menu options will always be printed in boldface.) This option can also be chosen by moving the highlight bar with the mouse to the first row of the menu and clicking. A third alternative is to move the mouse pointer out of the menu box, use the arrow keys to move the highlight bar, and then press ↩. After selecting this option you will see the Data Menu, from which you can make a further selection, e.g., `Load new data set`, in the same way. To return to the Main Menu, select the last item of the data menu (e.g., by typing **R**). For the remainder of the book we shall indicate selection of menu items by typing the highlighted letter, but in all cases the other two methods of menu selection can equally well be used.

There are several distinct functions of the program PEST. The first is to plot, analyze, and possibly transform time series data, the second is to compute properties of time series models, and the third utilizes the previous two in fitting models to data. The latter includes checking that the properties of the fitted model match those of the data in a suitable sense. Having found an appropriate model, we can (for example) then use it in conjunction with the data to forecast future values of the series. Sections D.2–D.5 of this appendix deal with the modelling and analysis of data, while Section D.6 is concerned with model properties.

It is important to keep in mind the distinction between data and model properties and not to confuse the data with the model. At any particular time PEST typically stores one data set and one model (which can be identified using the option [`Current model and data file status`] of the Main Menu). Rarely (if ever) is a real time series generated by a model as simple as those used for fitting purposes. Our aim is to develop a model that mimics important features of the data, but is still simple enough to be used with relative ease.

The examples that follow constitute a tutorial session for PEST in serialized form. They lead you through a complete analysis of the well-known airline passenger series of Box and Jenkins (1976) filed as AIRPASS.DAT on the diskette.

D.2 Preparing Your Data for Modelling

The observed values of your time series should be available in a single-column ASCII file (or two columns for a bivariate series). You can then begin model fitting with PEST. The program will read your data from the file, plot it on the screen, compute sample statistics, and allow you to do a number of transformations designed to make your transformed data representable as a realization of a zero-mean stationary process.

Example D.2.1 To illustrate the analysis we shall use the file AIRPASS.DAT, which contains the number of international airline passengers (in thousands) for each month from Jan. '49 to Dec. '60. □

D.2.1 Entering Data

From the Main Menu of PEST select the first option (**D**`ata entry; statistics; transformations`) by typing **D**. The Data Menu will then appear. Choose Option 1 and you will be asked to confirm that you wish to enter new data. Respond by typing **Y**. A list of data files will then appear. To select a data file for analysis, move the highlight bar to the name of the required file and press ↩. The program PEST will then read in your data and display on the screen the number of observations in the data as well as the first three and last data points.

A new data file can always be imported using Option 1 of the Data Menu. Note, however, that the previous data file is eliminated from PEST each time a new file is read in.

Example D.2.2 Go through the above steps to read the airline passenger data into PEST. The file name is AIRPASS.DAT. □

D.2.2 Filing Data

You may wish to change your data using PEST and then store it in another file. At any time before or after transforming the data in PEST, the data can be filed by choosing Option 10 from the Data Menu. Do not use the name of a file that already exists or it will be overwritten.

D.2.3 Plotting Data

The first step in the analysis of any time series is to plot the data. With PEST the data can be plotted by selecting Option 2 from the Data Menu. This will first produce a histogram of the data; pressing any key then causes a graph of the data vs time to appear on the screen.

Under the histogram several sample statistics are printed. These are defined as follows:

Mean:

$$\bar{X} = \frac{1}{n}\sum_{i=1}^{n} X_i$$

Standard Deviation:

$$s = \sqrt{\frac{1}{n}\sum_{i=1}^{n}(X_i - \bar{X})^2} = \sqrt{\frac{1}{n}\left(\sum_{i=1}^{n} X_i^2 - n\bar{X}^2\right)}$$

Coefficient of Skewness:

$$\hat{\nu}_3 = \sqrt{\frac{1}{n}\sum_{i=1}^{n}(X_i - \bar{X})^3} = \sqrt{\frac{1}{n}\left(\sum_{i=1}^{n} X_i^3 - 3\bar{X}\sum_{i=1}^{n} X_i^2 + 2n\bar{X}^3\right)}$$

Example D.2.3 Continuing with our analysis of the data file AIRPASS.DAT, choose Option 2 from the Data Menu. The first graph displayed is a histogram of the data. Then press any key to obtain a time-series graph of the data. (The data is shown in Figure 9.4.) Finally press any key and type **C** to return to the Data Menu. □

D.2.4 Transforming Data

Transformations are applied in order to produce data which can be successfully modelled as "stationary time series". In particular we need to eliminate trend and cyclic components and to achieve approximate constancy of level and variability with time.

Example D.2.4 The airline passenger data (see Figure 8.3) are clearly not stationary. The level and variability both increase with time, and there appears to be a large seasonal component (with period 12). □

Nonstationary data must be transformed before attempting to fit a stationary model. PEST provides a number of transformations that are useful for this purpose.

Box-Cox Transformations

Box-Cox transformations can be carried out by selecting Option 5 of the Data Menu. If the original observations are $Y_1, Y_2, \ldots, Y_n$, the Box-Cox transformation f_λ converts them to $f_\lambda(Y_1), f_\lambda(Y_2), \ldots, f_\lambda(Y_n)$, where

$$f_\lambda(y) = \begin{cases} \dfrac{y^\lambda - 1}{\lambda}, & \lambda \neq 0, \\ \log(y), & \lambda = 0. \end{cases}$$

These transformations are useful when the variability of the data increases or decreases with the level. By suitable choice of λ, the variability can often be made nearly constant. In particular, for positive data whose standard deviation increases linearly with level, the variability can be stabilized by choosing $\lambda = 0$.

The choice of λ can be made by trial and error, using the graphs of the transformed data that can be plotted using Option 2 of the Data Menu. (After inspecting the graph for a particular λ you can invert the transformation using Option 5 of the Data Menu, after which you can then try another value of λ.) Very often it is found that no transformation is needed or that the choice $\lambda = 0$ is satisfactory.

Example D.2.5 For the series AIRPASS.DAT the variability increases with level and the data are strictly positive. Taking natural logarithms (i.e., choosing a Box-Cox transformation with $\lambda = 0$) gives the transformed data shown in Figure D.1. You can perform this transformation and plot the graph (starting in the Data Menu with the data file AIRPASS.DAT) by typing **5** 0↩ (to transform the data), then **2** ↩ (to plot the graph).

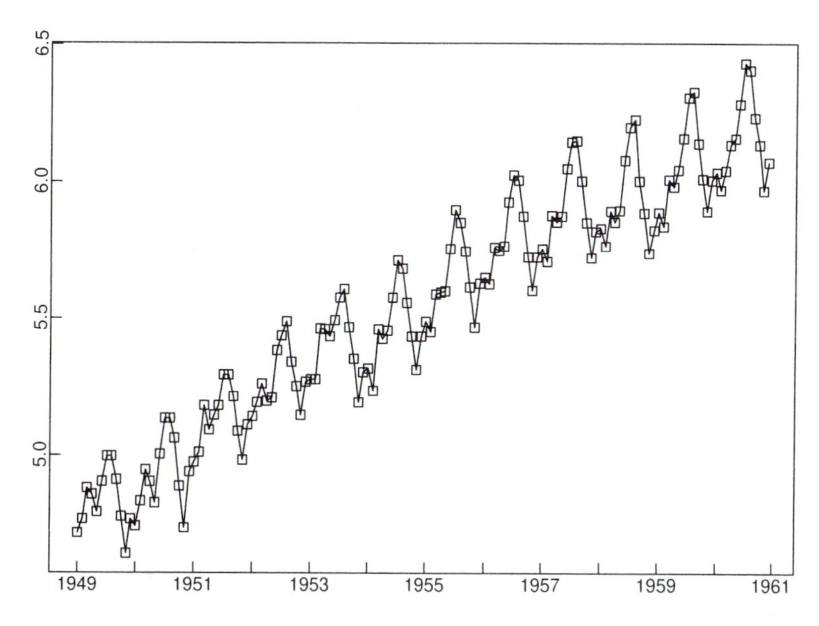

Figure D-1
The series AIRPASS.DAT after taking logs.

Notice how the variation no longer increases. The seasonal effect remains, as does the upward trend. These will be removed shortly. Since the log transformation has stabilized the variability, it is not necessary to consider other values of λ. Note that the data stored in PEST now consists of the natural logarithms of the original data. □

Classical Decompositon

There are two methods provided in PEST for the elimination of trend and seasonality. These are

i. "classical decomposition" of the series into a trend component, a seasonal component, and a random residual component, and
ii. differencing.

Classical decomposition of the series $\{X_t\}$ is based on the model

$$X_t = m_t + s_t + Y_t,$$

where X_t is the observation at time t, m_t is a "trend component," s_t is a "seasonal component," and Y_t is a "random noise component," which is stationary with mean zero. The objective is to estimate the components m_t and s_t and subtract them from the data to generate a sequence of residuals (or estimated noise) that can then be modelled as a stationary time series.

To achieve this, select Option 6 then Option 7 from the Data Menu. (You can also estimate trend only or seasonal component only by selecting the appropriate option separately.)

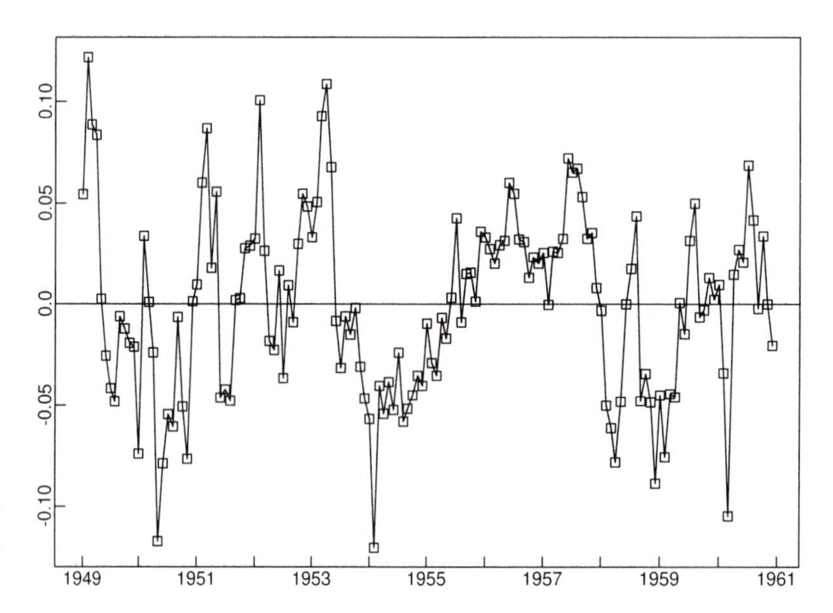

Figure D-2
The logged AIRPASS.DAT series after removal of trend and seasonal components by classical decomposition.

The estimated noise sequence automatically replaces the previous data stored in PEST.

Example D.2.6 The logged airline passenger data has an apparent seasonal component of period 12 (corresponding to the month of the year) and an approximately quadratic trend. A plot of the data together with the estimated seasonal and quadratic trend components can be displayed by typing **6** 12↩ ↩ ↩ ↩ ↩ **C 7 q** ↩ ↩ (starting from the Data Menu).

Type ↩ **C N** to return to the data menu and then **2** ↩ to view a plot of the transformed data (or residuals) Y_t, as in Figure D.2, obtained by removal of trend and seasonality from the logged AIRPASS.DAT series using the classical decomposition. $\{Y_t\}$ shows no obvious deviations from stationarity and it would now be reasonable to attempt to fit a stationary time series model to this series. We shall not pursue this approach any further here, but turn instead to the **differencing** approach. (You should have no difficulty in later returning to this point and completing the classical decomposition analysis by fitting a stationary time series model to $\{Y_t\}$.)

Restore the original airline passenger data into PEST by using Option 1 of the Data Menu and reading in the file AIRPASS.DAT. □

Differencing

Differencing is a technique that can also be used to remove seasonal components and trends. The idea is simply to consider the differences between pairs of observations with appropriate time separations. For example, to remove a seasonal component of period 12 from the series $\{X_t\}$, we generate the transformed series

$$Y_t = X_t - X_{t-12}.$$

It is clear that all seasonal components of period 12 are eliminated by this transformation, which is called **differencing at lag 12**. A linear trend can be eliminated by differencing at lag 1, and a quadratic trend by differencing twice at lag 1 (i.e., differencing once to get a new series, then differencing the new series to get a second new series). Higher-order polynomials can be eliminated analogously. It is worth noting that differencing at lag 12 eliminates not only seasonal components with period 12 but also any linear trend.

Repeated differencing can be done with PEST by selecting Option 8 from the Data Menu.

Example D.2.7 At this stage of the analysis we have restored the original data set AIRPASS.DAT into PEST with the Data Menu displayed on the screen. Type **5** `0↩` to replace the stored observations by their natural logs. The transformed series can now be deseasonalized by differencing at lag 12. To do this type **8** `12↩`. Inspection of the graph of the deseasonalized series suggests a further differencing at lag 1 to eliminate the remaining trend. To do this type **8** `1↩`. Then type **2** `↩` and you should see the transformed and twice differenced series shown in Figure D.3. □

Subtracting the Mean

The term *ARMA model* is used in PEST to denote a zero-mean ARMA process (see Definition 3.1.1). To fit such a model to data, the sample mean of the data should

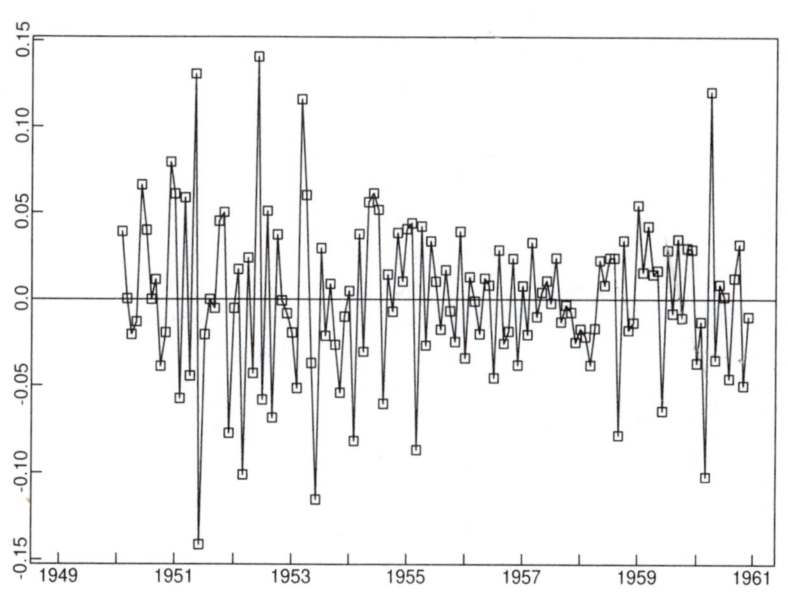

Figure D-3 The series AIRPASS.DAT after taking logs and differencing at lags 12 and 1.

therefore be small. (An estimate of the standard error of the sample mean is displayed on the screen just after reading in the data file.) Once the apparent deviations from stationarity of the data have been removed, we therefore (in most cases) subtract the sample mean of the transformed data from each observation to generate a series to which we then fit a zero-mean stationary model. Effectively we are estimating the mean of the model by the sample mean, then fitting a (zero-mean) ARMA model to the "mean-corrected" transformed data. If we know a priori that the observations are from a process with zero mean, then this process of mean correction is omitted. PEST keeps track of all the transformations (including mean correction) that are made. You can check these for yourself by going to Option 10 of the Main Menu. When it comes time to predict the original series, PEST will invert all these transformations automatically.

Example D.2.8 Subtract the mean of the transformed and twice differenced AIRPASS.DAT series by typing **9**. Type **R** to return to the Main Menu, then **C** to check the status of the data and model that currently reside in PEST. You will see in particular that the default white noise model (ARMA(0,0)) with variance 1 is displayed since no model has yet been entered. □

D.3 Finding a Model for Your Data

After transforming the data (if necessary) as described above, we are now in a position to fit an ARMA model. PEST uses a variety of tools to guide us in the search for an appropriate model. These include the sample ACF (autocorrelation function), the sample PACF (partial autocorrelation function), and the AICC statistic (a bias-corrected form of Akaike's AIC statistic (see Section 5.5.2).

D.3.1 The Sample ACF and PACF

Option 3 of the Data Menu computes and plots the sample ACF and PACF for values of the lag h from 1 up to 40. Values that decay rapidly as h increases indicate short-term dependency in the time series, while slowly decaying values indicate long-term dependency. For ARMA fitting it is desirable to have a sample ACF that decays fairly rapidly. A sample ACF that is positive and very slowly decaying suggests that the data may have a trend. A sample ACF with very slowly damped periodicity suggests the presence of a periodic seasonal component. In either of these two cases you may need to transform your data before continuing.

When you select Option 3 of the Data Menu you will be prompted to specify the maximum lag required. This is restricted by PEST to be less than or equal to 40. (As a rule of thumb, the estimates are reliable for lags up to about $\frac{1}{3}$ of the sample size.

It is clear from the definition of the sample ACF, $\hat{\rho}(h)$, that it will be a very poor estimator of $\rho(h)$ for h close to the sample size n.)

Once you have specified the maximum lag, M, the sample ACF and PACF values will be plotted on the screen for lags h from 1 to M. The horizontal lines on the graph display the bounds $\pm 1.96/\sqrt{n}$, which are approximate 95% bounds for the autocorrelations of a white noise sequence. If the data are a (large) sample from an independent white noise sequence, approximately 95% of the sample autocorrelations should lie between these bounds. Large or frequent excursions from the bounds suggest that we need a model to explain the dependence and sometimes suggest the kind of model we need (see below). The numerical values of the sample ACF and PACF are printed below the graphs. Press any key to return to the Data Menu.

The graphs of the sample ACF and PACF sometimes suggest an appropriate ARMA model for the data. As a rough guide, if the sample ACF falls between the plotted bounds $\pm 1.96/\sqrt{n}$ for lags $h > q$ then an MA(q) model is suggested, while if the sample PACF falls between the plotted bounds $\pm 1.96/\sqrt{n}$ for lags $h > p$, then an AR(p) model is suggested.

If neither the sample ACF nor PACF "cuts off" as in the previous two paragraphs, a more refined model selection technique is required (see the discussion of the AICC statistic in Section 5.5.2). Even if the sample ACF or PACF does cut off at some lag, it is still advisable to explore models other than those suggested by the sample ACF and PACF values.

Example D.3.1 Figure D.4 shows the sample ACF of the AIRPASS.DAT series after taking logarithms, differencing at lags 12 and 1, and subtracting the mean. Figure D.5 shows the corresponding sample PACF. These graphs suggest we consider an MA model of order 12 (or perhaps 23) with a large number of zero coefficients, or alternatively an AR model of order 12. □

D.3.2 Entering a Model

To do any serious analysis with PEST, a model must be entered. This can be done either by specifying an ARMA model directly using the option [**E**`ntry of an ARMA(p,q) model`] or (if the program contains a data file that is to be modelled as an ARMA process) by using the option [**P**`reliminary estimation of ARMA parameters`] of the Main Menu. If no model is entered, PEST assumes the default ARMA(0,0) or white noise model,

$$X_t = Z_t,$$

where $\{Z_t\}$ is an uncorrelated sequence of random variables with mean zero and variance 1.

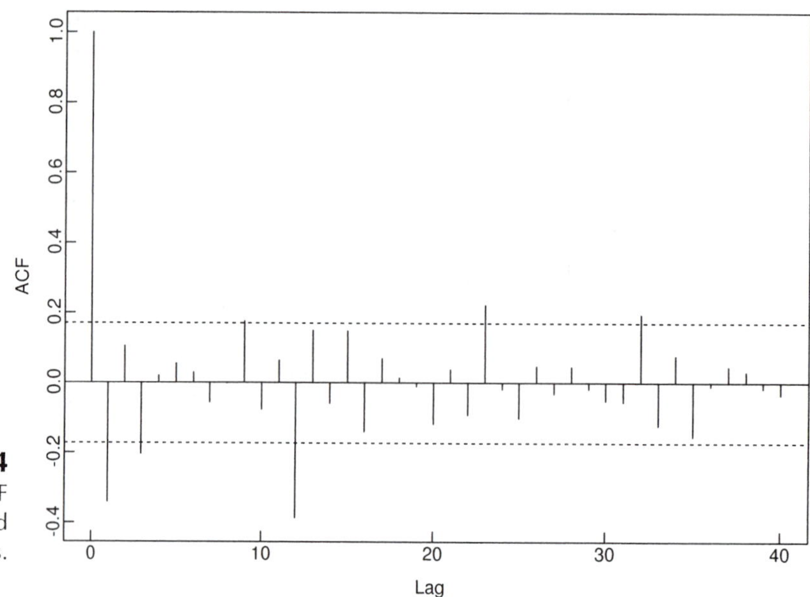

Figure D-4
The sample ACF of the transformed AIRPASS.DAT series.

If you have data and no particular ARMA model in mind, it is best to let PEST find the model by using the option [**P**reliminary estimation of ARMA parameters].

Sometimes you may wish to try a model used in a previous session with PEST or a model suggested by someone else. In that case use the option [**E**ntry of an ARMA(p,q) model]. A useful feature of this option is the ability to import a model stored in an earlier session. PEST can read the stored model, saving you the trouble of repeating an optimization or entering the model coefficient by coefficient.

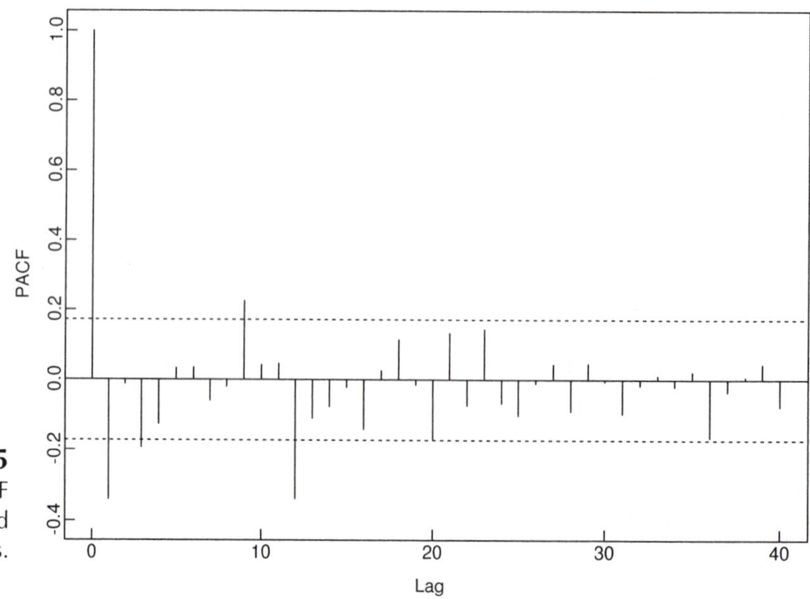

Figure D-5
The sample PACF of the transformed AIRPASS.DAT series.

To enter a model directly, specify the order of the autoregressive and moving average polynomials as requested. You will then be required to enter the coefficients. Initially PEST will set the white noise variance to 1. To enter a model stored in a file, choose the autoregressive order to be -1. You will then be asked to specify the name of the file in which the model has previously been saved by PEST.

After you have entered the model, you will see the Model Menu, which gives you the opportunity to make any required changes. To alter a specific coefficient in the model, you must enter the index number of the coefficient. The autoregressive coefficients are numbered $1, 2, \ldots, p$ and the moving average coefficients are numbered $p+1, p+2, \ldots, p+q$. For example, to change the second moving average coefficient in an ARMA(3,2) model, type **C** to change a coefficient and then type `5`↵.

D.3.3 Preliminary Estimation

The option [`Preliminary estimation of ARMA parameters`] of the Main Menu contains fast (but somewhat rough) model-fitting algorithms. These are useful for suggesting the most promising models for the data, but they should be followed by the more refined maximum likelihood estimation procedure in the option [`ARMA parameter estimation`] of the Main Menu. The fitted preliminary model is generally used as an initial approximation with which to start the nonlinear optimization carried out in the course of maximizing the (Gaussian) likelihood.

The AR and MA orders p and q of the model to be fitted must be entered first (see Section 3.1). For pure AR models, the preliminary estimation option of PEST offers you a choice between the Burg and Yule-Walker estimates. The Burg estimates frequently give higher values of the Gaussian likelihood than the Yule-Walker estimates. For the case $q > 0$, PEST will also give you a choice between the two preliminary estimation methods based on the Hannan-Rissanen procedure and the innovations algorithm. If you choose the innovations option by typing **I**, a default value of m will be displayed on the screen. The parameter m was defined in Section 5.1.3. The standard choice is the default value computed by PEST.

Once the values of p, q, and m have been entered, PEST will quickly estimate the parameters of the specified model and display a number of diagnostic statistics.

The estimated parameters are given with the ratio of each estimate to 1.96 times its standard error. The denominator ($1.96 \times$ standard error) is the critical value for the coefficient. Thus, if the ratio is greater than 1 in absolute value, we may conclude (at level .05) that the corresponding coefficient is different from zero. On the other hand, a ratio less than 1 in absolute value suggests the possibility that the corresponding coefficient in the model may be zero. (If the innovations option is chosen, the ratios of estimates to $1.96 \times$ standard error are displayed only when $p = q$.)

After the estimated coefficients are displayed on the screen, press any key and PEST will then do one of two things depending on whether or not the fitted model is causal (see Section 3.1).

If the model is causal, PEST will give an estimate $\hat{\sigma}^2$ of the white noise variance, $\mathrm{Var}(Z_t)$, and some further diagnostic statistics. These are $-2\ln L(\hat{\phi}, \hat{\theta}, \hat{\sigma}^2)$, where L denotes the Gaussian likelihood, (5.2.9), and the AICC statistic,

$$-2\ln L + 2(p+q+1)n/(n-p-q-2),$$

(see Section 5.5.2).

Our eventual aim is to find a model with as small an AICC value as possible. Smallness of the AICC value computed in the option [`Preliminary estimation`] is indicative of a good model, but should be used only as a rough guide. Final decisions between models should be based on maximum likelihood estimation computed in the option [`ARMA parameter estimation`], since for fixed p and q, the values of $\boldsymbol{\phi}$, $\boldsymbol{\theta}$, and σ^2 that minimize the AICC statistic are the maximum likelihood estimates, not the preliminary estimates. In the option [`Preliminary estimation`] of the Main Menu, it is possible to minimize the AICC for pure autoregressive models fitted either by Burg's algorithm or the Yule-Walker equations by entering -1 as the selected autoregressive order. Autoregressions of all orders up to 26 will then be fitted by the chosen algorithm and the model with smallest AICC value will be selected.

If the preliminary fitted model is noncausal, PEST will set all coefficients to .001 to generate a causal model with the specified values of p and q. Further investigation of this model must then be done with the option [`ARMA parameter estimation`].

After completing the preliminary estimation, PEST will store the fitted model coefficients and white noise variance. The stored estimate of the white noise variance is the sum of squares of the residuals (or one-step prediction errors) divided by the number of observations.

At this point you can try a different model, file the current model, or return to the Main Menu. When you return to the Main Menu, the most recently fitted preliminary model will be stored in PEST. You will now see a large number of options available on the Main Menu.

Example D.3.2 Let us first find the minimum-AICC AR model for the logged, differenced, and mean-corrected AIRPASS.DAT series currently stored in PEST. From the Main Menu type **P** and then type `-1`↩ for the order of the autoregression. Type **B** to select the Burg estimation procedure. The minimum-AICC AR model is of order 12 with an AICC value of -458.57. Now let us fit a preliminary MA(25) model to the same data set. Select the option [`Try another model`] and type `0`↩ for the order of the autoregressive polynomial and `25`↩ for the order of the moving average polynomial. Choose the Innovations estimation procedure by typing **I** and type **N** to use the default value for m, the number of autocovariances used in the estimation procedure.

The ratios, (estimated coefficient)/(1.96×standard error), indicate that the coefficients at lags 1 and 12 are nonzero, as we suspected from the ACF. The estimated coefficients at lags 3 and 23 also look substantial even though the corresponding ratios are less than 1 in absolute value. The displayed values are as follows:

```
MA COEFFICIENTS
    -.3568      .0673     -.1629     -.0415      .1268
     .0264      .0283     -.0648      .1326     -.0762
    -.0066     -.4987      .1789     -.0318      .1476
    -.1461      .0440     -.0226     -.0749     -.0456
    -.0204     -.0085      .2014     -.0767     -.0789
RATIO OF COEFFICIENTS TO (1.96*STANDARD ERROR)
   -2.0833      .3703     -.8941     -.2251      .6875
     .1423      .1522     -.3487      .7124     -.4061
    -.0353    -2.6529      .8623     -.1522      .7068
    -.6944      .2076     -.1065     -.3532     -.2147
    -.0960     -.0402      .9475     -.3563     -.3659
<Press any key to continue>
```

Press any key to see the value of the white noise variance. Press ↩ once again to display the values of $-2 \ln L$ and the AICC. After pressing ↩, you can return to the Main Menu by typing **R** with the fitted MA(25) model now stored in PEST. Note that at this stage of the modelling process the fitted AR(12) model has a smaller AICC value than the MA(25) model. Later we shall find a subset MA(25) model that has an even smaller AICC value. □

D.3.4 The AICC Statistic

The AICC statistic for the model with parameters p, q, ϕ, and θ, is defined (see Section 5.2.2) as

$$\mathrm{AICC}(\phi, \theta) = -2 \ln L(\phi, \theta, S(\phi, \theta)/n) + 2(p+q+1)n/(n-p-q-2),$$

and a model chosen according to the AICC criterion minimizes this statistic.

Model selection statistics other than AICC are also available in PEST. A Bayesian modification of the AIC statistic, known as the BIC statistic,is also computed in the option [`ARMA parameter estimation`]. It is used in the same way as the AICC.

An exhaustive search for a model with minimum AICC or BIC value can be very slow. For this reason the sample ACF and PACF and the preliminary estimation techniques described above are useful in narrowing down the range of models to be considered more carefully in the maximum-likelihood estimation stage of model fitting.

D.3.5 Changing Your Model

The model currently stored by the program and the status of the data file can be checked at any time using the option [Current model and data file status] of the Main Menu. Any parameter can be changed with this option, including the white noise variance, and the model can be filed for use at some other time.

Example D.3.3 We shall now set some of the coefficients in the current model to zero. To do this choose the option [Current model and data file status] from the Main Menu by typing **C**. The preliminary estimation suggested that the most significant coefficients in the fitted MA(25) model were those at lags 1, 3, 12, and 23. Let us therefore try setting all the other coefficients to zero. To change the lag-2 coefficient, select [Change a coefficient] from the menu and enter the number 2 (to indicate which coefficient is to be changed) followed by the new value of the coefficient (i.e., 0). To make this change press return and type **C** 2↩ 0↩. Repeat these steps for each coefficient to be changed. You should then see the screen message

THE ARMA(0,25) MODEL IS $X(t) = Z(t)$
+(-.357)*Z(t-1) +(-.163)*Z(t-3) +(-.499)*Z(t-12) +(.201)*Z(t-23)

White noise variance = .115169E-02
<Press any key to continue> □

D.3.6 Maximum Likelihood Estimation

Once you have specified values of p and q and possibly set some coefficients to zero, you can carry out efficient parameter estimation by using the option [ARMA parameter estimation] of the Main Menu.

From the Main Menu type **A** to obtain the ARMA Estimation Menu.

Much of the information displayed in this menu concerns the optimization settings. For most purposes you will need to use the default settings only. (With the default settings, any coefficients that are set to zero will be treated as fixed values and not as parameters. If you wish to include a particular coefficient in the parameters to be optimized, you must therefore not set its initial value equal to zero.)

To find the maximum likelihood estimates of your parameters choose the option [Optimize with current settings] by typing **O**. PEST will then try to find the parameters that maximize the likelihood of your model with respect to all the nonzero coefficients in the model currently stored by PEST.

If you wish to compute the Gaussian likelihood (or one-step predictors and residuals) without doing any optimization, type **L** to select the option [Likelihood of Model (no optimization)].

Example D.3.4. Let us find the maximum likelihood estimates of the parameters in the current model for the logged, differenced, and mean-corrected airline passenger data stored in PEST.

Starting from the Main Menu type **A** ⟵ and you will see the Estimation Menu. Choose the default option by typing **O**. After a short delay the iterations will cease and you will see the the message

```
THE ARMA(0,25) MODEL IS X(t) = Z(t)
+(-.355) * Z(t - 1) + (-.201) * Z(t - 3) + (-.523) * Z(t - 12) + (.242) * Z(t - 23)

MOVING AVERAGE PARAMETERS :
THETA( 1)=                    -.35528650   THETA( 3)=   -.20131570
THETA(12)=                    -.52297480   THETA(23)=    .24152310

WHITE NOISE VARIANCE    =      .125024E-02

BIC STATISTIC           =      -487.613800
-2 ln(LIKELIHOOD)       =      -496.517400
AICC STATISTIC          =      -486.037400     LAST= -440.962900

 # FUNCTION CALLS = 46 ;# ITERATIONS = 5;ACCURACY PARAM =.002053
 STOPPING VALUE 2 : WITHIN ACCURACY LEVEL
 <Press any key to continue>
```

The stopping value of 2 indicates that the minimum of $-2 \ln L$ has been located with the specified accuracy. If you see the message

```
STOPPING VALUE  4 : ITERATION LIMIT EXCEEDED
```

then the minimum of $-2 \ln L$ could not be located with the number of iterations (50) allowed. You can continue the search (starting from the point at which the iterations were interrupted) by typing **C** to return to the Estimation Menu, then typing **O** as before. □

D.3.7 Optimization Results

After running the optimization algorithm, PEST displays the model parameters (coefficients and white noise variance), the values of $-2 \ln L$, AICC, and BIC, information regarding the computations, and the Results Menu.

Example D.3.5 The next stage of the analysis is to consider a variety of competing models and to select the most suitable. The following table shows the AICC statistics for a variety of subset moving average models of order less than 24.

Lags						*AICC*
1	3		12		23	−486.04
1	3		12	13	23	−485.78
1	3	5	12		23	−489.95
1	3		12	13		−482.62
1			12			−475.91

The best of these models from the point of view of AICC value is the one with nonzero coefficients at lags 1, 3, 5, 12, and 23. To substitute this model for the one currently stored in PEST (starting from the Results Menu), type

C ↵ **C C** 1↵ 5↵ **R** ↵ **O**

and optimization will begin as before. You should obtain the noninvertible model

$$X_t = Z_t - .439Z_{t-1} - .302Z_{t-3} + .242Z_{t-5}$$
$$- .656Z_{t-12} + .348Z_{t-23}, \quad \{Z_t\} \sim \text{WN}(0, .00103).$$

For future reference, store the model under the filename AIRPASS.MOD using the option [`Store the Model`] of the Results Menu. □

We have seen in our example how, when optimizing iterations cease, the stopping code indicates whether or not more iterations are required. The display of results that includes this information also contains a menu that allows us to investigate properties of the fitted model, including its goodness of fit. The options available in the Results Menu are described below.

D.4 Testing Your Model

Once we have a model, it is important to check whether it is any good or not. Typically this is judged by comparing observations with corresponding predicted values obtained from the fitted model. If the fitted model is appropriate then the prediction errors should behave in a manner that is consistent with the model. The **residuals** are the rescaled one-step prediction errors,

$$\hat{W}_t = (X_t - \hat{X}_t)/\sqrt{r_{t-1}},$$

where $\hat{X}_t$ is the best linear mean-square predictor of X_t based on the observations up to time $t-1$, $r_{t-1} = E(X_t - \hat{X}_t)^2/\sigma^2$ and σ^2 is the white noise variance of the fitted model.

If the data were truly generated by the fitted ARMA(p, q) model with white noise sequence $\{Z_t\}$, then for large samples the properties of $\{\hat{W}_t\}$ should reflect those of $\{Z_t\}$. To check the appropriateness of the model we therefore examine the residual series $\{\hat{W}_t\}$, and check that it resembles a realization of a white noise sequence.

PEST provides a number of tests for doing this in the Residuals Menu, which is obtained by selecting the option [`File and Analyze Residuals`] of the Results Menu.

To examine the residuals from a *specified* model without doing any optimization, enter the data and model, then use the option [`Likelihood of Model`] of the Estimation Menu followed by [`File and Analyze Residuals`] of the Results Menu. You will then see the following menu:

RESIDUALS MENU: File residuals
Plot rescaled residuals
ACF/PACF of residuals
File ACF/PACF of residuals
Tests of randomness of residuals
Return to display of estimation results

D.4.1 Plotting the Residuals

The **rescaled residuals** are defined as

$$\hat{W}_t^{(r)} = \sqrt{n}\hat{W}_t/(\sum_{j=1}^{n} \hat{W}_t^2).$$

From the Residuals Menu type **P** and you will see a histogram of the rescaled residuals.

If the fitted model is appropriate, the histogram of the rescaled residuals should have mean close to zero. If the fitted model is appropriate and the data is Gaussian, this will be reflected in the shape of the histogram, which should then resemble a normal density with mean zero and variance 1.

Press any key after inspecting the histogram and you will see a graph of $\hat{W}_t^{(r)}$ vs t. If the fitted model is appropriate, this should resemble a realization of a white noise sequence. Look for trends, cycles, and nonconstant variance, any of which suggest that the fitted model is inappropriate. If substantially more than 5% of the rescaled residuals lie outside the bounds ± 1.96 or if there are rescaled residuals far outside these bounds, then the fitted model should not be regarded as Gaussian.

Example D.4.1 After selecting the option [`Plot rescaled residuals`] of the Residuals Menu you will see the histogram of the rescaled residuals as shown in Figure D.6. The mean is close to zero and the shape suggests that the assumption of Gaussian white noise is not unreasonable in our proposed model for the transformed airline passenger data.

Press any key to see the graph shown in Figure D.7. A few of the rescaled residuals are greater in magnitude than 1.96 (as is to be expected), but there are no obvious indications here that the model is inappropriate. Press any key then enter ↩ and type **C** to return to the Residuals Menu. □

D.4.2 ACF/PACF of the Residuals

If we were to assume that our fitted model is the true process generating the data, then the observed residuals would be realized values of a white noise sequence.

In particular, the sample ACF and PACF of the observed residuals should lie within the bounds $\pm 1.96/\sqrt{n}$ roughly 95% of the time. These bounds are displayed on the graphs of the ACF and PACF. If substantially more than 5% of the correlations are outside these limits, or if there are a few very large values, then we should look for a better-fitting model. (More precise bounds, due to Box and Pierce, can be found in TSTM Section 9.4.)

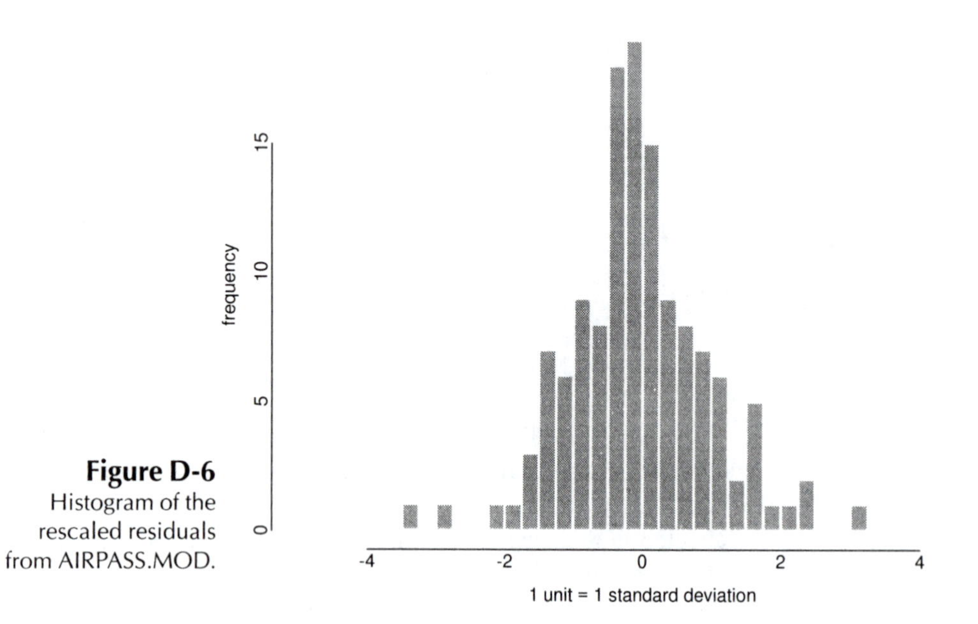

Figure D-6 Histogram of the rescaled residuals from AIRPASS.MOD.

Example D.4.2 Choose the option [`ACF/PACF of residuals`] of the Residuals Menu. After entering **N**, the sample ACF and PACF of the residuals will then appear as shown in Figures D.8 and D.9. No correlations are outside the bounds in this case. They appear to be compatible with the hypothesis that the residuals are in fact observations of a white noise sequence. Type ↩ to return to the Residuals Menu. □

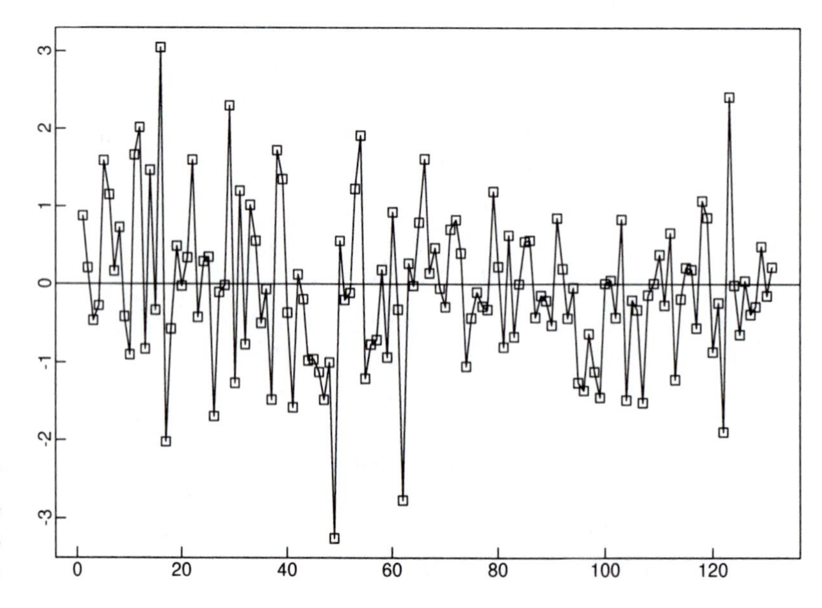

Figure D-7 Time plot of the rescaled residuals from AIRPASS.MOD.

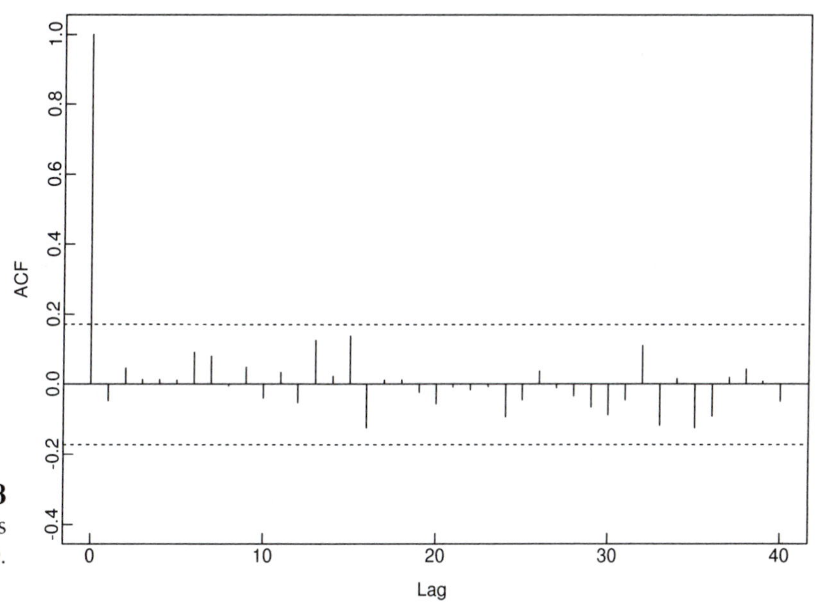

Figure D-8
Sample ACF of the residuals from AIRPASS.MOD.

D.4.3 Testing for Randomness of the Residuals

The option [`Tests of randomness of the residuals`] in the Residuals Menu carries out the six tests for randomness of the residuals described in Section 5.3.3.

Example D.4.3 Type **T** from the Residuals Menu to test the residuals for randomness. You will then be prompted to input the value of h, the total number of sample autocorrelations of the residuals used to compute the portmanteau statistics. After entering the suggested

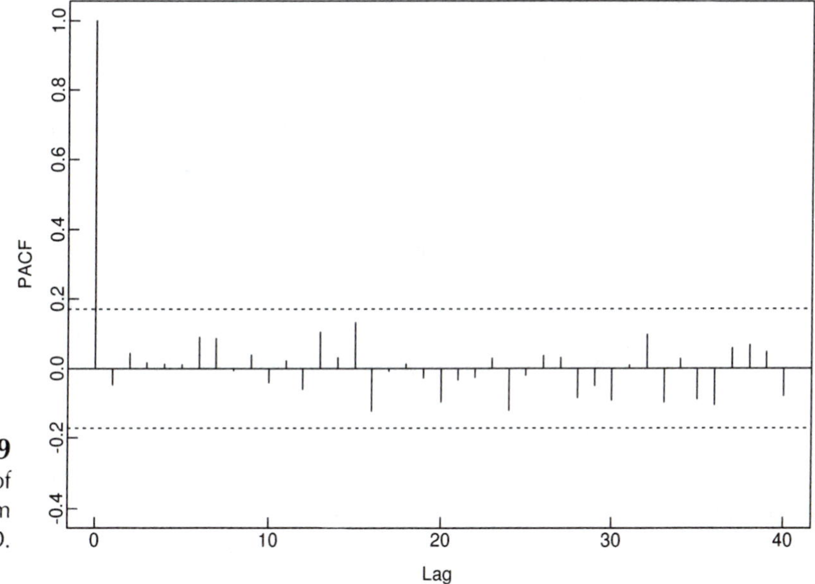

Figure D-9
Sample PACF of the residuals from AIRPASS.MOD.

value of h (in this case PEST suggests using $h = 25$), the results of the six tests of randomness described above are as follows:

```
RANDOMNESS TEST STATISTICS (see Section 5.3.3)
______________________________________________
LJUNG-BOX PORTM.    =    13.76    CHISQUR(      20)
MCLEOD-LI PORTM.    =    17.37    CHISQUR(      25)
TURNING POINTS      =    87.      ANORMAL(  86.00,   4.79**2)
DIFFERENCE-SIGN     =    65.      ANORMAL(  65.00,   3.32**2)
RANK TEST           =    3935.    ANORMAL(4257.50, 753.91**2)
ORDER OF MIN AICC YW MODEL FOR RESIDUALS = 0
```

Every test is easily passed by our fitted model (with significance level $\alpha = .05$) and the order of the minimum-AICC AR model for the residuals supports the compatibility of the residuals with white noise. Press $\hookleftarrow$ to return to the Residuals Menu. For later use, save the residuals under the filename AIRRES.DAT using the option [**F**ile residuals]. □

D.5 Prediction

One of the main purposes of time series modelling is the prediction of future observations. Once you have found a suitable model for your data, you can predict future values using either the option [**F**orecasting] of the Main Menu or (equivalently) the option [**P**redict Future Data] of the Results Menu.

D.5.1 Forecast Criteria

Given observations $X_1, \ldots, X_n$ of a series which we assume to be appropriately modelled as an ARMA(p, q) process, PEST predicts future values of the series X_{n+h} from the data and the model by computing the linear combination $P_n(X_{n+h})$ of $X_1, \ldots, X_n$ that minimizes the mean squared error $E(X_{n+h} - P_n(X_{n+h}))^2$.

D.5.2 Forecast Results

Assuming that you have data stored in PEST that have been adequately fitted by an ARMA(p, q) model, also stored in PEST, choose [**F**orecasting] from the Main Menu, after which you will be asked for the number of future values you wish to predict.

After the model is displayed on the screen you will be asked if you wish to change the white noise variance. (This will not affect the predictors but only their mean squared errors.) The predicted values of the fitted ARMA process will then be displayed in the column labelled XHAT. In the column labelled SQRT(MSE) you will see the square roots of the estimated mean squared errors of the corresponding predictors. These are calculated under the assumption that the observations are truly generated by the current model. They measure the uncertainty of the corresponding forecasts. A smaller value indicates a more reliable forecast. As is to be expected, the mean squared error of $P_n(X_{n+h})$ increases with the lead time h of the forecast.

If your data was mean-corrected, the third column of the PEST output will show the predicted values in Column 1 plus the previously subtracted sample mean. If there has been no mean-correction, the third column will be the same as the first.

D.5.3 Inverting Transformations

The predictors and mean squared errors calculated so far do not pertain to your *original* time series unless you have made no data transformations other than mean-correction (in which case the relevant predictors are those in the third column). What we have found are predictors of your *transformed* series. To predict the original series, you will need to invert all the data transformations that you have made in order to fit a zero-mean stationary model. PEST will do this for you. In fact one transformation, mean-correction, has already been inverted to generate the predicted values that were displayed in Column 3.

If you used differencing transformations, you will see the Prediction Menu displayed following the printing of the ARMA predictors. Type **U** to select the option [`Undo differencing`]. The predictors of the undifferenced data (still Box-Coxed if you made such a transformation) will then be printed on the screen together with the square roots of their mean squared errors. Type ↩ and the undifferenced data will be plotted. Type ↩ again and the predicted values will be added to the graph of the data. Type **C** and you will be asked if you wish to invert the Box-Cox transformation (if you made one). If so type **Y** and the original data will be plotted on the screen. Type ↩ again and the predictors of the original series will be added to the graph. Type **C** and you will be asked if you wish to file the predicted values, then returned to the Prediction Menu.

If you used classical decomposition rather than differencing, PEST will automatically add back the trend and/or seasonal component immediately after listing the ARMA predictors. The resulting data values and corresponding predictors will then be plotted, after which you will again be given the opportunity to invert the Box-Cox transformation (if any) as in the previous paragraph.

Example D.5.1 We left our logged, differenced, and mean-corrected airline passenger data stored in PEST as AIRPASS.DAT along with the fitted MA(23) model, AIRPASS.MOD. To predict the next 24 values of the original series AIRPASS.DAT, return to the Main Menu and select [`Forecasting`]. Type 24↩ to specify that 24 predicted values are required after the last observation. After a brief delay and a ↩, you will be asked if you wish to change the white noise variance. Type **N** and the predictors of the

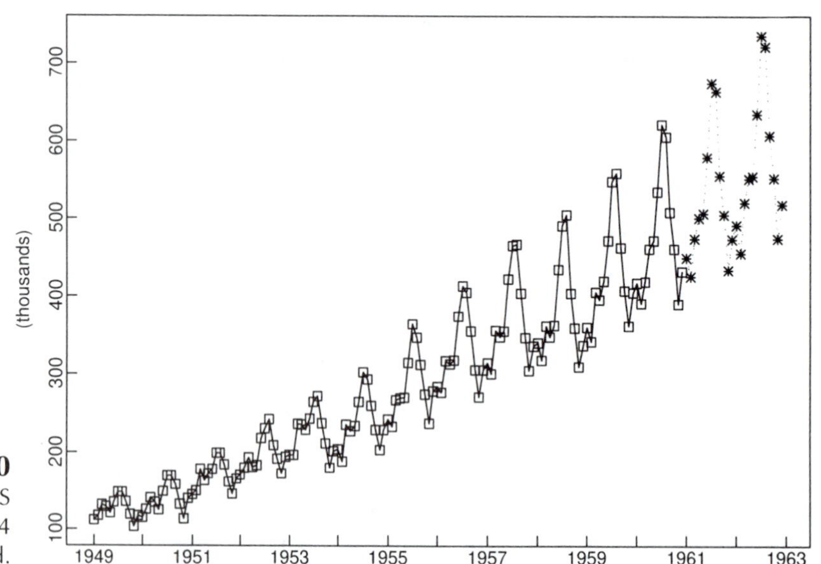

Figure D-10 The original AIRPASS data with 24 forecasts appended.

transformed data together with the square roots of the corresponding mean squared errors will be printed on the screen.

To obtain forecasts of the undifferenced series, choose the option [`Undo differencing`] from the Prediction Menu and follow the program prompts to obtain the graph of the undifferenced series with the forecasts of the next 24 values appended.

To undo the Box-Cox transformation and recover the *original* data and predictors, type ↩ **C Y** ↩ and a graph of the original AIRPASS data will be plotted on the screen. Type ↩ and the 24 predicted values will be added, giving the graph shown in Figure D.10. □

D.6 Model Properties

PEST can be used to analyze the properties of a specified ARMA process without reference to any data set. This enables us in particular to compare the properties of potential ARMA models for a given data set to see which of them best reproduces particular features of the data.

PEST allows you to look at the autocorrelation function and spectral density, to examine MA(∞) and AR(∞) representations and to generate realizations for any specified ARMA process. The use of these options is described in this section.

Example D.6.1 We shall illustrate the use of PEST for model analysis using the model AIRPASS.MOD that is currently stored in the program. □

D.6.1 ARMA Models

For modelling zero-mean stationary time series, PEST uses the class of ARMA processes. PEST enables you to compute characteristics of the causal ARMA model defined by

$$X_t = \phi_1 X_{t-1} + \phi_2 X_{t-2} + \cdots + \phi_p X_{t-p} + Z_t + \theta_1 Z_{t-1} + \theta_2 Z_{t-2} + \cdots + \theta_q Z_{t-q},$$

where $\{Z_t\} \sim \text{WN}(0, \sigma^2)$ and $\phi_1, \ldots, \phi_p, \theta_1, \ldots, \theta_q$ and σ^2 are numerically specified values of the parameters.

PEST works exclusively with causal ARMA models. It will not permit you to enter a model for which $1 - \phi_1 z - \cdots - \phi_p z^p$ has a zero inside or on the unit circle, nor does it generate fitted models with this property. From the point of view of second-order properties this represents no loss of generality (Section 3.1). If you are trying to enter an ARMA(p, q) model manually, the simplest way to ensure that your model is causal is to set all the autoregressive coefficients close to zero (e.g., .001). PEST will not accept a noncausal model.

PEST does not restrict models to be invertible; however, if the current model is noninvertible, i.e., if the moving average polynomial $1 + \theta_1 z + \cdots + \theta_q z^q$ has a zero inside or on the unit circle, you will be informed by the program. (You can check the model status by choosing the option [**C**`urrent model and data file status`] of the Main Menu.) A noninvertible model can always be converted to an invertible model with the same autocovariance function by choosing [**A**`RMA parameter estimation`] of the Main Menu, then option [**L**`ikelihood of Model`] of the Estimation Menu, then option [**N**`ONINVERTIBLE MODEL`] from the Results Menu.

D.6.2 Model ACF, PACF

The *model* ACF and PACF can be obtained using the selection [**M**`odel ACF/PACF, AR/MA infinity representations`] of the Main Menu. They can be calculated for lags up to 1150. Normally you should not need more than about 40.

After typing **M** from the Main Menu, the Model ACF/PACF Menu consisting of seven items will appear. It allows you to reset the maximum lag (the default value is 40), to plot the ACF and PACF, to file the values, to plot the sample ACF and PACF with the model ACF and PACF, to change the white noise variance and to compute the coefficients in the MA(∞) and AR(∞) representations of the process (see Section 3.1).

Example D.6.2 Starting from the Main Menu, the ACF and PACF for the current model AIRPASS.MOD may be plotted by typing **M A**. To compare the sample ACF/PACF with the model ACF/PACF, press ↩ to return to the Model ACF/PACF Menu and type **S**. The sample and model ACF and PACF are shown in Figures D.11 and D.12. The vertical lines represent the model values and the squares are the sample ACF/PACF. These graphs show that the data and the model ACF both have large values at lag 12,

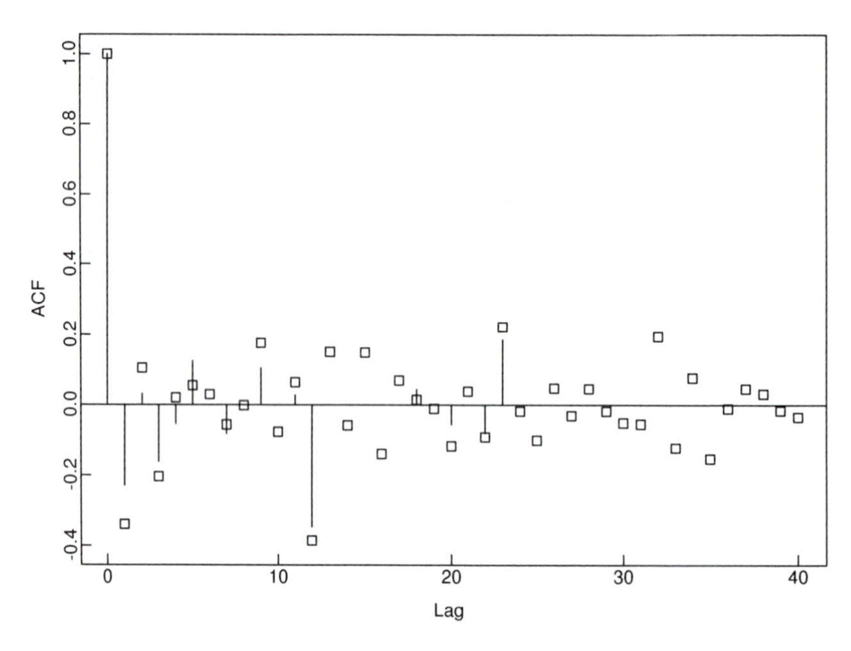

Figure D-11
The ACF of AIRPASS.MOD together with the sample ACF of the transformed AIRPASS.DAT series.

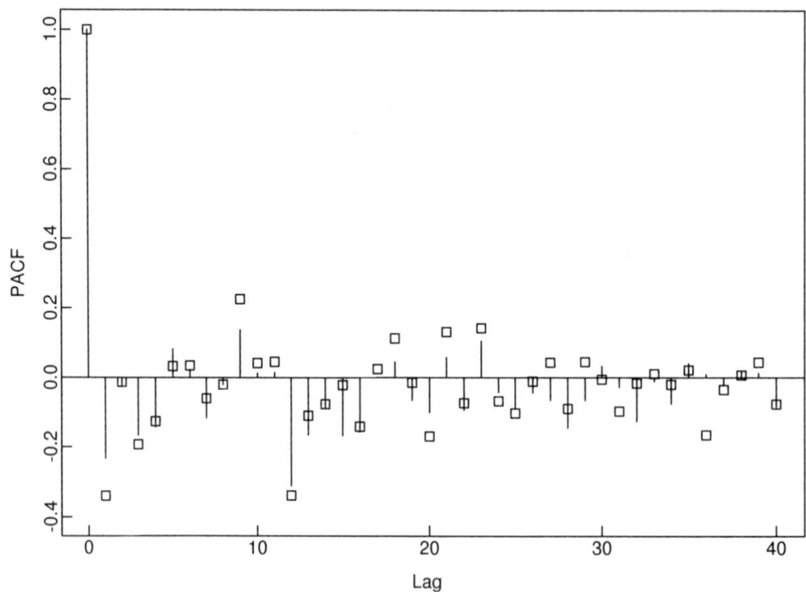

Figure D-12
The PACF of AIRPASS.MOD together with the sample PACF of the transformed AIRPASS.DAT series.

while the sample and model partial autocorrelation functions both tend to die away geometrically after the peak at lag 12. The similarities between the graphs indicate that the model is capturing some of the important features of the data. □

D.6.3 Model Representations

As indicated in Section 3.1, if $\{X_t\}$ is a causal ARMA process, then it has as an MA(∞) representation

$$X_t = \sum_{j=0}^{\infty} \psi_j Z_{t-j}, \quad t = 0, \pm 1, \pm 2, \ldots,$$

where $\sum_{j=0}^{\infty} |\psi_j| < \infty$ and $\psi_0 = 1$.

Similarly, if $\{X_t\}$ is an invertible ARMA process, then it has an AR(∞) representation

$$Z_t = \sum_{j=0}^{\infty} \pi_j X_{t-j}, \quad t = 0, \pm 1, \pm 2, \ldots,$$

where $\sum_{j=0}^{\infty} |\pi_j| < \infty$ and $\pi_0 = 1$.

For any specified ARMA model you can determine the coefficients in these representations by selecting the option [`MA or AR infinity representations`] from the Model ACF/PACF Menu. Starting from the Main Menu type **M M**. You will then be asked to choose between the [`MA-Infinity Representation`] and [`AR-Infinity Representation`] of the model. (If the model is not invertible, the AR(∞) choice will not be possible.) After entering the maximum lag required, PEST will then print the desired coefficients (ψ_j or π_j) on the screen. They can be stored either as they appear on the screen or in model format for later use in PEST.

Example D.6.3 AIRPASS.MOD does not have an AR(∞) representation since it is not invertible. However, we can convert AIRPASS.MOD to an equivalent invertible model and then find an AR(∞) representation for it. To convert to an invertible model, start from the Main Menu and type **A** ↩ **L** ↩ ↩ **N** ↩ ↩ **R**. To find the AR(∞) representation, type **M M A** 50↩. This gives 50 coefficients, the first 19 of which are shown below.

AR-infinity coeffs up to lag 50

j	pi(j)
0	1.0000000
1	.3613552
2	.1173539
3	.3052337
4	.2738022

```
 5        -.0051080
 6         .0538636
 7         .1687211
 8         .1005408
 9         .0208243
10         .0813016
11         .0672451
12         .5841435
13         .4196610
14         .2307828
15         .3741191
16         .3943860
17         .1059944
18         .0887224
19         .2328299
<Press any key to continue>
```

There is little point in using PEST to find the MA(∞) representation of this model. What is it? □

D.6.4 Generating Realizations of a Random Series

PEST can be used to generate realizations of a random time series defined by the currently stored model.

To generate such a realization, select the option [**G**`eneration of simulated data`] from the Main Menu. You will be asked if you wish to continue with the simulation (type **Y**) and if you wish to change the white noise variance. Next you will be prompted for the number of data points you wish to generate and then you will be asked to enter a random number seed. This should be an integer with fewer than 10 digits. By using the same random number seed you can reproduce the same realization of the process at any other time.

Once the values of an ARMA process have been generated, you will be given the opportunity to add any specified mean to the observations. If you have previously mean-corrected a data set, the subtracted mean will have been stored by PEST and it will be displayed so that you may choose to add this value to the simulated ARMA data. (If you have previously performed a classical decomposition on a data set, you will also be given the opportunity to add the stored trend and seasonal components. This allows you to simulate the *original data*, not just the random noise component. If, however, you transform your original data by differencing, PEST allows you to simulate the differenced data only.)

The simulated data will be stored in PEST, overwriting any data previously stored in the program.

Example D.6.4 To generate 135 data points using the model AIRPASS.MOD, type **G** from the Main Menu. Then type

G Y N `135↩ 1327↩ 0↩` **N**.

(The number 1327 is the random number seed.) You can plot the sample ACF and PACF of the generated data using Option 3 of the Data Menu. These should be compared with those in Figures D.11 and D.12. By computing the sample ACF and PACF for a variety of different realizations you can get a feeling for the magnitude of the random fluctuations in these functions. □

D.6.5 Spectral Properties

Spectral properties of both data and fitted ARMA models can also be computed and plotted with the aid of PEST. The spectral density of the *model* is determined by selecting the option [**S**`pectral density of MODEL on (-pi,pi)`] of the Main Menu. Estimation of the spectral density of a stationary series can be carried out in two ways, either by fitting an ARMA model as already described and computing the spectral density of the fitted model (Section 4.4) or by computing the periodogram of the data and smoothing (Section 4.2). The latter method is applied by selecting the option [**N**`onparametric spectral estimation`] from the Main Menu of PEST. Examples of both approaches are given in Chapter 4.

Bibliography

Akaike, H. (1969), Fitting autoregressive models for prediction, *Annals of the Institute of Statistical Mathematics*, *21*, 243–247.

Akaike, H. (1973), Information theory and an extension of the maximum likelihood principle, *2nd International Symposium on Information Theory*, B.N. Petrov and F. Csaki (eds.), Akademiai Kiado, Budapest, 267–281.

Akaike, H. (1978), Time series analysis and control through parametric models, *Applied Time Series Analysis*, D.F. Findley (ed.), Academic Press, New York.

Anderson, T.W. (1971), *The Statistical Analysis of Time Series*, John Wiley, New York.

Anderson, T.W. (1980), Maximum likelihood estimation for vector autoregressive moving average models, *Directions in Time Series*, D.R. Brillinger and G.C. Tiao (eds.), Institute of Mathematical Statistics, 80–111.

Ansley, C.F. (1979), An algorithm for the exact likelihood of a mixed autoregressive-moving average process, *Biometrika*, *66*, 59–65.

Ansley, C.F. and Kohn, R. (1985), On the estimation of ARIMA models with missing values, *Time Series Analysis of Irregularly Observed Data*, E. Parzen (ed.), Springer Lecture Notes in Statistics, *25*, 9–37.

Aoki, M. (1987), *State Space Modeling of Time Series*, Springer-Verlag, Berlin.

Atkins, Stella M. (1979), Case study on the use of intervention analysis applied to traffic accidents, *J. Opns Res. Soc.*, *30*, 7, 651–659.

Barndorff-Nielsen, O. (1978), *Information and Exponential Families in Statistical Theory*, John Wiley, New York.

Bergstrom, A.R. (1990), *Continuous Time Econometric Modelling*, Oxford University Press, Oxford.

Bhattacharyya, M.N. and Layton, A.P. (1979), Effectiveness of seat belt legislation on the Queensland road toll – an Australian case study in intervention analysis, *J. Amer. Stat. Assoc.*, *74*, 596–603.

Bloomfield, P. (1976), *Fourier Analysis of Time Series: An Introduction*, John Wiley, New York.

Bollerslev, T. (1986), Generalized autoregressive conditional heteroskedasticity, *J. Econometrics*, *31*, 307–327.

Box, G.E.P. and Cox, D.R. (1964), An analysis of transformations (with discussion), *J. R. Stat. Soc. B*, *26*, 211-252.

Box, G.E.P. and Jenkins, G.M. (1976), *Time Series Analysis: Forecasting and Control, Revised Edition*, Holden-Day, San Francisco.

Box, G.E.P. and Pierce, D.A. (1970), Distribution of residual autocorrelations in autoregressive-integrated moving average time series models, *J. Amer. Stat. Assoc.*, *65*, 1509–1526.

Box, G.E.P. and Tiao, G.C. (1975), Intervention analysis with applications to economic and environmental problems, *J. Amer. Stat. Assoc.*, *70*, 70–79.

Breidt, F.J. and Davis, R.A. (1992), Time reversibility, identifiability and independence of innovations for stationary time series, *J. Time Series Anal.*, *13*, 377–390.

Brockwell, P.J. (1994), On continuous time threshold ARMA processes, *J. Statistical Planning and Inference*, *39*, 291–304.

Brockwell, P.J. and Davis, R.A. (1988), Applications of innovation representations in time series analysis, *Probability and Statistics, Essays in Honor of Franklin A. Graybill*, J.N. Srivastava (ed.), Elsevier, Amsterdam, 61–84.

Brockwell, P.J. and Davis, R.A. (1991), *Time Series: Theory and Methods, 2nd Edition*, Springer-Verlag, New York.

Brockwell, P.J. and Davis, R.A. (1994), *ITSM for Windows*, Springer-Verlag, New York.

Chan, K.S. and Ledolter, J. (1995), Monte Carlo EM estimation for time series models involving counts, *J. Amer. Stat. Assoc.*, *90*, 242–252.

Chan, K.S. and Tong, H. (1987), A note on embedding a discrete parameter ARMA model in a continuous parameter ARMA model, *J. Time Series Anal.*, *8*, 277–281.

Chung, K.L. and Williams, R.J. (1990), *Introduction to Stochastic Integration, 2nd Edition*, Birkhäuser, Boston.

Cochran, D. and Orcutt, G.H. (1949), Applications of least squares regression to relationships containing autocorrelated errors, *J. Amer. Stat. Assoc.*, *44*, 32–61.

Davis, M.H.A. and Vinter, R.B. (1985), *Stochastic Modelling and Control*, Chapman and Hall, London.

Davis, R.A., Chen, M., and Dunsmuir, W.T.M. (1995), Inference for MA(1) processes with a root on or near the unit circle, *Probability and Mathematical Statistics*, *15*, 227–242.

Davis, R.A., Chen, M., and Dunsmuir, W.T.M. (1996), Inference for seasonal moving average models with a unit root. *Preprint, Colorado State University*.

Davis, R.A. and Dunsmuir, W.T.M. (1996), Maximum likelihood estimation for MA(1) processes with a root on or near the unit circle, *Econometric Theory*, *12*, 1–29.

Dempster, A.P., Laird, N.M. and Rubin, D.B. (1977), Maximum likelihood from incomplete data via the EM algorithm, *J. R. Stat. Soc. B*, *39*, 1–38.

Dickey, D.A. and Fuller, W.A. (1979), Distribution of the estimators for autoregressive time series with a unit root, *J. Amer. Stat. Assoc.*, *74*, 427–431.

Duong, Q.P. (1984), On the choice of the order of autoregressive models: a ranking and selection approach, *J. Time Series Anal.*, *5*, 145–157.

Engle, R.F. (1982), Autoregressive conditional heteroscedasticity with estimates of the variance of UK inflation, *Econometrica*, *50*, 987–1007.

Engle, R.F. and Granger, C.W.J. (1987), Co-integration and error correction: representation, estimation and testing, *Econometrica*, *55*, 251–276.

Engle, R.F. and Granger, C.W.J. (1991), *Long-run Economic Relationships*, Advanced Texts in Econometrics, Oxford University Press, Oxford.

Fuller, W.A. (1976), *Introduction to Statistical Time Series*, John Wiley, New York.

de Gooijer, J.G., Abraham, B., Gould, A., and Robinson, L. (1985), Methods of determining the order of an autoregressive-moving average process: A survey, *Int. Stat., Review*, *53*, 301–329.

Granger, C.W.J. (1981), Some properties of time series data and their use in econometric model specification, *J. Econometrics*, *16*, 121–130.

Gray, H.L., Kelley, G.D., and McIntire, D.D. (1978), A new approach to ARMA modelling, *Comm. Stat.*, *B7*, 1–77.

Graybill, F.A. (1983), *Matrices with Applications in Statistics*, Wadsworth, Belmont, California.

Grunwald, G.K., Hyndman, R.J., and Hamza, K. (1994), Some properties and generalizations of nonnegative Bayesian time series models, *Technical Report*, Statistics Dept., Melbourne Univeristy, Parkville, Australia.

Grunwald, G.K., Raftery, A.E., and Guttorp, P. (1993), Prediction rule for exponential family state space models, *J. R. Stat. Soc. B*, *55*, 937–943.

Hannan, E.J. (1980), The estimation of the order of an ARMA process, *Ann. Stat.*, *8*, 1071–1081.

Hannan, E.J. and Deistler, M. (1988), *The Statistical Theory of Linear Systems*, John Wiley, New York.

Hannan, E.J. and Rissanen, J. (1982), Recursive estimation of mixed autoregressive moving average order, *Biometrika*, *69*, 81–94.

Harvey, A.C. (1990), *Forecasting, Structural Time Series Models and the Kalman Filter*, Cambridge University Press, Cambridge.

Harvey, A.C. and Fernandes, C. (1989), Time Series models for count data of qualitative observations, *J. Business and Economic Statistics*, *7*, 407–422.

Holt, C.C.(1957), Forecasting seasonals and trends by exponentially weighted moving averages, *ONR Research Memorandum 52*, Carnegie Institute of Technology, Pittsburgh, Pennsylvania.

Hurvich, C.M. and Tsai, C.L. (1989), Regression and time series model selection in small samples, *Biometrika 76*, 297–307.

Jones, R.H. (1975), Fitting autoregressions, *J. Amer. Stat. Assoc.*, *70*, 590–592.

Jones, R.H. (1978), Multivariate autoregression estimation using residuals, *Applied Time Series Analysis*, David F. Findley (ed.), Academic Press, New York, 139–162.

Jones, R.H. (1980), Maximum likelihood fitting of ARMA models to time series with missing observations, *Technometrics*, *22*, 389–395.

Karatzas, I. and Shreve, S.E. (1991), *Brownian Motion and Stochastic Calculus, 2nd Edition*, Springer-Verlag, New York.

Kendall, M.G. and Stuart, A. (1976), *The Advanced Theory of Statistics, Vol. 3*, Griffin, London.

Kitagawa, G. (1987), Non-Gaussian state-space modelling of non-stationary time series, *J. Amer. Stat. Assoc.*, *82* (with discussion), 1032–1063.

Kuk, A.Y.C. and Cheng, Y.W. (1994), The Monte Carlo Newton-Raphson algorithm, *Technical Report S94-10*, Department of Statistics, U. New South Wales, Sydney, Australia.

Lehmann, E.L. (1983), *Theory of Point Estimation*, John Wiley, New York.

Lehmann, E.L. (1986), *Testing Statistical Hypotheses, 2nd Edition*, John Wiley, New York.

Liu, J. and Brockwell, P.J. (1988), The general bilinear time series model, *J. Appl. Probability 25*, 553–564.

Ljung, G.M. and Box, G.E.P. (1978), On a measure of lack of fit in time series models, *Biometrika*, *65*, 297–303.

Lütkepohl, H. (1993), *Introduction to Multiple Time Series Analysis, 2nd Edition*, Springer-Verlag, Berlin.

McCullagh, P., and Nelder, J.A. (1989), *Generalized Linear Models, 2nd Edition*, Chapman and Hall, London.

McLeod, A.I. and Li, W.K. (1983), Diagnostic checking ARMA time series models using squared-residual autocorrelations, *J. Time Series Anal.*, *4*, 269–273.

Mage, D.T. (1982), An objective graphical method for testing normal distributional assumptions using probability plots, *Amer. Stat.*, *36*, 116–120.

Makridakis, S., Andersen, A., Carbone, R., Fildes, R., Hibon, M., Lewandowski, R., Newton, J., Parzen, E., and Winkler, R. (1984), *The Forecasting Accuracy of Major Time Series Methods*, John Wiley, New York.

Makridakis, S., Wheelwright, S.C., and McGee, V.E. (1983), *Forecasting: Methods and Applications*, John Wiley, New York.

May, R.M. (1976), Simple mathematical models with very complicated dynamics, *Nature*, *261*, 459–467.

Mendenhall, W., Wackerly, D.D., and Scheaffer, D.L. (1990), *Mathematical Statistics with Applications, 4th Edition*, Duxbury, Belmont.

Mood, A.M., Graybill, F.A., and Boes, D.C. (1974), *Introduction to the Theory of Statistics*, McGraw-Hill, New York.

Newton, H.J. and Parzen, E. (1984), Forecasting and time series model types of 111 economic time series, *The Forecasting Accuracy of Major Time Series Methods*, S. Makridakis et al. (eds.), John Wiley and Sons, Chichester.

Nicholls, D.F. and Quinn, B.G. (1982), *Random Coefficient Autoregressive Models: An Introduction*, Springer Lecture Notes in Statistics, *11*.

Oksendal, B. (1992), *Stochastic Differential Equations: An Introduction with Applications, 3rd Edition*, Springer-Verlag, New York.
Pantula, S. (1991), Asymptotic distributions of unit-root tests when the process is nearly stationary, *J. Business and Economic Statistics*, *9*, 63–71.
Parzen, E. (1982), ARARMA models for time series analysis and forecasting, *J. Forecasting*, *1*, 67–82.
Pole, A., West, M., and Harrison, J. (1994), *Applied Bayesian Forecasting and Time Series Analysis*, Chapman and Hall, New York.
Priestley, M.B. (1988), *Non-linear and Non-stationary Time Series Analysis*, Academic Press, London.
Rosenblatt, M. (1985), *Stationary Sequences and Random Fields*, Birkhäuser, Boston.
Said, S.E. and Dickey, D.A. (1984), Testing for unit roots in autoregressive moving-average models with unknown order, *Biometrika*, *71*, 599–607.
Schwert, G.W. (1987), Effects of model specification on tests for unit roots in macroeconomic data, *J. Monetary Economics*, *20*, 73–103.
Shapiro, S.S. and Francia, R.S. (1972), An approximate analysis of variance test for normality, *J. Amer. Stat. Assoc.*, *67*, 215–216.
Shibata, R. (1976), Selection of the order of an autoregressive model by Akaike's information criterion, *Biometrika*, *63*, 117–126.
Shibata, R. (1980), Asymptotically efficient selection of the order of the model for estimating parameters of a linear process, *Ann. Stat.*, *8*, 147–164.
Silvey, S.D. (1975), *Statistical Inference*, Halsted, New York.
Smith, J.Q. (1979), A Generalization of the Bayesian steady forecasting model, *J. R. Stat. Soc. B*, *41*,, 375–387.
Sorenson, H.W. and Alspach, D.L. (1971), Recursive Bayesian estimation using Gaussian sums, *Automatica*, *7*, 465–479.
Stramer, O., Brockwell, P.J., and Tweedie, R.L. (1995), Existence and stability of continuous time threshold ARMA processes, *Statistica Sinica*.
Subba-Rao, T. and Gabr, M.M. (1984). *An Introduction to Bispectral Analysis and Bilinear Time Series Models*, Springer Lecture Notes in Statistics, *24*.
Tam, W.K. and Reinsel, G.C. (1995), Tests for seasonal moving average unit root in ARIMA models, *Preprint, University of Wisconsin-Madison.*
Tanaka, K. (1990), Testing for a moving average unit root, *Econometric Theory*, *9*, 433–444.
Tong, H. (1990), *Non-linear Time Series: A Dynamical Systems Approach*, Oxford University Press, Oxford.
Venables, W.N. and Ripley, B.D. (1994), *Modern Applied Statistics with S-Plus*, Springer-Verlag, New York.
West, M. and Harrison, P.J. (1989), *Bayesian Forecasting and Dynamic Models*, Springer-Verlag, New York.
Whittle, P. (1963). On the fitting of multivariate autoregressions and the approximate canonical factorization of a spectral density matrix, *Biometrika*, *40*, 129–134.

Wichern, D. and Jones, R.H. (1978), Assessing the impact of market disturbances using intervention analysis, *Management Science*, *24*, 320–337.

Wu, C.F.J. (1983), On the convergence of the EM algorithm, *Ann. Stat.*, *11*, 95–103.

Zeger, S.L. (1988), A regression model for time series of counts, *Biometrika*, *75*, 621–629.

Index

Springer Texts in Statistics *(continued from page ii)*

Jobson	Applied Multivariate Data Analysis, Volume II: Categorical and Multivariate Methods
Kalbfleisch	Probability and Statistical Inference, Volume I: Probability, Second Edition
Kalbfleisch	Probability and Statistical Inference, Volume II: Statistical Interference, Second Edition
Karr	Probability
Keyfitz	Applied Mathematical Demography, Second Edition
Kiefer	Introduction to Statistical Inference
Kokoska and Nevison	Statistical Tables and Formulae
Kulkarni	Modeling, Analysis, Design, and Control of Stochastic Systems
Lehmann	Elements of Large-Sample Theory
Lehmann	Testing Statistical Hypotheses, Second Edition
Lehmann and Casella	Theory of Point Estimation, Second Edition
Lindman	Analysis of Variance in Experimental Design
Lindsey	Applying Generalized Linear Models
Madansky	Prescriptions for Working Statisticians
McPherson	Statistics in Scientific Investigation: Its Basis, Application, and Interpretation
Mueller	Basic Principles of Structural Equation Modeling: An Introduction to LISREL and EQS

continued on next page

Springer Texts in Statistics *(continued from previous page)*

Nguyen and Rogers	Fundamentals of Mathematical Statistics, Volume I: Probability for Statistics
Nguyen and Rogers	Fundamentals of Mathematical Statistics, Volume II: Statistical Inference
Noether	Introduction to Statistics: The Nonparametric Way
Peters	Counting for Something: Statistical Principles and Personalities
Pfeiffer	Probability for Applications
Pitman	Probability
Rawlings, Pantula, and Dickey	Applied Regression Analysis
Robert	The Bayesian Choice: A Decision-Theoretic Motivation
Robert and Casella	Monte Carlo Statistical Methods
Santner and Duffy	The Statistical Analysis of Discrete Data
Saville and Wood	Statistical Methods: The Geometric Approach
Sen and Srivastava	Regressions Analysis: Theory, Methods, and Applications
Shao	Mathematical Statistics
Shumway and Stoffer	Time Series Analysis and its Applications
Terrell	Mathematical Statistics: A Unified Introduction
Whittle	Probability via Expectation, Third Edition
Zacks	Introduction to Reliability Analysis: Probability Models and Statistical Methods

ITSM2000

To install the package ITSM2000 in a subfolder ITSM2000 of drive C:
Place the ITSM CD in your CD-ROM drive.
Choose the RUN command from the START Menu.
Type D:INSTALL2000 D C
(where D: is the CD-ROM drive indicator)
Click OK
To run ITSM
Open the folder ITSM2000 and double-click on the ITSM icon,
OR Choose the RUN command from the START Menu
Type C:\ITSM2000\ITSM.EXE
Click OK
Once the ITSM2000 window is open, further information on running the program can be obtained by clicking on Help>Contents>GETTING STARTED.

ITSM96

To install the package ITSM96 in a subfolder ITSM96 of drive C:
Place the ITSM CD in your CD-ROM drive.
Choose the RUN command from the START Menu.
Type D:INSTALL96 D C
(where D: is the CD-ROM drive indicator)
Click OK
Consult the text file README.DOC in the ITSM96 folder to complete the installation and for information on running the program.